REVIEWS in MINERALOGY

Volume 5

ORTHOSILICATES

SECOND EDITION

PAUL H. RIBBE, Editor

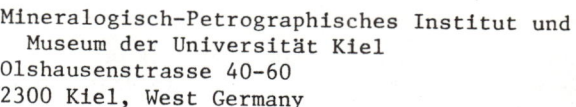

The Authors

Friedrich Liebau

> Mineralogisch-Petrographisches Institut und
>   Museum der Universität Kiel
> Olshausenstrasse 40-60
> 2300 Kiel, West Germany

E. Patrick Meagher

> Department of Geological Sciences
> University of British Columbia
> Vancouver, B.C. Canada V6T 2B4

J. Alexander Speer

Paul H. Ribbe

> Department of Geological Sciences
> Virginia Polytechnic Institute & State University
> Blacksburg, Virginia 24061

Gordon E. Brown, Jr.

> Geology Department
> Stanford University
> Stanford, California 94305

Series Editor   Paul H. Ribbe

> Department of Geological Sciences
> Virginia Polytechnic Institute and State University
> Blacksburg, Virginia 24061

MINERALOGICAL SOCIETY OF AMERICA

PRINTED BY

BookCrafters, Inc.
Chelsea, Michigan 48118

REVIEWS IN MINERALOGY

(Formerly: SHORT COURSE NOTES)
ISSN 0275-0279

VOLUME 5 : ORTHOSILICATES
First Edition   (1980) ISBN 0-939950-05-7
Second Edition (1982) ISBN 0-939950-13-8

Additional copies of this volume as well as those
listed below may be obtained at moderate cost from

Mineralogical Society of America
2000 Florida Avenue, NW
Washington, D.C.   20009

ORTHOSILICATES   SECOND EDITON

FOREWORD

The Mineralogical Society of America sponsored its first Short Course in conjunction with its annual meetings in November 1974. Contributions of the lecturers were published in a paperback book entitled *Sulfide Mineralogy*. In 1975 the Short Course and book were entitled *Feldspar Mineralogy*, in 1976 *Oxide Minerals*, and in 1977 *Mineralogy and Geology of Natural Zeolites*. In 1978 the Short Course Committee decided to forego activities because the annual meeting of the M.S.A. was held together with the Mineralogical Association of Canada, who sponsored a Short Course in Uranium Deposits and published a book by the same title. A number of mineralogists expressed regret at the potential loss of momentum in M.S.A.'s production of this series and encouraged several authors of this book to press on with their idea of publishing Volume 5 — *Orthosilicates*. Work was begun in 1978; however, without the pressure of a deadline associated with presenting the material to students of a short course at the annual meeting, procrastination set in and the first edition of this volume was not completed until September 1980 (with the exception of Chapters 1 and 2 which were submitted in their present form in 1978). In the meantime Volume 6, *Marine Minerals*, appeared in time for the annual meeting of the Society and a Short Course in San Diego in November 1979.

In 1980 the Council of the M.S.A. changed the name of the published volumes from "SHORT COURSE NOTES" to "REVIEWS in MINERALOGY" in order to more aptly describe the material contained in this now highly successful series. The First Edition of *Orthosilicates* was the first volume to appear under the "REVIEWS" banner. Subsequently Volumes 7, 8, 9A, 9B, and 10 have appeared (see p. ii); Volume 2 is being totally revised, Volume 11 is planned to be a monograph on *Fluid Inclusions*, and Volume 12 will be entitled *Carbonates*.

This is the Second Edition of *Orthosilicates*. It contains an updating and minor revisions of Chapters 3 through 10 (only) and two new chapters originally intended for the First Edition. Chapter 12 contains very brief descriptions of the paragenesis and crystal chemistry of many orthosilicates that fit the description stated in the Preface (p. iv). It may be used as an index, because all orthosilicates are listed alphabetically, including those discussed in Chapters 2 through 11. Minerals which have individual SiO_4 groups polymerized to other cations (Be, B, Al, Zn, etc.) in tetrahedral coordination are described in Chapter 13, together with some whose structures are unknown but are thought to be orthosilicates and some which have been classified as orthosilicates in the past but are now known not to be such.

Paul H. Ribbe
Series Editor
Blacksburg, VA

ACKNOWLEDGMENTS

The editors of journals and publishers of books from which figures have been reproduced are gratefully acknowledged for their cooperation.

Ramonda Haycocks and Margie Strickler patiently and skillfully typed the manuscript, and Alex Speer shared proof-reading responsibilities with the editor. The Department of Geological Sciences at Virginia Polytechnic Institute and State University provided the facilities at which much of the manuscript writing and preparation were undertaken. The editor acknowledges the financial support of the National Science Foundation (Grant EAR 77-23114 to G.V. Gibbs and PHR) during the four years in which this book has been in progress.

PREFACE

The intent of this volume is to emphasize the crystal chemistry and related physical properties of the major rock-forming orthosilicates. Though in some chapters more attention is given to phase equilibria and paragenesis than in others, these are for the most part cursorily treated with references to the more important papers and to review articles (also see Deer, Howie and Zussman, 1962, *Rock-forming Minerals, Vol. 1, Ortho- and Ring Silicates*).

Some confusion will inevitably result from the definition of the term used as the title for this volume. In Chapter 1 Liebau (p. 14) says that "silicates containing [SiO$_4$] groups should be called *monosilicates* rather than orthosilicates or nesosilicates." The editor chose *not* to adopt Liebau's terminology for the title, because *monosilicate* is not yet widely accepted (although it might well be). To set manageable boundaries for the scope of the First Edition of *Orthosilicates*, an editorial option was exercised in rejecting as "orthosilicates" those minerals with both [SiO$_4$] tetrahedra *and* [Si$_2$O$_7$] groups (zoisite, epidote, vesuvianite, etc.), as well as those with [SiO$_4$] tetrahedra that are polymerized to other tetrahedra by sharing corners with [BeO$_4$], [BO$_4$], [AlO$_4$], [ZnO$_4$], etc. However, as mentioned in the Foreword, Chapter 13 has been added to the Second Edition to correct for the latter omission. Also, Chapter 12 serves as an alphabetical index for the previous ten chapters and includes most of the more obscure minerals that fit our restricted definition of orthosilicate.

ORTHOSILICATES

TABLE of CONTENTS

CHAPTER 11, continued

CHAPTER 12.   <u>MISCELLANEOUS ORTHOSILICATES</u>   J.A. Speer & P.H. Ribbe

INTRODUCTION   393

    This chapter contains brief descriptions of minerals in which $[SiO_4]$ groups
are *not* polymerized to other $[SiO_4]$ groups, nor are they polymerized to
other tetrahedral radicals containing such cations as Be, B, Al or Zn.

    In addition, this may be used as an INDEX to Chapters 2-11, since all miner-
al names of *orthosilicates* as defined in the Preface (p. iv) are listed here
<u>alphabetically</u>, with reference to the place they are described in the text.

CHAPTER 13.   <u>ORTHOSILICATES with SiO_4 Polymerized to</u>
        <u>Other Tetrahedral Polyanions</u>   J.A. Speer & P.H. Ribbe

INTRODUCTION   429

    This chapter contains an alphabetical listing of all the silicates the
authors could locate in major compendia which have been classified as
ortho-, neso-, or monosilicates but in which "isolated" $[SiO_4]$ groups
are polymerized by corner-sharing with other tetrahedral groups, such
as $[BeO_4]$, $[BO_4]$, $[AlO_4]$, and $[ZnO_4]$. Included are some minerals whose
structures are unknown but are suspected to be orthosilicates.

Chapter 1

CLASSIFICATION of SILICATES   F. Liebau

INTRODUCTION

Silicates form the largest single group of minerals. In order
to deal efficiently with the large variety of silicates it is necessary to put them in a suitable order, $i.e.$, to classify them. As is
indicated by the word 'suitable,' classification has no end in itself,
but it does fulfill a purpose. Because there are different purposes,
there will be different classifications. The best classification is
that which fulfills the particular purpose best.

Dealing with the structural chemistry of silicates requires a
classification based on both the atomic structure and the chemistry
of the silicates. In this regard chemistry may be applied directly
as chemical composition and/or as character of the chemical bonds
(covalent, ionic) occurring in the silicates. Since the atomic structure of a substance is controlled by the chemical properties of its
constituents, a classification that is based on atomic structure in
such a way that it reflects the chemistry of the substance would be
a very suitable one.

It is common practice to classify silicates as well as other
minerals according to the kinds of their coordination polyhedra and
the way these polyhedra are linked. A silicate may be represented
as $M'_r, M''_{r'}, M'''_{r''}, \ldots Si_s O_t$, for there is a wide variety of $[MO_n]$ polyhedra in silicates. By contrast, $[SiO_4]$ tetrahedra and $[SiO_6]$
octahedra are the only $[SiO_n]$ polyhedra known to exist in silicates.
This makes $[SiO_n]$ polyhedra the most suitable for a basic classification of silicates. Classification schemes based on the kind and
degree of polymerization of $[SiO_4]$ tetrahedra have been developed
by Bragg (1930) and Náray-Szabó (1930) and extended by Zoltai (1960)
and Liebau (1962, 1972, 1978). In the following a concise description of the present state of this crystal chemical classification of
silicates is given; a more detailed description together with a
discussion of the crystal chemical interpretation is in preparation
(Liebau, 1980).

1

Silicon is either tetrahedrally coordinated by four oxygen atoms or octahedrally by six. Although the number of phases containing $[SiO_6]$ groups is still small ($ca.$ 30) it increases as new high-pressure methods become available.

In principle, $[SiO_4]$ and $[SiO_6]$ groups can either be isolated or share corners, edges or faces (Table 1). Due to strong repulsive forces between the silicon atoms only one example -- fibrous SiO_2 (Weiss and Weiss, 1954) -- with edge-sharing tetrahedra and none with

Table 1. Very broad division of silicate anions; one or more example known, +; no examples, o.

	$[SiO_4]$ tetrahedra	$[SiO_6]$ octahedra
isolated	+	+
corner-shared	+	+
edge-shared	+	+
face-shared	o	o

face-sharing $[SiO_4]$ has been observed. By contrast, for $[SiO_6]$ octahedra edge-sharing (as in stishovite) seems to be as common as corner-sharing. Face-sharing has not been observed in either of the two kinds of $[SiO_n]$ polyhedra. While the large number of silicates containing $[SiO_4]$ tetrahedra require further subdivision, it is premature to further subdivide phases with $[SiO_6]$ octahedra.

Treatment of Tetrahedrally Coordinated Cations

Whenever, under any conditions, there is isomorphous replacement of silicon by cations M in a tetrahedrally coordinated site T, the corresponding $[TO_4]$ tetrahedron is regarded to be part of the silicate anion. Such replacement is most common for Al^{3+} but is also observed for Fe^{3+}, Ge^{4+}, Ti^{4+}, B^{3+}, P^{5+}, Ga^{3+}, and Be^{2+}. If a tetrahedrally coordinated cation M does not replace Si, not even in small amounts and at temperatures and pressures near to the stability limits of the silicate, its $[MO_4]$ tetrahedra are not regarded as part of the silicate anion.

2

According to this convention potassium feldspar is a framework silicate, $K[AlSi_3O_8]$, even in completely ordered maximum microcline, while petalite is a layer silicate of formula $Li^{[4]}Al^{[4]}[Si_4O_{10}]$ since it transforms into a new phase before statistical redistribution (disorder) of Si/Al or Si/Li is achieved.

Dimensionality of Silicate Anions

If a finite number of $[TO_4]$ tetrahedra is linked the resulting complex anion is finite or, in other words, infinite in zero dimensions. Condensation of an infinite number of tetrahedra leads either to one-dimensionally infinite chains or to two-dimensionally infinite layers or to three-dimensionally infinite frameworks. The *dimensionality* of such anions is, therefore, said to be d = 0, 1, 2, or 3.

Unbranched, Branched and Hybrid Silicate Anions

An isolated $[TO_4]$ group that does not share a corner with another $[TO_4]$ group is called a *singular* or *single tetrahedron*. With increasing number of corners shared with other $[TO_4]$ groups the tetrahedra are called *primary*, *secondary*, *tertiary* and *quaternary* $[TO_4]$ *tetrahedra*. Silicate anions containing only primary and secondary $[TO_4]$ tetrahedra are called *linear silicate anions*. They are either multiple tetrahedra or single chains or single rings. Figure 1 presents several typical linear anions.

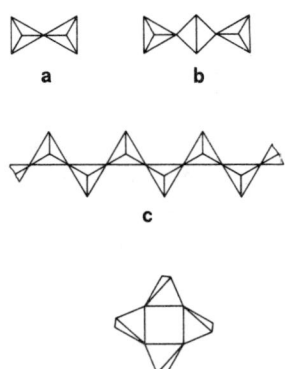

a   b

c

d

Figure 1. Several fundamental linear anions.
(a) Double tetrahedron $[Si_2O_7]$.
(b) Triple tetrahedron $[Si_3O_{10}]$.
(c) Linear single chain $\{\frac{1}{\infty}\}[Si_2O_6]$.
(d) Single ring $\{c\}[Si_4O_{12}]$.

3

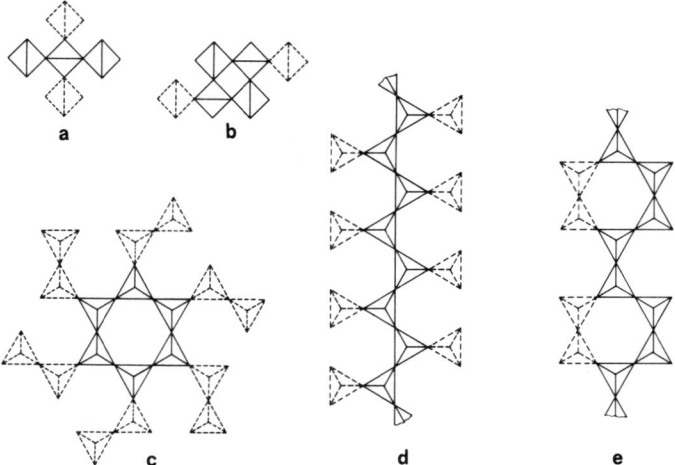

Figure 2. Several fundamental branched silicate anions.

(a) Open branched triple tetrahedron $\{oB\}[Si_5O_{16}]$ of zunyite.

(b) Open branched *vierer* ring $\{oB,c\}[^4Si_6O_{18}]$ of eakerite.

(c) Open branched *sechser* ring $\{oB,c\}[^6Si_{18}O_{54}]$ of tienshanite.

(d) Open branched *zweier* single chain $\{oB, \frac{1}{\infty}\}[^2Si_4O_{12}]$ of astrophyllite.

(e) Loop branched *vierer* single chain $\{\ell B, \frac{1}{\infty}\}[^4Si_6O_{17}]$ of deerite.

Here and in the following figures the tetrahedra regarded as branches
are drawn with broken lines.

Within the last decade more and more silicates have been found
that contain silicate anions in which additional $[TO_4]$ tetrahedra are
linked to a linear silicate anion.   Several such anions, which are
called *branched silicate anions*, are illustrated in Figure 2.   When
the additional $[TO_4]$ tetrahedra (branches) are linked to the linear
part of the silicate anion by only one corner, the anion is called
open branched.   When a branch is bonded by more than one corner to
the linear part, a *loop branched* silicate anion is formed.

The linear as well as the branched anions described so far are
the *fundamental silicate anions*.   They can be linked to larger
composed anions.   All silicate anions that can be regarded as formed
by linking branched fundamental anions are called *branched silicate
anions*.   Silicate anions that can be regarded as formed by linking
branched with unbranched anions are called *hybrid anions*.   All other
silicate anions are called *unbranched anions*.   Figures 3, 4 and 5
present a few anions of each of these groups, which are compiled in
Table 2.

4

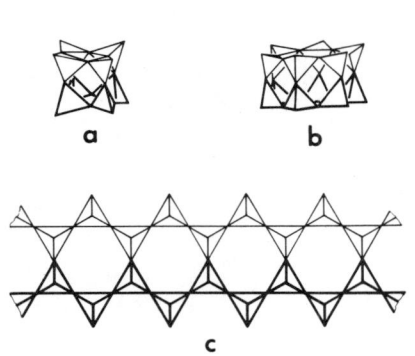
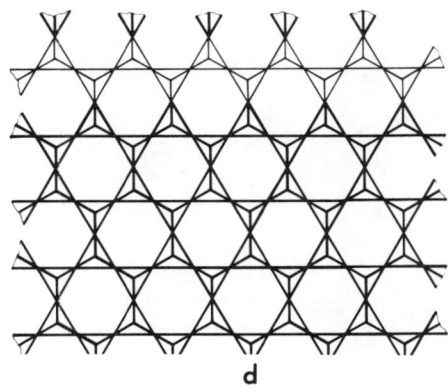
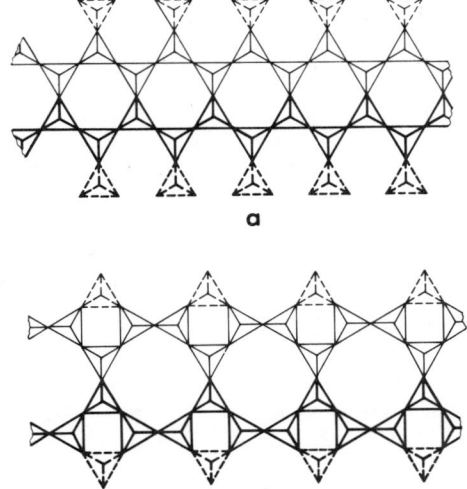
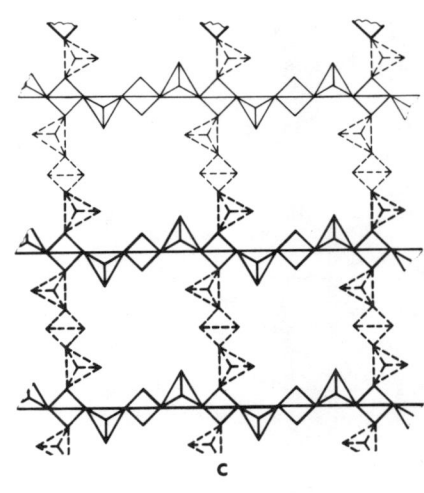

Figure 3. Several composed linear silicate anions.

(a) Unbranched *vierer* double ring $\{U, 2c\}[Si_8O_{20}]$ of ekanite.

(b) Unbranched *sechser* double ring $\{U, 2c\}[Si_{12}O_{30}]$ of milarite.

(c) Unbranched *zweier* double chain $\{U, 2\frac{1}{\infty}\}[^2Si_4O_{11}]$ of amphiboles.

(d) Unbranched *zweier* single layer $\{U, \frac{2}{\infty}\}[^2Si_4O_{10}]$ of micas.

In each composed anion of Figure 3 the fundamental anion is drawn with slightly thinner lines.

Figure 4. Several composed branched silicate anions.

(a) Open branched *zweier* double chain $\{oB, 2\frac{1}{\infty}\}[^2Si_6O_{17}]$, hypothetical.

(b) Loop branched *dreier* double chain $\{lB, 2\frac{1}{\infty}\}[^3Si_8O_{21}]$, hypothetical.

(c) Open branched *vierer* single layer $\{oB, \frac{2}{\infty}\}[^4Si_{14}O_{40}]$ of meliphanite.

The linear part of each anion of Figures 4 and 5 is drawn with solid heavy lines, the branched part with thinner lines, the branches with broken lines.

5

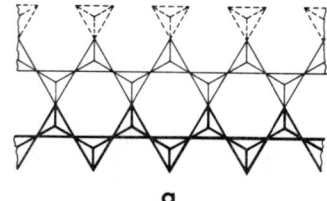 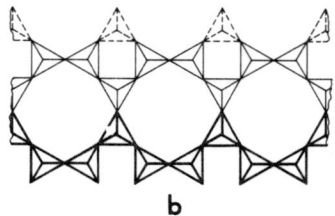

a   b

Figure 5. Hybrid silicate anions.
(a) Hybrid *zweier* double chain $\{H, 2_\infty^1\}[^2Si_5O_{14}]$, hypothetical.
(b) Hybrid *dreier* double chain $\{H, 2_\infty^1\}[^3Si_7O_{19}]$ of tinaksite.

Multiplicity of Silicate Anions

A finite number of $[TO_4]$ tetrahedra can be linked to a multiple
tetrahedron. So far multiple tetrahedra with *multiplicities* m = 2, 3,
4, 8, 9, and 10 have been observed. Several $[TO_4]$ tetrahedra can form
a single ring; a few silicates are known in which two rings are linked
to a double ring. Cyclic silicate anions with multiplicities $m > 2$
have not yet been discovered. An infinite number of tetrahedra can
form a single chain. Crystalline silicates having multiple chains
with m = 2, 3, 4, and 5 have been described. An infinite number of
single chains form a single layer. Several silicates containing double

Table 2. The various classes of silicate anions.

		single tetrahedron		
		UNBRANCHED ANIONS	BRANCHED ANIONS	HYBRID ANIONS
FUNDAMENTAL ANIONS		LINEAR ANIONS multiple single single tetrahedra chains rings	branched branched branched multiple single single tetrahedra chains rings	
COMPOSED ANIONS		multiple multiple chains rings	branched branched multiple multiple chains rings	hybrid multiple hybrid multiple chains rings
		single layers	branched single layers	hybrid single layers
		multiple layers	branched multiple layers	hybrid multiple layers
		frame- works	branched frame- works	hybrid frame- works

6

layers have been discovered but none yet with $m > 2$. Condensation
of an infinite number of single layers, eventually, leads to a three-
dimensional framework of $[TO_4]$ tetrahedra.

Consideration of the multiplicity results in a periodic system
of silicate anions as can be seen from Table 3. For each of the
classes -- unbranched, branched, and hybrid silicate anions -- such
a periodic table can be set up. However, at the present time the
very small number of silicates with hybrid anions makes such a subdivision for this class unnecessary.

Periodicity of Silicate Anions

In crystalline silicates containing linear single chains, the
structural motif of the silicate chain repeats after every p tetrahedra
where p is the *periodicity* of the chain. So far linear single chains
with p = 2, 3, 4, 5, 6, 7, 9, and 12 have been discovered.

The periodicity of a branched chain is the number of tetrahedra
of the linear portion of one period of the chain. According to their
periodicity single chains with p = 1, 2, 3, ... are called either *einer*
single chain, *zweier* single chain, *dreier* single chain .... Figures 6
and 7 show one example of each type of known linear and branched single
chains.

Multiple chains, layers and frameworks are subdivided according
to the fundamental single chain type from which they can be generated
by continuous linking of the chains. Figures 8 to 13 present a few
examples in which the fundamental chain type is set off with open tetrahedra while the other tetrahedra are given as solid polyhedra. The
subdivision according to chain periodicities of silicate anions with
dimensionalities $d \geqslant 1$ is presented in Table 4.

In the same way as chains are classified by their periodicity,
single rings are subdivided by their numbers of $[TO_4]$ tetrahedra. So
far silicate rings with p = 3, 4, 6, 8, 9, and 12 and double rings
with p = 3, 4, and 6 have been found in crystalline silicates (Figures
14 and 15).

7

Table 3. Classification of silicate anions based on multiplicity and dimensionality.

Dimensionality \ Multiplicity		UNBRANCHED											
		single	double	triple	4-fold	5-fold	6-fold	7-fold	8-fold	9-fold	10-fold	......	∞-fold
0	tetrahedra	+	+	+	+				+	+	+		+
	rings	+	+										+
1	chains	+	+	+	+	+							+
2	layers	+	+										+
3	frameworks	+											

Dimensionality \ Multiplicity		OPEN BRANCHED						LOOP BRANCHED					
		single	double	triple	4-fold	......	∞-fold	single	double	triple	4-fold	......	∞-fold
0	tetrahedra	0	0	+			+						+
	rings	+		+									
1	chains	+	+				+	+	+				+
2	layers	+					+	+	+				+
3	frameworks	+						+					

See Table 4 for symbols.

Table 4. Periodic table of silicate anions with dimensionalities $d > 1$ subdivided according to their periodicities. + : At least one example is known. o : Such anions are theoretically not possible.

Periodicity \ Dimensionality (Multiplicity)	UNBRANCHED ANIONS			BRANCHED ANIONS		
	1 (chains)	2 (layers)	3 (frameworks)	1 (chains)	2 (layers)	3 (frameworks)
	12345 ... ∞	123 ... ∞		1234 ... ∞	123 ... ∞	1
1 (einer)	+	oo	o	o	oo	o
2 (zweier)	++++ +	++ +	+	++ +		
3 (dreier)	++ +	++ +	+	+	+ +	+
4 (vierer)	++ +	++ +	+	+ +	+ +	+
5 (fünfer)	++			+	+	
6 (sechser)	++ +	+	+	+	+ ++	
7 (siebener)	+					
8 (achter)				+		
9 (neuner)	+ + +					
10 (zehner)				+		
11 (elfer)						
12 (zwölfer)	+					

8

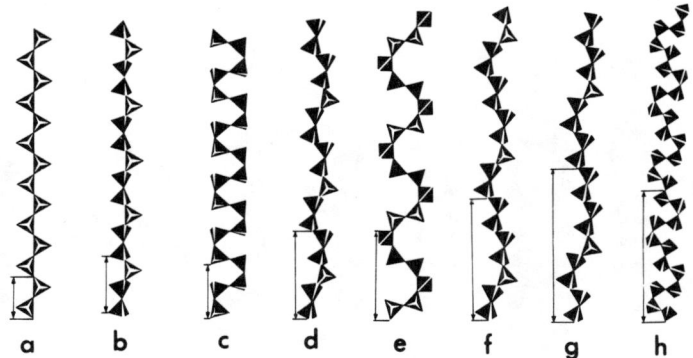

Figure 6. Unbranched single chain silicate anions.

(a) Unbranched *zweier* single chain of the pyroxenes, $M(1)M(2)\{U,\frac{1}{\infty}\}[^2Si_2O_6]$.

(b) Unbranched *dreier* single chain of wollastonite, $Ca_3\{U,\frac{1}{\infty}\}[^3Si_3O_9]$.

(c) Unbranched *vierer* single chain of haradaite, $Sr_2(VO)_2\{U,\frac{1}{\infty}\}[^4Si_4O_{12}]$.

(d) Unbranched *fünfer* single chain of rhodonite, $(Mn,Ca)_5\{U,\frac{1}{\infty}\}[^5Si_5O_{15}]$.

(e) Unbranched *sechser* single chain of stokesite, $Ca_2Sn_2\{U,\frac{1}{\infty}\}[^6Si_6O_{18}]\cdot4H_2O$.

(f) Unbranched *siebener* single chain of pyroxferroite, $(Fe,Ca)_7\{U,\frac{1}{\infty}\}[^7Si_7O_{21}]$.

(g) Unbranched *neuner* single chain of ferrosilite III, $Fe_7\{U,\frac{1}{\infty}\}[^9Si_9O_{27}]$.

(h) Unbranched *zwölfer* single chain of alamosite, $Pb_{12}\{U,\frac{1}{\infty}\}[^{12}Si_{12}O_{36}]$.

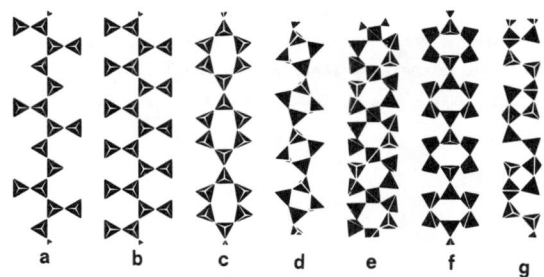

Figure 7. Branched single chain silicate anions.

(a) Open branched *vierer* single chain of aenigmatite, $Na_2Fe_5Ti\{oB,\frac{1}{\infty}\}[^4Si_6O_{18}]O_2$.

(b) Open branched *zweier* single chain of astrophyllite, $KNa_2Mg_2(Fe,Mn)_5Ti_2\{oB,\frac{1}{\infty}\}[^2Si_4O_{12}]_2(O,OH,F)_5$.

(c) Loop branched *vierer* single chain of deerite, $Fe_6^{2+}Fe_3^{3+}\{\ell B,\frac{1}{\infty}\}[^4Si_6O_{17}]O_3(OH)_5$.

(d) Loop branched *sechser* single chain of vlasovite, $Na_4Zr_2\{\ell B,\frac{1}{\infty}\}[^6Si_8O_{22}]$.

(e) Loop branched *sechser* single chain of lemoynite, $(Na,K)CaZr_2\{\ell B,\frac{1}{\infty}\}[^6Si_{10}O_{26}]$.

(f) Loop branched *achter* single chain of pellyite, $Ba_4Ca_2(Fe,Mg)_4\{\ell B,\frac{1}{\infty}\}[^8Si_{12}O_{34}]$.

(g) Loop branched *zehner* single chain of nordite, $Na_4(Na,Mn)_2(Sr,Ca)_2RE_2(Zn,Fe,Mg,Mn)\{\ell B,\frac{1}{\infty}\}[^{10}Si_{12}O_{34}]$.

Structural Formulae of Silicate Anions

The structural formulae of silicates should give as much information as necessary about its structure and should be as self-explanatory as possible. If the property of being unbranched, branched or hybrid is called the *branchedness* and given the symbol b, then the general

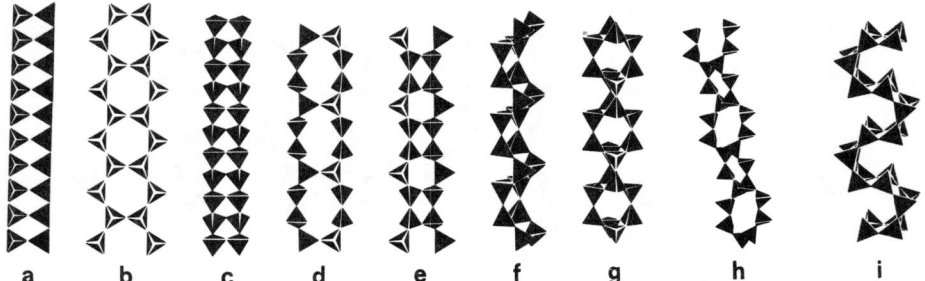

Figure 8. Unbranched double chain silicate anions.

(a) Unbranched *einer* double chain of sillimanite, $Al\{U, 2_\infty^1\}[^1(SiAl)O_5]$.

(b) Unbranched *zweier* double chain of amphibole, $Ca_2Mg_5\{U, 2_\infty^1\}[^2Si_4O_{11}]_2(OH)_2$.

(c) Unbranched *zweier* double chain of synthetic $Li_4\{U, 2_\infty^1\}[^2(SiGe_3)O_{10}]$.

(d) Unbranched *dreier* double chain of xonotlite, $Ca_6\{U, 2_\infty^1\}[^3Si_6O_{17}](OH)_2$.

(e) Unbranched *dreier* double chain predicted in devitrite, $Na_2Ca_3\{U, 2_\infty^1\}[^3Si_6O_{16}]$.

(f) Unbranched *dreier* double chain of synthetic $Na_2Be_2H\{U, 2_\infty^1\}[^3Si_6O_{15}](OH)$.

(g) Unbranched *vierer* double chain of narsarsukite, $Na_4(TiO)_2\{U, 2_\infty^1\}[^4Si_8O_{20}]$.

(h) Unbranched *fünfer* double chain of inesite, $(Mn,Ca)_9\{U, 2_\infty^1\}[^5Si_{10}O_{28}](OH)_2 \cdot 5H_2O$.

(i) Unbranched *sechser* double chain of tuhualite, $(Na,K)_2Fe_2^{2+}Fe_2^{3+}\{U, 2_\infty^1\}[^6Si_{12}O_{30}] \cdot H_2O$.

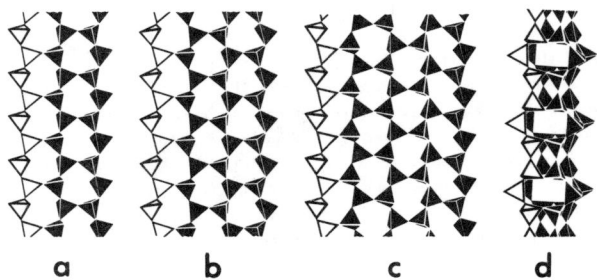

Figure 9. Unbranched multiple chain silicate anions.

(a) Unbranched *zweier* triple chain of synthetic $Ba_4\{U, 3_\infty^1\}[^2Si_6O_{16}]$.

(b) Unbranched *zweier* fourfold chain of synthetic $Ba_5\{U, 4_\infty^1\}[^2Si_8O_{21}]$.

(c) Unbranched *zweier* fivefold chain of synthetic $Ba_6\{U, 5_\infty^1\}[^2Si_{10}O_{26}]$.

(d) Unbranched *dreier* fourfold chain of miserite, $K_2Ca_{10}[Si_2O_7]_2\{U, 4_\infty^1\}[^3Si_{12}O_{30}](OH)_2F_2$.

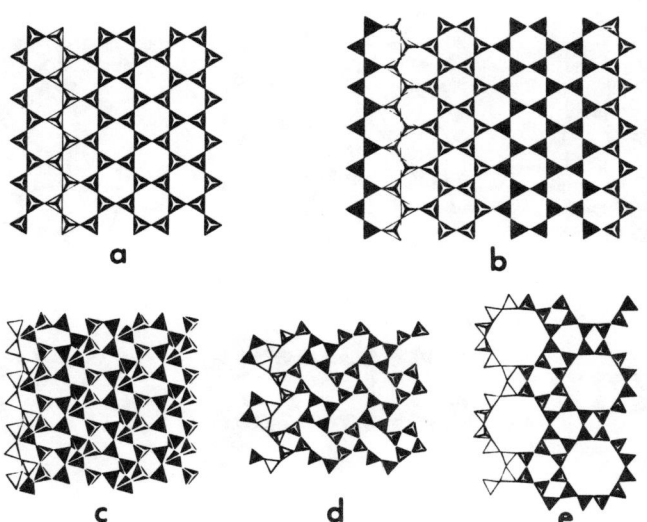

Figure 10. Several unbranched single layer silicate anions.

(a) Unbranched *zweier* single layer of mica, e.g. muscovite, $KAl_2\{U,^2_\infty\}[^2(Si_3Al)O_{10}](OH)_2$.

(b) Unbranched *zweier* single layer of sepiolite, $(Mg,Fe,Al)_8\{U,^2_\infty\}[^2Si_4O_{10}]_3(O,OH)_4\cdot 8H_2O$.

(c) Unbranched *dreier* single layer of dalyite, $K_2Zr\{U,^2_\infty\}[^3Si_6O_{15}]$.

(d) Unbranched *vierer* single layer of apophyllite, $K\,Ca_4\{U,^2_\infty\}[^4Si_8O_{20}]F\cdot 8H_2O$.

(e) Unbranched *sechser* single layer of manganese pyrosmalite, $Mn_{16}\{U,^2_\infty\}[^6Si_{12}O_{30}](OH)_{18}Cl_2$.

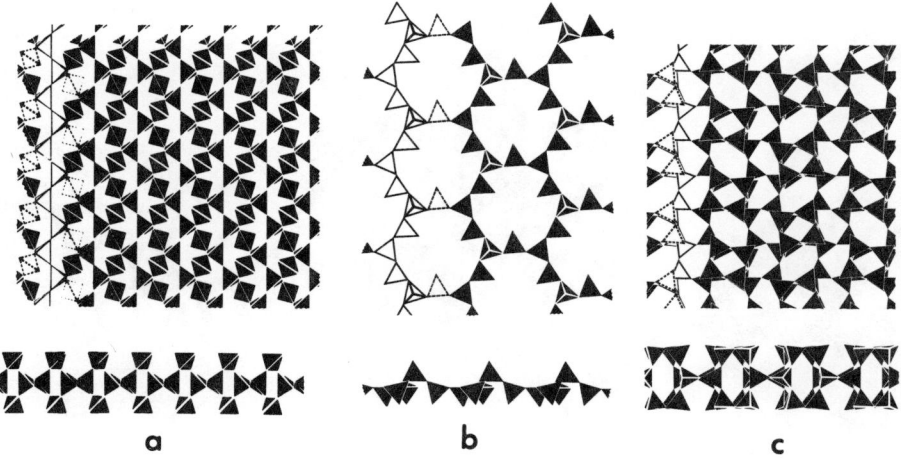

Figure 11. Several branched single layer silicates.

(a) Open branched *zweier* single layer of prehnite, $Ca_2(Al,Fe)\{oB,^2_\infty\}[^2(Si_3Al)O_{10}](OH)_2$.

(b) Open branched *vierer* single layer of zeophyllite, $Ca_{13}\{oB,^2_\infty\}[^4Si_5O_{14}]_2F_8(OH)_2\cdot 6H_2O$.

(c) Loop branched *vierer* single layer of synthetic $NaPr\{lB,^2_\infty\}[^4Si_6O_{14}]$.

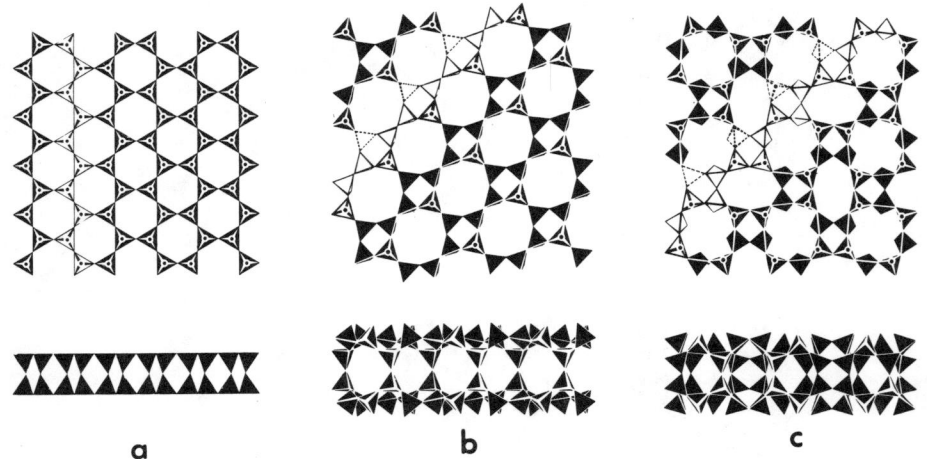

Figure 12. Several double layer silicate anions.

(a) Unbranched *zweier* double layer of hexacelsian, $Ba\{U, {}^2_\infty\}[{}^2(Si_2Al_2)O_8]$.

(b) Loop branched *dreier* double layer of delhayelite, $Ca_4(Na_3Ca)K_7\{lB, {}^2_\infty\}[{}^3_\infty Si_{14}Al_2)O_{38}]Cl_2F_4$.

(c) Loop branched *sechser* double layer of carletonite, $K_2Na_8Ca_8\{lB, {}^2_\infty\}[{}^6Si_{16}O_{36}][CO_3]_8(OH,F)_2 \cdot 2H_2O$.

Two sublayers of each of the kind shown are linked via the oxygen atoms marked by dots and lying on mirror or pseudomirror planes.

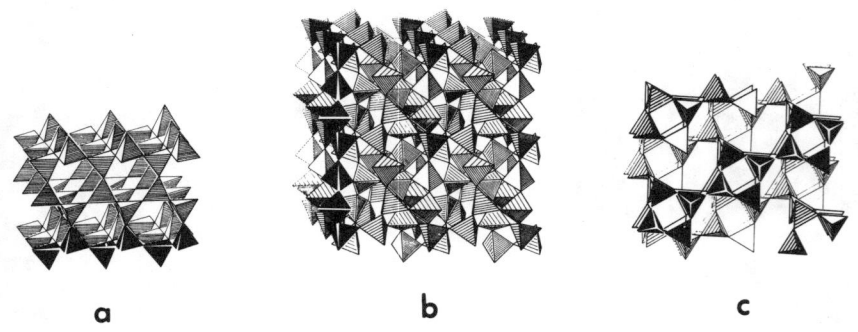

Figure 13. A few silicate frameworks.

(a) Unbranched *zweier* framework of tridymite, $\{U, {}^3_\infty\}[{}^2Si_4O_8]$.

(b) Unbranched *dreier* framework of keatite, $\{U, {}^3_\infty\}[{}^3Si_6O_{12}]$.

(c) Loop branched *dreier* framework of feldspars, e.g. orthoclase, $K_2\{lB, {}^3_\infty\}[{}^3(Si_6Al_2)O_{16}]$.

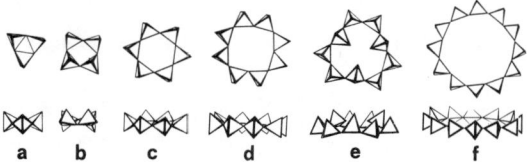

Figure 14. Unbranched single ring silicate anions.

(a) *Dreier* single ring of benitoite, $BaTi\{U,c\}[Si_3O_9]$.

(b) *Vierer* single ring of taramellite, $Ba_2(Fe,Ti,Mg)_2\{U,c\}[Si_4O_{12}]$.

(c) *Sechser* single ring of beryl, $Al_2Be_3\{U,c\}[Si_6O_{18}]$.

(d) *Achter* single ring of muirite, $Ba_{10}(Ca,Mn,Ti)_4\{U,c\}[Si_8O_{24}](Cl,O,OH)_{12}\cdot4H_2O$.

(e) *Neuner* single ring of eudialyte,

$(Fe,Mn,Mg)_3Zr_3(Zr,Nb)_x(Ca,R.E.)_6Na_{12}\{U,c\}[Si_3O_9]_2\{U,c\}[Si_3O_{27-y}(OH)_y]_2Cl_z$.

(f) *Zwölfer* single ring of traskite,

$Ba_{24}(Fe,Ti)_2(Fe,Mn)_2(Ca,Sr)(Ti,Fe,Mg,Al)_{12}\{U,c\}[Si_{12}O_{36}][Si_2O_7]_6(O,OH)_{30}Cl_6\cdot14H_2O$.

structural formula of a silicate anion can be described as

$$\{b,md\}[^pSi_xO_y]$$

where m, d and p are the multiplicity, dimensionality and periodicity of the anion. The branchedness is specified as U for unbranched, oB for open branched, ℓB for loop branched and H for hybrid anions. For anions having multiplicities $m \geqslant 2$ the multiplicity is indicated by the appropriate ciphers 2, 3, 4, *etc*. The dimensionality is designated as $\frac{1}{\infty}$ for one-dimensionally infinite chains, as $\frac{2}{\infty}$ for 2-dimensionally infinite layers and as $\frac{3}{\infty}$ for 3-dimensional frameworks. The zero-dimensional cyclic anions are indicated as c, and the likewise zero-dimensional oligosilicate anions, *i.e.* the multiple tetrahedra, as t.

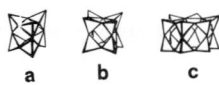

a   b   c

Figure 15
Unbranched double ring silicate anions.

(a) *Dreier* double ring of synthetic
$[Ni(H_2N.CH_2.CH_2.NH_2)_3]_3$
$\{U,2c\}[^3Si_6O_{15}]\cdot26H_2O$.

(b) *Vierer* double ring of ekanite,
$K(Ca,Na)_2Th\{U,2c\}[^4Si_8O_{20}]$.

(c) *Sechser* double ring of milarite
$KCa_2AlBe_2\{U,2c\}[^6Si_{12}O_{30}]\cdot1/2H_2O$.

For unbranched single rings and unbranched single chains the number x of silicon atoms is equal to the number of tetrahedra in the ring or in the repeat unit of the chain, that is equal to the periodicity p. For branched single rings and branched single chains as well as for all silicate anions with higher condensation degrees the periodicity can be indicated by a left-sided

13

superscript to the silicon atoms within the square bracket. Examples of these formulae are to be found in the legends to the figures.

Nomenclature of Silicates

Unfortunately, mineralogists, chemists, and ceramists each use different nomenclatures for silicates. The most commonly used nomenclature is the chemical one which is also the most systematic one. In addition it has the advantage to be not only valid for silicates but for any other inorganic substance and offers more information than the mineralogical nomenclature. Table 5 presents the modern chemical notation for the main groups of silicates together with the notation usually used by mineralogists. Note that the classification is the same as that of Table 3.

Table 5. Nomenclature of silicates *

OLIGOSILICATES (——)	MONOSILICATES (Nesosilicates)	DISILICATES	TRISILICATES	TETRASILICATES	
		(S o r o s i l i c a t e s)			
CYCLOSILICATES	MONOCYCLOSILICATES (——)	DICYCLOSILICATES (——)			
POLYSILICATES (Inosilicates)	MONOPOLSILICATES (——)	DIPOLYSILICATES (——)	TRIPOLYSILICATES (——)	TETRAPOLYSILICATES (——)	PENTAPOLY- SILICATES (——)
PHYLLOSILICATES	MONOPHYLLO- SILICATES (——)	DIPHYLLO- SILICATES (——)			
TECTOSILICATES	TECTOSILICATES				

*Where the nomenclature used by mineralogists differs from the general one it is given in parentheses. (——) : no distinction is made by mineralogists.
Note that the classification is the same as that of the Periodic Table of Silicate Anions in Table 3

Although the chemical nomenclature could be used to describe silicates with even more complicated anions unequivocally this would lead to rather long and inconvenient names. If needed the same information can more conveniently be given by the structural formula of the silicate. It is recommended to use the nomenclature set in capital letters in Table 5. This means that silicates containing $[SiO_4]$ groups should be called *monosilicates* rather than orthosilicates or nesosilicates, those containing $[Si_2O_7]$ double tetrahedra should be called *disilicates* rather than pyrosilicates. Particularly the name metasilicate should be dropped and replaced by *monocyclosilicate* or *monopolysilicate* depending on whether the anion is a single ring or a single chain.

Mixed Anion Silicates

The classification of silicate anions described in the previous section can directly be applied to silicates that contain only one kind of silicate anions. However, a steadily increasing number of silicates that contain more than one kind of anion are being discovered. Table 6 contains a selection of such *mixed anion silicates*. These can be filed into separate groups just after the corresponding anion with the higher condensation degree. Then, for example, epidote, $Ca_2(Al,Fe)_3[SiO_4][Si_2O_7]O(OH)$, would be listed just after the silicates containing double tetrahedra, and chesterite,

$(Mg,Fe)_{17}\{U,2\frac{1}{\infty}\}[Si_4O_{11}]_2\{U,3\frac{1}{\infty}\}[Si_6O_{16}]_2(OH)_6$, just after the triple

Table 6. Mixed anion silicates

silicate	silicate anions			
zoisite, epidote, ganomalite, vesuvian serendibite, rustumite	$[SiO_4]$		$[Si_2O_7]$	
kilchoanite, ardennite	$[SiO_4]$		$[Si_3O_{10}]$	
joesmithite	$[SiO_4]$		$\{U,\frac{1}{\infty}\}$	$[^2Si_2O_6]$
meliphanite	$[SiO_4]$		$\{oB,\frac{2}{\infty}\}$	$[^4Si_7O_{20}]$
traskite	$[Si_2O_7]$		$\{U,c\}$	$[Si_{12}O_{36}]$
miserite	$[Si_2O_7]$		$\{U,4\frac{1}{\infty}\}$	$[^3Si_{12}O_{30}]$
bavenite	$[Si_3O_{10}]$		$\{oB,2\frac{1}{\infty}\}$	$[^2Si_6O_{16}]$
eudialyte	$\{U,c\}$	$[Si_3O_9]$	$\{U,c\}$	$[Si_9(O,OH)_{27}]$
vinogradovite	$\{U,\frac{2}{\infty}\}$	$[^2Si_2O_6]$	$\{U,2\frac{1}{\infty}\}$	$[^2Si_4O_{10}]$
chesterite	$\{U,2\frac{1}{\infty}\}$	$[^2Si_4O_{11}]$	$\{U,3\frac{1}{\infty}\}$	$[^2Si_6O_{16}]$
reyerite	$\{U,\frac{2}{\infty}\}$	$[^4Si_8O_{20}]$	$\{U,2\frac{2}{\infty}\}$	$[^4(Si_{14}Al_2)O_{38}]$
high-pressure "garnet"	$[SiO_4]$		$[SiO_6]$	
synthetic $Si_5(PO_4)_6O$	$[SiO_4]$		$[SiO_6]$	
high-pressure $K_2Si_4O_9$	$\{U,c\}$	$[Si_3O_9]$	$[SiO_6]$	

chain silicates. A few silicates, e.g., some high-pressure garnets such as $Mg_3(Si_{0.5}AlMg_{0.5})^{[6]}[SiO_4]_3$, contain both tetrahedrally and octahedrally coordinated silicon.

Further Subdivision of Silicates

For some of the groups of silicates classified so far, the number of known members is so large that further subdivision is necessary. Such subdivision can either be based on properties of the silicate anions or by taking into account the $[MO_n]$ polyhedra and the way they are linked. The first of these two methods is particularly suitable for silica-rich silicates having an atomic ratio $Si:O > 1:3$. The latter method is highly recommended for more basic silicates having $Si:O < 1:3$, since these silicates contain in general more $[MO_n]$ poly-hedra than $[SiO_4]$ tetrahedra.

Due to the large variation in size of the M cations their coor-dination numbers vary from 1 and 2 for hydrogen up to about 12 for large cations such as K^+, Cs^+ or Ba^{2+}. The coordination polyhedra exhibit an even larger diversity from very small, rather regular and rigid polyhedra to the large very irregular and readily deformable polyhedra. This makes a general classification based on $[MO_n]$ poly-hedra which is similarly systematic as the one based on $[SiO_4]$ difficult and rather inconvenient.

The large variation of cation valence from 1 to 4 if only the normal silicates are considered or even to 7 if synthetic phases with P^{5+}, S^{6+} and Cl^{7+} are also taken into consideration adds to the un-suitability of $[MO_n]$ polyhedra for a consequent subclassification. Instead, for each of the groups of silicates classified according to their silicate anions the most appropriate way to subdivide has to be chosen separately. Nevertheless, $[MO_n]$ polyhedra in some way or other should definitely be used to subclassify silicates with $Si:O < 1:3$.

Shortcomings of the Crystal Chemical Classification of Silicates

Like any other classification the crystal chemical classification of silicates described here has some shortcomings. One of these is that the structure of the silicate to be classified has to be known because the classification is based on atomic structure. For

crystalline silicates, i.e., for the majority of silicate minerals,
this is no major problem since the development of powerful diffraction
methods for structure determination. For vitreous and molten silicates
as well as for silicates in aqueous solution the situation is less
favorable.

Classifying into groups means assembling as well as separating
according to some rules chosen. It is, therefore, unavoidable that
in some cases two species, which are so similar that one would prefer
to have them belonging to the same group, are separated into different
groups. This is true no matter where the limits are drawn. One of
the classification rules used is that cations, which at least under
certain conditions show isomorphous replacement with silicon are,
together with their oxygen atoms, regarded as part of the silicate
anion. Here one has to decide up to which Si content a $[TO_4]$ tetrahedron is to be regarded as belonging to the silicate anion.

Anorthite, $CaAl_2Si_2O_8$, and the very calcium-rich plagioclases,
$Ca_{1-x}Na_xAl_{2-x}Si_{2+x}O_8$, suggest that the limiting Si content of a given
tetrahedral site should be chosen well below the 1 percent level.
Otherwise both anorthite as well as, for example,
$Ca_{0.99}Na_{0.01}Al_{1.99}Si_{2.01}O_8$, would have to be classified as monosilicates $CaAl_2{}^{[4]}[SiO_4]_2$ and $(Ca_{0.99}Na_{0.01})(Al_{0.995}Si_{0.005})_2[SiO_4]_2$,
respectively, rather than as framework silicates with the feldspars.
However, no matter which Si content of a tetrahedral site is chosen,
the limit between monosilicates $(Ca,Na)(Si,Al)_2{}^{[4]}[SiO_4]_2$ and tectosilicates $(Ca,Na)\{{}^3_\infty\}[(Si,Al)_4O_8]$ has to be set somewhere, separating
very similar silicates into very different structural classes.
On the other hand, it seems more reasonable to classify the synthetic
pyroxene derivative $(Mg_{38.46}Sc_{3.11})(Li_{1.16}Si_{0.18}Si_{40})O_{124}$ (Takéuchi et $al.$,
1977), in which one of the T sites is statistically occupied by
0.58 Li and 0.09 Si and is to 33 percent vacant, as a decasilicate
$(Mg_{19.23}Sc_{1.555})(Li_{0.58}Si_{0.09}[]_{0.33}){}^{[4]}[Si_{10}O_{31}]_2$ rather than as a
loop branched $zweiundzwanziger$ single chain
$(Mg_{38.46}Sc_{3.11})\{lB,{}^1_\infty\}[{}^{22}(Li_{1.16}Si_{40.18}[]_{0.66})O_{124}]$. This, however,
would require one not to regard the lithium containing tetrahedral

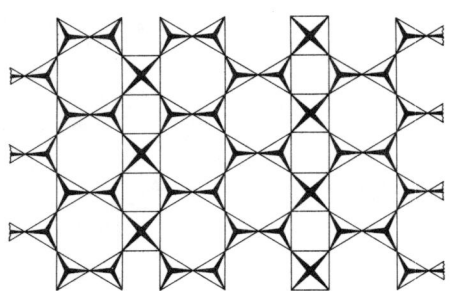

Figure 16. In astrophyllite open branched silicate single chains are linked with [TiO$_6$] octahedra to form composite single layers.

site as part of the silicate anion, even though it contains 9 percent silicon.

Another problem arises sometimes with branched silicates. Almost all branched silicates contain strongly electronegative cations (Liebau, 1978) such as Fe^{2+}, Fe^{3+}, Mn^{2+}, Ti^{4+}, Cu^{2+}, B^{3+}, etc. which form rather rigid cation-oxygen polyhedra. The bonds between these cations and oxygen are to a considerable extent covalent, some even more covalent than the Si-O bond, so that these polyhedra together with the silicate anion might be regarded as a composite anion. In fact, in some cases the branched silicate anion is complemented by these rigid cation-oxygen polyhedra to more complex anions. As an example Figure 16 demonstrates how the open branched *zweier* single chains of astrophyllite are complemented by [TiO$_6$] octahedra to an unbranched single layer.

The shortcomings mentioned are caused by the fact that the silicate anions have been chosen for the classification and that, therefore, a decision had to be made which atoms to consider as belonging to the silicate anions. If the limits as to which atoms to include into the anion were extended, the intricacies would only be shifted somewhere else. To reduce these shortcomings the non-silicon cations should also be taken into consideration, for instance by using them for subclassification, especially for silicates with strongly electronegative cations, with tetrahedrally coordinated cations and with high M:Si ratios. In any case, a classificiation should not be used thoughtlessly; it should not be a strait-jacket but an aid.

Kostov's Classification

Kostov (1975) has developed a classification of silicates to meet the needs of geoscientists in particular. Therefore, he based this classification only partly on crystal structure and to the other part on chemical composition and on crystal morphology of the silicates. As can be seen from Table 7 Kostov uses the degree of silification in form of the ratio (Si,Al):M' for a first broad classification of silicates. Here M' is equal to M_2^+, M^{2+}, $M_{0.67}^{3+}$ and $M_{0.5}^{4+}$ so that (Si,Al):M' is approximately inversely proportional to the Si:O atomic ratio, i.e. to the degree of condensation of the $[(Si,Al)O_4]$ tetrahedra.

Table 7. Kostov's classification of silicates

Cations	(Si,Al):M'		
	> 4 : 1	3:1 to 1:1	< 1 : 1
Be,Al,Mg(Fe)	axial, planar, isometric	axial, planar, isometric	axial, planar, isometric
Zr,Ti(Sn),Nb	axial, planar, isometric	axial, planar, isometric	axial, planar, isometric
Ca(RE),Mn,Ba	axial, planar, isometric	axial, planar, isometric	axial, planar, isometric
Zn,Cu,Pb(U)	axial, planar, isometric	axial, planar, isometric	axial, planar, isometric

Since some elements often occur together and very often replace each other isomorphously in natural silicates, each of these three classes is subdivided into four geochemically distinct groups. Silicates of the first of these groups contain Be, Al, Mg and Fe as principal cations, those of the second contain Zr, Ti, Sn and Nb, those of the third contain Ca, rare earth elements, Mn and Ba, and those of the fourth contain Zn, Cu, Pb and U.

In addition to these three main classes with (Si,Al):M' > 4:1, 3:1 $\geqslant$ (Si,Al):M' $\geqslant$ 1:1 and (Si,Al):M' < 1:1, there is a fourth class containing the borosilicates. The borosilicates, as well as each of the 3 x 4 groups of nonborosilicates, are then further subclassified according to their morphology into three families containing silicates with axial (elongated), planar, and isometric crystals.

Although the crystal structure is not used in this classification it is indirectly introduced, once by using the (Si,Al):M' ratio, and again by using the crystal morphology. Since the latter is controlled by the Si–O bonds as well as by the M–O bonds, one of the shortcomings of the crystal chemical classification, the over-emphasis of the Si–O bonds, is avoided. On the other hand, there are other inadequacies not observed in the crystal chemical classification. For example, in Kostov's classification the wollastonite family is listed as to contain silicates with double chains (xonotlite), single chains (wollastonite and foshagite), single rings (pseudowollastonite), triple tetrahedra (rosenhahnite) and $[SiO_6]$ octahedra (thaumasite). In another family, monosilicates (alleghanyite, leucophoenicite and sonolite), and mono-polysilicates (pyroxmangite and pyroxferroite) are grouped together, while the olivines Mg_2SiO_4 and Fe_2SiO_4 are separated from Mn_2SiO_4 and $CaMnSiO_4$, and albite, $NaAlSi_3O_8$, is separated from its boron analogue reedmergnerite, $NaBSi_3O_8$.

Zoltai's Classification

Zoltai (1960) starts by dividing silicates into classes in which the silicate anions have dimensionalities d = 0, 1, 2 or 3, respectively, and a fifth class of silicates containing anions of more than one of these four classes. The first three of these classes are subdivided according to the multiplicity of the silicate anions. For further subclassification a *sharing coefficient*

$$f_{sh} = 2n + 1 - \frac{A}{4T} (n^2 + n)$$

is introduced, in which A is the number of oxygen atoms and T the number of tetrahedrally coordinated cations in the silicate anion of general formula $[(Si,Al,...)_T O_A]$ and n is the integral part of 4T/A. This system is illustrated in Table 8.

In order to have a classification of substances containing tetrahedra that is universally applicable, Zoltai allows not only for corner, edge and face sharing of the tetrahedra but also for sharing of one tetrahedral corner by more than two tetrahedra. In fact, the sharing coefficient defined above is the average number of tetrahedra participating in the sharing of a corner in the particular structure.

20

Table 8. Zoltai's classification applied to silicate
anions containing only corner-shared tetrahedra

Types	Subtypes	f_{sh}	T : O
1) Isolated groups of tetrahedra	a) single tetrahedron	1.00	1 : 4
	b) double tetrahedron	1.25	1 : 3.5
	c) triple tetrahedron	1.33	1 : 3.33
	d) large groups	1.33-1.50	1:3.33 - 1:3
	e) mixed groups	1.00-1.50	1:3.67 - 1:3
2) 1-dimensionally non-terminated structures of tetrahedra	a) single chains	1.50	1:3
	b) single rings	1.50	1:3
	c) double chains	1.50 - 1.75	1:3 - 1:2.5
	d) double rings	1.50 - 1.75	1:3 - 1:2
	e) multiple chains	1.50 - 2.00	1:3 - 1:2
	f) multiple rings	1.50 - 2.00	1:3 - 1:2
	g) mixed chains and rings	1.50 - 2.00	1:3 - 1:2
3) 2-dimensionally non-terminated structures of tetrahedra	a) single layers	1.50 - 2.00	1:3 - 1:2
	b) double layers	1.50 - 2.00	1:3 - 1:2
	c) multiple layers	1.50 - 2.00	1:3 - 1:2
	d) mixed layers	1.50 - 2.00	1:3 - 1:2
4) 3-dimensionally non-terminated structures of tetrahedra		1.50 - 2.00	1:3 - 1:2

In addition Zoltai accepts all tetrahedrally coordinated cations in
the silicate anion regardless of isomorphous replacement and valence.
That means tetrahedral Li^+, Be^{2+} and Zn^{2+} are considered part of the
silicate anion just as tetrahedral Al^{3+} and Si^{4+}. This makes phenacite, Be_2SiO_4, and willemite, Zn_2SiO_4, tectosilicates in which each
oxygen atom is linking three tetrahedra, rather than monosilicates,
and Li_2SiO_3 would be classified as framework silicate rather than as
chain silicate.

For further subclassification Zoltai uses the chain periodicity
for single chain silicates as in the crystal chemical classification
of silicates. Structures containing cyclic anions, multiple chains,
layers, and frameworks, however, are subdivided according to the
number of tetrahedra in the loops that their anions contain. The
amphibole double chain then has loop size 6, that in inesite has loop

sizes 6 and 8 (Fig. 8), the single layer in mica has loop size 6, the double layer in hexacelsian 4 and 6 (Figs. 10, 12) and so on. Formally in a framework the number of loop sizes is infinite. Therefore, for classification purposes this number is limited to four different sizes with 12 being the largest loop size.

Like the other two classifications the one developed by Zoltai has certain shortcomings. One is that single chains are characterized by their periodicities, while already the corresponding double chains are described by the sizes of their loops, thus concealing their structural, crystal chemical and also geochemical similarities. Another deficiency is that weak and very ionic bonds such as Li^+-O are put on a level with strong and much more covalent bonds such as Si^{4+}-O, P^{5+}-O and even S^{6+}-O. As a consequence, e.g. $Li_6Si_2O_7$, Li_2SiO_3, $Li_2Si_2O_5$ and quartz, SiO_2, would all be classified as tectosilicates differing in sharing coefficient, while $Na_6Si_2O_7$ would be grouped with the disilicates having $[Si_2O_7]$ anions and $Li_6Si_2O_7$ with the tectosilicates.

COMPARISON OF DIFFERENT SILICATE CLASSIFICATIONS

Because of the different principles chosen, each of the three classification schemes described serves a particular purpose.

The Kostov classification uses geochemical relations of chemical elements as a main principle and is, therefore, suitable to meet the needs of geochemists and petrologists. It is exclusively applicable to silicates.

The crystal chemical classification, by comparison, is directly based on the way the $[TO_4]$ tetrahedra are linked, i.e. on the topology and conformation of the silicate anions. Since linkage is controlled by the chemical properties of the cations, pressure and temperature, and reaction kinetics (see e.g. Liebau 1972, 1978, 1980), the classification is indirectly based on these chemical and physical properties. On the other hand, thermodynamic stability, chemical reactivity and durability, structural similarities and dissimilarities are strongly correlated with the atomic structure of the silicates. This is not only true for natural silicate minerals but for synthetic silicates as well. The *crystal chemical classification* of silicates, therefore,

22

has a much wider field of application than the Kostov classification, ranging from earth science to materials science, wherever silicate properties are involved that are based on structure and chemical composition. It works better for silicates with Si:O $\geqslant$ 1:3 than for those with Si:O < 1:3. This classification can easily be adapted to classify other inorganic substances that contain other anionic [XO_4] tetrahedra and their condensation products such as germanates, phosphates, sulfates, etc.

The Zoltai classification has an even wider applicability as far as the chemical composition is concerned. It is directly applicable to all substances containing tetrahedrally coordinated cations. Unfortunately this is at the expense of the special nature of silicates. Since there does not seem to be the same sort of correlation between loop size and cation properties as there is between periodicity and cation properties (Liebau and Pallas, 1980), the otherwise very straightforward Zoltai classification does not reflect the crystal chemical correlations as well as the crystal chemical classification does.

Each of the three classifications has its own favorable fields of applicability as well as its own shortcomings. Before deciding which one to use, one has to become clear about the particular purpose to be fulfilled.

REFERENCES: SILICATE CLASSIFICATION

Bragg, W.L. (1930) The structure of silicates. *Z. Kristallogr.*, *74*, 237-305.

Kostov, I. (1975) Crystal chemistry and classification of silicate minerals. *Geochem. Mineral. Petrol.*, *1*, 5-41.

Liebau, F. (1962) Die Systematik der Silikate. *Naturwiss.*, *49*, 481-491.

―――― (1972) Crystal chemistry of silicon. *In*, K.H. Wedepohl, Ed., *Handbook of Geochemistry*, Volume II/3, chapter 14. Berlin, Springer-Verlag.

―――― (1978) Silicates with branched anions: a crystallochemically distinct class. *Am. Mineral.*, *63*, 918-923.

―――― (1980) *Structural Chemistry of Silicates*. Springer-Verlag, in preparation.

―――― and I. Pallas (1980) The influence of cation properties on the shape of silicate chains. *Z. Kristallogr.*, in press.

Náray-Szabó, S. (1930) Ein auf der Kristallstruktur basierendes Silicatsystem. *Z. Physik. Chem. Abt.*, *B9*, 356-377.

Takéuchi, Y., Y. Kudoh and J. Ito (1977) High-temperature derivative structure of pyroxene. *Proc. Japan Acad.*, *53*, 60-63.

Weiss, A. and A. Weiss (1954) Zur Kenntnis der faserigen Silicium-dioxyd-Modifikation. *Z. anorg. allg. Chem.*, *276*, 95-112.

Zoltai, T. (1960) Classification of silicates and other minerals with tetrahedral structures. *Am. Mineral.*, *45*, 960-973.

ACKNOWLEDGMENT

Financial support by the Deutsche Forschungsgemeinschaft is gratefully acknowledged.

Chapter 2

SILICATE GARNETS   E. P. Meagher

INTRODUCTION

The garnets are a diverse group of minerals chemically, physically
and in mode of occurrence. This diversity is brought about, in part, by
the ability of the garnet structure to accommodate cations with a wide
range of sizes and valence states. This varied chemistry results in a
succession of colors from white, through shades of red, brown, yellow
and green, to black. Hardness varies from $6\frac{1}{2}$ to $7\frac{1}{2}$ and specific gravity
from approximately 3.5 to 4.3. Despite variations in the physical properties of garnets, their crystal morphology is rather limited with the
predominant crystal forms being the dodecahedron {110} and the trapezo-
hedron {211} or combinations of the two.

Garnets can occur as stable phases in a wide range of pressures,
temperatures and chemical environments. Although they are most commonly
associated with contact and regional metamorphic rocks, they are also
found in igneous rocks ranging from granite to peridotite as well as in
felsic volcanics and pegmatites. Because of their resistance to chemical
and mechanical breakdown they occur as detrital grains in sedimentary
rocks.

In the following section the mineralogy and crystal chemistry of
silicate garnets will be discussed. Although the synthetic, non-silicate garnets constitute an unique and important group of compounds, they
are not considered in this volume.

CRYSTAL STRUCTURE

Menzer (1926) determined the crystal structure of garnet in a
powder diffraction investigation of grossular and later established
the isostructural relationship between grossular and the remaining common silicate garnets (Menzer, 1928). The general formula for the col-
lective garnet group is: $X_3Y_2Z_3O_{12}$ with eight formula units per unit
cell. The common natural garnets possess a body-centered cubic unit
cell with space group symmetry $Ia3d$. The X-, Y-, and Z-cations occupy
special sites which are fixed by the space group symmetry (Table 1)
while oxygen occupies a general position with variable coordinates,
xyz.

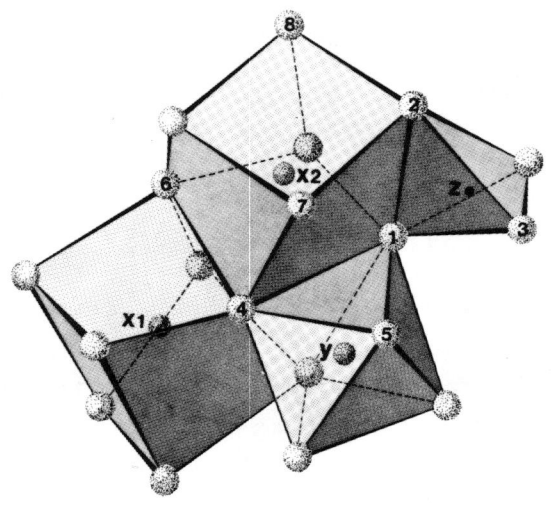

Figure 1. A portion of the garnet structure illustrating the
polyhedral configuration about the X-, Y-, and Z-cations.
After Novak and Gibbs (1971).

There are three kinds of coordination polyhedra in the garnet structure. The X-cation is coordinated by eight oxygens in a polyhedral configuration which is described as a distorted cube or, alternatively, as
a triangular dodecahedron (Fig. 1). Novak and Gibbs (1971) pointed out
that the XO_8 polyhedron is only slightly distorted from a regular triangular dodecahedron (Lippard and Russ, 1968) and should be referred to
as such. The Y-cation is coordinated by six oxygens in the polyhedral
configuration of a slightly distorted octahedron. The Z-cation is four-
coordinated by oxygens in a distorted tetrahedron or, more accurately, a
tetragonal disphenoid.

Table 1. Crystallographic data for the garnet structure.

Atom	Site	Point Symmetry	Coordinates	Coordination
X	24c	222	1/8 0 1/4	8
Y	16b	$\bar{3}$	0 0 0	6
Z	24d	$\bar{4}$	3/8 0 1/4	4
Oxygen	96h	1	x y z	4

26

If we consider only the ZO_4 tetrahedra and YO_6 octahedra in the garnet structure we see that a continuous linkage exists whereby tetrahedra share each corner with adjacent octahedra and vice versa (Fig. 2). Because no two tetrahedra share a common corner the structure is classified as an orthosilicate. Figure 3 illustrates the linkage between the tetrahedra and the triangular dodecahedra in which opposite edges of each tetrahedron are shared with dodecahedra forming chains which parallel the three mutually perpendicular a-axes. Finally, the octahedron shares six of its twelve edges with adjacent dodecahedra.

Each oxygen is coordinated by two X-, one Y-, and one Z-cation in a near tetrahedral configuration. Relative to other silicates the garnet structure is quite efficient from the standpoint of packing of oxygens. In the pyrope structure, assuming a spherical oxygen-ion with a radius of 1.38 Å (Shannon and Prewitt, 1969), the oxygens occupy 70% of the volume of the unit cell compared to 74% for a closest packed arrangement. This efficiency in packing is brought about by the lack of tetrahedral polymerization and the large number of shared polyhedral edges.

GARNET CHEMISTRY

Although the garnet structure can accommodate a variety of cations, the common garnets are composed of the most abundant di-, tri-, and tetravalent cations in the earth's crust; namely, Mg, Al, Si, Ca, Ti, Mn, and Fe. Table 2 summarizes the important X-, Y-, and Z-cations in the general formula for garnet.

Winchell (1933), in a study of the chemistry of garnets from different environments, divided the common garnets into two groups within which complete solid solution exists but between which a compositional gap seemed evident. The names of the two groups (ugrandite and pyralspite) are acronyms derived from the abbreviations of the members in each group:

Ugrandite group		Pyralspite group	
Uvarovite	$Ca_3Cr_2Si_3O_{12}$	Pyrope	$Mg_3Al_2Si_3O_{12}$
Grossular	$Ca_3Al_2Si_3O_{12}$	Almandine	$Fe_3Al_2Si_3O_{12}$
Andradite	$Ca_3Fe_2Si_3O_{12}$	Spessartine	$Mn_3Al_2Si_3O_{12}$

Figure 2. A portion of the garnet structure illustrating the linkage of the ZO_4 tetrahedra and YO_6 octahedra. The X-cation positions are not shown. After Gibbs and Smith (1965).

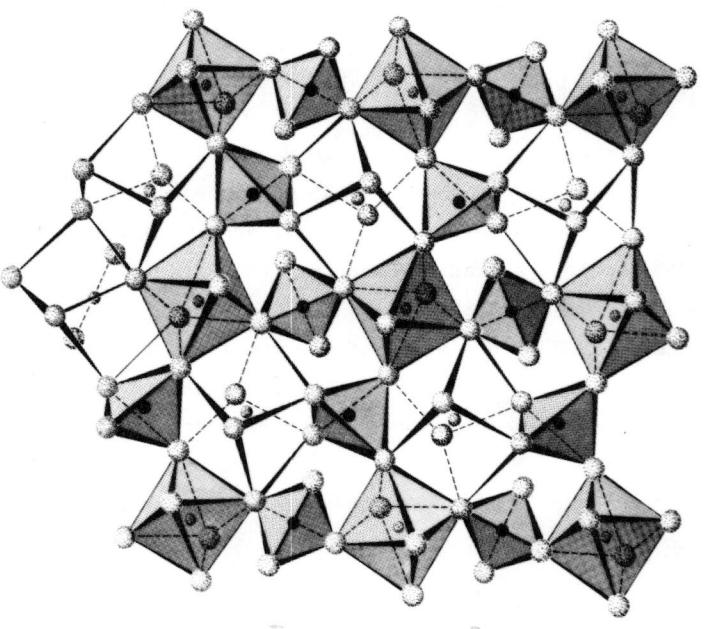

Figure 3. A section of the garnet structure as viewed down the a_3-axis. Note the chain of edge-shared tetrahedra and triangular dodecahedra. After Novak and Gibbs (1971).

Table 2. List of major, minor and trace substituents in garnet with the general formula, $X_3Y_2Z_3O_{12}$.

	X	Y	Z
Major:	Ca Mn Fe^{2+} Mg	Al Cr Fe^{3+}	Si
Minor:	Zn Y^{3+} Na	$Ti^{3+,4+}$ V^{3+} Fe^{2+} Zr Sn	Al Ti^{4+} $Fe^{3+,2+}$ P
Trace concentrations reported:	Li Be B F Sc Cu Ga Ge Sr Nb Ag		
	Cd In La Ce Pr Nd Gd Dy Ho Er Yb		

As the number of chemical analyses increases, more natural garnets are being found which bridge the proposed compositional gap between the ugrandite and pyralspite groups. For example, numerous garnets have been reported which have compositions intermediate between grossular and each member of the pyralspite group (Nemec, 1967; Brown, 1969; Heritsch, 1973; Reid *et al.*, 1976; Shimazaki, 1977).

Theoretical considerations led Ganguly and Kennedy (1974) to conclude that in the series pyrope-grossular and spessartine-almandine, positive excess free energies of mixing occur. The highest critical mixing temperature was proposed for the pyrope-grossular mixture; however, they were unable to experimentally confirm that a miscibility gap exists in this system. A similar theoretical analysis by Ganguly (1976) indicated possible unmixing of andradite and uvarovite, but experimental verification was not attempted. A study of the activity-composition relationships in pyrope-grossular by Hensen *et al.* (1975) indicates that a solvus occurs in the solid solution with a temperature of critical mixing at approximately 630°, in close agreement with the work of Ganguly and Kennedy (1974).

Cressy *et al.* (1978), in a study of the thermodynamic properties of almandine-grossular solid solution, demonstrates that a negative excess free energy of mixing occurs in the series at the $Fe_{85}Ca_{15}$ compositional range. In agreement with this study, Cressy (1978) has observed exsolution in a high temperature metamorphic almandine-pyrope-grossular garnet. Transmission electron microscope analyses of ion-thinned garnets have revealed exsolution of Fe-rich, Ca-poor granules in a more calcic garnet matrix. The two garnets have a measured cell edge difference of approximately 0.12Å.

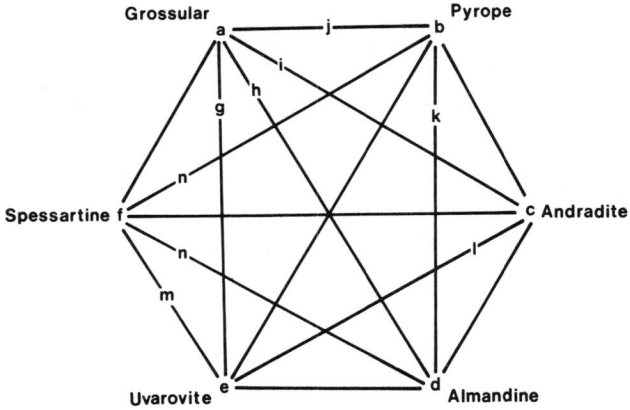

Figure 4. Phase-equilibria studies of the common end-member garnets.
a. Hays (1967); Huckenholz
   et al. (1975a)
b. Boyd & England (1959);
   Schreyer & Seifert (1969)
c. Huckenholz & Yoder (1971)
d. Hsu (1968); Keesmann
   et al. (1971)
e. Huckenholz (1975)
f. Snow (1943); Mottana (1974)

g. Huckenholz & Knittel (1975)
h. Hariya & Nakano (1972)
i. Huckenholz *et al.* (1974)
j. Yoder & Chinner (1960)
k. Hsu & Burnham (1969)
l. Isaacs (1965)
m. Naka *et al.* (1975)
n. Matthes (1961)

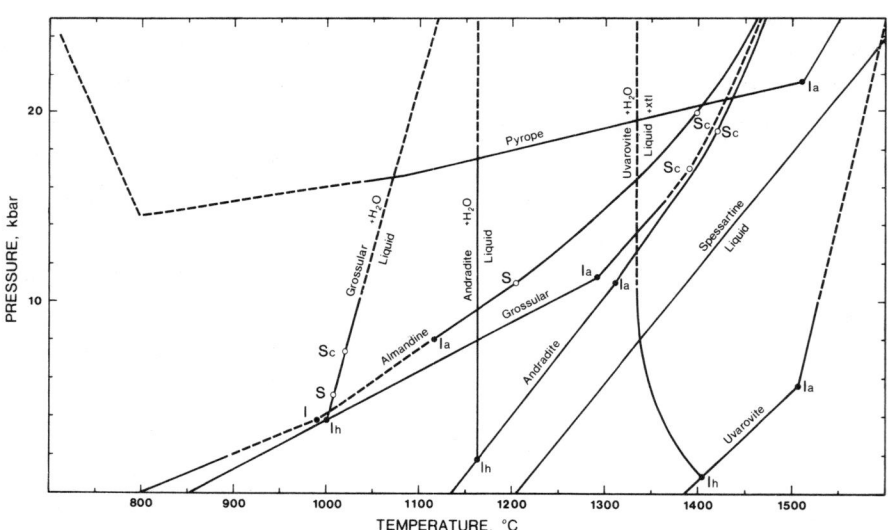

Figure 5. A total pressure-temperature plot of reaction curves for the six common end-member garnets. I, invariant points; S, singular points; I_h, hydrous melting of grossular, andradite and uvarovite; I_a, anhydrous melting of pyrope, almandine, grossular, andradite, spessartine and uvarovite; Sc, congruent melting of garnets under hydrous and anhydrous conditions. Dashed lines are estimates. Taken from Huckenholz (1975).

30

Table 3. Garnet group members.

Name		
Pyrope	$Mg_3Al_2Si_3O_{12}$	Deer, Howie and Zussman (1962)
Knorringite	$Mg_3Cr_2Si_3O_{12}$	Nixon and Hornung (1968)
Khoharite	$Mg_3Fe_2Si_3O_{12}$	McConnell (1966)
Majorite	$Mg_3(Fe,Al,Si)Si_3O_{12}$	Smith and Mason (1970)
Almandine	$Fe_3Al_2Si_3O_{12}$	Deer *et al.* (1962)
Spessartine	$Mn_3Al_2Si_3O_{12}$	Deer *et al.* (1962)
Calderite	$Mn_3Fe_2Si_3O_{12}$	Dunn (1979)
Grossular	$Ca_3Al_2Si_3O_{12}$	Deer *et al.* (1962)
Andradite	$Ca_3Fe_2Si_3O_{12}$	Deer *et al.* (1962)
Uvarovite	$Ca_3Cr_2Si_3O_{12}$	Deer *et al.* (1962)
Goldmanite	$Ca_3V_2Si_3O_{12}$	Moench and Meyrowitz (1964)
Kimzeyite	$Ca_3(Zr,Ti)_2(Al,Fe,Si)_3O_{12}$	Milton, Ingram and Blade (1961)
Schorlomite	$Ca_3(Fe,Ti)_2(Si,Fe)_3O_{12}$	Ito and Frondel (1967)
Melanite (Ti andradite)		Deer *et al.* (1962)
Hydrogrossular series	$Ca_3Al_2(SiO_4)_{3-x}(OH)_{4x}$	Zabinski (1966); Shoji (1974)
Hydroandradite series	$Ca_3Fe_2(SiO_4)_{3-x}(OH)_{4x}$	Zabinski (1966); Shoji (1975)
Henritermierite	$Ca_3(Mn,Al)_2(SiO_4)_2(OH)_4$	Fleischer (1969)

Experimental phase equilibria studies have been completed for all six of the common end-member garnets and on several of the joins between these members. These investigations are referenced in Figure 4. A pressure-temperature plot of the reaction curves for these six garnets is presented in Figure 5. It is important to bear in mind that compositionally some natural garnets cannot be expressed in terms of these six end members alone. For example, one cannot express the composition of certain Cr-bearing pyropes without utilizing the end member $Mg_3Cr_2Si_3O_{12}$ (knorringite). Rickwood (1968) considers the problem of recasting garnet compositions in terms of end members.

In addition to the common cations which constitute most of the silicate garnet compositions, there are several elements which occur with sufficient frequency to warrant the following brief discussions. A summary of garnet group end members is presented in Table 3.

Titanium. Ti-rich garnets are most commonly Ca-garnets which are referred to as melanites or schorlomites and are typically associated with undersaturated alkaline igneous rocks. Deer *et al.* (1962) describe melanite as an andradite containing 1 to 5 percent TiO_2 while schorlomite can contain up to 20 percent TiO_2. Howie and Woolley (1968) have proposed a division between the two at 8 percent TiO_2. In a study of synthetic titanium garnets Ito and Frondel (1967) proposed that a solid solution series exists between andradite, $Ca_3Fe_2Si_3O_{12}$ and $Ca_3Ti_2(Fe_2Si)O_{12}$ and suggested that the term schorlomite be applied to calcium garnets with Ti^{4+} dominant in atomic percent at the octahedral site and either Fe or Al entering the tetrahedral site to maintain charge balance. Garnets are known to contain titanium in both the trivalent and the tetravalent states (Huggins *et al.*, 1977a). A brief discussion on Ti-garnets is included in the section on crystal chemistry.

Zirconium. The Zr-rich garnets reported to date seem to be the calcic garnets which frequently contain Ti (Dowty, 1971). The Zr-garnet kimzeyite, discovered in a carbonatite at Magnet Cove, Arkansas, has the ideal end-member composition $Ca_3Zr_2(Al_2Si)O_{12}$ (Milton *et al.*, 1961). Ti replaces Zr and Fe replaces Al in the mineral. Ito and Frondel (1967) suggest that the name kimzeyite refers to calcium garnets with Zr dominant in atomic percent when referred to the octahedral position and either Fe or Al as the charge-compensating ion in the tetrahedral site. Based upon hydrothermal syntheses they proposed a solid solution between

32

$Ca_3Zr_2(Fe_2Si)O_{12}$ and $Ca_3Zr_2(Al_2Si)O_{12}$.

Phosphorus-Sodium. Minor concentrations of Na (Smith and Mason, 1970; Ringwood and Lovering, 1970; Sobolev and Lavrent'ev, 1971) and P and Na (Thompson, 1975; Reid *et al.*, 1976; Bishop *et al.*, 1976, 1978) have frequently been observed in garnets from peridotites, eclogites and diamond-bearing kimberlites. Although the amount of P_2O_5 and Na_2O in garnets is usually less than one weight percent, these elements show promise as pressure indicators (Sobolev and Lavrent'ev, 1971; Thompson, 1975; Bishop *et al.*, 1978). Synthetic high pressure garnets of composition $NaCa_2(AlSi)Si_3O_{12}$ and $Na_2CaTi_2Si_3O_{12}$ have been reported by Ringwood and Major (1971).

Yttrium. Y_2O_3 concentrations in excess of one weight percent are reported to occur in spessartines associated with pegmatitic environments (Jaffe, 1951, Wakita *et al.*, 1969). Jaffe proposed a coupled substitution in these garnets of the type $Y^{3+}Al^{3+} \rightleftarrows Mn^{2+}Si^{4+}$ which agrees with the later work of Yoder and Keith (1951) who demonstrated a $Mn_3Al_2Si_3O_{12}$-$Y_3Al_2(AlO_4)_3$ solid solution.

Vanadium. In the few natural garnets which contain appreciable amounts of vanadium, the valence state appears to be V^{3+}. The garnets are predominantly calcic and the only known end member is goldmanite, $Ca_3V_2Si_3O_{12}$, which was discovered in a metamorphosed uranium-vanadium deposit (Moench and Meyerowitz, 1964). The garnet yamatoite, $Mn_3V_2Si_3O_{12}$, has been discredited on grounds that the specimen is actually a manganoan goldmanite and, therefore, a member of the $Ca_3V_2Si_3O_{12}$-$Mn_3V_2Si_3O_{12}$ series (Fleischer, 1965). A vanadium grossular has recently been reported in association with clinopyroxene (Gubelin and Weibel, 1975).

Tin. As much as four to five weight percent SnO_2 has been reported in andradite-grossular garnets (Mulligan and Jambor, 1968; McIver and Mihálik, 1975). The replacement $Sn^{4+} + Fe^{2+} \rightleftarrows 2\ Fe^{3+}$ in the octahedral site has been proposed by the latter authors.

RELATION BETWEEN PHYSICAL PROPERTIES AND CHEMISTRY

Ford (1915) was one of the first to quantitatively relate the physical properties of garnets to chemical composition. His study was limited to the use of specific gravity and refractive index. However, with the availability of x-ray diffraction data, investigations subsequently included unit cell dimensions (Fleischer, 1937; Sriramadas, 1957;

Winchell, 1958), and Biswas (1974) has added magnetic susceptibility to the list of physical properties.

Winchell's (1958) graphical method has been a widely used technique for estimating garnet compositions from physical properties. The underlying assumption in this method is that specific gravity, unit cell dimension and refractive index are each additive functions of the substituent atoms in a compound. Of these three properties, Winchell believed that specific gravity was the most difficult to determine accurately; therefore, his charts are constructed with the unit cell parameter and refractive index as independent variables. Hutchinson (1974) has recently modified Winchell's determinative charts by including Skinner's (1956) data for synthetic end-member garnets and has constructed a new chart which includes data for uvarovite. The disadvantage to this type of analysis is that one must often use additional data of a compositional nature in order to unambiguously estimate the chemistry of a garnet. Recent efforts to improve on Winchell's method by computing multiple regression equations relating chemistry to garnet composition have met with limited success (Biswas, 1974).

McConnell (1964) demonstrated that the refractive index of a garnet can be estimated with reasonable accuracy utilizing the Lorentz-Lorentz theory. This theory relates refractive index to unit cell dimension, unit cell chemistry and ionic refractivities; the latter being treated as empirical constants. Errors between observed and calculated refractive indices for a variety of garnet compositions ranged to 0.011. The relationship used by McConnell reveals that at least 70% of the refractive index in a garnet is contributed to by the characteristics of the packing of oxygen.

Biswas (1973) has studied the dispersion of the refractive index and reflectivity of garnets as a function of chemistry. His results indicate that except for those end members with the highest and lowest indices and reflectivities (schlorlomites and pyrope), the dispersion curves are not diagnostic due to overlap.

The great variety of color in garnets is mainly a result of the variety of transition elements which occur either as minor impurities or in major concentrations in the 8-, 6-, and 4-coordinated sites. Readers interested in the origin of color in garnets are referred to Loeffler and Burns (1976).

34

Table 4.   Physical constants for synthetic end members of the common garnets

Name	Composition	a_o (Å)	n(Na)	Bulk modulus[†]	ρ calc (gm./cc.)	Reference
Pyrope	$Mg_3Al_2Si_3O_{12}$	11.459(1)	1.714(2)	1.730(9)	3.582	Skinner (1956)
Almandine	$Fe_3Al_2Si_3O_{12}$	11.528(1)	1.829(3)	1.779(8)	4.315	Hsu (1968)
Spessartine	$Mn_3Al_2Si_3O_{12}$	11.614(1)	1.799(3)	1.742(9)	4.197	Hsu (1968)
Grossular	$Ca_3Al_2Si_3O_{12}$	11.851(1)	1.734(2)	1.691(8)	3.594	Skinner (1956)
Andradite	$Ca_3Fe_2Si_3O_{12}$	12.048(1)	1.887(2)	1.379(17)	3.859	Skinner (1956)
Uvarovite	$Ca_3Cr_2Si_3O_{12}$	11.996(2)	1.865(3)	1.43	3.850	Huckenholz & Knittel (1975)
Goldmanite	$Ca_3V_2Si_3O_{12}$	12.070(5)	1.834(3)		3.765	Strens (1965)

† Bulk modulus, in Mbar, taken from Babuska et al. (1978). The value for uvarovite is estimated.

Physical constants for synthetic end members in the garnet group are given in Table 4.

CRYSTAL CHEMISTRY

The interrelationship between chemistry and the crystal structure of garnets has been the subject of a considerable amount of research in the past two decades within two fields of interest. In the earth sciences the silicate garnets are studied because of their importance as rock-forming minerals in the earth's crust and upper mantle. In solid state physics the synthetic non-silicate garnets are investigated because of their ferrimagnetic and their laser properties. An excellent review of the crystal chemistry of garnets, particularly the non-silicate garnets, is given by Geller (1967). Studies of silicate garnet crystal chemistry have been published by Zemann (1962) and Novak and Gibbs (1971).

Although the Si-O bond has approximately 50% covalent character in garnets, the common 6- and 8-coordinated cations (Ca,Mn,Fe,Mg,Al) form predominantly ionic bonds. Consequently, trends between cation chemistry and crystal structure variations can be rationalized successfully when based on an empirical hard-sphere ionic model (Novak and Gibbs, 1971). Quantitative applications of simple ionic bonding theory to silicate garnets have been few in number. In a crystal chemical investigation, Born and Zemann (1964) used an ionic model which included Coulombic interaction and closed-shell repulsive energies to compute

the shift in position of a rigid SiO_4 tetrahedron as a function of increasing X-cation radius. Bloomfield *et al.* (1961) and Runciman and
Sengupta (1974) used a point charge model to calculate crystal field
splitting for Fe^{2+} in silicate garnets.

Structural response to chemical variations

Several high precision structure refinements on end-member silicate
garnets have been completed (Table 5) which, in combination with Mössbauer and infra-red spectroscopic information, make up the bulk of the
data base from which an understanding of the crystal chemistry of garnets
is derived. We will first consider the response of each of the polyhedra to chemical variations and then consider the total garnet structure.

SiO_4 tetrahedron. As mentioned previously, the SiO_4 tetrahedron is
distorted in all garnets to a tetragonal disphenoid with the elongation
parallel to the $\bar{4}$-axis. The O(1)-O(2) edge (Fig. 1) is shared with the
XO_8 polyhedron and is shortened relative to the unshared edge. Electrostatic repulsions between Si and the X-cation appear to play a role in
shortening of the shared edge in agreement with Pauling's rules (Pauling,
1929); however, it has been argued that geometric constraints brought
about by neighboring cation-oxygen and oxygen-oxygen interactions also
play an important role in the SiO_4 shared edge shortening in garnets
(*cf.* Born and Zemann, 1964; Meagher, 1975). Distortion of the tetrahedron
is also affected by the size of the X-cation, and in the aluminum silicate
garnets Novak and Gibbs (1971) found the shared and unshared edges increase and decrease in length, respectively, as the radius of the X-
cation increases. As a result the tetrahedron is slightly more distorted
in pyrope than in grossular.

The Si-O bond length, d(Si-O), varies with garnet composition; however, no clear correlation between d(Si-O) values and the X- or Y-cation
substitution is apparent. Novak and Gibbs (1971) have recorded a range
in d(Si-O) values between 1.628(2) Å in almandine and 1.655(2) Å in
goldmanite with an overall mean d(Si-O) value of 1.641 Å for the nine
garnets analyzed.

In addition to variations within the SiO_4 tetrahedron, there is a
rotation of the tetrahedron about the $\bar{4}$-axis in the aluminum silicate
garnets in response to the mean radius of the X-cations, $\langle r_X \rangle$, occupying

36

Table 5. Summary of structure refinements on silicate garnets.

Garnet	STP	High Temperature	High Pressure
Pyrope	a,b	c	d
Cr-pyrope	e		
Fe-pyrope	f		
Almandine	g		
Spessartine	g	h	
Uvarovite	g		
Grossular	g,d,i,j	c	d
Mn-grossular	g		
Andraditic-grossular	k		
Andradite	g,l	h	
Goldmanite	g		
Hydrogrossular	m,n		
Henritermierite	o		

a. Zemann and Zemann (1961)
b. Gibbs and Smith (1965)
c. Meagher (1975)
d. Hazen and Finger (1978)
e. Novak and Meyer (1970)
f. Euler and Bruce (1965)
g. Novak and Gibbs (1971)
h. Rakai (1975)

i. Abrahams and Geller (1957)
j. Prandl (1966)
k. Takéuchi and Haga (1976)
l. Quareni and dePiere (1966)
m. Cohen-Addad *et al.* (1964)
n. Foreman (1968)
o. Aubry *et al.* (1969)

Figure 6. A portion of the garnet structure illustrating the position angle γ and the rotation, about the $\bar{4}$ axis, of the SiO_4 tetrahedra with increasing $<r_x>$. Note the change indicated by arrows on the shared and unshared octahedral edges with tetrahedral rotation. After Meagher (1975). →

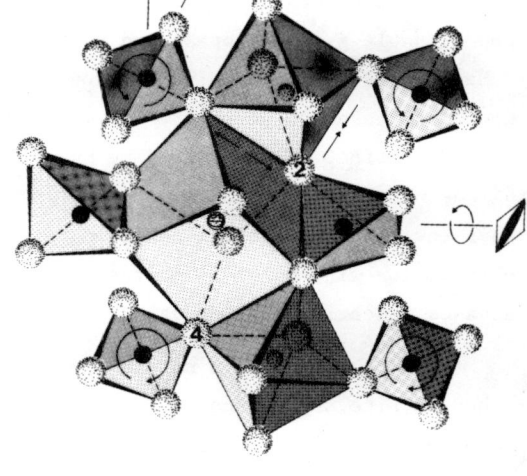

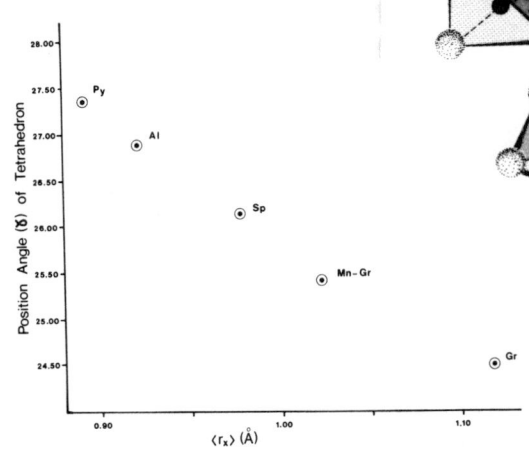

Figure 7. A plot of the position angle γ versus mean radius of the X-cation, $<r_x>$, for Y = Al garnets: pyrope (Py), almandine (Al), spessartine (Sp), manganese-grossular (Mn-Gr) and grossular (Gr). After Meagher (1975).

the 8-coordinated site. Born and Zemann (1964) were the first to discuss this rotation and defined a "position angle" for the tetrahedron
as the smaller of the two angles between the tetrahedral O-O edge normal
to the $\bar{4}$-axis and the two crystallographic axes normal to the $\bar{4}$-axis
(Fig. 6). In Figure 7 the positional angle γ is plotted against $\langle r_X \rangle$
for the aluminum garnets to illustrate the continuous rotation of the
tetrahedron to smaller position angles with increasing $\langle r_X \rangle$. This rotation increases the size of the triangular dodecahedron and causes a
change in octahedral distortion which will be discussed in the following
section.

Minor amounts of Fe^{2+}, Fe^{3+}, Ti^{4+}, Al and P have been reported to
occur in the tetrahedral site of natural garnets (Huggins *et al.*, 1977a;
Thompson, 1975). Based on an investigation of synthetic garnets,
Huggins *et al.* (1977b) proposed a tetrahedral site preference of Al $\geq$
Fe^{3+} > Ti^{4+} which is in agreement with the conclusions of Hartman (1969).
They further proposed that this preference can only be rationalized on
the basis of a combination of factors which include cation size, charge
and electronegativity. The tetrahedral site preference, Fe^{3+} > Al, Ti^{4+}
was determined in a Mössbauer spectroscopic study of natural Fe-Ti garnets
by Schwartz and Burns (1978). Bishop *et al.* (1976) proposed that in
upper-mantle garnets containing P the coupled substitution Ca^{VIII} + Si^{IV}
$\rightleftarrows$ Na^{VIII} + P^{IV} is operative and found it unnecessary to place Si in 6-
coordination as suggested by Sobolev and Lavrent'ev (1971).

The effect on the structure by cation substitution for tetrahedral
Si has not been studied in detail at this time. Nevertheless it appears
that the mean tetrahedral cation-oxygen distance follows a simple linear
correlation with $\langle r_Z \rangle$ (Hawthorne, 1978) and the unit cell parameter increases linearly with increasing $\langle r_Z \rangle$ (Novak and Colville, pers. comm.).

YO_6 *octahedron*. In natural garnets the Y-cations occupying the octahedral site are typically Al, Fe^{3+}, and Cr with less common occupancies
reported for V^{3+}, Zr, Ti, and Fe^{2+}. The silicate garnet structure can
accommodate cations in the octahedral site with a range in radii between
approximately 0.5 and 1.05 Å; however, within this range there is a
general interdependence between the radius of the Y-cation and that of
the accompanying X-cation.

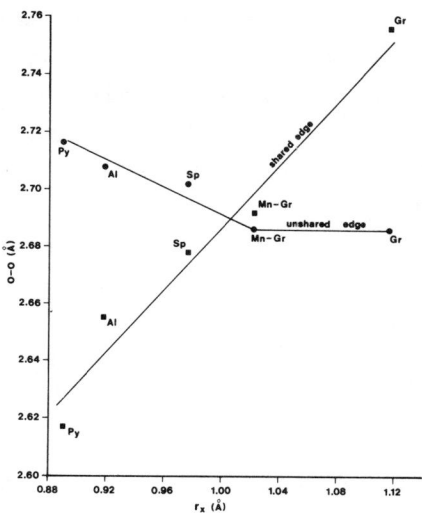

Figure 8. A plot of the unshared, O(1)-O(5), and shared, O(1)-O(4), octahedral edge versus mean X-cation radius, $\langle r_X \rangle$. Refer to Figure 7 for abbreviations. After Novak and Gibbs (1971).

The YO_6 octahedron is only slightly distorted in the garnet structure and is more regular than octahedra found in other orthosilicates such as the olivines and humites. The Y-cation occupies a position of point symmetry $\bar{3}$; therefore, all d(Y-O) distances are equivalent and there are only two non-equivalent O-Y-O angles. These angles vary no more than ±2° from the ideal octahedral angle in the nine garnets refined by Novak and Gibbs (1971). Distortion of the octahedron is, in part, dependent on the radius of the X-cation, and for Y = Al garnets the smallest distortion occurs for an intermediate $\langle r_X \rangle$. This is illustrated in Figure 8 where lengths of the shared and unshared octahedral edges are plotted as a function of $\langle r_X \rangle$. In pyrope, the shared octahedral edge, O(1)-O(4), is shorter than the unshared edge as one might expect by Pauling's (1929) electrostatic bonding principles. However, as $\langle r_X \rangle$ increases, the shared edge lengthens while the unshared edge shortens such that when $\langle r_X \rangle$ approaches 1.00 Å the edges will be approximately equal in length and a symmetric octahedron results. With further increase in $\langle r_X \rangle$ the shared edge continues to lengthen while no further

39

shortening of the unshared edge below ∿2.86 Å is apparent. In grossular
the shared edge is 0.03 Å longer than the unshared edge which is contrary to our expectations based upon crystal chemical principles. Novak
and Gibbs (1971) have proposed that the 2.68 Å O-O distance for an unshared octahedral edge is approaching a lower limit due to strong anion-
anion, non-bonded repulsions.

The Y = Al series of garnets also reveals the effect the effect
the X-cation radius has on the size of the octahedron. Pyrope, $\langle r_X \rangle =$
0.89 Å, possesses a mean Al-O distance of 1.886 Å while in grossular,
$\langle r_X \rangle = 1.12$ Å, the value has increased to 1.924 Å (Novak and Gibbs,
1971). For the X = Ca garnets a linear trend is evident between $\langle r_Y \rangle$
and the Y-O bond length.

Titanium-bearing garnets present an interesting crystal chemical
problem in that Ti can apparently occupy the tetrahedral as well as the
octahedral site and both Ti^{3+} and Ti^{4+} have been reported to occur in
garnets (Burns and Burns, 1971; Huggins et $al.$, 1977a). Inasmuch as
Ti-garnets are frequently high in Fe, contain some Al and are often Si
deficient (Howie and Woolley, 1968); assignment of Ti to the tetrahedral
and octahedral sites is often ambiguous. Based on a chemical and Mössbauer spectroscopic study of natural melanites and schorlomites, Huggins
et $al.$ (1977a) have proposed the following coupled substitutions for in-
corporating Ti into the octahedral sites: $Y^{2+} + Ti^{4+} \gtrless 2Y^{3+}$, $(Y^{2+}$ = Mg,
Fe) and $Ti^{3+} \gtrless Y^{3+}$. The first substitution is of further interest because of the reported occurrence of octahedral Mg and Fe^{2+} in Ti-garnets.
Huggins et $al.$ (1977a) estimate a maximum $Ti^{3+}/\Sigma Ti$ ratio of 0.25 in garnets they analyzed, which is in agreement with the maximum observed by
Schwartz and Burns (1978).

Among the transition elements that most commonly occupy the octahedral site in natural end-member garnets, only Cr^{3+} has the appropriate
electron configuration to achieve a crystal field stabilization energy
(CFSE). The CFSE for Cr^{3+} in uvarovite is relatively high at approximately 56.6 kcal (Burns, 1970). Of the Cr-bearing minerals associated
with garnet in upper mantle rocks the CFSE for 6-coordinated Cr^{3+} decreases as: spinel > garnet > pyroxene ≈ olivine (Burns, 1976).

XO$_8$ triangular dodecahedron. The major elements occupying the eight coordinated triangular dodecahedral site in silicate garnets are Mg, Fe^{2+}, Mn and Ca. Zemann (1962) proposed the permissible range of r_X in the natural garnets to be 0.8-1.1 Å. It appears, however, that the maximum permissible size of the X-cation is dependent on the size of the accompanying Y-cations. For synthetic silicate garnets Novak and Gibbs (1971) proposed that as the size of the Y-cation is increased, the upper allowed limit for r_X can be increased to approximately 1.5 Å.

There are two symmetry non-equivalent X-O distances which are labeled X(1)-O(4) and X(2)-O(4) (Fig. 1). In all silicate garnet structures refined to date, the X(2)-O(4) distance is significantly longer and appears to be the result of geometric constraints related to Y-O and O-O distances in the neighboring octahedra (Zemann, 1962; Gibbs and Smith, 1965). The average X-O distance, <X-O>, appears to increase linearly with $<r_X>$ in the Al-silicate garnets; however, the <X-O> value is also strongly dependent on $<r_Y>$. For example, in the Ca-garnets as $<r_Y>$ increases from 0.53 Å in grossular to 0.64 Å in andradite, <Ca-O> increases by 0.03 Å (Novak and Gibbs, 1971; *cf.* Higgins and Ribbe, 1977, Fig. 2).

Among the common X-cations only Fe^{2+} experiences a crystal field stabilization energy. Burns (1970) estimates that the CFSE in the pyrope-almandine series ranges from 12.4 kcal in alm$_1$ to 11.7 kcal in alm$_{100}$. Of the Fe^{2+}-containing minerals associated with garnet in upper mantle rocks, the CFSE decreases as: garnet (8-coordinated) > pyroxene M$_2$-site > pyroxene M$_1$ ≃ olivine M$_1$ and M$_2$, which is consistent with the relative enrichment of Fe^{2+} in the garnets (Burns, 1976).

Composite structure. In the preceding sections the geometric relations of the specific polyhedra have been discussed relative to chemical variations. It is apparent, however, that a strong interdependence exists between the geometry of a given polyhedron and the chemistry of adjacent polyhedra. For example, the size and shape of an octahedron is strongly dependent not only on the $<r_Y>$ for the cations occupying the octahedron but also on the radius of the cations occupying the neighboring triangular dodecahedral sites.

One of the first to investigate these interdependent relations in garnets was Zemann (1962). Zemann was curious about the geometric

41

constraints operative in pyrope and grossular which prohibit the occurrence of an undistorted SiO_4 tetrahedron and AlO_6 octahedron. Using ideal SiO_4 and AlO_6 polyhedra of appropriate dimensions for pyrope and grossular, he computed the resulting interatomic distances in the MgO_8 and CaO_8 polyhedra and found an abnormally short unshared 0-0 edge of 2.44 Å present in each. Upon expanding the distance to 2.75 Å (which Zemann felt was a minimum allowable distance for the unshared edge of an XO_8 polyhedron), distortion in the tetrahedron and octahedron resulted.

A more powerful means of determining the interrelationships between various bond lengths and angles is now available in the form of the Distance Least Squares (DLS) computer program (Meier and Villiger, 1969). With this program, interatomic distances are prescribed, and a least-squares adjustment of independent atomic parameters and unit cell parameters is computed, to minimize the difference between prescribed and calculated distances. A distance least-squares (DLS) investigation of aluminum silicate garnets was undertaken by Meagher (1975) to evaluate the response of the structure to increasing values of r_X. The calculations indicate that geometric constraints brought about by cation-anion and anion-anion interactions in adjacent polyhedra play an important role in determining that the shared tetrahedral edge is shorter than the unshared edge in the Y = Al garnets. In addition, the observed trend in lengths of shared and unshared AlO_6 octahedral edges (Fig. 8) as a function of $<r_X>$ can be reproduced with this method.

Calculation of cell parameter and atomic coordinates. Regression equations have been computed by McConnell (1964) and Novak and Gibbs (1971) with which one can estimate the unit cell parameter, a, knowing the mean radius of the X- and Y-cations in a silicate garnet. The multiple regression analysis by Novak and Gibbs (1971) was completed for 56 silicate garnets and led to the following relation:

$$a = 9.04(2) + 1.61(4) <r_X> + 1.89(8) <r_Y>$$

The numbers in parentheses refer to one estimated standard deviation in the last place quoted. A subsequent analysis by Novak and Colville (pers. comm.) extended the multiple regression to include $<r_Z>$ for over 1100 natural and synthetic garnets and yielded the relation:

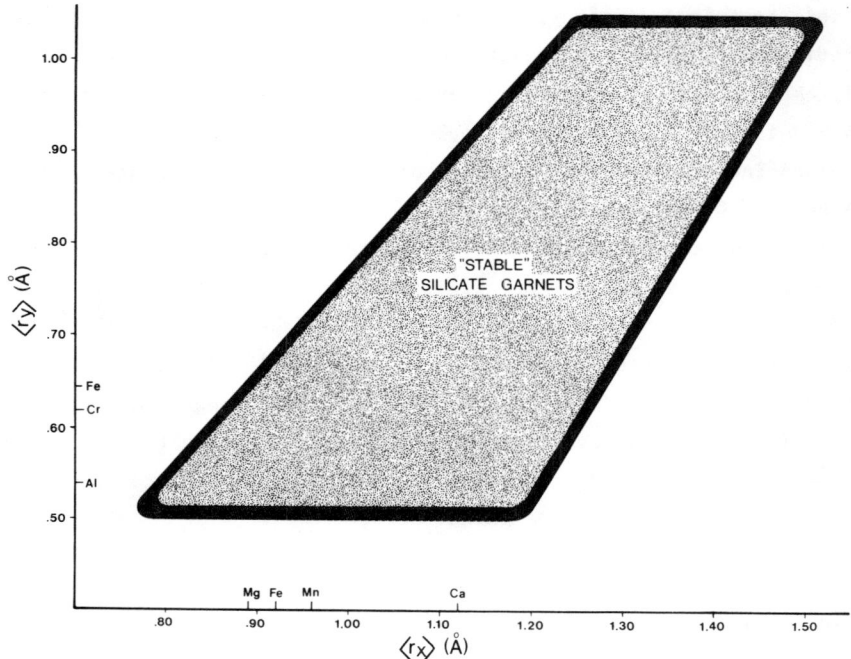

Figure 9. The proposed "stable" silicate garnets for which compatible combinations of $\langle r_X \rangle$ and $\langle r_Y \rangle$ occur. Effective ionic radii of some common X- and Y-cations (Shannon and Prewitt, 1969) are shown. The radius of Fe is for 2+ along $\langle r_X \rangle$ and 3+ along $\langle r_Y \rangle$. After Novak and Gibbs (1971).

$$a = 8.44 + 1.71(1)\ \langle r_X \rangle + 1.78(1)\ \langle r_Y \rangle + 2.17(1)\ \langle r_Z \rangle$$

In both equations the effective ionic radii of Shannon and Prewitt (1969) are used with the exception of some rare-earth ionic radii in the Novak and Colville investigation.

Novak and Gibbs (1971) completed a multiple linear regression analysis of the oxygen atomic coordinates versus $\langle r_X \rangle$ and $\langle r_Y \rangle$ using the data from their nine silicate garnet refinements. The following equations are obtained:

$$x = 0.0059(5) + 0.022(2)\ \langle r_X \rangle + 0.014\ \langle r_Y \rangle$$
$$y = 0.0505(4) - 0.023(2)\ \langle r_X \rangle + 0.037\ \langle r_Y \rangle$$
$$z = 0.6431(7) - 0.009(3)\ \langle r_X \rangle + 0.034\ \langle r_Y \rangle$$

Using the above relations to compute the cell parameter and oxygen atomic coordinates, the crystal structure of a hypothetical silicate garnet can be approximated since the X- and Y-cations, as well as Si, occupy special positions in the space group $Ia3d$. Novak and Gibbs (1971)

43

used this technique, along with crystal chemical data obtained from garnet refinements, to estimate what limitations might exist in the chemistry of silicate garnets. To do this they proposed a series of guidelines which a "stable" silicate garnet could not exceed. In general, these guidelines included the maximum allowed length of the calculated Si-O distance, the O-O distance in the unshared octahedral edge and the calculated size of the octahedral and triangular dodecahedral sites compared to the $r_Y + r_O$ and $r_X + r_O$ radii sums. Accordingly, an area was mapped out on an r_X versus r_Y plot within which the crystal-chemically stable silicate garnets occur. This type of plot is illustrated in Figure 9 along with the Shannon-Prewitt radii for the more common X- and Y-cations. One can see that the common silicate garnets plot toward the lower portion of this diagram and much larger X- and Y-cations appear to be compatible with the garnet structure.

Structural response to elevated temperatures

To date, structure refinements of garnets at elevated temperatures have been completed for pyrope and grossular (Meagher, 1975) and for spessartine and andradite (Rakai, 1975). The structural adjustments with increasing temperature in these four garnets are similar in that no significant change in the Si-O interatomic distance occurs whereas the mean octahedral Y-O distances, <Y-O>, and <X-O> distances increase in a simple linear manner.

Although increased temperature does not significantly change the dimension of the SiO_4 tetrahedron, the tetrahedron plays more than a passive role in the thermal expansion of garnets. In pyrope and spessartine, the tetrahedron rotates to smaller values of γ with increasing temperature (as was illustrated in Fig. 7 for increasing $<r_X>$) which has the effect of increasing the size of the triangular dodecahedron and the unit cell. In the Y = Al garnets the rate of tetrahedral rotation with increasing temperature, $d\gamma/dT$, is highest in pyrope, intermediate in spessartine and insignificant in grossular.

An additional consequence of the tetrahedral rotation with heating is the change of distortion to the octahedron. In pyrope and spessartine, as the tetrahedron rotates about the $\bar{4}$ axis to a smaller γ angle, the shared AlO_6 octahedral edge increases while the unshared edge remains

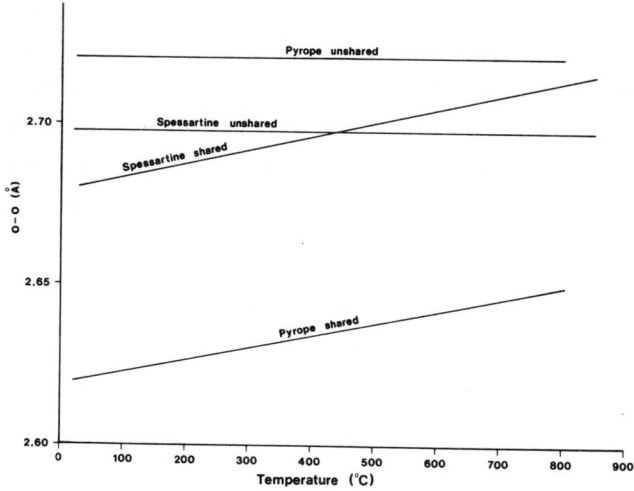

Figure 10. A plot of the shared and unshared octahedral edge
lengths versus temperature for pyrope and spessartine. Data
taken from Meagher (1975) and Rakai (1975), respectively.

essentially unchanged. Figure 10 illustrates this change and shows that
the crossover occurs in spessartine at approximately 425°C resulting in
an undistorted AlO_6 octahedron. In grossular, which represents the
upper limit for $\langle r_X \rangle$ in an aluminum silicate garnet, no rotation of the
tetrahedron occurs and distortion of the AlO_6 octahedron is unchanged
with heating.

Thermal expansion of the interatomic distance within the octahedron
is dependent not only upon the cation occupancy of the octahedron but
also upon the mean cation radius, $\langle r_X \rangle$, in the dodecahedral site. For
example, the rate of thermal expansion in the Y–O interatomic distance
is substantially less in the non-calcium garnets (pyrope and spessartine)
than in the Ca-garnets (grossular and andradite). In addition, for the
Y = Al garnets, the mean bond length expansion coefficients, $\alpha_{\langle X-O \rangle}$, for
the XO_8 polyhedra vary inversely with the magnitude of the $\langle X-O \rangle$ distance
such that $\alpha_{\langle Mg-O \rangle} > \alpha_{\langle Mn-O \rangle} > \alpha_{\langle Ca-O \rangle}$ (Rakai, 1975). Both of these observations are consistent with the fact that grossular has one of the
lowest unit cell thermal expansions of all the Y = Al garnets.

Thermal expansion of the unit cell parameter in the common silicate garnets is, to a first approximation, linear (Skinner, 1956; Rakai, 1975). To date, however, an interpretation of the garnet cell expansion based on structural considerations has not been made. Perhaps this is not too surprising in light of the preceding discussion on the interrelationship between chemistry and the structural response to increased temperature.

Structural response to elevated pressures

Due to the technological problems of obtaining single crystal x-ray intensity data at high pressure, investigations on the response of the garnet structure to elevating pressures have only recently been attempted. Hazen and Finger (1978) have determined the structures of pyrope and grossular to 60 kbars using a diamond anvil apparatus with a water and glycerin hydrostatic pressure medium. In general, they observe that longer bonds are more compressible and, accordingly, the polyhedral compressibilities in the two garnets increase in the order: SiO_4 tetrahedron < AlO_6 octahedron < MgO_8 < CaO_8 triangular dodecahedron. Distortions in the tetrahedra and octahedra remain unchanged within the pressure range of their investigation, although the MgO_8 and CaO_8 polyhedra become somewhat more regular with pressure. This is in agreement with the Mössbauer work of Huggins (1976) who attributed the change in quadrupole splitting of Fe^{2+} in almandine with increasing pressure to a decrease in site distortion.

The electronic absorption spectra of Fe^{3+} in andradite and Cr^{3+} in uvarovite have been recorded at pressures to 100 kbars and 177 kbars, respectively, by Abu-Eid (1976). Negligible shifts in the spectra led Abu-Eid to conclude that the ionic bonding character of Fe^{3+} and Cr^{3+} in the octahedral environment changes little with increasing pressure. In a high pressure Mössbauer study of Fe-bearing minerals Huggins (1976) observed a general decrease in the isomer shift of both Fe^{2+} and Fe^{3+} with increasing pressure. Changes in the isomer shift with pressure will reflect compressibility of the metal-oxygen bond as well as changes in covalency of the bond, and in garnet Huggins interpreted the general decrease to be the result of an increase in the contribution of 3s-electrons to the electron density at the nucleus which he attributed to an increased delocalization of the 3d-electrons. This increased delocalization with higher pressures is equivalent to an increased covalent character in the iron-oxygen bond; nevertheless, there does not appear to be a large change in bond character for Fe or Cr in garnets up to pressures of approximately 175 kbars.

Because of the importance of garnets in mantle rocks, considerable effort has been devoted to determining elastic constants in garnets of various compositions (for a review of past investigations refer to Babuska *et al.*, 1978). The bulk moduli for end-member garnets in Table 4 were extrapolated from members of solid solution series by Babuska *et al.* (1978). These values indicate that the effect of composition on garnet compressibility is somewhat more complex than with simple oxides.

Anderson and Anderson (1970) proposed that in simple oxides which are isostructural, the product of the bulk modulus (K) and molar volume (Vm) would be a constant. The trend does not follow for the bulk moduli listed in Table 4, and an interesting anomaly appears to exist between pyrope and almandine. Pyrope, which has a smaller molar volume than almandine, would be expected by the K·Vm = constant relationship to possess a larger bulk modulus than almandine. However, as observed by Babuska *et al.* (1978) the reverse is true, and in the pyrope-almandine solid solution Bonczar *et al.* (1977) found the bulk modulus to decrease with increasing Mg-content. The anomaly may be contributed to, in part, by the effect of the crystal field stabilization energy for Fe^{2+} on the bulk modulus of almandine (Weaver *et al.*, 1976).

The study of high pressure phase transitions in those minerals which occur within the earth's lower crust and upper mantle is an important aspect of current geophysical research. Field and experimental petrology studies indicate that garnets in upper mantle peridotites and eclogites are predominantly almanditic-pyropes. Accordingly, high pressure investigations in the system $Mg_3Al_2Si_3O_{12}$-$Fe_3Al_2Si_3O_{12}$ are of interest. Boyd (1964) proposed and Liu (1977a) experimentally confirmed that pyrope transforms to an ilmenite-like structure at high pressure. Liu achieved the transformation at a loading pressure between 240 and 250 kbars at 1000°-1400°C and proposed that a transformation at still higher pressures to a perovskite structure would occur. In the system $Mg_3Al_2Si_3O_{12}$-$Fe_3Al_2Si_3O_{12}$ Liu (1975) indicated that successive phase transitions occur with increasing pressure which are dependent on the Mg/Fe ratio. The true nature of these phase transitions, is, as yet,

47

uncertain. Ahrens and Graham (1972), in a study of shock-induced phase changes in Mg-almandine, reported a transition to an ilmenite-like structure.

A unique high-pressure garnet called majorite, discovered in the Coorara Meteorite, is reported to contain Si in 6- as well as 4-coordination (Smith and Mason, 1970). Assuming total iron as divalent, the structural formula is given as

$$(Mg_{2.98}Na_{0.10})^{VIII}(Fe_{1.02}Si_{0.78}Al_{0.23}Cr_{0.03})^{VI}Si_3^{IV}O_{12}.$$

Majorite's discovery appeared to corroborate the suggestion by Ringwood (1967, 1970) that $(Mg,Fe)SiO_3$ pyroxenes transform to the garnet structure at high pressure. Majorite has been synthesized by Ringwood and Major (1971) at a pressure between 250-300 kbars and 1000°C.

To illustrate the high-pressure phase transitions in the earth's mantle Liu (1977b) proposed a model with an enstatite-peridotite crust made up of a 1:1 molecular ratio of forsterite (Mg_2SiO_4) and orthopyroxene (90% $MgSiO_3$·10% Al_2O_3). He further assumed no interaction between these two phases. With increasing depth, in the region 70-420 km, the aluminous enstatite disproportionates to a mixture of pyrope plus an aluminum-poor enstatite (Boyd and England, 1959), resulting in a decreasing ratio of pyroxene to garnet. At the 620 to 660 km discontinuity garnet transforms to the ilmenite-like structure, with six-coordinated Si, and spinel transforms to a mixture of perovskite and MgO with a halite structure.

GARNET-HYDROGARNET SERIES $X_3Y_2(SiO_4)_{3-p}(H_4O_4)_p$ $(0 \leq p \leq 3)$

Although the hydrogen content of garnet is reported as weight percent H_2O+, x-ray and neutron diffraction studies, in agreement with infra-red absorption spectroscopy, indicate that H_2O does not occur as a molecular group in hydrogarnets to any appreciable extent. The substitution $4H^+ \gtrless Si^{4+}$ appears to be the operative mechanism whereby each of the four oxygens about the tetrahedral site is strongly bonded to one hydrogen. In garnets containing minor amounts of H_2O+ (<1%) the nature of the water is not known.

Until recently the term hydrogarnet has been used almost synony-mously with hydrogrossular; however, natural hydrogarnets of varied compositions are continually being discovered and it seems appropriate to use a general formula to describe the hydrogarnet group of minerals. As in the general formula for anhydrous garnets, X refers to the eight-coordinated cation and Y to the octahedrally-coordinated cation. Al-though numerous end-member hydrogarnets (p = 3) have been synthesized, natural garnets beyond intermediate compositions are not known to occur.

Hydrogarnets display significant changes in physical properties with increasing hydrogen content. Characteristically, the unit cell size increases and the refractive index decreases as the silica content decreases. A decrease in specific gravity would also be expected with decreasing silica content. In the hydrogrossular series specific gravity varies from 3.59 in grossular to approximately 2.53 in $Ca_3Al_2(H_4O_4)_3$ (Zabinski, 1966). In addition, the hydrogarnets commonly crystallize in the {111} octahedral form which is not a common form for anhydrous garnets.

Hydrogrossular. Grossular-rich garnets were the first garnets ob-served to contain hydrogen. Foshag (1920) described a hydrous calcium aluminum silicate from the Crestmore Quarry, California which he named plazolite (from the Greek word *plazo*, to perplex) because of his difficulty in determining the composition. His choice of this name was prophetic in that the CO_2 he reported was later found to re-sult from included impurities. He believed the mineral was related to sodalite, but Winchell (1933) later recognized that the mineral was closely related to a ugrandite garnet. Subsequently, Pabst (1937) dis-regarded the CO_2 reported in the composition by Foshag and upon com-paring the powder diffraction patterns of plazolite and grossular con-cluded that they were isostructural but with OH replacing oxygen and 16 silicons randomly distributed at the 24d site in plazolite.

Belyankin and Petrov (1941), in investigating the mineral hibschite (originally described by F. Cornu, 1906), found the reported chemistry in error and the actual chemistry to be similar to that of plazolite. They proposed that both minerals were members of the same mineral group which they called the grossularoid group. Pabst (1942) subsequently

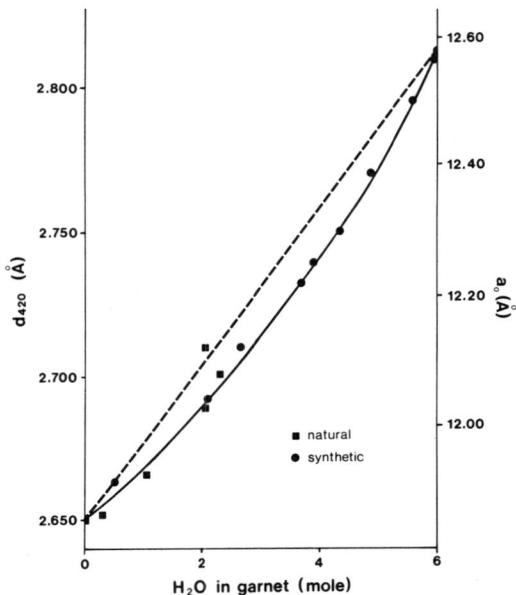

Figure 11. Unit cell parameter, a_0, and d-spacing of the (420) plane for natural and synthetic garnets in the series $Ca_3Al_2Si_3O_{12}-Ca_3Al_2(H_4O_4)_3$. The solid line was used by Shoji (1974) to estimate H_2O content of his synthetic hydrogrossulars. After Shoji (1974).

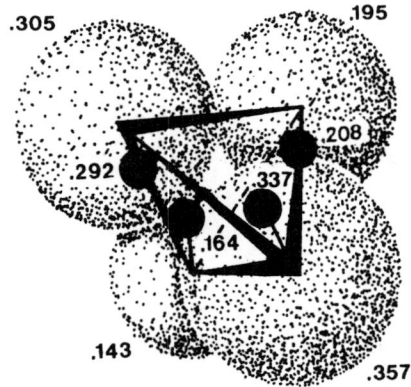

Figure 12. The D_4O_4 tetrahedral group as viewed down the a_3-axis in the structure of $Ca_3Al_2(D_4O_4)_3$. The deuterated hydrogens are represented as solid circles. The z-fractional coordinates are given. After Foreman (1968).

reinvestigated hibschite and confirmed its similarity to plazolite.
Hibschite is preferred over plazolite as the mineral name and has the
ideal formula: $Ca_3Al_2(SiO_4)_2(H_4O_4)$.

Most hydrogrossulars can be classified as occurring in (1) calcsilicate rocks which are the products of thermal metamorphosed marls or
marl-limestones or (2) hydrothermally altered basic igneous rocks.
Natural hydrogrossulars are most often compositionally intermediate
to grossular and hibschite (Deer et $al.$, 1962; Zabinski, 1966) although
varieties more hydrous than hibschite have been reported. Based only
upon cell edge data, Mason (1957) reported a hydrogrossular of approximate composition $Ca_3Al_2(SiO_4)(H_4O_4)_2$. In a hydrothermal and x-ray
study of the grossular-hydrogrossular series, Flint et $al.$ (1941) concluded that a continuous solid solution exists between $Ca_3Al_2Si_3O_{12}$ and
$Ca_3Al_2(H_4O_4)_3$. A more recent investigation of this series led Shoji
(1974) to conclude that above 200°C the series is not continuous but
is composed of a limited hydrogrossular series $Ca_3Al_2(SiO_4)_{3-p}(H_4O_4)_p$
$0 \leq p \leq 1$ or 2 and a hydrogarnet series $Ca_3Al_2(SiO_4)_p(H_4O_4)_{3-p}$ $0 \leq p$
≤ 0.4. Shoji used the relation between the d-spacing of the (420) plane
and the H_2O content illustrated in Figure 11 to determine the compositions of his fine-grained synthetic hydrogrossulars.

Silica-free $Ca_3Al_2(H_4O_4)_3$ decomposes above approximately 250°C,
and the hydrogen content in the grossular-hydrogrossular series decreases with increasing temperature (Shoji, 1974). $Ca_3Al_2(SiO_4)_2(H_4O_4)$
was found to persist up to a temperature of 300°C (1 kbar), while
$Ca_3Al_2(SiO_4)_{2.5}(H_4O_4)_{0.5}$ is reported to approximately 425°C. The
stability of the phases to these temperatures is uncertain,

$Crystal$ $structure$ of $Ca_3Al_2(H_4O_4)_3$. The crystal structure of the
hydrogarnet end member was determined by Cohen-Addad et $al.$ using neutron diffraction and NMR methods. A subsequent structure refinement
of a deuterated hydrogarnet, $Ca_3Al_2(D_4O_4)_3$ was completed using neutron
and x-ray diffraction data (Foreman, 1968). Both investigators found
the Ca hydrogarnet end member to be isostructural with grossular in $Ia3d$
space group symmetry and confirmed the $4H^+ \rightleftarrows Si^{4+}$ substitution. Each
hydrogen is strongly bonded to one of four oxygens about the tetrahedral
void as illustrated in Figure 12. The O-H distances are 0.95 Å, and

each hydrogen is 1.25 Å from the tetrahedron centroid with resulting H-H distances of 1.82 and 2.42 Å (Foreman, 1968). The centroid-oxygen distance within the tetrahedron is 1.931 Å which compares with a value of 1.645 Å in grossular illustrating the significant increase in the size of the tetrahedron as the hydrogens replace silicon. As in grossular, the shared tetrahedral edge (3.02 Å) is shorter than the unshared edge (3.21 Å). The CaO_8, AlO_6, and D_4O_4 polyhedra are larger in hydrogarnet than the respective polyhedra in grossular which results in a significantly larger unit cell.

The structure refinement of henritermierite, $Ca_3(Mn,Al)_2(SiO_4)_2(H_4O_4)$, also reveals substitution of the type $4H^+ \rightleftarrows Si^{4+}$. In addition, this mineral is unique because of the reported occurrence of Mn^{3+} and its tetragonal symmetry (Aubry *et al.*, 1969).

Hydroandradite. In a study of the synthesis of Ca,Al- and Ca-Fe-hydrogarnets Flint *et al.* (1941) proposed that a continuous solid solution extends between andradite and $Ca_3Fe_2(H_4O_4)_3$ but they were unable to synthesize the hydrogarnet end member. Ito and Frondel (1967) were able to synthesize hydroandradites from precipitated gels with compositions up to $Ca_3Fe_2(SiO_4)_{0.25}(H_4O_4)_{2.75}$. Likewise, Shoji (1975) was unable to hydrothermally synthesize the silica-free end member and proposed that above 200°C the composition of hydroandradite is limited to $Ca_3Fe_2(SiO_4)_{3-p}(H_4O_4)_p$, $0 \le p \le 1.5$.

Few natural hydroandradites have been reported to date. Peters (1965) discovered hydroandradite occurring in an ophicalite which contained one weight percent H_2O+ while a hydroandradite discovered in a serpentinized peridotite possessed 5.3 weight percent H_2O+ (Zabinski, 1966). A range of 0.15 to 0.67 weight percent H_2O+ was reported in four andradites by Wilkens and Sabine (1973); however, it is difficult to be certain of the nature of water in these garnets.

Hydrospessartine. Although the number of natural hydrospessartines reported to date is sparse, they apparently do occur. Wilkens and Sabine (1973) have analyzed two hydrospessartines from Sterling Hill, N. J. which contain approximately 2.5 weight percent H_2O+. Matthes (1961) and Hsu (1968) have reported the synthesis of hydrospessartine.

Figure 13. Photomicrograph (crossed nicols, X30) of a contact metamorphic garnet of bulk composition $An_{95}Gr_5$ from Oro Grande, New Mexico. Isotropic zones alternate with birefringent zones which display "dodecahedral" twinning. Taken from Bloom (1975).

BIREFRINGENT GARNETS

Optical birefringence has been reported in garnets since before the turn of the century. Although this anomaly has been the subject of numerous studies, no general consensus has been reached regarding its origin. Within the garnet group, the calcic garnets most commonly display a weak birefringence and it is especially common in grossular-andradite garnets associated with contact metamorphic or metasomatic deposits.

Figure 13 is a photomicrograph of a contact metamorphic garnet of bulk composition $An_{95}Gr_5$ under crossed nicols which·shows alternating isotropic and birefringent zones. These zones are contained within sectors which converge toward the center of the crystal resulting in a pattern which is typical of birefringent skarn garnets.

Birefringent grossular-andradite garnets have been irreversibly transformed to optically isotropic garnets through heating. The transition temperature of 860°C was determined for two different andradite-rich garnets by Stose and Glass (1938) and by Allen and Fahey (1957) in their attempt to gain some insight about the temperature of garnet

53

crystallization. On the other hand, for a garnet of intermediate
grossular-andradite composition the birefringence weakened but was still
observable after quenching from a temperature of 1060°C (Ingerson and
Barksdale, 1943). Hariya and Kimura (1978) undertook a series of heating experiments under atmospheric pressure on several optically anisotropic garnets and succeeded in reversing the anisotropic to isotropic
transformation. Garnets of composition Gr_8An_{92} and $Gr_{84}An_{16}$ which became isotropic after heating at 945°C for 44 hours were again rendered
birefringent upon annealing at 800°C for 72 hours.

Various proposals have been put forth regarding the origin of birefringence in garnets. These include substitution of H_4 for Si in the
hydrogarnets, anisotropy due to residual strain in the structure (Chase
and Lefever, 1960; Kitamura and Komatsu, 1978), substitution of rare
earth ions in the X-cation site giving rise to magneto-optic effects
(Blanc and Maisonneuve, 1973) and ordering of octahedral cations thereby
reducing the symmetry (Takéuchi and Haga, 1976). Kalinin (1967) has
demonstrated that anisotropy cannot be ascribed to the water content of
garnets and suggests that in synthetic grossular-andradite series garnets the pure end members are characteristically not birefringent,
whereas intermediate compositions are.

Winchell and Winchell (1951) classified the twinning in anisotropic garnets on the basis of the isometric external form that is retained
by the twinned birefringent segments. They report the most common type
to be the dodecahedral twin (Fig. 13) with hexoctahedral, trapezohedral
and octahedral twins also occurring. Little is known about the origin
of twinning and the nature of the twin laws in anisotropic garnets
because little is known about the true nature of the birefringent garnets themselves.

Cation ordering. Takéuchi and Haga (1976), through a single-
crystal x-ray investigation of a birefringent garnet of composition
$Ca_{2.91}Al_{1.36}Fe_{0.68}Si_{2.97}O_{12}$, were the first to substantiate that cation
ordering does occur in the silicate garnets thereby lowering their symmetry. The structure was refined in orthorhombic space group symmetry
Fddd which yields two symmetry non-equivalent octahedral sites. Site
refinements yielded an occupancy of M(1) = 0.424(3)Fe + 0.576Al and M(2)
= 0.244Fe + 0.776Al. Marezio *et al.* (1978) have proposed that all oxide
garnets are non-cubic based upon comparisons of single-crystal x-ray intensities and the occurrence of reflections forbidden by the cubic space

group *Ia3d*. They propose that trigonal space group $R\bar{3}$ represents the true symmetry of the garnets they have analyzed, including a natural manganese and chromium-rich silicate garnet. As precise structure refinements of these garnets are completed the true nature of their symmetry will emerge.

COMPOSITIONAL ZONING IN GARNETS

Since the advent of the electron microprobe the study of compositional zoning in minerals has been greatly facilitated. This is particularly true in the garnets because of their isotropic optics. Numerous electron microprobe analyses have now been published and several theories have recently been developed to model the origin and nature of zoning in garnets. Because compositional zoning is particularly common in metamorphic garnets most models relate to them.

Metamorphic garnets. Zoning in metamorphic garnets most commonly involves the major X-cations Ca, Mn, Fe^{2+} and Mg which occupy the large eight-coordinated site within the structure. One or more of the following processes are usually incorporated in the models which have been proposed to date to explain compositional zoning developed during or subsequent to crystal growth: (1) preferential fractionation and segregation of ions during growth; (2) volume diffusion of ions within the garnet; and (3) intergranular diffusion between garnet and matrix.

Hollister (1966) utilized the Rayleigh fractionation model to interpret the origin of Mn zoning in a metamorphic almandine. The bell shape of the Mn profile illustrated in Figure 14 suggested to Hollister that a progressive depletion of Mn from the surrounding rock was a factor in the zoning. This model requires that each layer of the garnet be isolated as the crystal grows and that volume diffusion of Mn is inconsequential during and subsequent to crystal growth. In addition, the matrix about the garnet must act as a homogeneous reservoir for Mn. The equation derived by Hollister (1966) is:

$$M_G = \lambda M_o \left(1 - \frac{W^G}{W^o}\right)^{\lambda - 1}$$

M_G = weight fraction of element at garnet edge

λ = fractionation factor

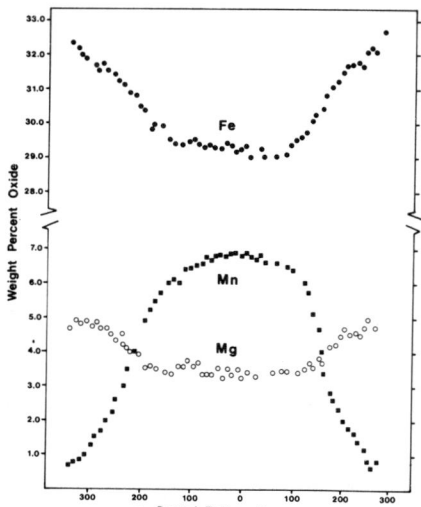

Figure 14. The Mn, Fe and Mg distribution in a zoned metamorphic garnet. The points represent electron microprobe analyses across the crystal. After Hollister (1966).

M_o = initial weight fraction of element in rock prior to garnet crystallization

W^G = total weight of crystallized garnet

W^o = initial weight of rock prior to garnet crystallization.

The weight percent of Mn calculated with this relation along a diameter of the crystal compares favorably to the bell-shaped profile illustrated in Figure 14. Atherton (1968) utilized a somewhat different segregation model to explain a similar zonation in garnets in medium-grade pelitic rocks.

A theoretical approach by Anderson and Buckley (1973), which considered only diffusion models, was also successful in reproducing such a bell-shaped Mn-profile. Significant diffusion rates of X-cations in garnets are evident through observations of homogeneity of zoning with increasing metamorphic grade (Anderson and Olimpio, 1977; Woodsworth, 1977). Loomis (1975) has demonstrated that compositional zoning is possible through diffusion-exchange or diffusion-reaction mechanisms involving garnets and accompanying phases as partition coefficients vary in response to changing environments. Loomis (1978, 1978a) has recently formulated a multicomponent diffusion model for garnets which includes cross diffusion coefficients that arise due to coupling of the concentration gradient of one ion with the flux of another ion. Crawford (1977) has attributed the irregularities in Ca-zoning to the discontinuous nature of plagioclase solid solution in moderate-grade metamorphic rocks.

   Igneous, hydrothermal garnets. Fewer reports of zoned garnets in igneous rocks are available; nevertheless, compositional zoning does occur and is, in many respects, quite similar to that observed in metamorphic garnets. Leake (1967) studied Ca, Mn, Fe, Mg and Ti zoning in garnets which occur in granites and aplites and concluded that the zoning is best described by a Rayleigh fractionation model. The Mn distribution in his garnets has the same bell-shaped profile as illustrated in Figure 14 for the metamorphic garnets. Gomes (1969) reports the occurrence of Fe/Ti zoning in melanite from a nepheline syenite with Ti-enrichment at the center of the garnet.

   Calcic garnets of hydrothermal and skarn deposits frequently possess compositional zoning. The zoning can be gradational (Shimazaki, 1977) or oscillatory with sharp boundaries. Oscillatory compositional zoning in these garnets is often correlated with isotropic and birefringent regions such as those shown in Figure 13. In some grossular-andradite series garnets the zoning correlates with variation in Al/Fe^{3+} with the Al-rich zones being isotropic (Lessing and Standish, 1973) while in others the isotropic zones are Fe^{3+} rich (Murad, 1976). The origin of oscillatory zoning is unknown although cyclical variations in the composition of the hydrothermal solutions (Lessing and Standish, 1973) and of the temperature during crystal growth (Murad, 1976) have been proposed.

   NOTE ADDED IN PROOF: "Recent work on the pyrope-grossular and almandine grossular solid solution series has shown that at concentrations of Ca up to 15 mol.% the garnets show negative excess volumes and negative excess free energies of mixing. It has been suggested that this results from a change in structure caused by Ca [Mg] ordering." Preliminary X-ray studies indicate there is such a structural change, possibly to space group $I2_13$, though further work must be undertaken to confirm the cause. (From M.J. Dempsey (1980) Evidence for structural changes in garnet caused by calcium substitution. *Contrib. Mineral. Petrol. 71*, 281-282.)

Abrahams, S.C. and S. Geller (1957) Refinement of the structure of a grossularite garnet. *Acta Crystallogr., 11*, 437-441.

Abu-Eid, R.M. (1976) Absorption spectra of transition metal-bearing minerals at high pressures. In R.G.J. Strens, Ed., *The Physics and Chemistry of Minerals and Rocks*, John Wiley and Sons, 641-675.

Ahrens, T.J. and E.K. Graham (1972) A shock-induced phase change in iron-silicate garnet. *Earth Planet. Sci. Lett., 14*, 87-90.

Allen, V.T. and J.J. Fahey (1957) Some pyroxenes associated with pyro-metasomatic zinc deposits in Mexico and New Mexico. *Bull. Geol. Soc. Amer., 68*, 881.

Anderson, D.L. and O.L. Anderson (1970) The bulk modulus-volume relationship for oxides. *Jour. Geophys. Res., 75*, 3494-3500.

Anderson, D.E. and G.R. Buckley (1973) Zoning in garnets—diffusion models. *Contrib. Mineral. Petrol., 40*, 87-104.

—— and J.C. Olimpio (1977) Progressive homogenization of metamorphic garnets, South Morar, Scotland: evidence for volume diffusion. *Can. Mineral., 15*, 205-216.

Atherton, M.P. (1968) The variation in garnet, biotite and chlorite composition in medium grade pelitic rocks from the Dalradian, Scotland, with particular reference to the zonation in garnet. *Contrib. Mineral. Petrol., 18*, 347-371.

Aubry, A., Y. Dusausoy, A. Laffaille and J. Protas (1969) Détermination et étude de la structure cristalline de l'henritermiertie, hydrogrenat de symétrie quadratique. *Bull. Soc. Fr. Mineral. Cristallogr., 92*, 126-133.

Babuska, V., J. Fiala, M. Kumazawa and I. Ohno (1978) Elastic properties of garnet solid-solution series. *Phys. Earth Planet. Inter., 16*, 157-176.

Belyankin, D.S. and V.P. Petrov (1941) The grossularoid group (hibschite, plazolite). *Am. Mineral., 26*, 450-453.

Bishop, F.C., J.V. Smith and J.B. Dawson (1976) Na, P, Ti and coordination of Si in garnet from peridotite and eclogite xenoliths. *Nature, 260*, 696-697.

——, —— and —— (1978) Na, K, P and Ti in garnet, pyroxene and olivine from peridotite and eclogite xenoliths from African kimberlites. *Lithos, 11*, 155-173.

Biswas, D.K. (1973) Refractive indices (n) and reflectivity (R) of natural garnets, *Indian Mineral., 14*, 74-79.

——, (1974) Quantitative relationship between chemical composition and physical properties of natural garnets. *Indian Jour. Earth Sci., 1*, 141-147.

Blanc, Y. and J. Maisonneuve (1973) Sur la biréfringence des grenats calciques. *Bull. Soc. Fr. Mineral. Cristallogr., 96*, 320-321.

Bloom, M.S. (1975) Mineral paragenesis and contact metamorphism in the Jarilla Mountains, Orogrande, New Mexico. M.Sc. Thesis, New Mexico Institute of Mining and Technology, Socorro, New Mexico.

Bloomfield, P., A.W. Lawson and C. Rey (1961) Crystal field splitting and covalent bonding in Fe^{2+} silicate garnets. *Jour. Chem. Phys.*, *34*, 749-756.

Bonczar, L.J., E.K. Graham and H. Wang (1977) The pressure and temperature dependence of the elastic constants of pyrope garnet. *Jour. Geophys. Res.*, *82*, 2529-2534.

Born, L. and J. Zemann (1964) Abstrandsberechnungen und gitterenergitische berechnungen an granaten. *Contrib. Mineral. Petrol.*, *10*, 2-23.

Boyd, F.R. (1964) Geological aspects of high-pressure research. *Science*, *145*, Num. 3627, 13-20.

—— and J.L. England (1959) Pyrope. *Carnegie Inst. Wash. Year Book*, *58*, 83-87.

Brown, E.W. (1969) Some zoned garnets from the greenschist facies. *Am. Mineral.*, *54*, 1662-1677.

Burns, R.G. (1970) *Mineralogical Applications of Crystal Field Theory*. Cambridge Univ. Press.

—— (1976) Partitioning of transition metals in mineral structures of the mantle. In R.G.J. Strens, Ed., *The Physics and Chemistry of Minerals and Rocks*. John Wiley and Sons, New York, 555-572.

—— and V.M. Burns (1971) Study of the crystal chemistry of titaniferous garnets by Mössbauer spectroscopy. *Geol. Soc. Am. Abstr. Programs*, *3*, 519-520.

Chase, A.B. and R.A. Lefever (1960) Birefringence of synthetic garnets. *Am. Mineral.*, *45*, 1126-1129.

Chinner, G.A., F.R. Boyd and J.L. England (1960) Physical properties of garnet solid solutions. *Carnegie Instit. Year Book*, *59*, 76-78.

Coes, L., Jr. (1955) High pressure minerals. *Jour. Am. Ceram. Soc.*, *38*, 298.

Cohen-Addad, C., P. Ducros, A. Durif, E.R. Bertaut and A. De La Palme (1964) Détermination de la position des atomes d'hydrogène dans l'hydrogranat Al_2O_3, 3CaO, $6H_2O$ par résonance magnétique nucléaire et diffraction neutronique. *Le Jour. Phys.*, *25*, 478-483.

Coleman, R.G., D.E. Lee, L.B. Beatty and W.W. Brannock (1965) Eclogites and eclogites: their differences and similarities. *Geol. Soc. Am. Bull.*, *76*, 483-508.

Cornu, F. (1906) Hibschit, ein neues kontaktmineral. *Tscher. Min. Petr. Mitt.*, *26*, 457-468.

Crawford, M.L. (1977) Calcium zoning in almandine garnet, Wissahickon formation, Philadelphia, Pennsylvania. *Can. Mineral.*, *15*, 243-249.

Cressey, G. (1978) Exsolution in almandine-pyrope-grossular garnet. *Nature, 271,* 533.

————, R. Schmid and B.J. Wood (1978) Thermodynamical properties of almandine-grossular garnet solid solution. *Contrib. Mineral. Petrol., 67,* 397-404.

Dadák, V. and F. Novák (1965) Tin-containing andradite from Plavno mine in the Krušné Hory Mts., Czechoslovakia. *Mineral. Mag., 35,* 379-385.

Deer, W.A., R.A. Howie and J. Zussman (1962) Rock-forming minerals, Vol. 1: Ortho-and ring silicates. John Wiley and Sons, New York.

Dowty, E. (1971) Crystal chemistry of titanian and zirconian garnet: I review and spectral studies. *Am. Mineral., 56,* 1983-2009.

Dunn, P.J. (1979) On the validity of calderite. *Can. Mineral., 17,* 569-571.

Euler, F. and J.A. Bruce (1965) Oxygen coordinates of compounds with garnet structures. *Acta Crystallogr., 19,* 971-978.

Fleischer, M. (1937) The relation between chemical composition and physical properties in the garnet group. *Am. Mineral., 22,* 751-759.

———— (1965) New mineral names. *Am. Mineral., 50,* 810.

———— (1969) New mineral names, henritermierite. *Am. Mineral., 54,* 1739.

———— (1975) Glossary of mineral species. *Mineralogical Record, Inc.,* Bowie, Maryland.

Flint, E.P., H.F. McMurdie and L.S. Wells (1941) Hydrothermal and x-ray studies and the relationship of the series to hydration products of portland cement. *Jour. Res. Natl. Bur. Stand., 26,* 13-33.

Ford, W.E. (1915) A study of the relations existing between the chemical, optical and other physical properties of the members of the garnet group. *Am. Jour. Sci., 40,* 33-49.

Foreman, D.W., Jr. (1968) Neutron and x-ray diffraction of Ca, $Al_2(O_4D_4)_3$, a garnetoid. *Jour. Chem. Phys., 48,* 3037-3041.

Foshag, W.F. (1920) Plazolite, a new mineral. *Am. Mineral., 5,* 183-185.

Fuchs, L.H. (1971) Occurrence of wollastonite, rhönite and andradite in the allende meteorite. *Am. Mineral., 56,* 2053-2068.

Ganguly, J. (1976) The energetics of natural garnet solid solution II. mixing of the calcium silicate end-members. *Contrib. Mineral. Petrol., 55,* 81-90.

———— and G.C. Kennedy (1974) The energetics of natural garnet solid solution I. mixing of the aluminosilicate end-members. *Contrib. Mineral. Petrol., 48,* 137-148.

Geller, S. (1967) Crystal chemistry of the garnets. *Zeit. für Kristal., 125,* 1-47.

Gibbs, G.V. and J.V. Smith (1965) Refinement of the crystal structure of synthetic pyrope. *Am. Mineral., 50,* 2023-2039.

Gomes, C.B. (1969) Electron microprobe analysis of zoned garnets. *Am. Mineral., 54,* 1654-1661.

Gubelin, E. and M. Weibel (1975) Vanadium-grossular von lualenyi bei voi. Kenja. *Neues Jahrb. Mineral. Abh., 123,* 191-197.

Hariya, Y. and K. Nakano (1972) Garnet solid solution between grossular and almandine. *Jour. Mineral. Soc. Japan, 10,* 373. (in Japanese)

—————— and M. Kimura (1978) Optical anomaly garnet and its stability field at high pressures and temperatures. *Jour. Fac. Sci. Hokkaido Univ., Ser. IV, 18,* 611-624.

Hartman, P. (1969) Can Ti^{4+} replace Si^{4+} in silicates? *Mineral Mag., 37,* 366-369.

Hawthorne, F. (1978) Personal Communication.

Hays, J.F. (1967) Lime-alumina-silica. *Carnegie Inst. Wash. Year Book, 65,* 234-239.

Hazen, R.M. and L.W. Finger (1978) Crystal structures and compressibilities of pyrope and grossular to 60 kbars. *Am. Mineral., 63,* 297-303.

Hensen, B.J., R. Schmid and B.J. Wood (1975) Activity-composition relationships for pyrope-grossular garnet. *Contrib. Mineral. Petrol., 51,* 161.

Heritsch, H. (1973) Noch einmal: granat mit einer zusammensetzung zwischen almandin und grossular. *Contrib. Mineral. Petrol., 40,* 83-85.

Hollister, L.S. (1966) Garnet zoning: an interpretation based on the Rayleigh fractionation method. *Science, 154,* 1647-1650.

Howie, R.A. and A.R. Woolley (1968) The role of titanium and the effect of TiO_2 on the cell-size, refractive index and specific gravity in the andradite-melanite-schorlomite series. *Mineral. Mag., 36,* 775-790.

Hsu, L.C. (1968) Selected phase relations in the system Al-Mn-Fe-Si-O-H: A model for garnet equilibria. *Jour. Petrol., 9,* 40-83.

—————— and C.W. Burnham (1969) Phase relations in the system $Fe_3Al_2Si_3O_{12}$-$Mg_3Al_2Si_3O_{12}$-H_2O at 2.0 kilobars. *Geol. Soc. Am. Bull., 80,* 2392-2408.

Huckenholz, H.G. (1975) Uvarovite stability in the $CaSiO_3$-Cr_2O_3 join up to 10 kbar. *Neues Jahrb. für Mineral., 8,* 337-360.

——————, E. Hölzi and W. Lindhuber (1975) Grossularite, its liquidus and subsolidus relations up to 10 kbar. *Neues Jahb. Mineral. Abh., 124,* 1-47.

—————— and D. Knittel (1975) Stability of grossularite-uvarovite solid solutions. *Contrib. Mineral. Petrol., 49,* 211-232.

———, W. Lindhuber and J. Springer (1974) The join $CaSiO_3$-Al_2O_3-Fe_2O_3 of the CaO-Al_2O_3-Fe_2O_3-SiO_2 quaternary system and its bearing on the formation of granditic garnets and fassaitic pyroxenes. *Neues Jahrb. Mineral. Abh., 121,* 160-207.

——— and H.S. Yoder, Jr. (1971) Andradite stability relations in the $CaSiO_3$-Fe_2O_3 join up to 30 kb. *Neues. Jahrb. Mineral. Abh., 114,* 246-280.

Huggins, F.E. (1976) Mössbauer studies of iron minerals under pressures of up to 200 kilobars. In R.G.J. Strens, Ed., *The Physics and Chemistry of Minerals and Rocks,* John Wiley and Sons, 613-640.

———, D. Virgo and H.G. Huckenholz (1977a) Titanium containing silicate garnets II. The crystal chemistry of melanites and schorlomites. *Am. Mineral., 62,* 646-665.

———, D. Virgo and H.G. Huckenholz (1977b) Titanium containing silicate garnets I. The distribution of Al,Fe^{3+}, and Ti^{4+} between octahedral and tetrahedral sites. *Am. Mineral., 62,* 475-490.

Hutchison, C.S. (1974) Laboratory handbook of Petrographic Techniques. John Wiley and Sons, New York, 209-214.

Ingerson, E. and J.D. Barksdale (1943) Iridescent garnet from the Adelaide Mining District, Nevada. *Am. Mineral., 28,* 303.

Isaacs, T. (1965) A study of uvarovite. *Mineral. Mag., 35,* 38-45.

Ito, J. and C. Frondel (1967) Synthetic zirconium and titanium garnets. *Am. Mineral., 52,* 773-781.

Jaffe, H.W. (1951) The role of yttrium and other minor elements in the garnet group. *Am. Mineral., 36,* 133-155.

Kalinin, D.V. (1967) Anisotropism in garnets in relation to composition and chemical background of their synthesis. Doklady A'kad. *Nauk SSR 172,* 128-130.

Keesman, I., S. Matthes, W. Schreyer and F. Seifert (1971) Stability of almandine in the system FeO-(Fe_2O_3)-Al_2O_3-SiO_2-(H_2O) at elevated pressures. *Contrib. Mineral. Petrol., 31,* 132-144.

Kitamura, K. and H. Komatsu (1978) Optical anisotropy associated with growth striation of yttrium garnet, $Y_3(Al,Fe)_5O_{12}$. *Kristall. und Technik, 13,* 811-816.

Kohn, J.A. and D.W. Eckart (1962) X-ray study of synthetic diamond and associated phases. *Am. Mineral., 47,* 1422-1430.

Leake, B.E. (1967) Zoned garnets from the Galway Granite and its aplites. *Earth Planet Sci. Lett., 3,* 311-316.

Lessing, P. and R.P. Standish (1973) Zoned garnet from Crested Butte, Colorado. *Am. Mineral., 58,* 840-842.

Lippard, S.J. and B.J. Russ (1968) Comments on the choice of an eight-coordinated polyhedron. *Inorg. Chem., 9,* 1686-1688.

Liu, L. (1975) High pressure reconnaissance investigation in the system $Mg_3Al_2Si_3O_{12}$-$Fe_3Al_2Si_3O_{12}$. *Earth Planet. Sci. Lett.*, *26*, 425–433.

———(1977a) First occurrence of the garnet-ilmenite transition in silicates. *Science 195*, 990–991.

——— (1977b) The system enstatite-pyrope at high pressures and temperatures and the mineralogy of the earth's mantle. *Earth Planet. Sci. Lett.*, *36*, 237–245.

Loeffler, B.M. and R.G. Burns (1976) Shedding light on the color of gems and minerals. *Am. Scientist.*, *64*, 636–647.

Loomis, T.P. (1975) Reaction zoning of garnet. *Contrib. Mineral. Petrol.*, *52*, 285–305.

——— (1978) Multicomponent diffusion in garnet: I. Formulation of isothermal models. *Am. Jour. Sci.*, *278*, 1099–1118.

——— (1978a) Multicomponent diffusion in garnet: II. Comparison of models with natural data. *Am. Jour. Sci.*, *278*, 1119–1137.

Manning, P.G. (1967) The optical absorption spectra of some andradites and the identification of the $^6A_1 \rightarrow {}^4A_1{}^4E(G)$ transition in octahedrally bonded Fe^{3+}. *Can. Jour. Earth Sci.*, *4*, 1039–1047.

——— (1972) Optical absorption spectra of Fe^{3+} in octahedral and tetrahedral sites in natural garnets. *Can. Mineral.*, *11*, 826–839.

——— and M.J. Tricker (1977) A Mössbauer spectral study of ferrous and ferric ion distributions in grossular crystals: evidence for local crystal disorder. *Can. Mineral.*, *15*, 81–86.

Marezio, M., J. Chenavas and J.C. Joubert (1978) On the symmetry of the garnet structure. (Abstr.) Winter Meeting, *Am. Crystallogr. Assoc., Norman, Okla.*, *6*, 23.

Mason, B. (1957) Larnite, scawtite and hydrogrossular from Tokatoka, New Zealand. *Am. Mineral.*, *42*, 379–392.

Matthes, S. (1961) Ergebnisse zur granatsynthese und ihre beziehungen zur natürlichen granatbildung innerhalb der pyralspit-gruppe. *Geochim. Et Cosmochim. Acta, 23*, 233–294.

McConnell, D. (1964) Refringence of garnets and hydrogarnets. *Can. Mineral.*, *8*, 11–22.

McIver, J.R. and P. Mihálik (1975) Stannian andradite from "David Ost" Southwest Africa. *Can. Mineral.*, *13*, 217–221.

Meagher, E.P. (1975) The crystal structures of pyrope and grossularite at elevated temperatures. *Am. Mineral.*, *60*, 218–228.

Meier, W.M. and H. Villiger (1969) Die methode der abstandsverfeinerung zur bestimmung der atomkoordinaten idealisierter ger üststrukturen. *Zeit. Kristallogr.*, *129*, 411–423.

Menzer, G. (1926) Die kristallstruckture von granat. *Zeit Kristallogr.*, *63*, 157–158.

──── (1928) Die kristallsruckture der granate. *Zeit. Kristallogr.*, *69*, 300–396.

Milton, C., B.L. Ingram and L.V. Blade (1961) Kimzeyite, a zirconium garnet from Magnet Cove, Arkansas. *Am. Mineral.*, *46*, 533–548.

Moench, R.H. and R. Meyrowitz (1964) Goldmanite, a vanadium garnet from Laguna, New Mexico. *Am. Mineral.*, *49*, 644–655.

Mottana, A. (1974) Melting of spessarite at high pressure. *Neues Jahb. Mineral. Mh.*, *6*, 256–271.

Müller, G. and A. Schneider (1971) Chemistry and genesis of garnets in metamorphic rocks. *Contrib. Mineral. Petrol.*, *31*, 178–200.

Mulligan, R. and J.L. Jambor (1968) Tin bearing silicates from skarns in the Cassiar District, Northern British Columbia. *Can. Mineral.*, *9*, 358–370.

Murad, E. (1976) Zoned, birefringent garnets from Thera Island, Santorini Group (Aegean Sea). *Mineral. Mag.*, *40*, 715–719.

Naka, S., Y. Suwa and T. Kameyama (1975) Solid solubility between uvarovite and spessartite. *Am. Mineral.*, *60*, 418–422.

Němec, D. (1967) The miscibility of the pyralspite and grandite molecules in garnets. *Mineral. Mag.*, *36*, 389–402.

Nishizawa, H. and M. Koizumi (1975) Synthesis and infrared spectra of $Ca_3Mn_2Si_3O_{12}$ and $Cd_3B_2Si_3O_{12}$(Bi, Al, Ga, Cr, V, Fe, Mn). *Am. Mineral.*, *60*, 84–87.

Nixon, P.H. and G. Hornung (1968) A new chromium garnet end member, knorringite, from kimberlite. *Am. Mineral.*, *53*, 1833–1840.

Novak, G.A. and A. Colville (1975) A linear regression analysis of garnet chemistries versus cell parameters. (Abstr.) *Geol. Sco. Amer. S.W. Section Meeting, Los Angeles*, *7*, 359.

──── and G.V. Gibbs (1971) The crystal chemistry of the silicate garnets. *Am. Mineral.*, *56*, 791–825.

──── and H.O.A. Meyer (1970) Refinement of the crystal structure of a chrome pyrope garnet: an inclusion in natural diamond. *Am. Mineral.*, *55*, 2124–2127.

Pabst, A. (1937) The crystal structure of plazolite. *Am. Mineral.*, *22*, 861–868.

──── (1942) Reexamination of hibschite. *Am. Mineral.*, *27*, 783–792.

Pauling, L. (1929) The principles determining the structure of complex ionic crystals. *Jour. Am. Chem. Soc.*, *51*, 1011–1026.

Peters, T.J. (1965) A water-bearing andradite from the Totalp serpentine (Dauos, Switzerland). *Am. Mineral.*, *50*, 1482–1486.

Prandl, W. (1966) Vereinerung der kristallstrukture des grossulars mit neutronen und rontgenstrahlbeugung. *Zeit. Kristallogr.*, *123*, 81–116.

Quareni, S. and R. dePieri (1966) La struttura dell'andradite. *Mem. Accad. Patavina, Sci. Mat. Nat., 78,* 153-164.

Rakai, R.J. (1975) Crystal structure of spessartine and andradite at elevated temperatures. M.Sc. thesis, Univ. British Columbia.

Reid, A.M., R.W. Brown, J.B. Dawson, G.G. Whitfield and J.C. Siebert (1976) Garnet and pyroxene compositions in some diamondiferous eclogites. *Contrib. Mineral. Petrol., 58,* 203-220.

Rickwood, P.C. (1968) On recasting analyses of garnet into end-member molecules. *Contrib. Mineral. and Petrol., 18,* 175-198.

Ringwood, A.E. (1967) The pyroxene-garnet transformation in the earth's mantle. *Earth and Planet. Sci. Lett., 2,* 255-263.

────── (1970) Phase transformation and the constitution of the mantle. *Phys. Earth Planet. Int., 3,* 109-155.

────── and J. F. Lovering (1970) Significance of pyroxene-ilmenite intergrowths among kimberlite xenoliths. *Earth and Planet. Sci. Lett., 7,* 371-375.

────── and A. Major (1971) Synthesis of majorite and other high pressure garnets and perovskites. *Earth and Planet. Sci. Lett., 12,* 411-418.

Runciman, W.A. and D. Sengupta (1974) The spectrum of Fe^{2+} ions in silicate garnets. *Am. Mineral., 59,* 563-566.

Schreyer, W. and F. Seifert (1969) High pressure phases in the system $MgO-Al_2O_3-SiO_2-H_2O$. *Am. Jour. Soi., 267-A,* 407.

Schwartz, K.B. and R.G. Burns (1978) Mössbauer spectroscopy and crystal chemistry of natural Fe-Ti garnets. *Trans. Am. Geophys. Union, 59,* No. 4, 395-396.

Shannon, R.D. and C.T. Prewitt (1969) Effective ionic radii in oxides and fluorides. *Acta Crystallogr., B25,* 925-946.

Shimazaki, H. (1977) Grossular-spessartine-almandine garnets from some Japanese scheelite skarns. *Can. Mineral., 15,* 74-80.

Shoji, T. (1974) $Ca_3Al_2(SiO_4)_3-Ca_3Al_2(O_4H_4)_3$ series garnet: composition and stability. *Jour. Mineral. Soc. Japan 11,* 359-372 (in Japanese).

────── (1975) Phase relations in the system $(CaO-Fe_2O_3-SiO_2-H_2O)$. *Jour. Mineral. Soc. Japan 12,* 143-156 (in Japanese).

Skinner, B.J. (1956) Physical properties of end-members of the garnet group. *Am. Mineral., 41,* 428-436.

Smith, J.V. and B. Mason (1970) Pyroxene-garnet transformation in Coorara Meteorite. *Science 168,* 832-833.

Snow, R.B. (1943) Equilibrium relationships on the liquidus surface in part of the $MnO-Al_2O_3 - SiO_2$ system. *Jour. Am. Ceram. Soc., 16,* 11.

Sobolev, N.V., Jr. and Ju. G. Lavrent'ev (1971) Isomorphic sodium admixture in garnets formed at high pressure. *Contrib. Mineral. Petrol.*, *31*, 1-12.

Sriramadas, A. (1957) Diagrams for the correlation of unit cell edges and refractive indices with the chemical composition of garnet. *Am. Mineral.*, *42*, 294-298.

Stose, G.W. and J.J. Glass (1938) Garnet crystals in cavities in metamorphosed Triassic conglomerate in York County, Pennsylvania. *Am. Mineral.*, *23*, 430-435.

Strens, R.G.J. (1965) Synthesis and properties of calcium vanadium garnet (goldmanite). *Am. Mineral.*, *50*, 260.

Takéuchi, Y. and N. Haga (1976) Optical anomaly and structure of silicate garnets. *Proc. Japan Acad.*, *52*, 228-231.

Thompson, R.N. (1975) Is upper-mantle phosphorus contained in sodic garnet? *Earth Planet. Sci. Lett.*, *26*, 417-424.

Wakita, H., K. Shibao and K. Nagashima (1969) Yttrian spessartine from Suishoyama, Fukushima Prefecture, Japan. *Am. Mineral.*, *54*, 1678-1683.

Weaver, Scott J. (1976) Isothermal compression of grossular garnets to 250 kbar and the effect of the calcium on the bulk modulus. *Jour. of Geophys. Res.*, *81*, 2475-2482.

Wilkens, R.W.T. and W. Sabine (1973) Water content of some nominally anhydrous silicates. *Am. Mineral.*, *58*, 508-516.

Winchell, A.N. (1933) *Elements of Optical Mineralogy*, part II, 3rd ed., John Wiley and Sons, New York, 183.

——— and H. Winchell (1951) *Elements of Optical Mineralogy*, part II, 4th ed., John Wiley and Sons, New York, 490.

Winchell, H. (1958) The composition and physical properties of garnet. *Am. Mineral.*, *43*, 595-600.

Woodsworth, G.J. (1977) Homogenization of zoned garnets from pelitic schists. *Can. Mineral.*, *15*, 230-242.

Yoder, H.S., Jr. (1950) Stability relations of grossularite. *Jour. Geol.*, *58*, 221-253.

——— and G.A. Chinner (1960) Grossularite-pyrope-water system at 10,000 bars. *Carnegie Instit. Year Book*, *59*, 78-81.

——— and M.L. Keith (1951) Complete substitution of aluminum for silicon: the system $3MnO \cdot 3SiO_2 - 3Y_2O_3 \cdot 5Al_2O_3$. *Am. Mineral.*, *36*, 519-533.

Zabinski, W. (1966) Hydrogarnets. Pol. Akad. Nauk Oddzial Krakowie Kom. *Nauk Mineral. No. 3*, 1-61.

Zemann, A. and J. Zemann (1961) Verfeinerung der kristallstruktur von synthetischeim pyrope, $Mg_3Al_2(SiO_4)_3$. *Acta Crystallogr.*, *14*, 835-837.

Zemann, J. (1962) Zur kristallchemie der granate. *Beitr. Mineral. Petrol.*, *8*, 180-188.

Chapter 3
ZIRCON
J. A. Speer

INTRODUCTION

If for no other reason, zircon is a remarkable mineral because
of its ubiquitous occurrence in crustal igneous, metamorphic and
sedimentary rocks, and in mantle xenoliths, lunar rocks, meteorites
and tektites. Zircons occur either as primary crystallization
products, xenocrysts, detrital grains, or recrystallized grains. They
are seldom abundant but can occur in significant amounts in heavy
mineral layers in sediments and as an essential constituent of some
alkaline igneous rocks such as zircon syenites. Discovery of radiometric dating in the beginning of the twentieth century added
importance to the widespread occurrence of zircon. Zircon contains
minor amounts of U and Th and can be dated by a variety of techniques
yielding ages of crystallization, cooling, and redistribution of
radioactive isotopes and their daughter products. Zircon as a source
or protolith indicator in igneous, metamorphic and sedimentary rocks
reflects its ability to survive one or more cycles of erosion,
sedimentary transport, diagenesis, metamorphism, and/or anatexis. As
an ore, zircon is recovered from unconsolidated sands (see Chapter 10,
Marine Placer Minerals, in Volume 6 of this series). Its most important commercial use is in foundary sands, refractories and ceramics
because of its thermal properties. Lesser amounts of zircon are used
for abrasives, gemstones and the production of zirconium and hafnium.

CRYSTAL STRUCTURE

Zirconium orthosilicate, $ZrSiO_4$, is polymorphous: zircon, the
naturally-occurring dimorph, is tetragonal, space group $I4_1/amd$,
$Z = 4$ with a structure-type named for the species. The scheelite-
type structure, with space group $I4_1/a$, $Z = 4$, is unknown in nature.
The two structures are generally similar, but $ZrSiO_4$ with the scheelite
structure is $\sim$ 11% more dense than zircon because of closer packing.
The discussion of scheelite-type $ZrSiO_4$ is included in a later section
on polymorphism of ABO_4 compounds.

Minerals having the zircon-structure are *hafnon* ($HfSiO_4$),
xenotime (YPO_4), *behierite* ($(Ta,Nb)BO_4$), and the actinide orthosilicates: *thorite* ($ThSiO_4$) and *coffinite* ($USiO_4$) (see the following
chapter). Other compounds include the synthetic lanthanide phosphates,
arsenates, chromates and vanadates, as well as Zr, Hf and Th-germanates.

Table 1a. Crystal data for zircon, $ZrSiO_4$, and hafnon, $HfSiO_4$. Space group: $I4_1/amd$. (Estimated standard deviations are given in parentheses and refer to the last decimal place.)

	1	2	3	4	5	6	7
a, Å	6.604	6.6164(5)	6.607(1)	6.612(2)	6.6042(4)	6.581	6.573
c, Å	5.979	6.0150(5)	5.982(1)	5.994(2)	5.9796(3)	5.967	5.964
Volume, Å^3	260.7	263.32	261.1	262.0	260.80	258.4	258.1
ρ, gm cm^{-3}	4.67	4.67	4.71	4.64	--	6.95	6.97

1. synthetic $ZrSiO_4$. (Salt *et al.*, 1967)
2. zircon, Ilmen Mt., Urals, Russia (Krstanovic, 1958)
3. zircon, Kragero, Norway. Hf ∿ 1 wt % (Robinson *et al.*, 1971)
4. zircon, North Bay, Ontario. 100 ppm U (Finger, 1974)
5. zircon, Finsch Pipe, Kimberley, South Africa. 1.2 wt % HfO_2, 26 ppm U (Hazen and Finger, 1979)
6. synthetic $HfSiO_4$. (Durif, 1961)
7. synthetic $HfSiO_4$. (Salt *et al.*, 1967)

Table 1b. Crystal data for scheelite-type polymorphs of $ZrSiO_4$ and $HfSiO_4$. Space group: $I4_1/a$.

	$ZrSiO_4$*	$HfSiO_4$**
a, Å	4.73	4.719
c, Å	10.48	10.433
Volume, Å^3	234.5	232.33
ρ, gm cm^{-3}	5.19	7.76

* Reid and Ringwood (1969)
** Caruba *et al.* (1975)

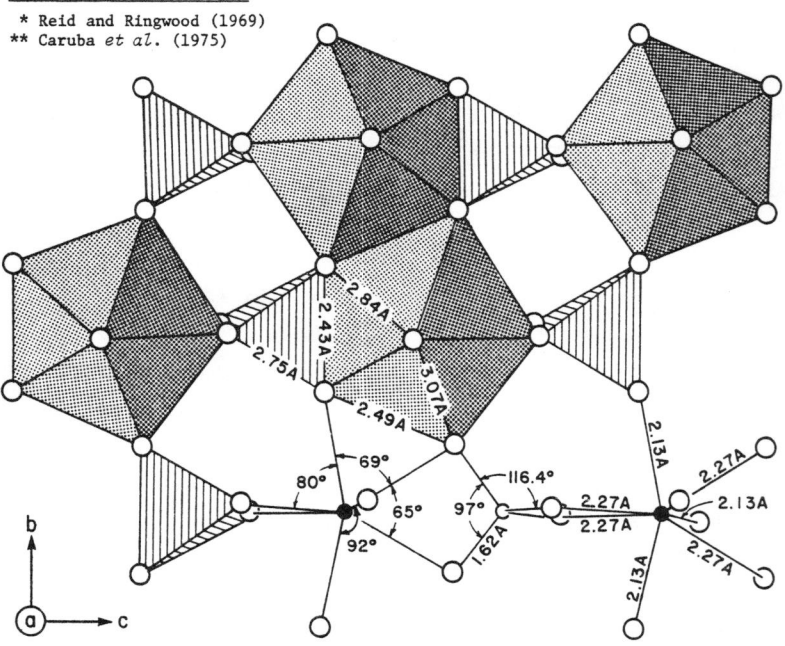

Figure 1. Perspective polyhedral and ball and stick representation of the zircon structure. Ruled tetrahedra are SiO_4 groups and shaded polyhedra are the ZrO_8 groups. Chains of alternating edge-sharing SiO_4 tetrahedra and ZrO_8 triangular dodecahedra extend parallel to c and are joined laterally by edge-sharing dodecahedra.

68

The Structure of Zircon

There have been at least nine determinations of the crystal structure of zircon since the initial works by Vegard (1926), Binks (1926), and Hassel (1926). Our discussion is based on recent refinements by Robinson *et al.* (1971), Finger (1974) and Hazen and Finger (1979), who discuss the earlier structural studies. Crystal data for the zircons used in these refinements, as well as that for synthetic $ZrSiO_4$ and $HfSiO_4$ are summarized in Table 1a.

The atoms in zircon occupy special positions and the two positional parameters of the oxygen atom are the only variables (Table 2). The principal structural unit in zircon is a chain of alternating, edge-sharing $[SiO_4]$ tetrahedra and $[ZrO_8]$ triangular dodecahedra extending parallel to c (Fig. 1). The chains are joined laterally by edge-sharing dodecahedra. "Similar chains occur in garnet [see Chapter 2], extending in three mutually perpendicular directions, but they are cross-linked by $[YO_6]$ octahedra as well as by $[XO_8]$ dodecahedra. In zircon octahedral voids are present but contain no cations. The structural similarities of zircon and garnet account for their similar hardness, density and high refraction indices" (Robinson *et al.*, 1971; see their Fig. 3).

The SiO_4 group is a tetragonal disphenoid, $\overline{4}2m$, elongate along the c-axis, parallel to the chain. The Si-O distance is 1.62 A (Table 2). The $O \cdots O$ edge shared with $[ZrO_8]$ polyhedra is 2.43 A subtending an O-Si-O angle of 97.0° whereas the unshared edge is 2.75 A opposite an O-Si-O angle of 116.1° (Fig. 1). The $[ZrO_8]$ polyhedron has been described as a triangular dodecahedron with symmetry $\overline{4}2m$. The two shortest $O \cdots O$ distances (2.43 A) and smallest O-Zr-O angles (65°) represent edges shared with SiO_4 tetrahedra (Fig. 1). The next shortest $O \cdots O$ distances (2.49 A) and O-Zr-O angles (69°) are for the four edges shared with other $[ZrO_8]$ dodecahedra. There are two sets of unshared edges, eight with $O \cdots O$ distances of 2.84 A and O-Zr-O angles of 80° and four with 3.07 A and 92°. The four Zr-O distances to edges shared between tetrahedra and dodecahedra are 2.27 A. The remaining four Zr-O distances are 2.13 A. The oxygen has site symmetry m and is coordinated by a planar array of one Si at 1.62 A and two Zr at 2.13 A and 2.27 A (*cf.* Fig. 4, Robinson *et al.*, 1971).

Table 2. Positional parameters, interatomic distances and interbond angles for zircon.
(Standard errors are 0.003 A or less for Si-O and Zr-O distances, 0.007 A or (usually)
less for O•••O distances, 0.2° or less for bond angles.)

Atom	Site	Multiplicity	Positional parameters	Site symmetry
Zr	a	4	0, 3/4, 1/8	$\bar{4}2m$
Si	b	4	0, 3/4, 3/8	$\bar{4}2m$
O	h	16	0, x, z	m

	Robinson *et al.* (1971)		Finger (1974)		Hazen & Finger (1979)	
Oxygen coordinates						
x	0.0661(1)		0.0660(5)		0.0660(4)	
z	0.1953(1)		0.1941(5)		0.1951(4)	
	Distance, A	Angle at cation	Distance, A	Angle at cation	Distance, A	Angle at cation
SiO$_4$ tetrahedron						
Si-O [4]*	1.622		1.630		1.623	
O•••O[1] [2]	2.430	97.0	2.435	96.6	2.431	97.0
O•••O [4]	2.752	116.1	2.769	116.2	2.749	116.1
ZrO$_8$ triangular dodecahedron						
Zr-O [4]	2.268		2.268		2.267	
Zr-O [4]	2.131		2.129		2.128	
mean	2.200		2.198		2.198	
O-Zr-O[1] [2]	2.430	64.8	2.435	65.0	2.431	64.8
O-Zr-O [8]	2.842	80.4	2.844	80.6	2.840	80.4
O-Zr-O[2] [4]	2.494	69.0	2.485	68.8	2.491	68.9
O-Zr-O [4]	3.071	92.2	3.068	92.2	3.068	92.2

* Bracketed numbers are bond multiplicities.
[1] Edge shared between tetrahedron and dodecahedron.
[2] Edge shared between two dodecahedra.

The refinement of Finger (1974) differs from those of Robinson
et al. (1971) and Hazen and Finger (1979). Finger discusses a few reasons for the discrepancies, but an additional cause may be that his
North Bay, Ontario zircon is slightly metamict. This suggestion is
based on its larger cell dimensions, as discussed in a later section.

A structure refinement of synthetic hafnon ($I4_1/amd$; a = 6.573,
c = 5.963 A) by Speer and Cooper (1982) yielded oxygen coordinates x =
0.0655, z = 0.1948. Similiarity of ionic radii of Hf (0.83 A) and Zr
(0.84 A) accounts for the interatomic distances and angles of hafnon and
zircon being identical within the stated errors, although hafnon has
systematically smaller distances.

Based on polarized infrared and Raman studies of zircon, Dawson *et
al.* (1971) calculated the force constants to be 0.94 and 0.96 mdyn/A,
respectively, for the 2.05 A and the 2.41 A Zr-O bonds. They assumed a
central force field model and estimated the Si-O force constant to be
9.01 mdyn/A. The effective charge on Si in zircon is between +1 and +2
based on calculated lattice energies (Sahl and Zemann, 1965) and +1.25
based on the Si$K\alpha$ line shift of the x-ray emission spectra (Marfunin,
1979).

ABO₄ Polymorphism

At ambient temperatures and pressures the zirconium, hafnium and
actinide (Th to Am) orthosilicates crystallize with the zircon structure.
The range in effective ionic radius for the eight-coordinated A-cation
is 0.83 A (Hf) to 1.05 A (Th) (Shannon, 1976). At higher temperatures
and pressures, orthosilicates with A-cations larger than 1.00 A [Th
(1.05 A) and Pa (1.01 A)] have the monoclinic, monazite-type structure
(see Chapter 10). Orthosilicates with A-cations smaller than 0.84 A
[Zr (0.84 A) and Hf (0.83 A)] transform to the scheelite structure at
high pressures. Reid and Ringwood (1969) have reported the synthesis
of the scheelite dimorph of $ZrSiO_4$ at 120 kbar and 900°C, and $HfSiO_4$
[scheelite] has been reported by Caruba _et al._ (1975). Their cell
parameters are listed in Table 1b. There is about an 11% increase in
density for the scheelite-type dimorphs as compared to zircon and hafnon,
which is accomplished by a more efficient packing of the coordination
polyhedra.

Although the actual structures of the scheelite-type polymorphs
have not been determined, they can be compared to that of the mineral,
scheelite, $CaWO_4$ (Burbank, 1965). As in zircon, Si is in tetrahedral
sites with four equal Si-O distances. Zr and Hf are in $(Zr,Hf)O_8$ polyhedra described as the interpenetration of two tetrahedra of unequal
size. There are eight unique oxygens, each belonging to one of the
eight different SiO_4 groups. Reid and Ringwood (1969) suggest that
$ZrSiO_4$ [scheelite] might be found in high pressure rocks, such as kimberlite pipes and that at still higher pressures, silicon will increase
in coordination to 6, and $ZrSiO_4$ will adopt the $KAlF_4$ structure. Hazen
and Finger (1979) predict that the [scheelite] → $KAlF_4$ structure transformation will be at about 170 kbar on the basis of their study of
zircon compressibility. Figure 2 summarizes the structural relationships involving zircon. Further details of the systematics of ABO_4
structural families are given by Muller and Roy (1974) and Zverzdinskaya
_et al._ (1977).

CHEMISTRY

Zircon is $ZrSiO_4$: 67.1% ZrO_2 and 32.9% SiO_2. More than 50 elements
have been reported in zircon analyses. Those elements which have been
best documented are listed in Table 3 along with their atomic radii.

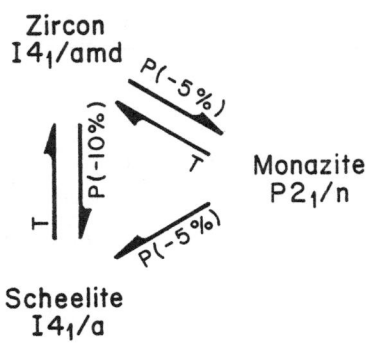

Zircon
I4₁/amd

Monazite
P2₁/n

Scheelite
I4₁/a

KAlF₄
P4/mmm

Figure 2. Phase relations among the structure types of ABO_4 orthosilicates. The large, A-cation ortho-silicates $ThSiO_4$ and $PaSiO_4$ follow the sequence zircon - monazite - scheelite whereas the small, A-cation orthosilicates $ZrSiO_4$ and $HfSiO_4$ follow the sequence zircon - scheelite - $KAlF_4$. Arrows indicate phase transitions between the structures as a function of temperature (T) and pressure (P). Numbers in parentheses are the approximate volume changes associated with the pressure transition.

Table 3. Ionic radii (in Angstroms, from Shannon, 1976) of elements substituting for eight-coordinated $Zr(A)$ and four-coordinated $Si(B)$ in zircon (ABO_4)

A	Radius	A	Radius	B	Radius
Zr^{+4}	0.84	Pb^{+2}	1.29	Si^{+4}	0.26
Hf^{+4}	0.83	Pb^{+4}	0.94	Al^{+3}	0.39
Y^{+3}	1.02	Ca^{+2}	1.12	P^{+5}	0.17
U^{+4}	1.00	Fe^{+3}	0.78	S^{+6}	0.12
Th^{+4}	1.05	Fe^{+2}	0.92		
La^{+3}	1.16	Na^{+1}	1.18		
↓		K^{+1}	1.51		
REE		Ti^{+4}	0.74		
↓		Nb^{+5}	0.74		
Lu^{+3}	0.98	Ta^{+5}	0.74		

In 8-fold coordination, tetravalent Zr is most closely related to tetravalent Hf, which is the major substituent for Zr in zircon. Because the element hafnium was discovered only as recently as 1923, it is not reported in older zircon analyses.

There are about 15 varietal names of zircon. They differ from zircon in their chemistry, water content and, because of radiation damage, their physical properties. The variation in physical properties intrigued and puzzled early workers until the discovery of radio-activity and idea of metamictization by Hamberg in 1914. The chemical varieties can be adequately described by Schaller's (1930) adjectival prefixes. Discussion of varieties of zircon is given in Vlasov (1966, pp. 350-353).

Many analyses of zircons have been reported for a bulk sample of small crystals separated from a large volume of rock. Because of the likelihood of contamination from similar-appearing minerals in the mineral separate, and because of abundant inclusions (Gorz, 1974; Ono, 1975) and occasional oriented intergrowths with xenotime and pyrochlore, the following discussion on the chemistry of zircon emphasizes analyses obtained by electron microprobe. Microprobe analyses also permit study of the strong compositional zoning present in zircons.

$ZrSiO_4$-$HfSiO_4$ Solid Solutions

Synthetic $HfSiO_4$ is isostructural with zircon and was called *hafnon* by Curtis *et al.* (1954), a name subsequently approved for the natural end-member by the International Mineralogical Association. Correria Neves *et al.* (1974) suggested the following IMA sanctioned nomenclature for natural $(Zr,Hf)SiO_4$ solid solutions:

mol % $HfSiO_4$:	0-10	10-50	50-90	90-100
	zircon	*hafnian zircon*	*zirconian hafnon*	*hafnon*

Hafnons from Zambézia, Mozambique, have compositional zones with up to 97 mol % Hf and bulk compositions up to 78 mol % (Correria Neves *et al.*, 1974). Ramakrishnan *et al.* (1969) synthesized intermediate compositions in the $ZrSiO_4$-$HfSiO_4$ system, demonstrating complete solid solution between zircon and hafnon.

Ahrens and Erlank (1969) compiled hafnium contents for 463 analyzed zircons (Fig. 3). In general, zircons contain less than 3% Hf with a range of 0.6 to 7.0 wt % and a mean of 1.71 %. Lunar zircons range from 0 to 3 wt % HfO_2 (Smith and Steele, 1976). A mean Hf content of 1.71 wt % represents a Zr:Hf ratio of 40:1, which closely corresponds to the ratio in which they are present in the earth's crust. This indicates that on the average an efficient process of geochemical separation of Hf and Zr does not occur in most geologic situations. Hf-rich rims on zircons in igneous rocks (Ono, 1975) and the uncommon enrichment of hafnium with differentiation of igneous rocks and their contained zircons (Ahrens and Erlank, 1969; Ehmann *et al.*, 1979) suggests some undefined and infrequent mechanism of Hf-enrichment. Most Hf-rich zircons are found in pegmatites containing tantalum and niobium minerals. Hf-enrichment associated with Ta and Nb mineralization is noted by von Knorring and Hornung (1961),

73

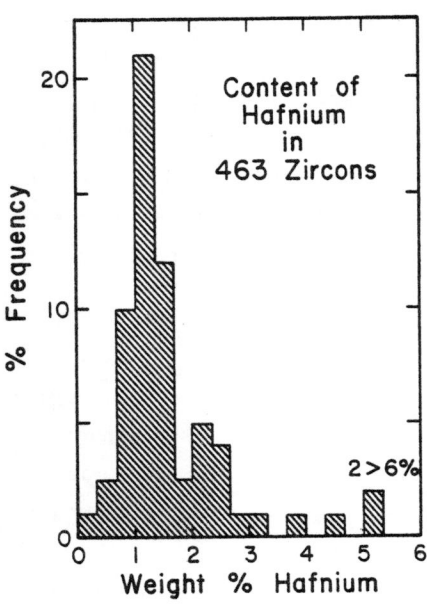

Figure 3. Frequency distribution diagram for weight % hafnium in 463 zircons compiled by Ahrens and Erlank (1969).

Su and Pan (1973), and Fontan *et al.* (1980). Fontan *et al.* (1980) noted that as Hf/(Zr+Hf) in zircons increased from 0.07 to 0.26, Ta/(Ta+Nb) in coexisting niobotantalates increased from 0.23 to 0.93. As might be anticipated, the process of Hf-enrichment causes increase of the even-odd pair [72]Hf-[73]Ta over the [40]Zr-[41]Nb pair.

$ZrSiO_4$-YPO_4-$REEPO_4$ Solid Solutions

Concentrations of rare earth elements in zircon can be as high as 25 wt % (Y + REE_2O_3), as reported by Medenbach (1976) in a microprobe study. Previous bulk analyses had reported as much as 15.89 wt % for the REE varities of zircon, but these have been subsequently shown to be zircon-xenotime intergrowths by single crystal x-ray and microprobe work (Robinson, 1978). The accompanying high phosphorus content of the REE-rich zircons suggests the substitution $REE^{3+} + P^{5+} = Zr^{4+} + Si^{4+}$, which represents a solid solution of zircon with its isostructural compound, xenotime. Often the amount of P in zircon is less than that of REE, requiring another charge balance mechanism. Romans *et al.* (1975) concluded that for the zircons they studied, the small quantities of Ca, Al, Fe and S detected share in the REE coupling substitution to establish charge balance. Caruba *et al.* (1974), Medenbach (1976), and Robinson (1978) suggest charge balance by substitution of $(OH)^-$: $(Zr_{1-y}REE_y)(SiO_4)_{1-x}(OH)_{4x-y}$. A hydroxyl zircon with about 10% Dy replacement for Zr was synthesized by Caruba *et al.* (1974). An electron paramagnetic resonance (EPR) study of zircon by Bershov (1971) found that the rare earths Tb and Tm had the unexpected charges of +4 and +2, respectively. At high concentrations of REE, when there is a high probability of substituting two REE's in adjacent A sites, charge balance can be achieved by formation of O^- centers.

These electron hole centers produce SiO_2^-, SiO_3^{3-}, and SiO_4^{5-} radicals
(Cainullina, 1971; Solntsev *et al.*, 1974). Solntsev *et al.* (1974) be-
lieve these radicals result from structural defects, such as Zr- and
O-vacancies, as well as non-isovalent substitution. Small amounts of
Ta- and Nb-pentoxide reported in zircon analyses suggest charge compensation by the couple $2Zr^{4+} = (Ta,Nb)^{5+} + REE^{3+}$ (Es'kova, 1959; Parker
and Fleischer, 1968).

Of the Y-REE group, Y predominates in zircon and can be present in
amounts of up to 16.5 wt % Y_2O_3 (Görz and White, 1970; Ono, 1975; Romans,
et al., 1975; Krasnobayev *et al.*, 1976; Medenbach, 1976). The highest Y
and P contents reported in zircon correspond to about 25 mol % xenotime.
Xenotime is reported to contain at most a few mol % $ZrSiO_4$ in solid solution (Vlasov, 1966). Coexisting xenotime and zircon are often reported
in rocks, as are epitaxial intergrowths, overgrowths and inclusions
(Vainshtain *et al.*, 1959; Cerny, 1956; Pigorini and Veniale, 1968;
Fujiwara *et al.*, 1965; Grünenfelder *et al.*, 1968; Ono, 1975; Romans *et
al.*, 1975; and Robinson, 1978). YPO_4 contents of zircon coexisting with
xenotime range from 0.1 to 10 wt %, but are usually less than 3 wt %.
This suggests that under most geologic conditions there is a wide miscibility gap on the $ZrSiO_4$-YPO_4 join.

Zircons usually contain less than 1 wt % REE_2O_3, but in more differentiated igneous rocks higher values are common, with a maximum near
10 wt %. It is often noted that the heavier REE predominate (Lyakhovich,
1962; Yes'kova and Ganzeyev, 1964; Khomyakov and Manukhova, 1970; Fielding,
1970; Nagasawa, 1970; Romans *et al.*, 1975; Medenbach, 1976; Gaudette *et al.*,
1981). In a survey of 128 REE analyses, Robinson (1978) found that
the REE distribution was Yb > Y = Dy > Tm > Lu = Er = Gd > Tb = Eu =
Ho = Ce > La > Sm > Nd = Pr. The distribution of REE in zircon can be
explained by the differing abundances of the odd-even REE (Oddo-Harkins
rule) and the ionic radii of these elements (see Table 3). The size of
Zr^{4+} is closer to the heavier than to the light REE's.

Enrichment of zircon in the lighter REE's has been reported by
Semenov and Barinskii (1958). In a study of the REE abundances in 62
zircons, Krasnobayev *et al.* (1976) found that zircon is not selective,
but will incorporate what is available. In addition, Krasnobayev *et al.*
(1976), Turovsky *et al.* (1967), and Khomyakov and Manukhova (1970) found

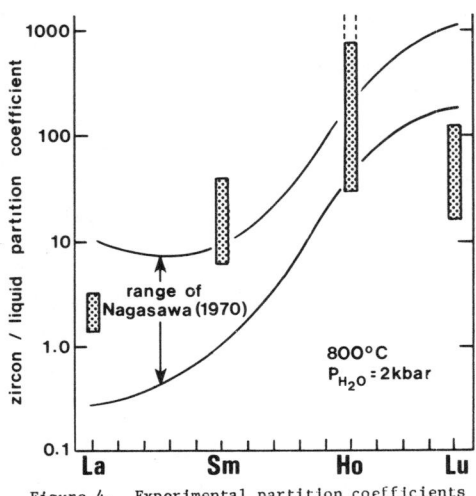

Figure 4. Experimental partition coefficients for La, Sm, Ho and Lu between synthetic zircon and a felsic, peralkaline silicate from Watson (1980) compared with the combined range of zircon/bulk rock and zircon/groundmass data of Nagasawa (1970). Eu data is excluded.

that the REE patterns change during crystallization. This is attributed to crystallization of other REE-selective minerals such as the ferromagnesian phases which change the REE pattern of the melt from which the zircons are continually crystallizing.

Better gauges of whether or not zircon is REE selective are the distribution coefficients between zircons and the melt from which they crystallized. Nagasawa (1970) calculated the partition coefficients from REE contents of zircons and their host dacites and granites and found heavy REE enrichment (Fig. 4). Experimental zircon/liquid partition coefficients obtained by Watson (1980) for a felsic, peralkaline liquid (SiO_2-Al_2O_3-Na_2O-K_2O-ZrO_2) at 800°C and P_{H_2O} = 2 kbars are: La 1.4-2.1, Sm 26-40, Ho 340+, Lu 72-126 (Fig. 4). These confirm the heavy REE enrichment of zircons over silicate melts found by Nagasawa (1970). The partitioning pattern is similar to garnet (Schnetzler and Philpotts, 1970), as expected in light of structural similarities to zircon. The concave downward pattern of the heavy REE end resembles some garnet/- and hornblende/liquid REE partitioning patterns (Watson, 1980).

Uranium, Thorium and Radiogenic Elements

The uranium contents of zircons range from 5 ppm in kimberlites (Kresten *et al.*, 1975; Davis, 1976) to 7 wt % U in a pegmatite (Muench, 1931). As much as 10 wt % Th has been reported in zircons from metasomatic rocks (Pavlenko *et al.*, 1957) and as little as 2 ppm in kimberlitic zircons (Ahrens *et al.*, 1967). Maximum values of U and Th determined by microprobe are 5.06 wt % UO_2 and 3.68 wt % ThO_2 (Medenbach, 1976). High U and Th zircons, which are often metamict and hydrous, are referred to as *crytolites* and *malacons*, varietal names still in common use. The concentrations of U and Th in zircon are usually much lower: 5-4000 ppm U and 2-2000 ppm Th (Gorz, 1974; Ahrens *et al.*, 1967), a fact which is understandable in terms of the large ionic radii of U and Th (Table 3).

A wide miscibility gap on the $ZrSiO_4$-$USiO_4$ and $ZrSiO_4$-$ThSiO_4$ joins is expected. In fact, Mumpton and Roy (1961) determined the limits of solid solution in zircon to be 4 ± 2 mol % $USiO_4$ and 4 mol % $ThSiO_4$. They believed that the phases which they synthesized with up to 20 mol % $USiO_4$ and 35 mol % $ThSiO_4$ were metastable. Caruba et al. (1975) were also able to synthesize a $(Zr_{0.8}U_{0.2})SiO_4$ composition but did not comment on its stability. Zircons grown in a flux with $1UO_2$:$2ZrO_2$ were found to contain 0.08 mol % UO_2 by Chase and Osmer (1966).

Natural zircons coexisting with U- and Th-silicates and -oxides are assumed to have incorporated maximum amounts of U and Th. These zircons are among those that have the highest U and Th contents but the contents are still low, indicative of wide miscibility gaps on the $ZrSiO_4$-$USiO_4$ and $ZrSiO_4$-$ThSiO_4$ joins. Pavlenko et al. (1957) found zircons coexisting with uraninite to contain as much as 0.5 wt % U and those coexisting with thorite and thorianite to contain up to 10 wt % Th. Steiger and Wasserburg (1966) and Silver and Deutsch (1963) found zircons coexisting with thorite to contain <300 ppm Th. Zircons coexisting with both uraninite and thorite in the Boulder Batholith, Idaho, have 0.07% U and Th (Effimoff, 1972).

The U and Th contents of rock-forming zircons are based on a large number of determinations because zircon is the most frequently used mineral for U-Th-Pb age dating. Zircons from granitoids have U and Th contents in the upper part of this range, >300 ppm U and >100 ppm Th (Ahrens et al., 1967). Zircons in kimberlites and mantle xenoliths have much lower U and Th contents: an average of 47 ppm U and a range of 4.8-440 ppm in 55 zircons and an average of 5 ppm Th for 6 zircons (Ahrens et al., 1967; Kresten, 1974; Davis, 1976, 1977). Zircons in lunar rocks typically have U contents of 100-400 ppm with a range of 10-1500 ppm (Roger and Adams, 1969). From data on U contents of rocks and their contained zircons, Kresten (1974) deduced a U_{zircon}/U_{liquid} partition coefficient of ≥ 100.

Ahrens (1965) and Ahrens et al. (1967) found that the Th/U ratio in zircons is less than 1, in contrast to the general value of 3.5-4 for igneous rocks. The reason for this is not known, but it could be accounted for by either preferential inclusion of U in zircon because it is closer to Zr in ionic radius than Th (Table 3), or cocrystallization with a Th-enriched phase such as allanite, monazite and thorite, or both.

Pb, Tl, He, Xe and possibly Ra and Bi are considered to be radiogenic and produced by decay of U and Th. The amounts of these daughter products depend on initial U and Th contents, duration of accumulation, and the ability of zircon to retain them. Zircons may contain up to 0.17 cc/gm He (Hurley, 1952; Hurley *et al.*, 1956) and small amounts (0.01-6.5 ppm) of nonradiogenetic or common lead (Krogh, 1971; Gulson and Krogh, 1975). Increasing amounts of common lead correlate with increasing numbers of inclusions.

Other Substituents

The remaining substitutions in zircons are usually minor with often only a singular high value reported. The problem of contamination is particularly acute in bulk samples separated from a large volume of rock. An added problem is that these elements are often confined to metamict and hydrated zones. However, as concluded by Romans *et al.* (1975), these elements are necessary for charge balance in zircon.

Nb and Ta contents are between 70 and 7000 ppm with the highest values from rocks containing coexisting Nb-Ta minerals (Pavlenko, 1957; Parker and Fleischer, 1968). Nb/Ta ratios range from 24 to 110 (Es'kova, 1959; Beus and Sitnin, 1968), reflecting the greater elemental abundance of Nb. The substitutional scheme is $(Nb,Ta)^{5+} + (REE^{3+},Fe^{3+}) = 2Zr^{4+}$.

Alkali and alkaline earths, Na, K, Mg, Ca, Sr, Ba and Ra, have been reported in zircon analyses (Görz, 1974). CaO is most often reported, with concentrations ranging up to 4.6% CaO (microprobe analyses by Roman *et al.*, 1975, and Medenbach, 1976). The highest values of CaO (9.00 wt %) and Na_2O (4.60 wt %) are for metamict zircons admixed with other minerals (Kopchenova *et al.*, 1974).

The only other elements reported in concentrations as high as 6 wt % by microprobe analysis are Fe and Al. Elements reported in amounts less than 1.0 wt % by a variety of techniques include Be, B, S, Sc, Ti, V, Cr, Mn, Ni, Cu, Zn, Ga, As, Mo, Ag, Sn, Sb, W and Au.

Water

Chemical analyses of zircons, particularly metamict zircons, often include water, with as much as 16.6 wt % reported by Coleman and Erd (1961). Agreement has not been reached on whether the water is essential or water that has been absorbed by metamict material. Frondel (1953)

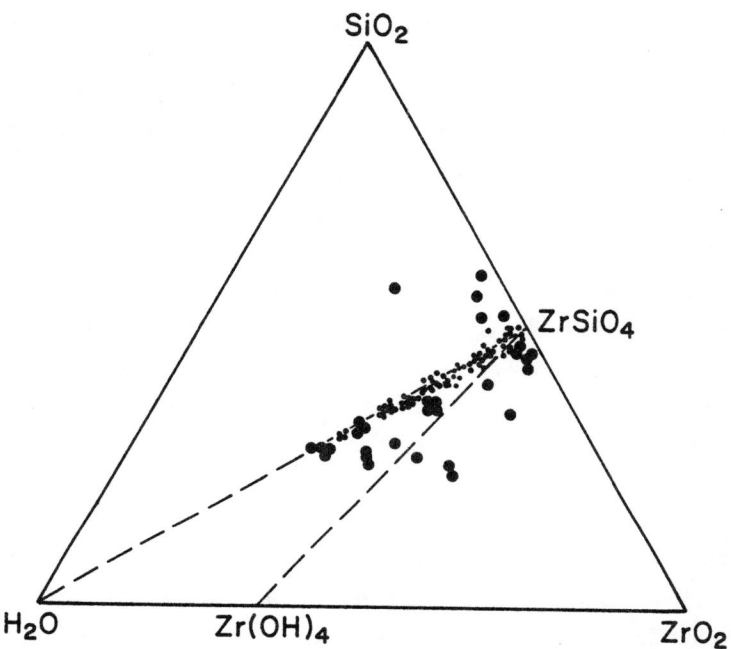

SiO₂ — at top (as SiO_2)

ZrSiO₄ (as $ZrSiO_4$)

H₂O (H_2O)   Zr(OH)₄ ($Zr(OH)_4$)   ZrO₂ (ZrO_2)

Figure 5. Compositions of natural zircons plotted on a molecular basis. Substituents for Zr are not specified. Zircons of the type $Zr(SiO_4)_{1-x}(OH)_{4x}$ would lie along the $ZrSiO_4$-$Zr(OH)_4$ join, whereas zircons with molecular water lie along the $ZrSiO_4$-H_2O join. Large circles are literature values compiled by Mumpton and Roy (1961). Small circles are based on microprobe analyses where (OH) is calculated from $((Si,P,Al)O_4)_{1-x} + (OH)_{4x}$ and molecular H_2O obtained by difference from the oxide sum (Sommerauer, 1976).

first suggested that the water in zircon represents $(OH)_4 \rightleftarrows (SiO_4)$ substitution. The water is often essential in compensating silicon deficiencies (Frondel, 1953; Pigorini and Veniale, 1966). Sommerauer (1976) found that for 137 microprobe analyses the amount of hydroxyl required to balance Si + P + Al deficiencies in the B-site was much less than the amount of molecular water needed to increase the low totals to 100%. Dymkov and Nazarenko (1962) suggested the solid solution series: $ZrSiO_4$-$Zr(SiO_4)_{1-x}(OH)_x$-$Zr(OH)_4$-ZrO_2. Mumpton and Roy (1961) pointed out that if natural zircons are $ZrSiO_4$-$Zr(OH)_4$ solid solutions they should lie along this join in a SiO_2-ZrO_2-H_2O plot. As evidenced in Figure 5, zircon analyses lie along the $ZrSiO_4$-H_2O join, indicating that water of natural zircons is present largely as molecular water, not hydroxyl.

Infrared, Raman and nuclear magnetic resonance spectroscopy show that both H_2O and $(OH)^-$ are present in zircon (Dawson et al., 1971; Rudnitskaya and Lipova, 1972; Krasnobaev and Ivonina, 1973). In addition, both H_2O and $(OH)^-$ are in several different sites in the structure, and are aligned

either parallel or perpendicular to the c axis. Not surprisingly,
Krasnobaev and Ivonina (1973) note that $(OH)^-$ is characteristic of
zircons with high water contents. Frondel and Collette (1957) synthesized hydroxyl zircon at 150°C, which was identified by its IR
spectra. Caruba *et al.* (1974) synthesized $Zr(SiO_4)_{.75}(OH)_{.25}$ with cell
dimensions slightly smaller than anhydrous zircon, although they are
identical within the errors stated. It dehydrated at 750°C, producing
zircon $+ ZrO_2 + H_2O$ (Caruba *et al.*, 1975).

From the chemical analyses and experimental and spectroscopic
data, it appears that substitution of $(OH)^-$ for SiO_4 in natural zircons is minor and that most water is absorbed by metamict material of
various compositions as either H_2O or $(OH)^-$.

ZONING

Optical microscopic examination, microprobe analysis, mass spectrometry, fission track mapping, color zoning and ultraviolet or cathodoluminescence have demonstrated that zircons are often inhomogeneous.
Several types of zoning can be distinguished on the basis of type or
origin: growth zoning, passive zoning, expitaxial overgrowths and outgrowths, chemical zoning, and sector zoning.

Chemical Zoning

Distribution of elements in zircon crystals have been studied by
electron and proton microprobe in works by Steiger and Wasserberg (1966),
Davis *et al.* (1968), Veniale *et al.* (1968), Gulson (1969), Görz and White
(1970), Köppel and Grünenfelder (1971), Köppel and Sommerauer (1974),
Sommerauer (1976), Romans *et al.* (1975) and Clack *et al.* (1979). Because of the sensitivity of the method, U distribution in zircons can
most effectively be studied by fission track mapping (Fielding, 1970;
Yeliseyeva *et al.*, 1974; Grauert *et al.*, 1974; Fleischer *et al.*, 1975).
Chemical zoning can also be studied by ultraviolet and cathodo-lumi-
nescence (Fielding, 1970; Ono, 1974, 1975; Romans *et al.*, 1975; Sommerauer, 1976). Because of the interaction of the large number of
substituting elements, no simple relationship between luminescent intensity, color and chemistry can be made (see PHYSICAL PROPERTIES -
Luminescence).

80

These studies have found that the elements which most often substitute for Zr and Si, *i.e.*, Hf, Y, P, Ca, Al, Fe, S, Th and U, usually
increase together. The zoning is best described as oscillatory, although
there is a general trend of enrichment of elements substituting for Zr
and Si toward the rim. These chemical substitutions produce zones of
nearly ideal zircon intergrown with zircon of variable composition and
different physical properties. Extreme cases of compositional zoning
are represented by intergrowths of xenotime. Because of the lack of
continuous solid solution between zircon and xenotime, their coexis-
tence in zoned crystals can be interpreted as epitaxial intergrowths of
two immiscible, isostructural phases (*cf*. Grunenfelder *et al.*, 1968).

Both xenotime and compositionally impure zircon differ in physical
properties from the purer zircon. They are more soluble in HF, more
susceptible to radiation damage, water-rich, unstable with increasing
temperature and diffuse Pb more easily. Strongly zoned zircons produce
discordant ages, and chronological data are explained by a mixing of
highly discordant and almost concordant systems. The highly discordant system consists of the compositionally impure zircon zones, which
were called "hot spots" by Steiger and Wasserberg (1966) because of
their elevated U and Th contents. The concordant system is the nearly
ideal zircon. Sommerauer (1976) described them as two-phase zircon
systems. It remains to be seen whether this phenomenon in zircon is
simply a result of strong compositional zoning, or the result of epitaxial intergrowths of immiscible phases or of exsolution.

Growth Zoning

Much of the chemical zoning in zircon results from growth in a
silicate melt, as evidenced by its euhedral habit. Growth zones result from physical and chemical variations in the melt or solutions
from which the zircon crystallizes and lack of subsequent reequilibra-
tion. The chemical zoning, seriate and hiatal textures of
zircons, and bulk compositional variation with grain size are cited
as evidence that zircon continually crystallizes in a cooling magma
(Silver and Deutsch, 1963; Gottfried and Waring, 1964; Köhler, 1970;
Veniale *et al.*, 1968). This is in contrast to the more classical ideas
advanced by Rosenbusch (1882) that small, euhedral accessory minerals
crystallize early. The zircon cores reflect the initial composition
of the melt and the zoning reflects the changes as crystallization pro-

CONTINUOUS GROWTH HISTORY

unzoned   zoned

DISCONTINUOUS GROWTH HISTORY

overgrowths (of zircon or xenotime)

euhedral
cores

                partial

                              aggregate crystals
total   also called synneusis
                              or parallel growth

anhedral or detrital
        cores

outgrowths (of zircon or xenotime)

pyramidal   hemispherical

Figure 6. Diagrammatic representation of the terminology applied
to zoning in zircons. The shaded
sections are the earlier generation zircons.

ceeds, eventually producing a rim rich
in Hf, U, Th, Y, REE and P, as might be
expected in the late-stage melts.

In Sri Lanka zircons Sahama (1981)
described three types of growth bands:
(1) *fine banding* ($\sim$3 μm wide) is defined
by differences in interference colors
(B = 0.001 - 0.002) and extinction
angle ($\sim$1°) and may be caused by strain
resulting from rhythmic fluctuations in
crystallization rate. Some fine bands
have strong relief and higher birefringence and densities. (2) These
high density bands are the most crystalline portions present. (3) *Coarse
banding* is superimposed on the fine and
is characterized by greater differences
in birefringence (B = 0.010 - 0.050)
correlated with varying degrees of radiation damage.

Passive Zoning

Zircons in which chemical zoning is caused by removal or addition
of elements without new growth is termed passive zoning. Grauert *et al.*
(1974), for example, found that zircons in a quartzite gained U on their
rims without new crystal growth.

Outgrowths and Overgrowths

Zircons in metamorphic and igneous rocks often contain distinct
cores which have been epitaxially overgrown by later generation zircon
or xenotime. Figure 6 presents the terminology that has been applied to
these zircons. The different generations can often be distinguished optically, but also can differ in chemical and isotopic composition. In
metamorphic rocks, the older zircon cores have been inherited from the
original igneous or sedimentary rock, so they are either euhedral or
well-rounded detrital grains. In igneous rocks the older cores are xenocrysts from either the adjacent country rock or the source area. It is

this persistence, evidenced by these older cores, from which zircon gets its reputation as a stable and refractory mineral.

There have been numerous reported examples of the overgrowth of zircon on prexisting, rounded detrital zircons or of the recrystallization of zircon under high grade contact or regional metamorphism (Poldervaart, 1950; Taubeneck, 1975; Schidlowski, 1963; Gastil *et al.*, 1967; Köppel and Grünenfelder, 1967; Ishizaka, 1967; Davis *et al.*, 1968; Tilton and Grünenfelder, 1968; Ishizaka and Yamaguchi, 1969; Köppel and Sommerauer, 1974; Gulson and Krogh, 1975). Occurrences in which zircon is described as a metamorphic mineral are uncommon (Gillson, 1925). Zircons with overgrowths have received attention because of their discordant ages. In the simplest case, the upper concordia intercept is the age of the older cores and the lower intercept the age of overgrowth. In addition to differences in isotopic composition, Köppel and Grünenfelder (1971) and Köppel and Sommerauer (1974) found the older, detrital cores to be nearly end-member $ZrSiO_4$ as compared to the overgrown rims. These authors consider this phenomenon to be the combined result of mechanical abrasion and removal of outer growth zones rich in other elements, the greater resistance of near end-member $ZrSiO_4$ to chemical and mechanical weathering, and the loss of substituting elements during annealing and recrystallization of zircon.

The chemical and isotopic changes that occur in zircons with increasing metamorphic grade were demonstrated by Davis *et al.* (1968). With increasing metamorphism in a contact aureole, they found there was loss of Pb and gain in U and Th, particularly on the rims, with a resulting change in apparent age (Fig. 7).

Overgrowths on zircon xenocrysts in igneous rocks have been reported by Poldervaart (1950), Veniale *et al.* (1968), and Kohler (1970). Because this new growth is readily recognized, problems in age dating can be avoided. However, the discordant age can be of great interest in identifying the age of the magmatic source region. Zircon xenocrysts in the Caledonian granites of Scotland (c. 400 Ma) have U-Pb systems yielding dates of 1000-2000 Ma and are believed to have been derived by partial melting of Proterozoic crust at depth (Pidgeon and Johnson, 1974; Pankhurst and Pidgeon, 1976; Pidgeon and Aftalion, 1978). The persistence of zircon in sedimentary and metamorphic processes is

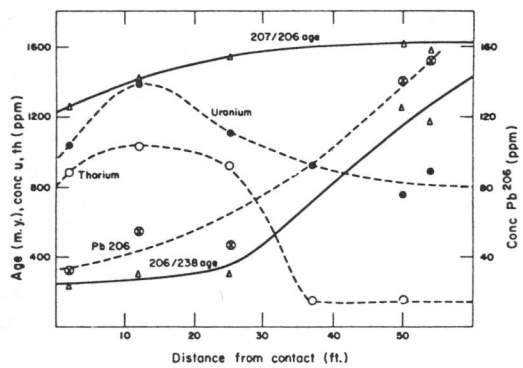

Figure 7. Change in $^{206}Pb/^{238}U$ and $^{207}Pb/^{206}Pb$ ages and in concentration of uranium, thorium and lead in zircons with increasing contact metamorphism as a function of distance from the contact at Eldora, Colorado. (After Davis *et al.*, 1968)

used by Halliday *et al.* (1979) to suggest the Caledonian granites could have also have come from sediments derived from the Proterozoic crust.

Sector Zoning

Sector zoning in both natural and synthetic zircon has been reported by Chase and Osmer (1966) and Fielding (1970). Preferential incorporation of U on the {111} growth faces results in conical zones of more intense color. Sector zoning indicates metastable conditions where growth rates of certain faces are too rapid for chemical equilibrium between the crystal and the growth medium (Dowty, 1976).

MORPHOLOGY

The crystal habit of zircon varies from long-prismatic, through prismatic, to dipyramidal. Structurally, it is the chains of edge-sharing tetrahedra and dodecahedra parallel to [100] that account for both its (110) cleavage and its dominant prismatic habit (Robinson *et al.*, 1971). The relationship of crystal structure and morphology can be examined by calculating the frequency and development of faces for zircon by the methods of Bravais, Donnay-Harker, Niggli or Schneer, but the dynamic behavior of the different crystal faces in varying physical and chemical environments is very important in accounting for the observed morphological variations in zircon as well.

Morphology as a Petrogenetic Indicator

Crystal morphology has received only limited attention in petrology except in the case of zircon. Morphology, when combined with other features such as color, size, inclusions, zoning, etc., has been used for both correlation and petrogenetic interpretations. Reviews are given by Poldervaart (1955, 1956), Pupin and Turco (1972), and Kostov (1972). The following is a list of the relationships of morphology and physical

84

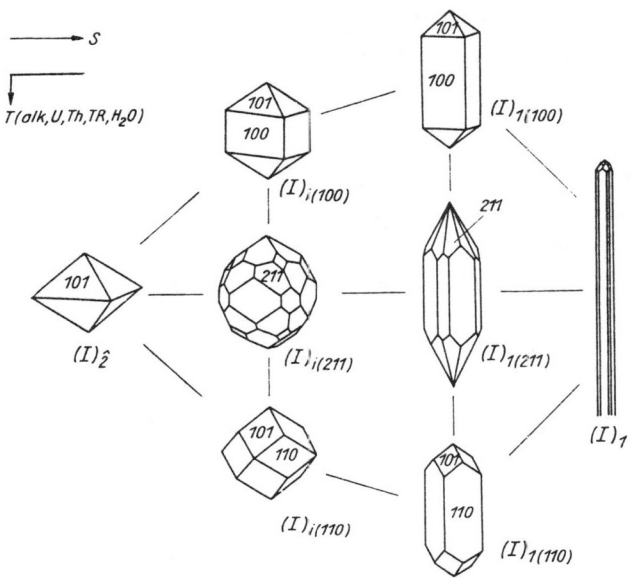

S

$T(alk,U,Th,TR,H_2O)$

$(I)_{1(100)}$

$(I)_{i(100)}$

$(I)_{\hat{2}}$

$(I)_{i(211)}$

$(I)_{1(211)}$

$(I)_1$

$(I)_{i(110)}$

$(I)_{1(110)}$

Figure 8a. Crystallogenetic diagram of zircon from Kostov (1973) showing the variation in habit with speed of crystallization and chemistry of the melt and zircons.

MILIEU	BASIQUE	NEUTRE	ACIDE				
COMPOSITION	ZrO_2 SiO_2 + $(K_2CO_3...)$	ZrO_2 + SiO_2	$ZrOCl_2.8H_2O$ + SiO_2	$Zr(NO_3)_4.5H_2O$ + SiO_2	$Zr(SO_4)_2.4H_2O$ + SiO_2	$Zr(SO_4)_2.4H_2O$ + SiO_2 + Na_2SO_4	$ZrF_4.H_2O$ + SiO_2
COMPOSES	ZIRCONO-SILICATES (Wadeite...)	ZIRCON					
MORPHOLOGIE							

Figure 8b. Effect of increasing acidity on the formation and morphology of zircon. (After Caruba *et al.*, 1975)

TEMPERATURE ± 50°C.	DEVELOPPEMENT RELATIF DES PRISMES.	EXEMPLES DE TYPES ET SOUS-TYPES (D'APRES CLASSIFICATION J.P.PUPIN, G.TURCO, 1971) PYRAMIDES (101) + (211)	(101) seule
1200° 900°	(100) seul	J2 J4	D
850°	(100) ≫ (110)	S22 S24	P5
800°	(100) > (110)	S17 S19	P4
750°	(100) = (110)	S12 S14	P3
700°	(100) < (110)	S7 S9	P2
650°	(100) ≪ (110)	S3 S4	P1
600°	110 seul	L2 L4	G1
550° 500°	O PRISME	AB2 AB4	A

Figure 8c. Changes in zircon morphology with estimated temperatures of crystallization of the host rock (see Pupin and Turco, 1972; 1975).

or chemical conditions of growth based on both experimental and petro-logic studies:

(1) Crystallization rate: rapid crystallization is found to favor long, prismatic crystals (see Fig. 8; Kostov, 1973; Larsen and Poldervaart, 1958).

(2) Acidity: increasing acidity favors more tabular crystals (see Fig. 8b; Caruba *et al.*, 1975).

85

(3) Agpaicity: related to (2) is the occurrence of bipyramidal zircons in agpaitic or highly alkaline rocks (K + Na > Al) (Caruba,
1978; Kostov, 1973; Pupin and Turco, 1975; Poldervaart, 1956;
Marchenko and Gurvov, 1966; see Fig. 8a).

(4) Temperature: the development of the {100} prism increases and
{110} prims decreases with increasing temperature (see Fig. 8c;
Pupin and Turco, 1972, 1975).

(5) Water content of the melt: {110} prisms dominate zircons in hydrous magmas whereas {100} prisms are dominant in dry magmas
(Pupin et al., 1978).

(6) Variation in zircon chemistry, substitution of U, Th, REE, P and
H_2O are presumed to favor bipyramids whereas Hf favors prismatic
habit (Kostov, 1973; see Fig. 8a).

(7) Crystal size: zircons change in crystal habit with growth, {110}
developing as the dominant face over {100} (Pupin and Turco, 1972;
Jocelyn and Pidgeon, 1974; Kohler, 1970). This may be the result
of decreasing temperature (Fig. 8c).

(8) Zircon overgrowths in igneous rocks are characteristic of autoch-
thonous granites (Poldervaart and Eckelman, 1955).

(9) Zircons from kimberlites are rounded (Kresten et al., 1975).

These conditions affecting zircon morphology are not independent of one
another and it is difficult to relate the morphology to any one condition. Pupin and Turco (1975) conclude that temperature and agpaicity
of the magma are the most important.

Using morphology as a petrogenetic indicator requires a method of
quantifying observations on a large number of crystals. Several methods
have been used:

(1) Statistical studies of crystal dimensions. Initially this was done
by graphical comparisons of frequency curves or histograms of
length (x), width (y) or elongation (x/y). This evolved into the
method of the reduced major axis (Alper and Poldervaart, 1957)
where the length and width of a number of unbroken crystals ($\sim$200)
are measured and the following calculated:

$\bar{x}$, mean length; S_x, standard deviation of x
$\bar{y}$, mean width; S_y, standard deviation of y

In a plot of length versus width, the point $(\bar{x}, \bar{y})$ is plotted and

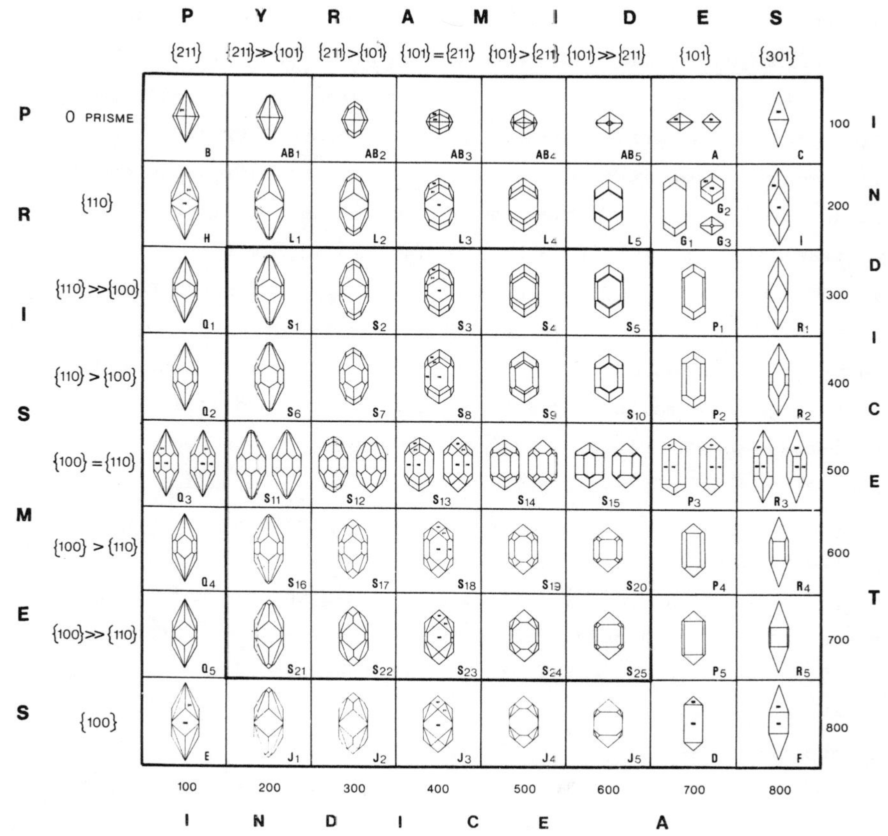

		{211}	{211}≫{101}	{211}>{101}	{101}={211}	{101}>{211}	{101}≫{211}	{101}	{301}		
P	0 PRISME	B	AB₁	AB₂	AB₃	AB₄	AB₅	A	C	100	I
R	{110}	H	L₁	L₂	L₃	L₄	L₅	G₁ G₂ G₃	I	200	N
I	{110}≫{100}	Q₁	S₁	S₂	S₃	S₄	S₅	P₁	R₁	300	D
S	{110}>{100}	Q₂	S₆	S₇	S₈	S₉	S₁₀	P₂	R₂	400	I
M	{100}={110}	Q₃	S₁₁	S₁₂	S₁₃	S₁₄	S₁₅	P₃	R₃	500	C
E	{100}>{110}	Q₄	S₁₆	S₁₇	S₁₈	S₁₉	S₂₀	P₄	R₄	600	E
S	{100}≫{110}	Q₅	S₂₁	S₂₂	S₂₃	S₂₄	S₂₅	P₅	R₅	700	T
	{100}	E	J₁	J₂	J₃	J₄	J₅	D	F	800	

100   200   300   400   500   600   700   800

I N D I C E A

Figure 9. Types and subtypes fundamental to the classification of zircons. (After Pupin and Turco, 1972)

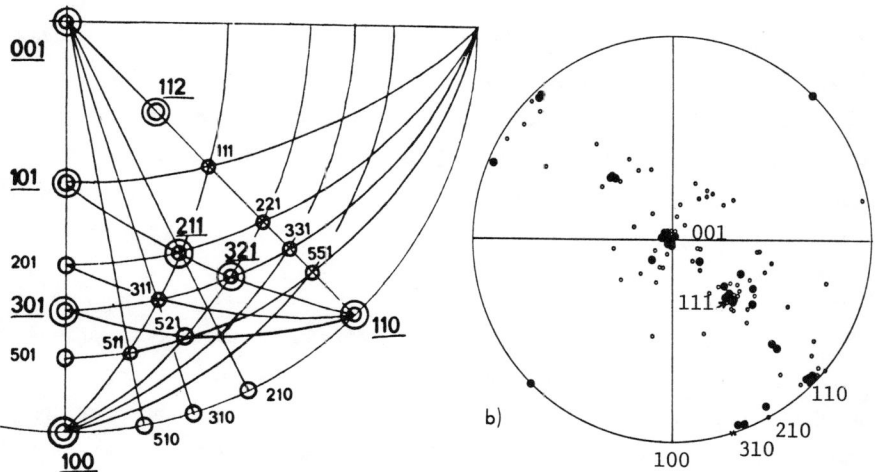

b)

Figure 10. (a) Stereographic projection of zircon (after Caruba and Turco, 1971); (b) poles to principal, cleavage planes (large filled circles) and subordinate fractures (small, open circles) in kimberlite zircons (after Kresten *et al.*, 1975).

a line, the slope of which is S_y/S_x, is drawn through the point. The end points of the line are obtained by omitting 2.5% each of the largest and shortest crystals measured. Zircon populations can be compared either visually or by statistical tests. The reduced major axis of a group of zircons in an igneous rock is supposed to approach the growth trend of the zircon crystals.

(2) Fourier shape analysis (Byerly *et al.*, 1975). The two-dimensional projection of zircons is resolved into a number of shape components or harmonics. The sum of the harmonics describe the grain shape. The harmonic amplitudes indicate the relative contribution of each harmonic to the total grain shape, and it is these amplitudes which are compared from one population to another.

(3) Predominance of principal prism and pyramidal crystal forms. The system of Pupin and Turco (1972) is the most extensive and incorporates the ideas of the earlier workers back to 1886. It is based on the relative development of pyramid and prism faces which are summarized on a grid reproduced in Figure 9 (from Pupin *et al.*, 1978). Statistical studies are made on unbroken zircon crystals which are grouped by similarity to the habits of the grid. The morphological setting, which is also the x-ray setting, is that suggested by Caruba and Turco (1971), who also presented a nonogram for identifying and classifying the faces.

Twinning

The commonly reported twin plane in zircon is {101}, which produces geniculate growth twins. This plane coincides with layers of isolated SiO_4 tetrahedra. A mirror plane would produce a twin-member whose structure would have a 180° rotation about [101] from a continuation of the untwinned crystal. Recent work and summary of previous work on zircon twinning is given by Jocelyn and Pidgeon (1974).

Cleavage

The most commonly reported cleavage in zircon is imperfect, prismatic and parallel to {110}, a plane which is parallel to the edge-sharing SiO_4-ZrO_8 chain and is also parallel to what is the nearest approximation to a close-packed layer of oxygen atoms in zircon. Kresten *et al.* (1975) found that in kimberlitic zircons {001} and {111} are the most frequently developed cleavages with {110}, {210}, {310}, {331} and {113} observed (Fig. 10b) in addition to a number of subordinate fractures and partings. Many of these cleavages are not equally developed as would be expected because of the symmetry, suggesting they are partings. Pronounced parting, traditionally attributed to twinning, may be a plane of weak adhesion such as between exsolved phases (White, 1979). Rather than exsolution, the weak adhesion in zircon may be between compositionally different growth zones which may differ in metamictization as well.

Optical Properties

Zircon is uniaxial positive, and synthetic zircon has refractive indices of ϵ = 1.984 and ω = 1.924. Dispersion is close to 0.04 for the interval of visible light, which is just slightly less than that of diamond, accounting for the fire of cut zircons and its use as a diamond substitute before the widespread availability of synthetic gems. Dispersion of the refractive indices of zircon have been measured by Il'inskiy and Krylova (1974) and their corrections for the dispersion relative to n at λ = 589.3 mμ are given in Table 4. With a mean refractive index of 1.98, zircon has a reflectivity of 11% which gives it a subadamantine luster.

Both the refractive indices and birefringence decrease with metamictization as measured by changes in density (Chudoba, 1935) or radioactivity (Morgan and Auer, 1941; Holland and Gottfried, 1955). The refractive index of metamict zircon approachs 1.81 as a limiting value (Fig. 11). This agrees well with the value of 1.83 calculated from the Gladstone-Dale relationship using a density of 3.96 for metamict zircon (see Density section below) and constants from Mandarino (1976). Some metamict specimens are biaxial with a $2V$ up to 10° (Krstanovic, 1964; Ueda, 1956), but most are isotropic.

Hafnian zircons with 21-31 wt % HfO_2 have refractive indices of ϵ = 1.97 and ω = 1.92 (von Knorring and Hornung, 1961). The slightly lower refractive indices of hafnon would be predicted from the Gladstone-Dale relationship which yields mean refractive indices of 1.980 for zircon and 1.946 for hafnon, using the densities in Table 1a for the synthetic material of Salt *et al.* (1967) and constants from

Table 4. Mean corrections for the dispersion of zircon (Il'inskiy and Krylova, 1974)

λ, mμ	ε	ω	ε-ω	λ, mμ	ε	ω	ε-ω
435.8	+0.0287	+0.0271	+0.0016	589.3	0.0000	0.0000	0.0000
486.1	+0.0166	+0.0155	+0.0011	620.0	-0.0033	-0.0032	-0.0001
500.0	+0.0138	+0.0130	+0.0008	656.3	-0.0068	-0.0066	-0.0003
546.1	+0.0058	+0.0055	+0.0003	700.0	-0.0101	-0.0097	-0.0004
578.0	+0.0013	+0.0012	+0.0001				

Mandarino (1976). Variation of optical properties with composition is often largely obscured by the effect of metamictization in zircon.

Color

Zircon is found in colorless, green, blue, red, orange, yellow, brown and grey hues. Colorless stones are often obtained by heating, although color can be sometimes restored by γ-radiation. Colors of transparent solids are a result of absorption bands in the visible spectrum, and zircons often have well-developed absorption spectra (up to 36 bands) which are sometimes distinctive enough to be used in identifying cut gemstones. Color in zircon has not been studied sufficiently to determine its origins in terms of all the possibilities outlined by Nassau (1978). Color centers may be associated with the presence of uranium and can be sector zoned, concentrated or most intense under (111) (Fielding, 1970; Chase and Osmer, 1966). Red is believed to result from color centers of Nb^{+4} ions produced by the radiation-induced reduction of Nb^{5+} substituting for Zr^{4+} (Fielding, 1970). Green color zoning of one specimen was found to be a result of differing amounts of U^{4+} substituting for Zr^{4+} (Chase and Osmer, 1966). Matumura and Koga (1962) concluded that color centers were related to Zr^{2+} produced by radiation-induced reduction of Zr^{4+}.

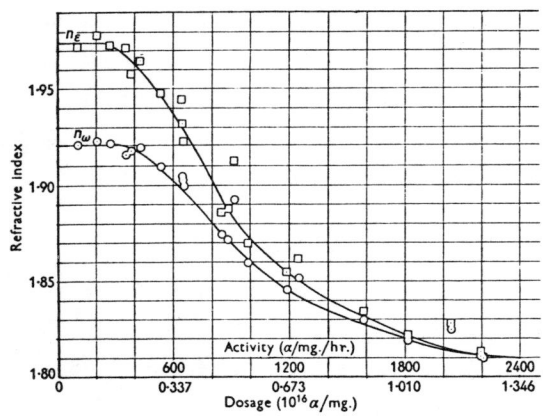

Figure 11. The indices of refraction of Ceylon zircons as a function of their present α activity and total α dosage, $i.e.$, metamictization. (After Holland and Gottfried, 1955.)

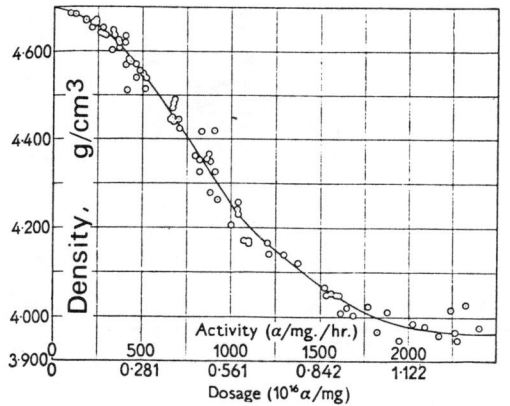

Figure 12. Density of Ceylon zircons as a function of
the present α activity and total α dosage, *i.e.*, metamictization. (After Holland and Gottfried, 1955.)

Density

The calculated density of zircon is 4.66 gm cm^{-3}. Substitution of hafnon (ρ = 6.97), thorite (ρ = 6.70), and coffinite (ρ = 7.16) components all increase the density whereas substitution with xenotime (ρ = 4.31) decreases the density. Correia Neves *et al.* (1974) observed a systematic increase of density with increasing Hf/(Hf + Zr) contents for natural Hf-rich zircons. However, these compositional effects are obscured in natural zircons by the decrease in density accompanying metamictization (Fig. 12) caused by the vacancies and fragmentation of the structure produced by radiation. The change in density with metamictization has led to the common designations:

low zircon: low density, metamict zircon;

high zircon: high density, nonmetamict zircon.

Lattice Parameters

The direction and magnitude of lattice parameter changes with composition can be predicted from Figure 13a, which is a plot of a versus c for silicates and phosphates with the zircon structure with which zircon could form solid solutions. However, natural zircons have a relatively narrow range of compositions, and consequently lattice parameters are not observed to change significantly. There is significant solid solution between zircon and hafnon, but because the dimensions of the two are close (Table 1 and Fig. 13a), lattice parameters will change only slightly. Gradual decreases in a and c were found by Ramakrishnan *et al.* (1969) with increasing Hf/(Hf + Zr) in synthesized $ZrSiO_4$-$HfSiO_4$ compounds (Table 5 and Fig. 14). The integrated intensities of the 211 and 101 reflections relative to 200 also increase as a result of the larger scattering factor of hafnium.

The limited variation in cell dimensions expected as a result of chemical variability will be obscured by metamictization. Holland and

Table 5. Lattice parameters and intensity ratios in the system ZrSiO₄-
HfSiO₄ (Ramakrishnan *et al.*, 1969)

Mole % HfSiO$_4$	Lattice parameters (Å)		Integrated intensity ratios	
	a	c	I_{211}/I_{200}	I_{101}/I_{200}
0	6.603	5.981	12	32
20	6.596	5.980	16	38
40	6.593	5.978	23	42
60	6.585	5.974	26	48
80	6.578	5.791	32	52
100	6.569	5.967	34	56

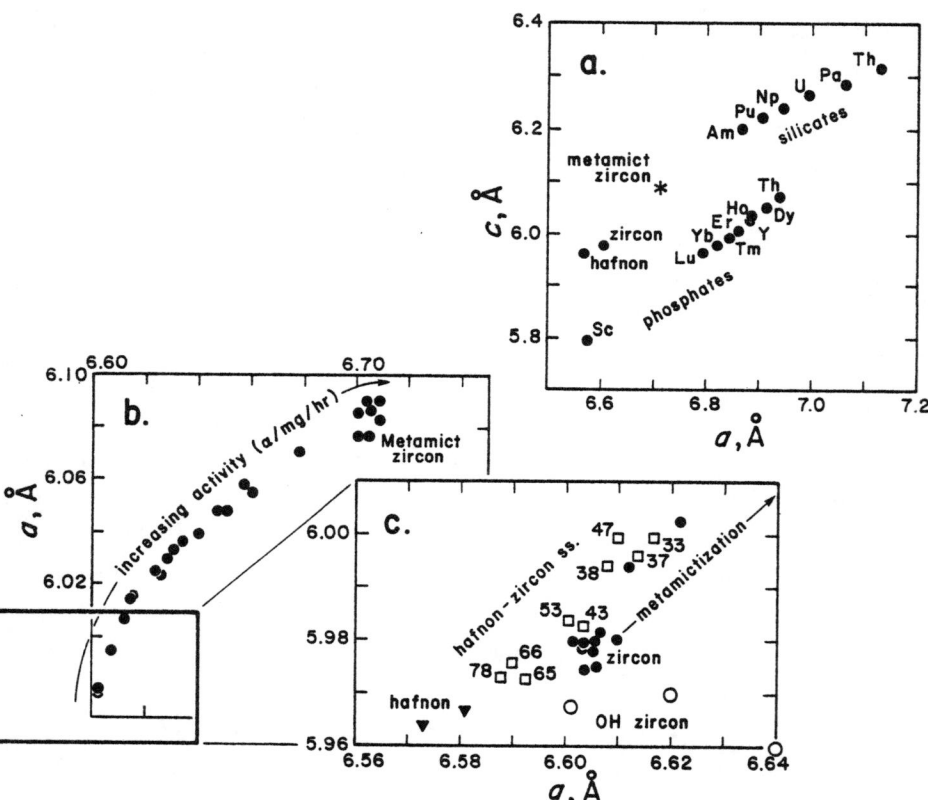

Figure 13.   (a) The cell parameter a as a function of c for silicates and phosphates with
zircon-type structures.   Lattice parameters for zircon and hafnon from Table 1.   References
for the lattice parameters of the actinide orthosilicates are given in Chapter 10; the REE
orthophosphates are from Muller and Roy (1974).   Metamict zircon is from Figure 13b.
(b) a as a function of c for Ceylon zircon showing trend of increasing α activity which
varies from 0 to 1000 α/mg/hr (data from Holland and Gottfried, 1955).   (c) a as a function
of c for synthetic zircon and hafnon and structurally analyzed zircons from Table 1.   Additional data from Subbarao and Gokhale (1968), Ozkan and Jamieson (1978) and Caruba *et al.*
(1974).   Synthetic OH-zircons from Frondel and Collette (1957) and Caruba *et al.* (1974).
Hafnon-zircon solid solutions from Correia Neves *et al.* (1974); the numbers are Hf/(Hf+Zr).

92

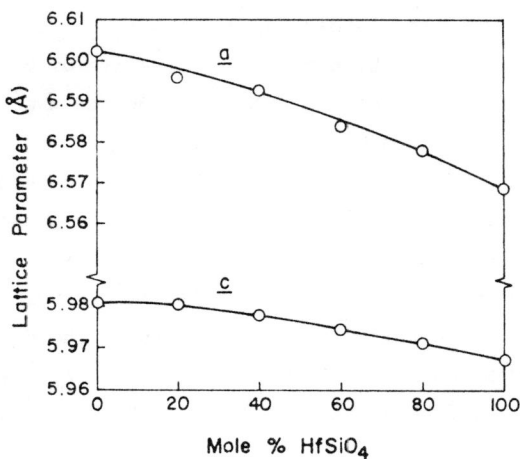

Figure 14. Variation of lattice parameters with composition in the system $ZrSiO_4$-$HfSiO_4$. (After Ramakrishnan *et al*., 1969.)

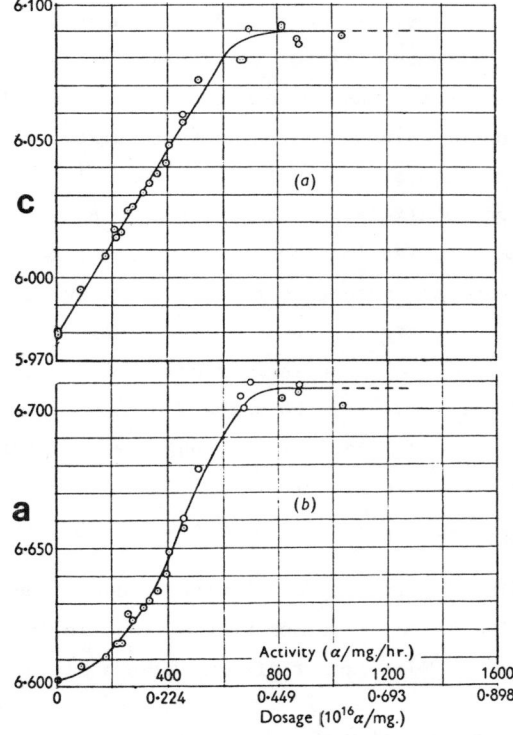

Figure 15. (a) The cell parameter c (in Angstroms) and (b) a (in Angstroms) for Ceylon zircon as a function of the present α activity and total α dosage, *i.e.*, metamictization. (After Holland and Gottfried, 1955.)

Gottfried (1955) found that the a and c cell dimensions rapidly increase with α activity and level off above 800 α/mg/hr at a = 6.708 A and c = 6.090 A (Fig. 15). The expansion of cell parameters with α activity coincides with a line drawn between normal zircon and the actinide orthosilicates (Fig. 13a). Subsequent determination of the composition of zircons used by Holland and Gottfried (1955) show no more than 0.006 wt % U (Pidgeon *et al.*, 1966; Gottfried *et al.*, 1956). This amount is much less than would be required to expand the cell dimensions, indicating that enlargement of the cell volume results from structural damage caused by radiation. Hurley and Fairbain (1953) suggested the degree of zircon metamictization can be estimated from the d_{112}-spacing, asymptomatically approaching 2θ = 35.1° from a value of 2θ = 35.635° for nonmetamict zircon.

Figure 13c is an a versus c plot of synthetic or chemically and crystalographically characterized zircons-hafnons from the literature. Most of the synthetic and structurally analyzed zircons fall in a cluster near a = 6.604 A and c = 5.979. Zircons used in structural determinations by Finger (1974) and Krstanovic (1958) (Table 1a) fall along the trend of zircon metamictization and may explain the differences of these structures from the others. Hydroxyl zircons synthesized by Frondel and Collette (1957) and Caruba *et al.* (1974) have smaller c dimensions but comparable or larger a dimensions than zircon. The hydroxyl zircons do not fall on the metamictization trend, suggesting that water-bearing metamict zircons have not undergone the substitution $(OH)_4 \rightleftarrows Si$. As would be predicted, the zircon-hafnon solid solutions described by Correia Neves *et al.* (1974) have increasingly smaller lattice parameters with Hf substitution. However, the variation does not lie between hafnon and zircon, but above it displaced along the metamictization curve. This suggests that they are metamict.

Hardness and Elastic Properties

Zircon has a hardness of $7\frac{1}{2}$ and a micro-indentation hardness of 841 to 1468 kg/mm^2. The hardness of metamict zircons is less, with a range of 485 to 841 kg/mm^2. Nonmetamict zircons have a hardness anisotropy of 1.22.

Elastic properties have been measured over a range of both temperatures of 25-300°C (Ozkan *et al.*, 1975) and pressures of 1 atm to 48 kbar

(Ozkan and Jamieson, 1978; Hazen and Finger, 1979). A range of values are reported but several conclusions can be drawn. Zircon has the lowest compressibility coefficients of any measured substance with tetrahedrally coordinated Si. The structure becomes unstable with pressure and the temperature and pressure derivatives of zircon's elastic constants are consistent with a zircon-scheelite structural transformation at higher pressures. Based on the Si-O bond compressibilities from the high pressure structure refinements of Hazen and Finger (1979), a zircon $\rightarrow$ KAlF$_4$-structure transformation with octahedrally-coordinated Si is predicted to occur at about 160 kbar (see STRUCTURE section).

The elastic constants obtained from synthetic zircons and nonmetamict zircon (a = 6.606(2) A, c = 5.980(2) A) by Ozkan et $al.$ (1974) are c_{11} = 4.237, c_{33} = 4.900, c_{14} = 1.136, c_{66} = 0.485, c_{12} = 0.703, and c_{13} = 1.495 (all times 10^{12} dyn/cm^2). A systematic and marked decrease of up to 69% in the elastic module of zircon is noted with increasing metamictization (Ozkan, 1976). All the longitudinal and the shear elastic moduli decrease with radiation damage as measured by the density and approach two common saturation values of 1.5 x 10^{12} and 0.49 x 10^{12} dyn/cm^2, respectively. The decrease in elastic moduli result from changes in interatomic bonding and lattice spacings caused by disruption of the structure.

In relating lattice-vibrational to thermodynamic properties, Kieffer (1979a,b,c; 1980) calculated the temperature dependence of the heat capacity and calorimetric Debye temperature between 0 and 1000°K. Her model uses published elastic, crystallographic and spectroscopic data for a number of minerals, including zircon.

Thermal Properties

The most recent determinations of thermal expansion are by Subbarao and Gokhale (1968) and Worlton et $al.$ (1972): $\alpha_c > \alpha_a$, and the bulk thermal expansion coefficient of zircon is small, 4-5 x 10^{-6}/°C between 25 and 1300°C, making it relatively insensitive to thermal shock. It is this property which makes zircon ideal for use as foundary sands and refractories.

The thermal conductivity of zircon is 120 $\pm$ 10 cal/°C cm sec (Crawford, 1965). Exposure to a radiation dosage of 30 x 10^{19} α/cm^2 decreases the thermal conductivity to 23 cal/°C cm sec because of the scattering of phonons by the radiation-induced defects.

Luminescence

Luminescence is the emission of visible light by a material in response to ultraviolet light (photoluminescence or fluorescence), energetic

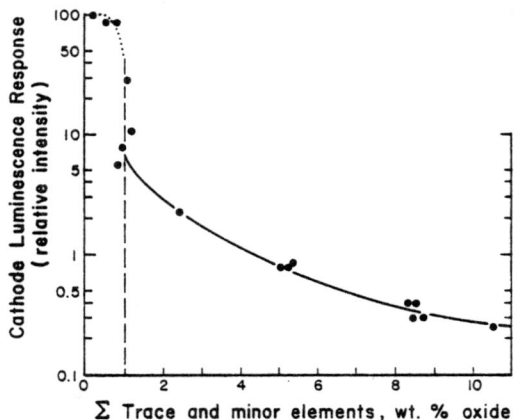

Figure 16. Variation in cathodoluminescent intensity with the amount of trace and minor element substitution in zircon (Sommerauer, 1976).

electrons (cathodoluminescence) or heat (thermoluminescence). Both chemistry and structure affect the luminescene properties of a material. Activators in zircon are one or more rare earth elements or uranium (Fielding, 1970; Caruba *et al.*, 1974; Trofimov, 1962). Interaction of the effects of REE's, as well as increasing REE concentrations, can lead to quenching of the luminescence (see Fig. 16). A rapid drop in cathodoluminescence intensity was observed at about 2 mol % REE (Sommerauer, 1976). The luminescent properties of zircon have been used (1) in studying compositional zoning in a qualitative or semi-quantitative manner (Sommerauer, 1976; Ono, 1975, 1974), (2) in correlating rock types based on similarity of the contained zircons, and (3) in thermoluminescent age dating, a subject discussed in the GEOCHRONOLOGY section. Materials scientists have investigated REE-doped zircons as potential phosphors.

METAMICTIZATION

Metamictization in minerals is generally considered to be the effect of radiation damage produced by radioactive decay of thorium and uranium. General reviews of metamictization are given by Pabst (1952), Billington and Crawford (1961), Crawford (1965), Mitchell (1973), and Chadderton (1965). Earlier summaries of zircon metamictization are given by Pellas (1954) and Holland and Gottfried (1955). Radiation damage in crystals consists of displaced atoms, vacancies and interstitial atoms (including helium) which result from elastic collisions between the atoms in the undisturbed structure and α particles, γ rays, and heavy charged particles and from nuclear recoil, and the ionization and high temperatures generated in the course of radioactive decay.

The intensity of radiation necessary to convert crystalline zircon into an optically amorphous structure is 10^{15}-10^{16} α/mg (Woodhead, 1978; Bursill and McLaren, 1966). Pellas (1954) and Holland and Gottfried (1955) calculated that zircon appears amorphous to x-rays when 20-30% of the atoms are displaced.

Bursill and McLaren's (1966) transmission electron microscopic study showed that no radiation damage is sustained by zircons exposed to $<10^{14}$ α/mg and that metamict zircons are composed of slightly misoriented zircon crystallites about 100 A in size. In other cases, metamict zircons have been shown to consist of cubic, tetragonal, or amorphous ZrO_2 and amorphous SiO_2 (Vance and Anderson, 1972; Lipova *et al.*, 1965; Sommerauer, 1979; Vance, 1975).

Transmission electron microscopy images of zircon taken by Yada *et al.* (1981) show that metamictization of zircon is principally the accumulated damage by fast-traveling nuclear particles in the form of fission tracks. Compared to the lattice images of undisturbed zircon (Fig. 17a), areas with fission tracks are extensively disordered. Intermediate zircon consists of similarly oriented lattice domains 50 to 100 Å in size separated by disordered areas tens of Ångströms wide (Fig. 17b). The individual domains are either rotated or tilted around 10^{-4} rad with respect to their neighbors. Low zircon appears entirely amorphous, containing abundant fission tracks (Fig. 17c). The fission tracks are 20-30 Å wide and 1000 Å long (Fig. 17d). The lattice is displaced on either side of the tracks. Adjacent to the tracks are bright and dark points, corresponding to vacancies and interstitial atoms - point defects caused by recoil particles.

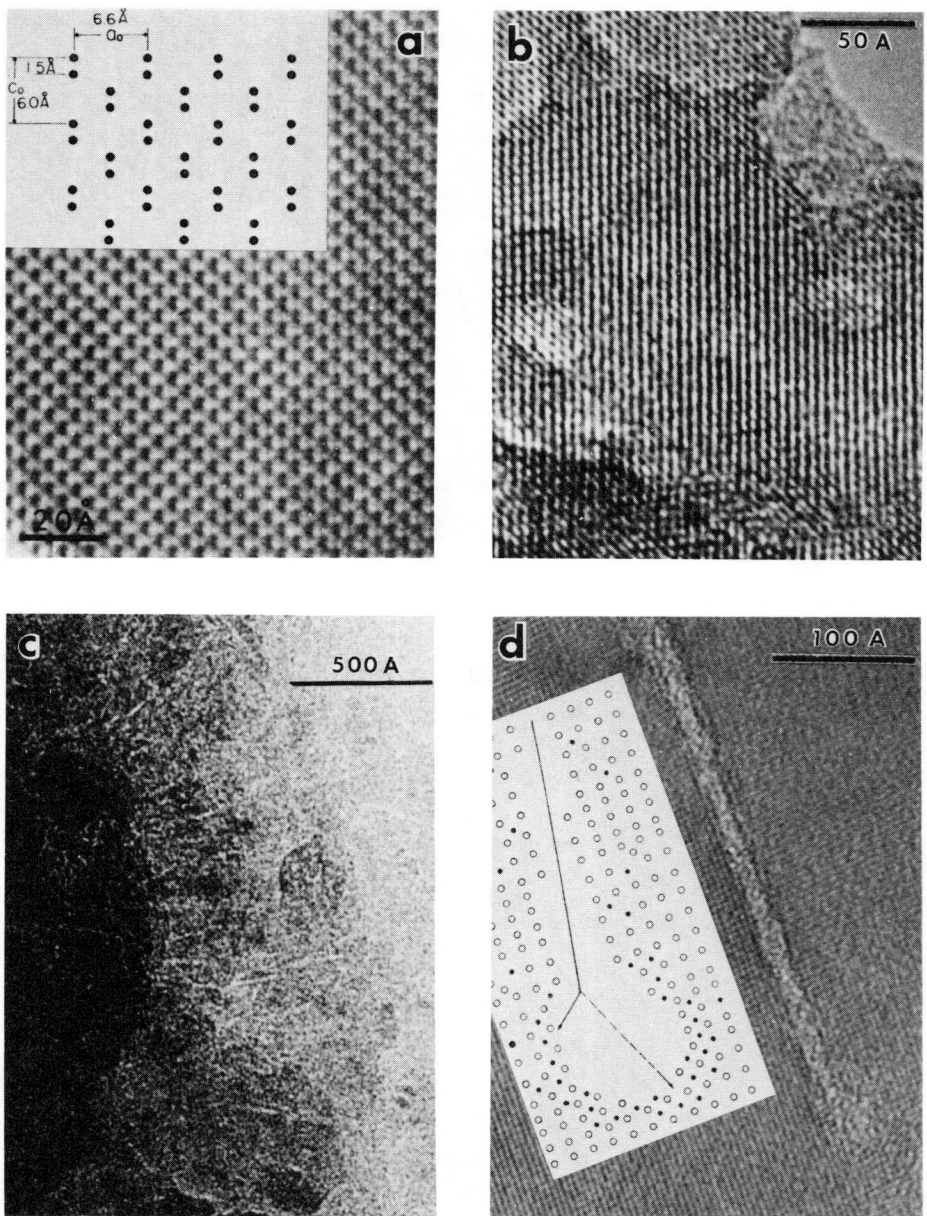

Figure 17. (a) Lattice image of high zircon viewed down the a-axis. The pattern is that of the zirconium atoms in zircon projected onto (100), which is a centered rectangular plane lattice of zirconium atom pairs as shown in the inset. (b) Lattice image of the domain structure of an intermediate zircon. (c) Lattice image of a metamict, low zircon. It is believed the white streaks are fission tracks. (d) Lattice image of a fission track and its termination. The inset drawing schematically shows the formation of fission tracks and point defects by the recoil particles. The photographs are from Yada *et al*. (1981).

The actual mechanism of metamictization in zircon is not known, but in general it is envisioned that crystalline zircon becomes saturated in vacancies and interstitial defects and ultimately decomposes to either microcrystalline zircon, zircon glass, crystalline and/or amorphous ZrO_2 and SiO_2, or all three. Holland and Gottfried (1955) believed the process to occur in stages whereas Pellas (1965) suggests continuous dissociation.

Physical properties of zircon are influenced by defects in the structure either as a result of the disrupted structure of the distorted electronic field. Radiation effects detected in zircon with increasing metamictization include:

1. absorption of H_2O; presence of Pb, Tl and He
2. decrease in refractive index and birefringence (Fig. 11)
3. broadening of x-ray peaks
4. increasingly larger cell parameters (Fig. 15)
5. decreasing density (Fig. 12)
6. decreasing hardness and loss of hardness anisotropy
7. increasing thermoluminescence
8. darker color
9. increasing susceptibility to chemical attack
10. increasing chemical diffusion
11. broadening and decreased intensity of the optical and infrared absorption spectra with loss of polarised spectra
12. decreasing thermal conductivity
13. decreasing elastic constants

Most of these properties could be used as measures of the degree of metamictization, but, in addition to those studies already discussed earlier in the sections on CHEMISTRY and PHYSICAL PROPERTIES, the following are considered the more important ones: refractive index and density (Vance and Anderson, 1972); solubility in acids (Bauer, 1939; Krasnobayev *et al.*, 1974); energy of recrystallization (Holland and Kulp, 1950; Kulp *et al.*, 1952); thermoluminescence (Vaz and Stenftle, 1971); change in x-ray spacing (Hurley and Fairbairn, 1953); and IR absorption (Rudnitskaya and Lipova, 1972; Delieus *et al.*, 1977; Woodhead *et al.*, 1978).

Radiation is an essential part in metamictization, but other aspects of chemistry, structure and age are important in determining the response of zircon to a radiation flux. This has been called the "predisposition" or "genetic factor" of metamictization (Krasnobaynev *et al.*, 1974; Lipova *et al.*, 1965). The amount of U and Th substituting for Zr will determine the magnitude of the radiation flux. Whereas zircon becomes amorphous at a flux of 10^{15}-10^{16} α/cm^2, thorite requires 10^{19} α/cm^2. The resistance of thorite to metamictization may be a result of increased bonding strength

or the fact that Th is a heavier atom. Zirconium has a small thermal neutron capture cross-section of 0.18 barn; hafnium has a much greater value
of 115 barn. Substitution of Hf as well as other elements for Zr will
raise the capture cross-section and increase its susceptibility to damage.

Energetic particles penetrating zircon do not encounter a random
collection of atoms but rather an orderly array of interdependent atoms.
Chances of colliding with atoms will vary with direction in the crystal
and thus skew the effective particle range. Energy may be removed from
the area of primary collision by vibrational propagation rather than displacement of atoms.

EXPERIMENTAL STUDIES

The stability of zircon at 1 atm has been studied in the ZrO_2-SiO_2
system by Butterman and Foster (1967). Zircon decomposes to tetragonal
ZrO_2 and cristobalite at 1676°C. Parfenenkov *et al.* (1969) found that
hafnon, in the HfO_2-SiO_2 system at 1 atm, melted incongruently to monoclinic HfO_2 + liquid at 1750 $\pm$ 15°C. Phase assemblages coexisting with
zircon in the SiO_2-ZrO_2-ThO_2-UO_2 system, the results of which have been
discussed in the CHEMISTRY section, is included in Chapter 10: *ACTINIDE
ORTHOSILICATES*. Hydrothermal experiments on the solubility of zircon in
the system SiO_2-Al_2O_3-Na_2O-K_2O were performed by Watson (1979). He found
that less than 100 ppm Zr is required for zircon saturation in peraluminous
melts. In peralkaline melts, zircon solubility increases with increasing
$(Na_2O + K_2O)/Al_2O_3$, with up to 3.9 wt % Zr at $(Na_2O + K_2O)/Al_2O_3 = 2.0$.
Larsen (1973) studied the crystallization of zircon in a water-saturated,
haplogranite system $(Q_{45}Ab_{15}Or_{40})$ at 830°C and 2 kbar and found saturation at 57 ppm Zr.

ALTERATION

Zircon is considered a very stable mineral, able to survive one or
more cycles of weathering, sedimentary transport, diagenesis, metamorphism and anatexis. Such a history is often detected from abrading and
subsequent overgrowths of zircon. Under some situations zircon can be
altered chemically.

Carroll (1953) found that zircons in soils are corroded in laterites
produced by alkaline leaching. This is consistent with the experimental
work of Maurice (1949) who found that in aqueous solutions at 400°C, zircon is not stable in alkaline solutions. Otherwise, zircons appear un-

affected by weathering. Zircons become rounded and frosted during sedimentary transport and usually remain that way during diagenesis and low
grade metamorphism. Saxena (1966) has presented evidence of authigenic
growth of zircon during diagenesis and has reviewed earlier work.

Kimberlite zircons are often coated with alteration products of
baddeleyite (monoclinic ZrO_2), tetragonal ZrO_2, or both (Kresten, 1975;
Davis, 1976). Raber and Haggerty (1979) identified Ti-baddeleyite, zirconolite, armalcolite, diopside and titanite at zircon-ilmenite or -rutile
interfaces. Zircon is interpreted as reacting with a carbonatitic fluid
phase:

ilmenite + zircon + calcite = baddeleyite + zirconolite + diopside + CO_2
zircon + rutile + calcite = baddeleyite + titanite + CO_2

at P < 12 kbar and T = 1450–1550°C.

Partial or complete breakdown of zircon to baddeleyite and silica
glass has been reported from several tektites and impact crater glasses
(Kleinmann, 1969, and included references). This must have occurred
above the zircon 1 atm decomposition temperature of 1676°C.

GEOCHRONOLOGY

The greatest geological interest in zircon stems from its use in
geochronology. This section reviews the geochronologic methods which
depend on chemical and physical properties of zircon. Complete discussions of geochronologic techniques and interpretation can be obtained
from Faure (1977) or Jäger and Hunziker (1979).

Isotopic U, Th-Pb method. ^{238}U, ^{235}U and ^{232}Th decay to ^{206}Pb,
^{207}Pb and ^{208}Pb with half lives of 4468, 704 and 14,010 Ma.   The amount
of daughter product Pb which accumulates during time t, given in terms
of the present amounts of parent isotope U or Th, is

$$\text{daughter Pb} = \text{parent U or Th isotope times } (e^{\lambda t} - 1),$$

where λ (= ln 2/half-life) is the decay constant of the parent isotope
U or Th. Zircon is used in U,Th-Pb dating because it is widely distributed in many rocks, contains isomorphous substitution of 10^2 to 10^3 ppm
U^{+4} and Th^{+4} for Zr^{+4} (whereas Pb^{+2} is essentially excluded during crystallization), and it is generally able to retain U, Th, Pb and intermediate
decay products over geologic time. The time t is the time elapsed since

closure of the zircon to interchange of U, Th, Pb, and intermediate daughters with the enclosing rock. In most crustal rocks this will represent the time of igneous and metamorphic crystallization. In the case of mantle zircons, the time represents a cooling age from the ambient temperatures of the mantle (Davis, 1976, 1977). If the dates given by the three decay schemes are the same, the date is *concordant*. Frequently, the dates are not the same and the zircon is said to give *discordant* dates. Discordancy results from loss or gain of Pb, U and Th; several models have been used to interpret these chemical and isotopic changes:

(1) Episodic loss of Pb (or gain of U) because of a thermal or chemical event such as metamorphism or weathering.

(2) Continuous loss of Pb by diffusion. Diffusion coefficients of Pb in zircon determined by Shestakov (1972) are too small to explain discordancy by low-temperature diffusion; however, diffusion of Pb at mantle temperatures is sufficient to prevent its accumulation until the zircons are brought to the surface (Davis, 1976, 1977).

(3) Dilatancy. Water contained in metamict zircon structure is lost with pressure decrease, together with accumulated Pb.

(4) Zoning and heterogeneous alteration. Chemical zoning in zircon results in zones differing in metamictization and susceptibility to chemical attack. The altered zones are found to be isotopically distinct from unaltered zones which causes discordancy (Krogh and Davis, 1975).

(5) Low-temperature ($\sim$300°C) annealing of metamict zircons, causing recrystallization of the structure with the exclusion of the accumulated radiogenic Pb (Gebauer and Grünenfelder, 1976). The low-temperature annealing is not an event but rather a response to ambient conditions.

(6) Mixing. A mixed zircon population is produced in a rock by incorporation of zircon xenocrysts, later overgrowths in igneous or metamorphic rocks, or hiatal crystallization.

Isotopic Xe-Xe method. The Xe-Xe dating technique was first described by Shukoljukov *et al.* (1974). It depends on the isotopic

102

differences of xenon produced by spontaneous fission of ^{238}U and xenon from induced ^{235}U fission by a thermal neutron fluence. The xenon is extracted from a mineral after irradiation in a number of release fractions at incrementally higher temperatures and is analysed by mass spectrometry. A series of apparent ages is calculated and plotted as a function of temperature, yielding an age spectra. The advantages to this are comparable to the better known ^{39}Ar/^{40}Ar dating in accounting for loss or extraneous accumulation of radiogenic Xe as well as yielding high accuracy because only isotopic ratios are measured. Limitations of the method for zircons are metamictization causing excessive xenon loss, presence of Th which also produces Xe daughters and absorption of neutrons by the REE. Shukoljukov *et al.* (1974) published one ^{136}Xe/^{131}Xe zircon age. More recently Shukoljukov *et al.* (1980) obtained ^{136}Xe/^{131}Xe age spectra of zircons showing that Xe given off at low temperatures, presumably from the more metamict portions of the crystal, had less radiogenic Xe whereas Xe yielded at higher temperatures corresponds to the age of crystallization.

Isotopic Sm-Nd method. The REE contents of zircon suggests their possible use in dating based on the alpha decay of ^{147}Sm to ^{143}Nd. To obtain a date, knowledge of the initial ^{143}Nd content or a zircon-whole rock isochron is required. The half life of ^{147}Sm is about 1.05×10^{11}a, which makes the technique useful only for "old" zircons. In a 3.4-3.8 Ga gneiss, Futa (1981, Geochimica Cosmochimica Acta *45*, 1245-1249) found that the Sm-Nd systematics for the zircons do not retain the original age or even a reliable age of a subsequent metamorphism of 1.75 Ga, but a much younger age.

Chemical methods. If only U and Th are incorporated in a crystallizing mineral, then an age can be calculated from the U and Th contents and one of the decay products such as Pb or He without isotopic analysis. This was the first geochronologic technique, but it is now done only in special cases on minerals high in U (such as uraninite). It assumes that Pb and/or He were not present initially and that no U, Th, Pb or He has been lost.

Fission track method. Spontaneous fission of ^{238}U produces charged particles which, in traveling through the crystal, leave a trail of damage about 7 microns long. The tracks are made visible by etching (Gleadow *et al.*, 1976), because the damaged areas are more readily attached by chemical etching than the undamaged zircon. Knowing the "density" of the tracks and the U content of the zircon permits a calculation of the length of time through which the tracks have accumulated. Temperature is the most important factor affecting fission track stability in minerals. At elevated temperatures, the damage caused by fission products is quickly annealed. The effective track retention temperature in zircon is 200 $\pm$ 50°C (Gleadow and Brooks, 1979). This low temperature means zircon fission track ages are cooling ages useable in studying the thermal history of rocks as well as in archaeometric dating.

Thermoluminescence dating. Thermoluminescence is the emission of light, in addition to the normal incandescence, which occurs when a material is heated. It is the energy emitted as electrons return to their stable positions in the structure from traps which had been gradually filled over time as a result of ionizing radiation. The thermoluminescence output, combined with measurements of the natural radiation dosage and the radiation sensitivity of zircon, can be used to calculate the time lapse since the zircon was last heated above 450°C (Vaz and Senftle, 1971). Thermoluminescent dating assumes that (1) the trapped electrons result from radiation alone, (2) the traps are not saturated, (3) there has been no loss of U and Th, and (4) there has been no radiation shielding. The technique has not been applied to rocks but it has been used in archeologic authentication. Thermoluminescent dating of zircons in the ceramic core of the Bronze Horse at the New York Metropolitan Museum of Art has shown that it was made in ancient times (Zimmerman *et al.*, 1974).

Degree of metamictization. The relation between degree of metamictization, radioactive element content, and age of zircons have long been recognized, but use in dating has been deterred by attempts to quantify the degree of metamictization and the effects of variable chemistry. Fission track dating is one measure of metamictization, and other proposals, such as IR-spectroscopy, x-ray diffraction, and DTA have been detailed in the section entitled METAMICTIZATION.

The factors complicating the use of the degree of metamictization as an age-determinative method in this complex mineral are many, including (among others) the variation in size of crystallites, chemical zoning, loss or gain of U, Th and Pb, and natural annealing at low temperatures over a long period of time. A suggested improvement is a pre-treatment with concentrated $HF-H_2SO_4$ to remove the more metamict portions of the zircon (Makeyev *et al*., 1981). The c dimension is used to calculate the age which, most importantly, is interpreted as a cooling or annealing age. If metamictization is an accumulation of fission tracks, as seems likely (Yada *et al*., 1981), this metamictization age is the time since the zircon cooled below the fission track retention temperature of 200 ±50°C (Gleadow and Brooks, 1979).

Abdel-Gawad, A.M. (1966) X-ray spectrographic determination of hafnium-zirconium ratio in zirconium minerals. Am. Mineral., 51, 464-473.

Ahrens, L. H. (1965) Some observations on the uranium and thorium distributions in accessory zircon from granitic rocks. Geochim. Cosmochim. Acta, 29, 711-716.

_____, R. D. Cherry, A. J. Erlank (1967) Observations on the Th-U relationship in zircons from granitic rocks and from kimberlites. Geochim. Cosmochim. Acta, 31, 2379-2387.

_____ and A. J. Erlank (1969) Hafnium, Sections B-O, *Handbook of Geochemistry II/5*. Springer-Verlag, New York.

Alper, A. and A. Poldervaart (1957) Zircons from the Animas stock and associated rocks, New Mexico. Econ. Geol., 52, 952-971.

Bauer, A. (1939) Untersuchungen zur Kenntnis der spezifisch leichten Zirkone. Neues Jahrb. Mineral., Beil. Bd., A75, 160-204.

Bershov, L. V. (1971) Isomorphism of Tb^{4+}, Tu^{2+} and Y^{3+} in zircon. Geochem. Int., 8/1, 24-28.

Billington, D.S. and J. H. Crawford (1961) *Radiation Damage in Solids*. Princeton University Press, Princeton, New Jersey.

Binks, W. (1926) The crystalline state of zircon. Mineral. Mag., 21, 176.

Burbank, R. D. (1965) Absolute integrated intensity measurement: application to $CaWO_4$ and comparison of several refinements. Acta Crystallogr., 18, 88-97.

Bursill, L. A. and A. C. McLaren (1966) Transmission electron microscope study of natural radiation damage in zircon ($ZrSiO_4$). Phys. Stat. Sol., 13, 331-343.

Butterman, W. C. and W. R. Foster (1967) Zircon stability and the ZrO_2-SiO_2 phase diagram. Am. Mineral., 52, 880-885.

Byerly, G. R., J. V. Mrakovich, and R. J. Malcuit (1975) Use of Fourier shape analysis in zircon petrogenetic studies. Geol. Soc. Am. Bull., 86, 956-958.

Carroll, D. (1953) Weatherability of zircon. J. Sed. Pet., 23, 106-116.

Caruba, R., A. Baumer, and G. Turco (1975) Nouvelles synthéses hydrothermales du zircon: substitutions isomorphiques; relation morphologie - milieu de croissance. Geochim. Cosmochim. Acta, 39, 11-26.

_____ and G. Turco (1971) Mise au point sur la notation des faces du zircon. Elaboration d'une méthode d'indexation rapide des faces des zircons accessoires des roches par utilisation d'abaques. Soc. Franc. Minéral. Crist. Bull., 94, 427-436.

_____, _____, P. Iacconi and P. Keller (1974) Solution solide d'éléments de transition trivalents dans le zircon et l'oxyde de zirconium. Étude par thermoluminescence artificielle. Bull. Soc. fr. Minéral. Crist., 97, 278-283.

Chadderton, L. T. (1965) *Radiation Damage in Crystals*. Methuen and Company, London, 202 p.

Chase, A. B. and J. A. Osmer (1966) Growth and preferential doping of zircon and thorite. J. Electrochem. Soc., 113, 198-199.

Chudoba, K. and M. V. Stackelberg (1936) Dichte und Struktur des Zirkons. Z. Kristallogr., 95, 230.

Crawford, J. H. (1965) Radiation damage in solids: A survey. Ceram. Bull., 44, 963-970.

Coleman, R. G. and R. C. Erd (1961) Hydrozircon from the Wind River formation, Wyoming. U. S. Geol. Surv. Bull., 424-C, 297-300.

Correia Neves, J. M., J. E. Lopes Nunes and Th. G. Sahama (1974) High hafnium members of the zircon-hafnon series from the granite pegmatites of Zambézia, Mozambique. Contrib. Mineral. Petrol., 48, 73-80.

Curtis, C. E., L. M. Doney and J. R. Johnson (1954) Some properties of hafnium oxide, hafnium silicate, calcium hafnate and hafnium carbide. J. Am. Ceram. Soc., 37, 458-465.

Davis, G. L. (1976) The ages and uranium contents of zircons from kimberlites and associated rocks. Carnegie Inst. Wash. Year Book, 62, 223-227.

_____ (1977) Zircons from the mantle. Carnegie Inst. Wash. Year Book, 77, 895-897.

_____, S. R. Hart and G. R. Tilton (1968) Some effects of contact metamorphism on zircon ages. Earth and Planet. Sci. Lett., 5, 27-34.

Dawson, P., M. M. Hargreave and G. R. Wilkinson (1971) The vibrational spectrum of zircon ($ZrSiO_4$). J. Phys. C: Sol. St. Phys., 4, 240-255.

Dennen, W. H. and R. Shields (1956) Yttria in zircon. Am. Mineral., 41, 655.

Doe, B. R. (1970) *Lead Isotopes*. Springer-Verlag, New York, 137 p.

Dowty, E. (1976) Crystal structure and growth: II. Sector zoning in minerals. Am. Mineral., 61, 460-469.

Durif, A. (1961) Structure du germanate d'hafnium. Acta Crystallogr., 14, 312.

Effimoff, I. (1972) *The Chemical and Morphological Variations of Zircons from the Boulder Batholith, Montana.* Ph.D. Dissertation, Univ. of Cincinnati.

Ehmann, W. D., L. L. Chyi, A. N. Garg and M. Z. Ali (1979) The distribution of zirconium and hafnium in terrestrial rocks, meteorites and the moon. In L. H. Ahrens, ed., *Origin and Distribution of the Elements.* Pergamon Press, New York, 247-259.

Faure, G. (1977) *Principles of Isotope Geology.* John Wiley and Sons, New York, 464 p.

Fielding, P. E. (1970) The distribution of uranium, rare earths; and color centers in a crystal of natural zircon. Am. Mineral., 55, 428-440.

Finger, L. W. (1974) Refinement of the crystal structure of zircon. Carnegie Inst. Wash. Year Book, 73, 544-547.

Fleischer, M. (1955) Hafnium content and hafnium-zirconium ratio in minerals and rocks. U. S. Geol. Surv. Bull., 1021-A, 1-13.

Fleischer, R. L., P. B. Price and R. M. Walker (1975) *Nuclear Tracks in Solids. Principles and Applications.* Univ. of California Press, Berkeley, 605 p.

Fontan, F., P. Monchous and F. Autefage (1980) Présence de zircons hafnifères dans des pegmatites granitiques des Pyrénées Ariégeoises; leur relation avec les niobo-tantalates. Bull. Soc. fr. Minéral. Cristallogr., 103, 88-91.

Frondel, C. (1953) Hydroxyl substitution in thorite and zircon. Am. Mineral., 38, 1007-1018.

_____ and R. L. Collette (1957) Hydrothermal synthesis of zircon, thorite and huttonite. Am. Mineral., 42, 759-765.

Fujiwara, S., K. Nagashima and M. Chiba (1975) Chemical investigations of minerals containing rare elements from the Far East LVIII. Zircon from Oro, Mineyama, Kyoto. Nippon Kagaku Zasski 86/6, 646-647. Chem. Abstr., 63, 11187e (1965).

Gastil, R. G., M. De Lisle and J. R. Morgan (1967) Some effects of progressive metamorphism on zircons. Bull. Geol. Soc. Am., 78, 879-906.

Gaudette, H. E., A. Vitrac-Michard, C. J. Allegre (1981) North American Precambrian history recorded in a single sample: high-resolution U-Pb systematics of the Potsdam sandstone detrital zircons, New York State. Earth Planet. Sci. Letters 54, 248-260.

Gebauer, D. and M. Grünenfelder (1979) U-Th-Pb dating of minerals. In E. Jäger and J. C. Hunziker, eds., *Lectures in Isotope Geology.* Springer-Verlag, Berlin, 105-131.

Gillson, J. L. (1925) Zircon, a contact metamorphic mineral in the Pend Oreille district, Idaho. Am. Mineral., 16, 187-194.

Gleadow, A. J. W., A. J. Hurford and R. Quaife (1976) Fission track dating of zircons: improved etching techniques. Earth and Planet. Sci. Lett., 33, 273-276.

_____ and C. K. Brooks (1979) Fission track dating, thermal histories, and tectonics of igneous intrusions in East Greenland. Contrib. Mineral. Petrol., 71, 45-60.

Gorz, H. (1974) Microprobe studies of inclusions and compilation of minor and trace elements in zircons from the literature. Chemie der Erde, 33, 326-357.

_____ and E. W. White (1970) Minor and trace elements in HF-soluble zircons. Contrib. Mineral. Petrol, 29, 180-182.

Gottfried, D. and C. L. Waring (1964) Hafnium content and Hf/Zr ratio in zircon from the Southern California batholith. U. S. Geol. Surv. Prof. Pap., 501, B88-91.

_____, F. E. Senftle and C. L. Waring (1956) Age determination of zircon crystals from Ceylon. Am. Mineral., 41, 157-161.

Grauert, B., M. G. Seitz and G. Soptrajanova (1974) Uranium and lead gain of detrital zircon studied by isotopic analyses and fission track mapping. Earth Planet. Sci. Lett., 21, 389-399.

Griffith, W. P. (1969) Raman studies on rock-forming minerals. Part 1. Orthosilicates and cyclosilicates. J. Chem. Soc. A, 9, 1372-1377.

Grünenfelder, M., G. N. Hanson, G. O. Brunner and E. Eberhard (1968) U-Pb discordance and phase unmixing in zircons (abstr.). Geol. Soc. Am. Spec. Pap., 101, 80-81.

Gulson, B. L. (1970) Electron microprobe determination of Zr/Hf ratios in zircon from the Yeoval diorite complex, N.S.W., Australia. Lithos, 3, 17-23.

_____ and T. E. Krogh (1975) Evidence of multiple intrusion, possible resetting of U-Pb ages, and new crystallization of zircons in the post-tectonic intrusions ('Rapakivi granites') and gneisses from South Greenland. Geochim. Cosmochim. Acta, 39, 65-82.

Halliday, A. M., M. Aftalion, O. van Breeman and J. Jocelyn (1979) Petrogenetic significance of Rb-Sr and U-Pb isotope systems in the c. 400 Ma old British Isles granitoids and their hosts. In A. L. Harris, C. H. Holland and B. E. Leake, eds., *The Caledonides of the British Isles - Reviewed.* Geol. Soc. Spec. Pub., London.

Hassel, O. (1926) Die Kristallstruktur einiger Verbindungen von der Zusammensetzung MRO_4 - I. Zirkon $ZrSiO_4$. Z. Kristallogr., 63, 247-254.

Hazen, R. M. and L. W. Finger (1979) Crystal structure and compressibility of zircon at high pressure. Am. Mineral., 64, 196-201.

Holland, H. D. and D. Gottfried (1955) The effect of nuclear radiation on the structure of zircon. Acta Crystallogr., 8, 291-300.

Hubin, R. and P. Tarte (1971) Etude infrarouge des orthosilicates et des orthogermanates - IV, structures scheelite et zircon. Spectrochim. Acta, 27A, 683-690.

Hurley, P. M. (1952) Alpha ionization damage as a cause of low helium ratios. Am. Geophys. Union Trans., 33, 174-183.

_____ and H. W. Fairbairn (1953) Radiation damage in zircon: a possible age method. Bull. Geol. Soc. Am., 64, 659-673.

_____, E. S. Larsen and D. Gottfried (1956) Comparison of radiogenetic helium and lead in zircon. Geochim. Cosmochim. Acta, 9, 98-102.

Jäger, E. and J. C. Hunziker (1979) Lectures in Isotope Geology. Springer-Verlag, Berlin, 329 p.

Kieffer, S. W. (1979a) Thermodynamics and lattice vibrations of minerals, 1, Mineral heat capacities and their relationships to simple lattice vibrational models, Rev. Geophys. Space Phys., 17, 1-19.

_____ (1979b) Thermodynamics and lattice vibrations of minerals, 2, Vibrational characteristics of silicates, Rev. Geophys. Space Phys., 17, 20-34.

_____ (1979c) Thermodynamics and lattice vibrations of minerals, 3, Lattice dynamics and an approximation for minerals with application to simple substances and framework silicates, Rev. Geophys. Space Phys., 17, 35-59.

_____ (1980) Thermodynamics and lattice vibrations of minerals: 4, Application to Chain and sheet silicates and orthosilicates. Rev. Geophys. Space Phys. 18, 862-886.

Kleinmann, B. (1969) The breakdown of zircon observed in the Libyan desert glass as evidence of its impact origin. Earth and Planet. Sci. Lett., 5, 497-501.

Knorring, O. von and G. Hornung (1961) Hafnian zircons. Nature, 190, 1098-1099.

Koppel, V. and M. Gruenenfelder (1971) A study of inherited and newly formed zircons from paragneisses and granitised sediments of the Strona-Ceneri-Zone (Southern Alps). Schweiz. Mineral. Petrogr. Mitt., 51, 385-409.

_____ and J. Sommerauer (1974) Trace elements and the behaviour of the U-Pb system in inherited and newly formed zircons. Contrib. Mineral. Petrol, 43, 71-82.

Köhler, H. (1970) Die Änderung der Zirkonmorphologie mit dem Differentiations-grad eines Granits. Neues Jahrb. Mineral. Mh., 9, 405-420.

Krasnobayev, A. A., Yu. M. Polezhayev, B. A. Yunikov and B. K. Novoselov (1974) Laboratory evidence on radiation and the genetic nature of metamict zircon. Geochem. Int., 11, 195-209.

Kresten, P., P. Fels and G. Berggren (1975) Kimberlitic zircons - A possible aid in prospecting for kimberlites. Mineral. Dep., 10, 47-56.

Krstanovic, I. R. (1958) Redetermination of the oxygen parameters in zircon ($ZrSiO_4$). Acta Crystallogr., 11, 896.

_____ (1964) X-ray investigation of zircon crystals containing OH-groups. Am. Mineral., 49, 1146-1148.

Larsen, E. S., C. L. Waring and J. Berman (1953) Zoned zircon from Oklahoma. Am. Mineral., 38, 1118-1125.

Larsen, L. (1973) Measurement of solubility of zircon ($ZrSiO4$) in synthetic granitic melts. EOS, 54, 479.

_____ and A. Poldervaart (1958) Measurement and distribution of zircons in some granitic rocks of magmatic origin. Mineral. Mag., 31, 544-564.

Levinson, A. A. and R. A. Borup (1959) High hafnium zircon from Norway (abstr.). Bull. Geol. Soc. Am., 70, 1638.

_____ and _____ (1960) High hafnium zircon from Norway. Am. Mineral., 45, 562-565.

Lipova, I. M. and M. M. Mayeva (1971) The relation of Zr/Hf ratio in zircon to crystal morphology. Geochem. Int., 8, 785-791.

_____, G. A. Kuznetsova and Ye.S. Makarov (1965) An investigation of the metamict state in zircons and cyrtolites. Geochem. Int., 2, 513-525.

Lyakhovich. V. V. (1962) Rare earth elements in the accessory minerals of granitoids. Geochem., 1, 39-51.

_____ (1967) Distribution of rare earths among the accessory minerals of granites. Geochem., 7, 691-696.

_____ and I. D. Shevaleevskii (1962) Zr/Hf ratio in the accessory zircon of granitoids. Geochem., 1, 508-524.

Lynd, L. E. (1980) Zirconium. *Mineral Commodity Summaries 1980*, U. S. Bureau of Mines, 186-187.

Makeyev, A. I., O. A. Levchenkov and R. S. Bubnova (1981) Radiation damage as an age measure for natural zircons. Geochem. International 18, 92-97.

Mamedov, Sh. A. (1970) Migration of radiogenetic products in zircon. Geochem. Int., 7, 203.

Marfunin, A. S. (1979) *Physics of Minerals and Inorganic Materials*. Springer-Verlag, Berlin, 340 p.

Matumara, O. and H. Koga (1962) On color centers in $ZrSiO_4$. J. Phys. Soc. Japan, 17, 409.

Maurice, O. D. (1949) Transport and deposition of the non-sulphide vein minerals: V. Zirconium Minerals. Econ. Geol., 44, 721-731.

Medenbach, O. (1976) *Geochemie der Elemente in Zirkon und ihre räumliche Verteilung - Eine Untersuchung mit der Electronenstrahlmikrosonde*. M. S. Thesis, Ruprecht Karl Universität, Heidelberg.

Mitchell, R. S. (1973) Metamict minerals: A review, Parts I and II. Mineral. Record, 4, 177-182, 214-223.

Morgan, J. H. and Auer, M. L. (1941) Optical, spectrographic, and radioactivity of zircon. Am. J. Sci., 239, 305-311.

Muench, O. B. (1931) The analysis of cyrtolite for lead and uranium. Am. J. Sci., 21, 350-357.

Muller, O. and R. Roy (1974) *The Major Ternary Structural Families*. Springer-Verlag, Heidelberg, 487 p.

Mumpton, F. A. and R. Roy (1961) Hydrothermal stability studies of the zircon-thorite group. Geochim. Cosmochim. Acta, 21, 217-238.

Nagasawa, H. (1970) Rare earth concentrations in zircons and apatites and their host dacites and granites. Earth Planet. Sci. Lett., 9, 359-364.

Nassau, K. (1978) The origins of color in minerals. Am. Mineral., 63, 219-229.

Nekrasova, R. A. and V. V. Gamjanina (1968) The composition of rare earth elements in kimberlite minerals. Dokl. Akad. Nauk SSSR, 182, 449-452.

_____ and I. V. Rozhdestvenskaya (1970) The ZrO_2/HfO_2 ratio in zircons from kimberlites and alluvial sediments. Geochem. Int., 7, 536-542.

Ono, A. (1974) Zircons from the Ryoke metamorphic rocks in the Takato-Shiojiri area, central Japan. J. Geol. Soc. Japan, 80, 187-191.

_____ (1975) Chemistry and zoning of zircons from some Japanese granitic rocks. J. Japan. Assoc. Min. Pet. Econ. Geol., 71, 6-17.

Özkan, H. (1976) Effect of nuclear radiation on the elastic moduli of zircon. J. Appl. Phys., 47, 4772-4779.

Pabst, A. (1952) The metamict state. Am. Mineral., 37, 137-157.

Pankhurst, R. J. and R. T. Pidgeon (1976) Inherited isotope systems and the source region prehistory of early Caledonian granites in the Dalradian Series of Scotland. Earth Planet. Sci. Lett., 31, 55-68.

Parfenenkov, V. N., R. G. Grebenschchikov and N. A. Toropov (1969) System HfO_2-SiO_2 Figure 4443 in E. M. Levin and H. F. McMurdle, eds., *Phase Diagrams for Ceramists*, 1975 Supplement, 165-166.

Pavlenko, A. S., E. E. Vainshtein and I. D. Shevaleefskii (1957) On the hafnium-zirconium ratio in zircons of igneous and metasomatic rocks. Geochem., 411-430.

Pellas, P. (1954) Sur la formation de l'état métamicte dans le zircon. Bull. Soc. Fr. Mineral. Cristallogr., 77, 447-460.

_____ (1965) Etude sur la récristallisation thermique des zircons métamictes. Mém. du Muséum Nat. d'Hist. Nat., ser. C, Sci. de la Terre Tome XII, Fascicule 5, 227-253.

Pidgeon, R. T. and M. R. W. Johnson (1974) A comparison of zircon U-Pb and whole-rock Rb-Sr systems in three phases of the Carn Chuinneag Granite, northern Scotland. Earth Planet. Sci. Lett., 24, 105-112.

_____ and M. Aftalion (1978) Cogenetic and inherited zircon U-Pb systems in granites: Paleozoic granites of Scotland and England. In D. R. Bowes and B. E. Leake, eds., *Crustal Evolution in Northwestern Britain and Adjacent Regions*. Geol. J. Spec. Issue, 10, 183-248.

_____, J. R. O'Neil and L. T. Silver (1966) Uranium and lead isotopic stability in a metamict zircon under experimental hydrothermal conditions. Science, 154, 1538-1540.

Pigorini, B., and F. Veniale (1966) Studio mediante microsonda elettronica dei diversi tipi di zircone accessorio nei graniti di Baveno, M. Orfano e Alzo. Atti., Soc. Ital. Sci. Nat., 105, 207-264.

Poldervaart, A. (1950) Statistical studies of zircon as a criterion in granitization. Nature, 165, 574-575.

_____ (1955) Zircons in rocks. 1. Sedimentary rocks. Am. J. Sci., 253, 433.

_____ (1956) Zircons in rocks. 2. Igneous rocks. Am. J. Sci., 254, 521.

Pupin, J. P., M. Boucarut, G. Turco and S. Gueirard (1969) Les zircons des granites et migmatites du Massif de l'Argentera-Mercantour et leur signification pétrogénétique. Bull. Soc. fr. Minéral. Cristallogr., 92, 472-483.

_____ and G. Turco (1970) Observations nouvelles sur les nuclei et le zonage des cristaux de zircon. Problèmes génétiques qui en découlent. Schweiz. Mineral. Petrogr. Mitt., 50, 527-538.

_____ and _____ (1972a) Une typologie originale du zircon accessoire. Bull. Soc. fr. Minéral. Cristallogr., 95, 348-359.

_____ and _____ (1972b) Application des données morphologiques du zircon accessoire en pétrologie endogéne. C. R. Acad. Sci., Paris, D, 275, 799-802.

_____ and _____ (1972c) Le zircon accessoire en géothermométrie. C. R. Acad. Sci., Paris, D, 274, 2121-2124.

_____ and _____ (1974a) Application à quelques roches endogenes du Massif franco-italien de l'Argentera-Mercantour d'une typologie originale du zircon accessoire et étude comparative avec la méthode des R.M.A. Bull. Soc. fr. Minéral. Cristallogr., 97, 59-69.

_____ and _____ (1974b) Contrôle thermique du développement de la muscovite dans les granitoides et morphologie du zircon. C. R. Acad. Sci., Paris, D, 278, 2719-2722.

Pyatenko, Yu. A. (1970) Behavior of metamict minerals on heating and the general problem of metamictization. Geochem. Int., 7/5, 758-763.

Quadrado, R. and J. Lima de Faria (1966) High hafnium zircon from Namacotche, Alto Ligonha, Mozambique. Garcia de Orta, 14, 311-315.

Raber, E. and S. E. Haggerty (1979) Zircon-oxide reactions in diamond-bearing kimberlites. In F. R. Boyd and H. O. A. Meyer, eds., *Kimberlites, Diatremes and Diamonds: Their Geology, Petrology, and Geochemistry.* Proc. Second Int'l Kimberlite Conf., 1, 229-240.

Ramakrishnan, S. S., K.V.G.K. Gokhale and E. C. Stubbarao (1969) Solid solubility in the system zircon-hafnon. Mat. Res. Bull., 4, 323-328.

Reid, A. F. and A. E. Ringwood (1969) Newly observed high pressure transformations in Mn_3O_4, $CaAl_2O_4$ and $ZrSiO_4$. Earth and Planet. Sci. Lett., 6, 205-208.

Reynolds, R. W., L. A. Boatner, C. B. Finch, A. Chatelain and M. M. Abraham (1972) EPR investigations of Er^{3+}, Yb^{3+}, and Cd^{3+} in zircon-structure silicates. J. Chem. Phys., 56, 5607.

Robinson, G. W. (1978) *The occurrence of rare earth elements in zircon.* Ph.D. Dissertation, Queen's University, Kingston, Ontario, Canada.

Robinson, K., G. V. Gibbs and P. H. Ribbe (1971) The structure of zircon: a comparison with garnet. Am. Mineral., 56, 782-790.

Rogers, J. J. W. and J. A. S. Adams (1969) 'Thorium - Uranium' In *Handbook of Geochemistry.* Springer-Verlag, Berlin.

Romans, P. A., L. Brown and J. C. White (1975) An electron microprobe study of yttrium, rare earth and phosphorous distribution in zoned and ordinary zircon. Am. Mineral., 60, 475-480.

Rosen, E. and A. Muan (1965) Stability of zircon in the temperature range 1180° to 1366°C. J. Am. Ceram. Soc., 48, 603-604.

Rosenbusch, H. (1882) Uber das Wesen der körnigen und porphyrischen Struktur bei Massengesteinen. Neues Jahrb., 2, 1-17.

Rudnitskaya, E. S. and M. Lipova (1972) Infrared spectroscopic and nuclear magnetic resonance study of metamict zircons. Izv. Vyssh. Ucheb. Zaved., Geol. Razved, 4, 43-50. Chem. Abstr., 77, 22810f.

Sahama, Th. G. (1981) Growth structure in Ceylon zircon. Bull. Mineral. 104, 89-94.

Sahl, K. and J. Zemann (1965) Gitterenergetische Berechungen an Zirkon. Ein Beitrag zur Ladungsverteilung in der Silikatgruppe. Tscherm. Min. und Petrol. Mitt., 10, 97-114.

Salt, D. J. and G. Hornung (1967) Synthesis and X-ray study of hafnium silicate. J. Am. Ceram. Soc., 50, 549-550.

Saxena, S. K. (1966) Evolution of zircons in sedimentary and metamorphic rocks. Sedimentol., 6, 1-33.

Schaller, W. T. (1930) Adjectival ending of chemical elements used as modifiers to mineral names. Am. Mineral., 15, 566-574.

Schidlawski, M.O.G. (1963) Recrystallization of zircon as an indication of contact metamorphism. Nature, 197, 68-69.

Schuiling, R. D., L. Vergouwen and H. van der Rijst (1976) Gibbs energies of formation of zircon ($ZrSiO_4$), thorite ($ThSiO_4$), and phenacite (Be_2SiO_4). Am. Mineral., 61, 166-168.

Semenov, E. I. and R. L. Barinskii (1958) The composition characteristics of the rare earths in minerals. Geochem., 4, 398-419.

Shannon, R. D. (1976) Revised effective ionic radii and systematic studies of interatomic distances in halides and chalcogenides. Acta Crystallogr., A32, 751-767.

Shestakov, G. I. (1972) Diffusion of lead in monazite, zircon, sphene and apatite. Geokhim., 10, 1197-1202.

Shimizu, N., M. F. Semet and C. J. Allègre (1978) Geochemical applications of quantitative ion-microprobe analysis. Geochim. Cosmochim. Acta 42, 1321-1334.

Shukoljukov, J., T. Kirsten and E. K. Jessberger (1974) The Xe-Xe spectrum technique, a new dating method. Earth Planet. Sci. Letters. 24, 271-281.

Shukolyukov, Yu. A., Ya. S. Kapusta, A. B. Vekhovskiy and M. Vaasjoki (1980) Neutron-induced fission xenon dating of zircon. Geochem. International 17, 122-133.

Silver, L. T. and S. Deutsch (1963) Uranium-lead isotopic variations in zircons: a case study. J. Geol., 71, 721-758.

_____ and A. A. Chodos (1966) Petrological and geochemical implications of composition variations in igneous zircons. Trans. Am. Geophys. Union, 47/3, 495-496.

Smith, J. V. and I. M. Steele (1976) Lunar mineralogy: a heavenly detective story. Part II. Am. Mineral., 61, 1059-1116.

Solntsev, V. P. and M. Ya. Shcherbakova (1974) Charge compensation mechanisms and the form in which Nb and Y are incorporated into the structure of zircon. Neorg. Mat., 10/10, 1834-1838.

Sommerauer, J. (1974) Trace element distribution patterns and the mineralogical stability of zircon — an application for combined electron microprobe techniques. Electron Micros. Soc. of Southern Africa Proc., vol. 4.

_____ (1976) *Die Chemisch-Physikalische Stabilität Natürlicher Zirkone und ihr U-(Th)-Pb System.* Ph.D. Disseration 5755, Swiss Federal Inst. of Technology, Zürich.

Speer, J.A. and B.N. Cooper (1982) Crystal structure of synthetic hafnon, $HfSiO_4$, comparison with zircon and the actinide orthosilicates. Am. Mineral. 67, 804-808.

Stern, T. W., S. S. Goldich and M. F. Nevell (1966) Effects of weathering on the U-Pb ages of zircon from the Morton Gneiss, Minnesota. Earth Planet. Sci. Lett., 1, 369-371.

Syme, R. W. G., D. J. Lockwood and H. J. Kerr (1977) Raman spectrum of synthetic zircon ($ZrSiO_4$) and thorite ($ThSiO_4$). J. Phys. C: Solid State Phys., 10, 1335-1348.

Taubeneck, W. H. (1957) Zircons in the metamorphic aureole of the Bald Mountain batholith, Elkhorn Mountains, north-eastern Oregon. Bull. Geol. Soc. Am., 68, 1803-1804.

Tomita, T. and Y. Karakida (1954) Effects of heat on the color and structure of hyacinth from Mamutu, Formosa. Japan. J. Geol. Geogr., 25, 145.

Trofimov, A. K. (1962) The luminescence spectrum of zircon. Geochem., 11, 1102-1108.

Tugarinov, A. I., E. E. Vainshtein and I. D. Shevalleeskii (1956) Hafnium-zirconium ratio in the zircons of igneous and metasomatic rocks. Geochem., 4, 361-374.

Ueda, T. (1956) On the biaxialization of zircon. Mem. Coll. Sci. Univ. Kyoto B. XXIII, No. 2, 297.

Vainshtein, E. E., A. I. Tugarinov, A. M. Tuzova and I. D. Shevaleefskii (1958) Hafnium-zirconium ratios in metamorphic and metasomatic rocks. Geochem., 3, 305-309.

_____, A. I. Ginzburg and I. D. Shevaleefskii (1959) The Hf/Zr ratio in zircons from granite pegmatites. Geochem., 2, 151-156.

Vance, E. R. (1975) α-recoil damage in zircon. Radiation Effects, 24, 1-6.

_____ and B. W. Anderson (1972) Study of metamict Ceylon zircons. Mineral. Mag., 38, 605-613.

Vaz, J. D. and F. E. Senftle (1971) Thermoluminescence study of the natural radiation damage in zircon. J. Geophys. Res., 76, 2038-2050.

Vegard, L. (1926) Results of crystal analysis, Part II, the zircon group. Phil. Mag., Ser. 7, 1, 1158-1168.

Veniale, F., B. Pigorini and F. Soggetti (1968) Petrological significance of the accessory zircon in the granites from Baveno, M. Orfano, and Alzo (North Italy). XXIII Int'l Geol. Congr., 13, 243-268.

Vinokurov, V. M., N. M. Gaynullina, N. M. Nizamutdinov and A. A. Krasnobayev (1972) Distribution of admixed Fe^{+3} ions in the single zircon crystals from the kimberlite pipe "Mir." Geokhim., 11, 1402-1405.

Vlasov, K. A., editor (1966) *Geochemistry and Mineralogy of Rare Elements and Genetic Types of their Deposits. Vol. II. Mineralogy of Rare Elements.* Israel Program for Scientific Translations, Jerusalem, 945 p.

Waring, C. L. (1964) Determination of hafnium content and Hf/Zr ratios in zircon with the direct reading emission spectrometer. U. S. Geol. Surv. Prof. Pap., 501, B146-147.

Watson, E. B. (1979) Zircon saturation in felsic liquids: experimental results and applications to trace element geochemistry. Contrib. Mineral. Petrol., 70, 407-419.

_____ (1980) Some experimentally determined zircon/liquid partition coefficients for the rare earth elements. Geochim. Cosmochim. Acta 44, 895-897.

White, J. S. (1979) Boehmite exsolution in corundum. Am. Mineral., 64, 1300-1302.

Willgallis, A. (1970) Zur Mikrosondenanalyse der U-Th-Minerale im Malsburger Granit. Neues Jahrb. Mineral. Abh., 114, 48-60.

Woodhead, J. A., G. R. Rossman and L. T. Silver (1978) X-ray and infrared studies of zircon met-amictization. EOS, 59, 394.

Yada, K., T. Tanji and I. Sunagawa (1981) Application of lattice imagery to radiation damage investigation in natural zircon. Phys. Chem. Minerals 7, 47-52.

Yes'kova, Ye. M. and A. A. Ganzeyev (1964) Rare earth elements in accessory minerals of the Vishnev Mountains. Geochm., 12, 1152-1163.

Zimmerman, D. W., M. P. Yuhas and P. Meyers (1974) Thermoluminescence authenticity measurements on core material from the Bronze Horse of the New York Metropolitan Museum of Art. Archaeometry, 16, 19-30.

Zvezdinskaya, L. V., N. L. Smironova and N. V. Belov (1977) System of polymorphic transitions of the structural types of compounds with the composition ABX_4. Sov. Phys. Crystallogr., 22, 439-442.

Chapter 4
The ACTINIDE ORTHOSILICATES   J. A. Speer

INTRODUCTION: OCCURRENCES

The lighter, tetravalent actinide elements form orthosilicates
with the general formula ABO_4, where B = Si and A = Th, Pa, U, Np, Pu
or Am. Each of these silicates is tetragonal and isostructural with
zircon and hafnon. Th and Pa also form monoclinic polymorphs that are
isostructural with monazite, $CePO_4$ (see Fig. 1). Three of the actinide orthosilicates are found in nature: tetragonal $ThSiO_4$, thorite; tetragonal $USiO_4$, coffinite; and monoclinic $ThSiO_4$, huttonite.
A summary of earlier work and historical information concerning these
minerals is given by Frondel (1958). He also includes varietal names
and discusses related, but incompletely described substances.

Although relatively rare in nature, the actinide orthosilicates
are widespread accessory minerals in igneous and metamorphic rocks
and are important ore minerals in some uranium and thorium deposits.
Because they are commonly primary in origin, geochronologic studies
of the actinide orthosilicates have been used to obtain ages of
mineralization. The minerals also provide material for studying the
electronic spectra of the
higher valence states of
rare earth and actinide
elements which are used
in phosphors. They con-
stitute crystalline hosts
which are optically
transparent, easily grown
in fluxes, readily doped
with lanthanides and
uranium, and which pro-
vide sites of tetragonal
symmetry for tetravalent
cations (Table 1).

Most of the natur-
ally occurring thorites
that have been described

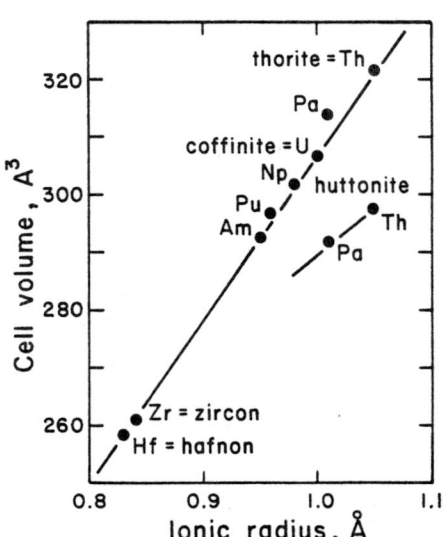

Figure 1. A plot of the unit cell volumes of
the actinide orthosilicates versus the effective ionic radius of the tetravalent actinide
in 8-fold coordination. Radii from Shannon
(1976).

113

are from pegmatites associated with granites, syenites or
nepheline syenites, or from placer concentrations derived from them.
Less commonly, thorites have been described from hydrothermal deposits (Phair and Shimamoto, 1952) and from skarns. The increasing
use of the electron microprobe has revealed that thorite is a widespread accessory mineral of igneous and metamorphic rocks which have
normal thorium concentrations (Silver and Deutsch, 1963; Willgallis,
1970; Renard, 1974; Speer *et al.*, 1980). A lunar occurrence is reported by Haines *et al.* (1972).

Coffinite occurs as a primary ore mineral of uranium in the
unoxidized, uranium-vanadium ores of the Colorado Plateau-type deposits. It is associated with or replaces carbonaceous material of
woody origin or asphaltite (Abdel-Gawad and Kerr, 1961). Coffinite
is found in hydrothermal, vein U deposits with pitchblende,
Co-Ni-Bi-As minerals or Mo (Ramdohr, 1961; Darnley *et al.*, 1963).

Table 1a. Crystallographic data for actinide orthosilicates
with the zircon structure (space group $I4_1/amd$).

Atom	Site	Point symmetry	Coordinates			Coordination
A	4a	$\bar{4}$2m	0	3/4	1/8	8
Silicon	4b	$\bar{4}$2m	0	3/4	5/8	4
Oxygen	16h	m	0	x	z	3

End-member	Cell dimensions		Calc. density $gm\ cm^{-3}$	Reference
	a	c		
$ThSiO_4$	7.1328(2)	6.3188(2)	6.70	Taylor & Ewing (1978)
$PaSiO_4$	7.068(7)	6.288(6)	6.832	Keller (1963)
$USiO_4$	6.994(5)	6.263(5)	7.164	Keller (1963)
$NpSiO_4$	6.950(7)	6.243(6)	7.254	Keller (1963)
$PuSiO_4$	6.906(6)	6.221(6)	7.415	Keller (1963)
$AmSiO_4$	6.87(1)	6.20(2)	7.561	Keller (1963)

Table 1b. Crystallographic data for actinide-orthosilicates with the
monazite structure (space group $P2_1/n$).

Atom	Site	Point symmetry	Coordinates			Coordination
A	4e	1	x	y	z	9(8)
Silicon	4e	1	x	y	z	4
Oxygen (1)	4e	1	x	y	z	4
Oxygen (2,3,4)	4c	1	x	y	z	3

	Cell dimensions				Calc. density $gm\ cm^{-3}$	Reference
	a	b	c	β		
$ThSiO_4$	6.784(2)	6.974(3)	6.500(3)	104.92(3)°	7.25	Taylor & Ewing (1978)
$PaSiO_4$	6.76(4)	6.92(4)	6.45(5)	104.83(25)°	7.36	Keller (1963)
cheralite (Ca,Ce,Th)(P,Si)O_4	6.717	6.920	6.434	103.83°	5.47	Rao & Finney (1965)

As in the case of thorite, increasing use of microtechniques has shown that coffinite also occurs as a disseminated accessory mineral of igneous and metamorphic rocks (Zavarzin, 1977; Speer *et al.*, 1980).

Huttonite is commonly encountered as an ignition product of metamict thorite and is rarely reported as a naturally occurring phase. Most published accounts list huttonite as a constituent of beach sands, presumed to have been derived from metamorphic rocks. Huttonite has also been found in a hydrothermal vein associated with a granite in a metamorphic terrain (Kosterin and Zuev, 1962) and an unspecified occurrence in metamorphic rocks at Sudbury, Ontario (Traill, 1969).

CRYSTAL STRUCTURE

Thorite

Thorite has long been accepted to be isostructural with zircon on the basis of the crystal habit and chemical composition. This was first conclusively established by Pabst (1951a) in an x-ray study. Thorite and the zircon group possess a body-centered, tetragonal unit cell with space group symmetry $I4_1/amd$. The general chemical formula is ABO_4 with Z=4. The A and B cations occupy special positions fixed by the space group symmetry, but the y and z positional parameters of the oxygen are variable (Table 1a). The O positional parameters have twice been obtained on synthetic material by powder methods: Fuchs and Gebert (1958): a = 7.142A, c = 6.327A, y = 0.084(10), z = 0.222(10); and Sinha and Prasad (1973): a = 7.148A, c = 6.309A, y = 0.081, z = 0.232. The most recent refinement was on a flux-grown single crystal by Taylor and Ewing (1978): a = 7.1332A, c = 6.319A, y = 0.0732(13), z = 0.2104(16). The following description of the structure is based on their work.

In thorite, Th is surrounded by eight oxygen atoms in triangular dodecahedral array, with four "equatorial" oxygens at distances 2.37A and another four "axial" oxygens at 2.47A which are denoted by primes in Figure 2b. Each pair of adjacent "axial" oxygens is bonded

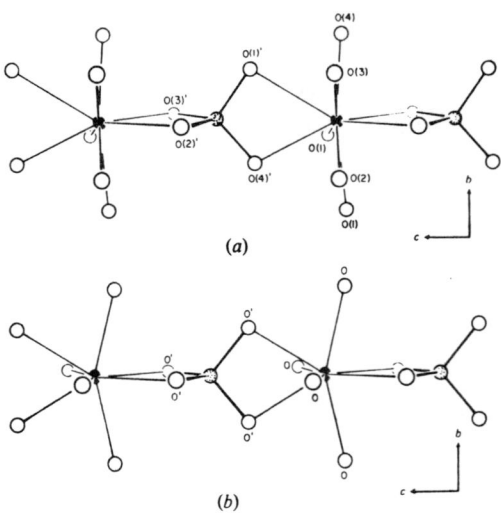

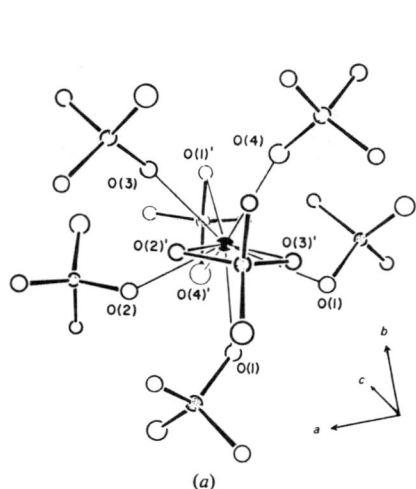

(a)

(a)

(b)

Figure 3. The c-axis chains in (a) huttonite and (b) thorite. After Taylor & Ewing (1978).

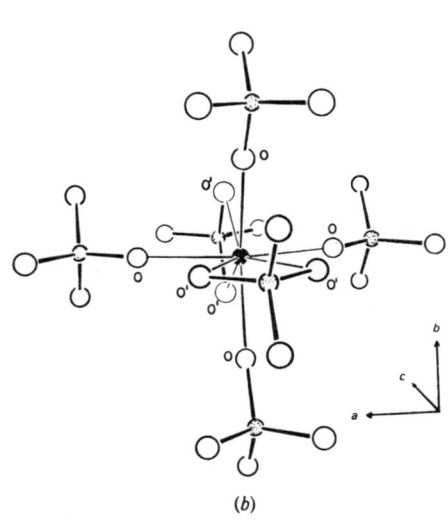

(b)

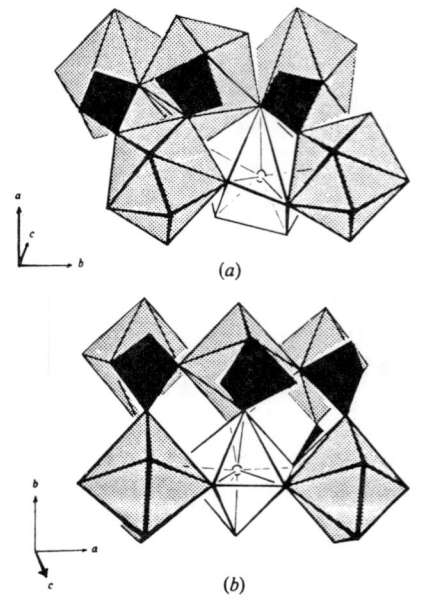

(a)

(b)

Figure 2. Th environments in (a) huttonite and (b) thorite. Open circles are O, gray circles are Si, and black ellipsoids are Th. Axial and equatorial O atoms have primed and unprimed labels respectively. After Taylor & Ewing (1978).

Figure 4. Perspective polyhedral representation of the (a) huttonite, and (b) thorite structures. Darker tetrahedra are SiO_4 groups and lighter polyhedra are (a) ThO_4 and (b) ThO_8 groups. After Taylor & Ewing (1978).

116

to the same Si, forming a chain of alternating edge-sharing SiO_4 tetrahedra and ThO_8 triangular dodecahedra extending parallel to c (Fig. 3b). These chains are joined in the a and b directions by edge-sharing ThO_8 dodecahedra (Fig. 4b). The SiO_4 tetrahedron is actually a tetragonal disphenoid and it has a smaller O-Si-O angle (101°) opposite the edge shared with the ThO_8 dodecahedron than that opposite the unshared-edge (113.9°). The Si-O bond length is 1.63A. Each O is coordinated by one Si and two Th. The silicate ion vibrations in the Raman spectrum are of lower frequency in zircon than in thorite, indicating a relatively weaker Si-O bond in thorite (Syme et al., 1977). The thorite structure has $14A^3$ voids centered on $\frac{1}{4}\frac{1}{4}\frac{1}{2}$ which are connected parallel to c. Mumpton and Roy (1963) have suggested that the zircon-like structures contain molecular water in these voids resulting in the composition $ABO_4 \cdot xH_2O$. The channels may also serve as pathways for water entering the metamict material.

Coffinite

In their description of coffinite, Stieff et al. (1956) concluded that the x-ray diffraction powder studies indicated that coffinite is tetragonal and isostructural with thorite and zircon. Subsequently, Fuchs and Gebert (1958) reported oxygen positional parameters of $y = 0.084(10)$ and $z = 0.222(10)$ obtained by powder methods on hydrothermally grown coffinite with lattice parameters $a = 6.995A$, $c = 6.263A$. Recently Nord (1977) examined a natural coffinite, $a = 6.938A$, $c = 6.291A$, with the transmission electron microscope and found single-crystal diffraction patterns consistent with tetragonal symmetry.

Huttonite

Pabst (1961b) suggested that huttonite should be isostructural with monazite, $CePO_4$. The structure determinations of monazite that have been done (Mooney, 1948; Veda, 1953; Mooney-Slater, 1962; Ghouse, 1965) are not in agreement. The structure determination of huttonite

by Taylor and Ewing (1978) and the huttonite-group mineral cheralite
by Finney and Rao (1967) differ in detail from each other and the
monazite structures. The huttonite used in the crystal structure
determination was a flux-grown crystal which has the space group
$P2_1/n$, $Z = 4$, lattice parameters $a = 6.784A$, $b = 6.974A$, $c = 6.500A$,
$\beta = 104.9°$. The cheralite is a mineral with an intermediate composition in the $REEPO_4$-$ThSiO_4$-$CaTh(PO_4)_2$ system. It has the space group
$P2_1/n$, $Z = 4$, lattice parameters $a = 6.717A$, $b = 6.920A$, $c = 6.434A$,
$\beta = 103.8°$.

In huttonite Th is surrounded by 9 oxygens: four "axial" O
atoms at distances of 2.43-2.81A and 5 "equatorial" O atoms at distances of 2.40-2.58A (Fig. 2a). In cheralite, the A atoms are bonded
to 8 oxygens at distances of 2.403 to 2.564A. The 4 next nearest
oxygen atoms occur at distances of 2.778 to 3.945A; three are at distances of 3.154 to 3.945A; and one, which is the ninth atom of Taylor
and Ewing's (1978) ThO_9 group, is at a distance of 2.778A. In cheralite, Finney and Rao (1967) say that each of the four oxygens is
bonded to three A-atoms and one B-atom, which is not consistent with
an AO_8 polyhedron. Two A-O bonds are short, 2.40-2.56A, and one is
long (2.78-3.95A). In huttonite, three oxygens are bonded to one Si
and two Th at distances of 2.40-2.52A, but one oxygen is bonded to one
Si and three Th with the additional Th-O bond (2.81A) being much
longer than the other two (2.50 and 2.58A). The difference in descriptions of huttonite and cheralite appears to be a choice of what
oxygens are included in the coordination polyhedron for the A-atom,
there being at least 8 at 2.5A or less and perhaps one additional
oxygen at a distance of $\sim$ 2.8A.

Similar to thorite, adjacent pairs of axial oxygens define edges
on opposite sides of the ThO_9 polyhedra which are shared with SiO_4
groups and form chains of alternating ThO_9 and SiO_4 parallel to c
(Fig. 3a). The remaining five equatorial oxygens form a nearly (001)
planar pentagonal array around the Th atom. These oxygens are corner-
shared with other SiO_4 and ThO_9 groups (Fig. 4a). As in the case of
thorite, O-Si-O angles opposite shared edges in huttonite, ranging

118

from 99 to 104°, and O-P-O angles in cheralite (104°) are smaller than unshared edges in huttonite (105-116°) and cheralite (106-114°). As expected, the mean tetrahedral bond length for the SiO_4 tetrahedron in huttonite (1.62A) is larger than that (1.54A) for the PO_4 tetrahedron in cheralite.

The huttonite structure forms a dense, space-filling network which, unlike the thorite-structure, cannot accomodate molecular water (Mooney-Slater, 1962). The structures of huttonite and cheralite are similar to one another and the monazite structures reported by Ueda (1953), Mooney (1948) and Mooney-Slater (1962).

Table 2. General actinide orthosilicate formula: ABX_4

	A	B	X
Major:	Th, U^{4+}, U^{6+}	Si	O
Minor:	Fe^{3+}, Ca	H, P	
Trace:	Mn, Ti, Nb, Ta, Fe^{2+} Al, Pb, Be, Mg, Ge Sn, Zr, Hf, Zn, K, Na Cu, Ce, La, Pr, Nd, Sm Eu, Gd, Tb, Dy, Ho, Er Tm, Yb, Lu, Y	S, As B	F, Cl (?) CO_2 (?)

CHEMISTRY

Forty-five elements have been reported as occurring in thorite, huttonite and coffinite, some of which may appear in various oxidation states (Table 2). Most published analyses actually list compositions of bulk samples which are usually metamict and altered to secondary minerals. In addition, several authors indicated that other minerals were included as contaminants. These factors make it difficult to determine the actual compositions of the actinide orthosilicates and rationalize the chemical substitutions. For these reasons the crystal chemistry of the synthetic systems is important in understanding the actinide orthosilicates.

Thorite

Substituents for thorium. End-member thorite is $ThSiO_4$. Major compositional variations involve the substitution of U, Fe, or rare earth elements (REE) for Th. U has been reported in amounts up to about 20 wt. % U_3O_8 (Robinson and Abbey, 1957), but this sample was suspected to contain uraninite as inclusions. A thorite with 15 wt. % UO_3 was found to contain thorian uraninite (Staatz *et al.*, 1976). Both U^{+4} and U^{+6} as UO_2 and UO_3 have been reported in thorite analyses, but total U is usually reported as either UO_2 or UO_3 or U_3O_8, without regard to the true valence state of uranium. Fuchs and Gebert (1958) synthesized intermediate $ThSiO_4$-$USiO_4$ solid solutions, but Mumpton and Roy (1961) could synthesize thorite with no more than 20 to 30 mole % $USiO_4$.

Although REE are generally limited to between 0.5 and 7 wt. %, they have been found in thorites in amounts up to 20 wt. % (Staatz *et al.*, 1976). Because of the metamict nature of the mineral and the uncertain presence of contaminants, the mechanism of charge balance of REE^{+3} for Th^{+4} is unknown. For thorites with minor REE content, small amounts of P substituting for Si could provide a coupled substitution: $REE^{+3} + P^{+5} \rightleftarrows Th^{+4} + Si^{+4}$. In thorites with REE in much greater amounts than P, U^{+6} substitution could provide a mechanism for charge balance: $2REE^{+3} + U^{+6} \rightleftarrows 3Th^{+4}$. In synthesized thorites with up to 1% Gd^{+3}, Er^{3+} and Yb^{3+}, electronic paramagnetic resonance (EPR) spectra of the REE's suggest that they occupy sites that have tetragonal and orthorhombic symmetries (Reynolds *et al.*, 1972). The tetragonal REE spectra are assigned to REE ions substituting at the tetragonal 4a site (Table 1) with the necessary charge compensation mechanism not disturbing the site symmetry. The orthorhombic spectra of REE at higher REE concentrations is proposed to result from substitution of REE at the 4a site with a nearest-neighbor oxygen vacancy distorting the site symmetry.

Thorite can apparently incorporate all REE's equally. Staatz *et al.* (1976) found that of the REE present, yttrium and yttrium-group lanthinides -- Gd, Dy, Er and Yb -- predominated in the thorites from the Seerie pegmatite, Colorado. In their review of thorite analyses from the literature that included the abundances of individual REE's, five analyses showed Ce as the most abundant REE and two contained abundant Nd. In these seven thorites, the cerium-group lanthinides dominated. Two other thorite analyses were similar to the Seerie thorites and were richer in yttrium-group lanthanides, except that Dy rather than Yb was the most abundant REE. From this data, Staatz *et al.* (1976) concluded that thorite does not selectively accomodate one group of REE's over another but rather incorporates whatever is available.

The third major substituent is Fe and it is present in amounts up to 12 wt. % Fe_2O_3. Staatz *et al.* reported a metamict thorite with 5 wt. % Fe_2O_3 containing small inclusions of goethite. Robinson and Abbey (1957) report pyrite and magnetite as contaminants in an iron-bearing thorite.

The other elements which have been cited as substituting for Th (Table 2) are usually reported in amounts of 1 wt. % or less. In rare cases, amounts of up to 7 wt. % are reported.

Any Pb present is believed to be radiogenic in origin. CO_2 is probably present in included carbonate minerals. Non-carbonate carbon may be present as microscopic films of hydrocarbon (Robinson and Abbey, 1957). Calcium (up to 6 wt. % CaO) is reported to substitute for U and Th in other minerals and may actually substitute for Th. A high manganese-plus-iron thorite with $\sim$ 13 wt. % Mn_2O_3 and 7 wt. % Fe_2O_3 (Krol, 1960) is believed to represent a mixture of thorite + Fe_2O_3 + Mn_2O_3. Reported amounts of Zr and Hf could represent solid solutions with zircon and hafnon. Mumpton and Roy (1961) found a miscibility gap on the $ZrSiO_4$-$ThSiO_4$ join with a maximum of 6 mole % $ZrSiO_4$ in thorite. A thorite with a high ZrO_2 content of 1.5 wt. % had a high SnO_2 content of 3.6 wt. % (Heinrich, 1963). The thorite was a stream sediment concentrate and may have contained zircon and cassiterite.

Substituents for silicon. Chemical substitutions for silicon have received more attention than those for thorium. Analyses of thorite have been reported with up to 4 wt. % P_2O_5 and 2.1 wt. % As_2O_5 (Krol, 1962), up to 0.15 wt. % B_2O_3 and up to 4.2 wt. % S. Substitution of P^{+5} and As^{+5} for Si^{+4} would help in charge balancing the substitution of di- and trivalent cations for tetravalent thorium. However, the amounts present are much less than those needed to compensate for the amounts of Ca, Fe, and REE's reported in thorites.

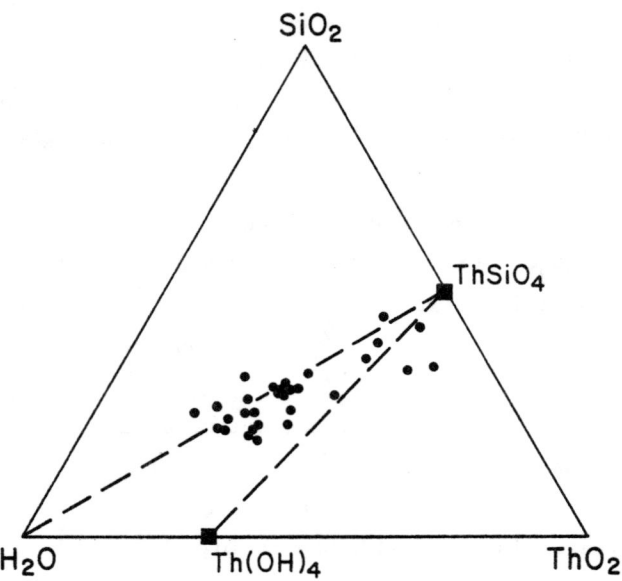

Figure 5. Compositions of natural thorites plotted on a molecular basis. Substituents for Th are not specified. It is evident that natural thorites are
not of the type $Th(SiO_4)_{1-x}(OH)_{4x}$ which would have compositions along
$ThSiO_4$-$Th(OH)_4$, but lie along the $ThSiO_4$-H_2O join, indicating that the water
is present as molecular water. After Mumpton and Roy (1961).

Water. Up to 15 wt. % water has been reported in thorite
analyses. Controversy has arisen as to whether the water is essential, or water that has been absorbed by metamict minerals. It is
generally accepted that the water is essential and that the minerals
are hydroxyl thorites (thorogummite) with the composition $Th(SiO_4)_{1-x}$
$(OH)_x$. Dymkov and Nazarenko (1962) have suggested a complete solid
solution: $ThSiO_4$ - $Th(SiO_4)_{1-x}(OH)_{4x}$ - $Th(OH)_4$ - ThO_2. Frondel
(1953) reviewed the evidence for such substitutions, and Frondel and
Collette (1957) hydrothermally synthesized at <400°C what they interpreted as hydroxyl thorite. It had larger cell dimensions (*a* 7.18A,
c 6.28A) and more diffuse x-ray reflections than higher temperature
thorite (*a* 7.08A, *c* 6.31A). Infrared studies showed strong (OH) absorption at 2.75 microns which disappeared with heating. Mumpton and
Roy (1961) similarly found that only well crystallized $ThSiO_4$ was produced above 400°C whereas thorite formed below 250°C had expanded cell
dimensions. They were reluctant to interpret the material as hydroxyl
thorite, based on a critical evaluation of published thorite analyses.

They reasoned that if hydroxyl thorite of composition $Th(SiO_4)_{1-x}$ $(OH)_{4x}$ exists, it represents solid solution on the join $ThSiO_4$, $x = 0$ and $Th(OH)_4$, $x = 1$. In a plot of the molecular compositions of thorite in Figure 5, the analyses lie along the $ThSiO_4-H_2O$ join, indicating that the water is present as molecular water, not $(OH)^{-1}$, and the formula is $ThSiO_4 \cdot xH_2O$.

Halogens. Up to 3 wt % of the halogens Cl and F have been reported in thorite analyses. These elements could be substituting for $[SiO_4]$ or for O and providing charge balance for the di- and trivalent cations substituting for Th. More than likely, halogens in the thorite analyses represent admixtures of other minerals.

Huttonite

Huttonite analyses are as rare as its occurrence in nature. The type huttonite contains, in addition to Th and Si, 1.2 wt % Fe_2O_3 and 2.6 wt % REE_2O_3 (Pabst, 1951b). The charge balance mechanism is not apparent. A metamict, hydrothermal huttonite described by Kosterin and Zuev (1962) has 12 wt % H_2O, 8.4 wt % Fe_2O_3, 2 wt % MnO and 5.5 wt % REE. Huttonite is isostructural with monazite, $REEPO_4$, and brabanite, $CaTh(PO_4)_2$, suggesting a potential for solid solutions. Monazites contain up to 12 wt % ThO_2. Silicon is usually present in smaller amounts, but several monazites have atomic ratios of Th/Si = 1, suggesting that the substitution is coupled: $Th^{+4} + Si^{+4} \rightleftarrows REE^{+3} + P^{+5}$ (Deer *et al.*, 1961). A monazite from Sri Lanka (Bowie and Horne, 1953) and a monazite-like mineral called "cerphosphorhuttonite" (Pavlenko *et al.*, 1965) are evidently intermediate compositions on the $ThSiO_4$-$REEPO_4$ join: $Ca_{0.02}REE_{2.71}Th_{1.09}Fe^{+3}_{0.21}[Si_{1.04}P_{2.91}]O_4$ and $REE_{0.41}Th_{0.51}U_{0.02}Fe_{0.14}[Si_{0.55}P_{0.46}]O_4$. Cheralite, a monazite isotype with high calcium content, $Ca_{1.08}Th_{1.15}REE_{1.62}U_{0.14}[Si_{0.34}P_{3.64}]O_4$ represents solid solution toward $CaTh(PO_4)_2$, and Bowie and Horne (1953) suggested that these monazite-type minerals are compositions in the ternary system $ThSiO_4$-$REEPO_4$-$CaTh(PO_4)_2$. A possible hydroxyl-bearing mineral of this group, analogous to thorogummite, is tombarthite (Neumann and Nilssen, 1968) which is a monazite isotype with ideal formula $(REE,Th,Ca,U,Mn)_4(Si+4H)_4O_{12-n}(OH)_{4+2n}$.

Karkhanavala (1956) synthesized huttonite and a monazite with 5% $ThSiO_4$ by sintering. Frondel and Collette (1957) hydrothermally synthesized huttonite with smaller interplanar spacings than dry synthesized huttonite, suggesting a hydroxyl huttonite.

Coffinite

Coffinite has been described as a uranous silicate that shows substitution of $(OH_4)^{-4}$ for $(SiO_4)^{-4}$ with the formula $U(Si,H_4)O_4$ or $U(SiO_4)_{4-x}(OH)_{4x}$ (Stieff et al., 1956). The type material could not be adequately characterized by chemical analysis because of admixed uraninite, vanadium minerals and organic material. Heating, leaching, x-ray and α-plate studies helped investigators arrive at the chemical composition. Determination of the essential nature of the OH was based on infrared studies. Subsequent analyses of coffinites from various localities revealed one with evidently admixed minerals (Arribas, 1966) but three other coffinites show only U and Si as major constituents; CaO and PbO are present in amounts of 2.5 wt. % or less, and Ti, Y, Fe, P, Th, Ce, Zn, V and Cr occur as trace elements (Zavarzin, 1977, Khalezov et al., 1974; Belova et al., 1969; Frenzel, 1975). Nord (1977) found by electron microprobe and qualitative x-ray energy-dispersive analyses in a transmission electron microscope that coffinite contains only U and Si; the elements Ca, Th and Pb were below the limits of detectability (<0.1 wt. %). By chemical, infrared, x-ray and DTA studies, Belova et al. (1969) and Abdel-Gawad and Kerr (1961) found natural coffinites that are uranous silicates lacking an essential OH component.

One problem in establishing the chemistry of the type material was the difficulty of synthesizing $USiO_4$. It has since been synthesized by hydrothermal techniques by Fuchs and Hoekstra (1959), Keller (1963), and Bayushkin and Dikov (1974). Fuchs and Gebert (1958) and Fuchs and Hoekstra (1959) found no evidence of essential OH in synthetic coffinite. Evidently, coffinite is not readily synthesized; Mumpton and Roy (1961) failed to synthesize it and assumed it to be metastable.

124

Table 3. Thermochemical data for actinide
orthosilicates at 298°K and 1 bar total pressure

Mineral	Formula Weight	Molar Volume (cm^{-3})	$\Delta H°F$ (kcal/mol)	$\Delta G°F$ (kcal/mol)	S° (cal/mol deg)	Reference
Zircon	183.3	39.26	−486	−458.63	20.1	Robie et al. (1978)
Zircon				−459.02		Schuiling et al. (1976)
Thorite	324.1	48.40		−489.67		Schuiling et al. (1976)
Huttonite	324.1	44.74				
Coffinite	330.1	46.12	−478	−452.0	28.4	Langmuir and Applin (1977)
Coffinite (amorphous)			−439.2			

PHYSICAL PROPERTIES

The physical properties which best characterize the actinide
orthosilicates are obtained from synthetic material. Synthetic
thorite is uniaxial (+) with ω = 1.827, ε = 1.885 (Finch et al., 1964).
Synthetic coffinite has an average refractive index of 1.83 (1.88
after heating to 1000°C; Fuchs and Hoekstra, 1959). Synthetic huttonite has α = 1.900 and γ = 1.930 (Finch et al., 1964). The type
huttonite has α = 1.898, β = 1.900 (calc), γ = 1.922, 2V = (+)25, and
r<v (Pabst, 1951b).

Calculated densities are given in Table 1, together with those
of other actinide orthosilicates.

Thermochemical data for the actinide orthosilicates are given in
Table 3. The molar volumes are calculated from the cell parameters of
Table 1. Values for zircon are included for comparison. An estimate
of the enthalpy of the thorite ⇄ huttonite transition is calculated
as 5.6(3) kcal mole at 298°K (Dachille and Roy, 1964).

The Raman spectra of synthetic thorite at 295°K have been
measured and the internal and lattice modes of the SiO_4^{4-} complex
assigned by Syme et al. (1977) and Griffith (1969). Ferraro (1975)
has performed a factor group analysis of the zircon-group minerals and
lists the selection rules of infrared and Raman activity and includes

the species of vibrations activity and the number of times each appears. The electronic paramagnetic resonance (EPR) spectra of $ThSiO_4$ doped with rare earth elements are reported by Reynolds *et al.* (1972). Application of EPR to the recrystallization of metamict thorite is given by Hubin (1974). In addition to the infrared work referenced in the section entitled *CHEMISTRY*, Hubin and Tarte (1970) have made detailed assignments of the infrared spectra of thorite and huttonite.

Metamict actinide orthosilicates.

Most natural thorites and coffinites have structural damage because of internally generated α-particles and nuclear recoil and, compared to their synthetic counterparts, exhibit decreased density, birefringence, refractive indices, hardness and x-ray diffraction intensities. The metamict minerals consist of microcrystalline or amorphous U- or Th-oxides, silica, iron oxides and galena. They also contain absorbed water. Huttonites are invariably crystalline, leading Ewing (1975) to suggest that the huttonite structure is more resistant to radiation damage than the thorite structure. However, in the absence of any compelling reason, all amorphous thorium silicates are labeled "thorite." Nevertheless, structure as well as the amount of radiation is important in the metamictization of crystalline solids. Considering their high α-particle flux, the actinide orthosilicates do appear to be more resistant to radiation damage than other minerals which should have lower fluxes because of their lower U and Th contents. Pellas (1951) found the minimum α-particle flux necessary to produce a metamict thorite is 1×10^{19} cm^{-2} compared to 2.3×10^{17} cm^{-2} for allanite. He attributed the greater stability of thorite to the more ionic character of its bonding.

With heating, metamict thorite and coffinite increase in crystallinity, density, refractive index and hardness. Metamict thorite produces an assemblage of thorite, huttonite or both in addition to thorianite ± uraninite ± SiO_2 ± Fe oxide. Pabst (1952) found that prolonged heating of thorite to 1200°C produces only huttonite. In addition to temperature, time and atmosphere also affect the

126

Table 4. Apparent ages of zircon, uranothorite and composite zircon-uranothorite fractions from the Johnny Lynn granodiorite (Silver and Deutsch, 1963)

Sample	Apparent Ages (Millions of Years)				Age Ratio
	Pb^{206}/U^{238}	Pb^{207}/U^{235}	Pb^{207}/Pb^{206}	Pb^{208}/Th^{232}	$\dfrac{(Pb^{206}/U^{238})}{(Pb^{208}/Th^{232})}$
R300 zircon acid-washed	1,030	1,230	1,610	1,390	0.74
R300 uranothorite wash no. 1	230	350	1,280	245	0.93
R-300 unwashed zircon and uranothorite composite.	710	950	1,570	360	1.97

products. Summaries of heating experiments are given by Pabst (1952) and Frondel (1953). Hubin (1972) conducted an infrared and x-ray diffraction study of the recrystallization of thorite and concluded that variation in bulk composition also affects the heating products. In an EPR study, Hubin (1974) found that in the recrystallization of metamict thorite at 800°C, Fe^{+3} substitutes for Th; above 1200°C, the iron forms a separate oxide.

Ramdohr (1961) found that the younger, Miocene to Recent coffinites of European mining districts are fresh and nonmetamict, whereas older, Precambrian-Miocene coffinites are metamict or pseudomorphed by quartz or pitchblende. The oldest coffinites are dusted by numerous inclusions of galena.

GEOCHRONOLOGY

In principle, the actinide orthosilicates would be ideal for radiometric age dating because of their high concentrations of radioactive isotopes. The minerals are rare in most rocks, however, and metamict minerals lose radiogenic lead easily, yielding younger apparent ages. Silver and Deutsch (1963) found that in the 1655(20) m.y. Johnny Lynn granodiorite, uranothorite made up less than 1% of the abundance but more than 50% of the activity in a zircon concentrate. The uranothorite could be eliminated by a hot acid wash. Silver and Deutsch concluded that in terms of U-Pb and presumably Th-Pb isotopic equilibrium, uranothorite constitutes a system separate from

127

zircon. Thorite apparent ages are in profound discordance as compared to zircon and can drastically affect the apparent age if not
removed from the zircon concentrate as illustrated in Table 4. The
behavior of uranothorite is attributed to the easy radiogenic lead
loss from the metamict grains.

Robinson and Abbey (1957) noted a low Pb/(U+0.38Th) ratio for
uranoan thorites from Bancroft, Ontario as compared to the coexisting
uraninites. They suggested that Pb was more easily lost from the
metamict thorites than the uraninites and formed the coexisting
galena.

Isotopic studies of coffinite and coexisting uraninite from
uranium deposits may be more widespread in the future because of the
chronological information they could yield about the formation, redistribution and weathering of the ores. Darnley *et al.* (1965) obtained $^{206}Pb/^{238}U$ ages of coffinite and pitchblende to define at
least three episodes of vein uranium mineralization in S.W. England.
Pavshukov *et al.* (1975) found by microprobe study that coffinite has
unevenly distributed radiogenic lead compared to uraninite. Total
lead dates on coffinite are one-tenth those of coexisting uraninite
or of Pb-isotopic ages. It is believed that the lead migrates from
the metamict coffinite and is incorporated in the coexisting sulfides.
By comparison, the uraninite structure is able to retain the radiogenic lead.

SYNTHESIS, PHASE STABILITY AND PHASE
ASSEMBLAGES OF THE ACTINIDE ORTHOSILICATES

End-member and intermediate composition actinide orthosilicates
can be prepared in crystals up to 10 mm long by direct reaction of an
actinide element and silicon oxides at high temperature, hydrothermal
synthesis from gels, or growth from a flux (Frondel and Colette, 1957;
Fuchs, 1958; Karkhanavala, 1956; Mumpton and Roy, 1961; Fuchs and
Hoekstra, 1959; Keller, 1963; Finch *et al.*, 1964; Dachille and Roy,
1964; and Sinha and Prasad, 1973).

Lunga (1966) has studied the liquidus relations of the

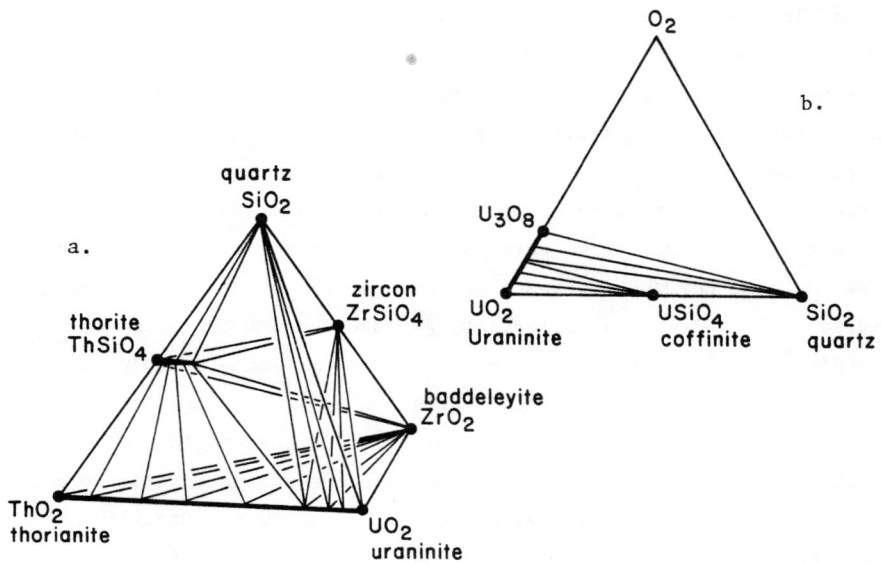

a.

quartz
SiO₂

b.

O₂

zircon
ZrSiO₄

thorite
ThSiO₄

baddeleyite
ZrO₂

ThO₂
thorianite

UO₂
uraninite

U₃O₈

UO₂
Uraninite

USiO₄
coffinite

SiO₂
quartz

Figure 6. *(a)* Subsolidus phase assemblages in the system SiO_2-ThO_2-UO_2-ZrO_2 based on work of Mumpton and Roy (1961). *(b)* Possible phase relations in coffinite-bearing assemblages.

ThO_2-SiO_2, UO_2-SiO_2 and ThO_2-UO_2-SiO_2 systems at one bar in a hydrogen atmosphere. Thorite incongruently melts to ThO_2 + liquid at 1975°C. No $USiO_4$ phase was found. The boundary curve in the ternary system lies near 1700°C at less than 5 mole % ThO_2 + UO_2. There is liquid immiscibility in the SiO_2-rich portion of the system. Mumpton and Roy (1961) studied the systems ThO_2-UO_2-SiO_2, ThO_2-ZrO_2-SiO_2 and UO_2-ZrO_2-SiO_2 at subsolidus conditions of 300 to 1350°C at one atmosphere. Their results, most of which have been discussed in the *CHEMISTRY* section, are summarized in the SiO_2-ZrO_2-ThO_2-UO_2 quadralateral of Figure 6a. $USiO_4$ is believed to be a metastable phase. Most rocks have quartz and Si >>(Zr+Th+U) and according to the phase diagram, they should contain the assemblage quartz + thorite + zircon ± thorian uraninite. This assemblage has been reported in studies of granitoids which include documentation of the U-Th minerals (Silver and Deutsch, 1963; Willgallis, 1970; Renard, 1974; Speer *et al.*, 1980). The presence or absence of uraninite depends on the U-content and whether it can be entirely incorporated in the uranoan thorite. The remaining assemblages depicted in the quadralateral are quartz-free: baddeleyite + uraninite - thorianite solid solution + thorite

129

and baddeleyite + zircon + thorite + thorianite - uraninite solid
solution. Thorianite, baddeleyite, or both commonly occur in quartz-
free igneous and metamorphic rocks, particularly carbonatites,
alkaline rocks and desilicified granitoids. Thorite and zircon are
also reported, but there is insufficient detail in the descriptions
to determine the coexisting accessory mineral assemblage.

Coffinite is considered metastable and does not appear in
Figure 6a. The common occurrence of the assemblage, quartz + thorite
+ zircon + uraninite, in quartz-bearing igneous rocks is consistent
with this view. The widespread occurrence of coffinite in Colorado
Plateau and vein uranium deposits suggests that it is stable under
some conditions. The occurrence of coffinite in the unoxidized
portions of the Colorado Plateau-type deposits suggests that it forms
in low temperature and pressure, reducing environments. Brookins
(1975, Abs. Bull. Am. Assoc. Petrol. Geol. *59*, 905) has estimated a
ΔG_F° for $USiO_4$ as -456 ±2.6 kcal/gfw (c.f. Table 3). This free energy
value indicates that the reaction $USiO_4 + 2H_2O = UO_2 + H_4SiO_4^\circ$ is at
equilibrium when dissolved silica is 10^{-69} mol/ℓ or 8 ppb SiO_2.
Langmuir (1978, Geochim. Cosmochim. Acta *42*, 547-569) points out that
the average silica concentration of 17 ppm in groundwaters invariably
exceeds this value and would make coffinite more stable than uraninite in nearly all natural waters at low temperatures. Langmuir
alternatively suggests that coffinite is stable relative to uraninite
at intermediate dissolved silica levels, above the levels of average
groundwaters and below saturation with amorphous silica. Coffinite
commonly coexists with uraninite and quartz, a metastable condition
unless an extra component is present. The uranium minerals of these
ore deposits are nearly pure because large radius cations such as Th,
Zr and REE are geochemically separated when U migrates, so that
Figure 6a does not apply. Perhaps coffinite is a hydrous phase which
permits it to coexist with uraninite + quartz. More likely it is the
uraninite which has a nonlinear composition with coffinite + quartz
because of the well-documented solid solution with O_2 between UO_2 and
U_3O_8. An oxidized uraninite could form a three-phase assemblage with
coffinite + quartz (Fig. 6b). Uraninite coexisting with coffinite has
cell dimensions of a = 5.45 A (Nord, 1977) which corresponds to a
composition of $UO_{2.15}$.

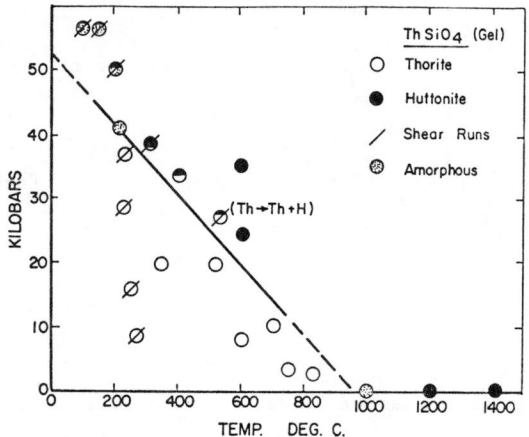

Figure 7. P-T diagram of the thorite $\rightleftarrows$ huttonite phase transition based on
static and shear runs using $ThSiO_4$ "gel" starting material (Dachille and
Roy, 1964). The negative slope explains the unexpected behavior of the
denser and lower symmetry, monoclinic huttonite being the higher temperature phase on heating tetragonal thorite at one atmosphere.

Most discussions of phase relations of the actinide orthosilicates have centered on the transition between the tetragonal thorite-
structure and the monoclinic huttonite-structure of $ThSiO_4$ and
$PaSiO_4$. Contrary to expectations, the high temperature huttonite
phase is denser and has a lower symmetry than thorite. Mumpton and
Roy (1961) suggested that huttonite is the stable phase and that
thorite is metastable. Finch *et al.* (1964) determined that the one
atmosphere equilibrium occurs at 1225°C based on growth synthesis.
Dachille and Roy's (1964) growth synthesis technique using shearing
stresses showed huttonite to be the high temperature phase at one
atmosphere, but more importantly the high pressure phase at any temperature. The thorite $\rightleftarrows$ huttonite univariant equilibrium has a
negative slope in P-T space (Fig. 7) which explains the apparent
behavior of the denser phase being the high temperature phase at one
atmosphere constant pressure. Because there were no reversals, Figure
7 is considered only a tentative curve.

Based on their thorite $\rightleftarrows$ huttonite phase diagram, Dachille and
Roy (1964) suggested that the rare occurrence of huttonite might be a
result of its instability in the pressures and temperatures normally
encountered in the earth's crust. The natural occurrences of huttonite have been in metamorphic terrains or sediments derived from
them. The huttonite stability field is probably increased with solid
solutions with monazite-group minerals such as $REEPO_4$ and $CaTh(PO_4)_2$,
brabantite.

131

Coffinite Analysis

Electron microprobe analysis of crystalline coffinite from the Woodrow Mine, New Mexico by Kim (1978) gave $UO_2 = 67.49$, $SiO_2 = 15.24$, $Al_2O_3 = 0.79$, $P_2O_5 = 2.18$, $V_2O_5 = 0.64$, $CaO = 2.34$, $FeO = 0.30$, $PbO = 0.38$ and, by difference, $H_2O = 10.65$ wt %. This yields the formula $(U_{0.85}Ca_{0.14})(Si_{0.86}P_{0.11}Al_{0.06}V_{0.02})O_2(OH)_4$ or, ideally, $(U,Ca)(Si,P)O_2(OH)_4$.

Brabantite

$CaTh(PO_4)_2$, which is isostructural with monazite, recently has been described as the mineral brabantite from a pegmatite on the Brabant farm, Karibib district, Namibia (Rose, 1980). The natural material contains about 10% huttonite solid solution, comparable to the cheralite described by Bowie and Horne (1953). Wang Xianjne (1978) described a brabantite of similar composition but with a small amount of $CaU(PO_4)_2$. See Figure 8.

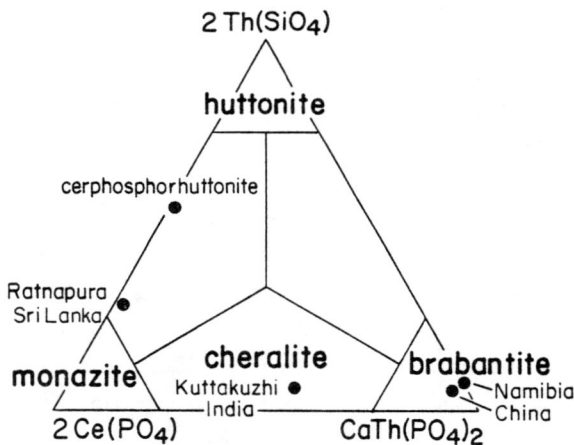

Figure 8. Nomenclature for the monazite-structure silicate and phosphate minerals. Plotted analyses are the recently described brabantites as well as the intermediate composition minerals from Kuttakuzhi and Ratnapura (Bowie and Horne, 1953) and the Russian "cerphosphorhuttonite" (Pavlenko *et al.*, 1965).

THE ACTINIDE ORTHOSILICATES: REFERENCES

Abdel-Gawad, A. M. and P. F. Kerr (1961) Urano-organic mineral association. Am. Mineral., 46, 402-419.

Arribas, A. (1966) New mineralogical and metallogenic data on coffinite. Estud. Geol., Inst. Invest. Geol. "Lucas Mallada" 22, 47-59. Chem. Abstr., 68, 51943w.

Bayushkin, I. M. and Yu. P. Dikov (1974) Uranium silicates in hydrothermal uranium mineralization. Geochem. Internat'l, 1974, 1162-1170.

Belova, L. N., G. A. Tananaeva and K. E. Frolova (1969) Coffinite. At. Energ., 27, 61-63 (Russian). Chem. Abstr., 71, 115090n.

Bowie, S.H.U. and J.E.T. Horne (1953) Cheralite, a new mineral of the monazite group. Mineral. Mag., 30, 93-99.

Chase, A. B. and J. A. Osmer (1966) Growth and preferential doping of zircon and thorite. J. Electrochem. Soc., 113, 198-199.

Dachille, F. and R. Roy (1964) Effectiveness of shearing stresses in accelerating solid phase reactions at low temperatures and high pressures. J. Geol., 72, 243-247.

Darneley, A. G., T. H. English, O. Sprake, E. R. Preece and D. Avery (1965) Ages of uraninite and coffinite from south-west England. Mineral. Mag., 34, 159-176.

Deer, W. A., R. A. Howie and J. Zussman (1963) *Rock-Forming Minerals, vol. 5, Non-silicates.* Longmans, London.

Dymkov, Yu. M. and N. G. Nazarenko (1962) Coffinite and the nature of pitchblende pseudocrystals. Geochem. Internat'l, 1962, 348-358.

Ewing, R. C. (1975) The crystal chemistry of complex niobium and tantalum oxides. IV. The metamict state: Discussion. Am. Mineral., 60, 728-733.

Ferraro, J. R. (1975) Factor group analysis for some common minerals. Appl. Spectros., 29, 418-421.

Finch, C. B., L. A. Harris and G. W. Clark (1964) The thorite-huttonite phase transformation as determined by growth of synthetic thorite and huttonite single crystals. Am. Mineral., 49, 782-785.

Frenzel, G., J. Ottemann and W. Kurtze (1975) Uran-Vererzungen und uranhaltige Rutile in einem permovulkanischen Tuffit von Boarezzo (Valganna, Varese). N. Jb. Mineral. Abh., 124, 75-102.

Frondel, C. (1953) Hydroxyl substitution in thorite and zircon. Am. Mineral., 38, 1007-1018.

_____ (1958) *Systematic mineralogy of uranium and thorium.* U. S. Geol. Survey Bull. 1064, 400 p.

_____ and R. L. Collette (1957) Hydrothermal synthesis of zircon, thorite and huttonite. Am. Mineral., 42, 759-765.

Fuchs, L. H. (1958) Formation and properties of synthetic thorite crystals. Am. Mineral., 43, 367-368.

Fuchs, L. H. and E. Gebert (1958) X-ray studies of synthetic coffinite, thorite and uranothorites. Am. Mineral., 43, 243-248.

_____ and H. R. Hoekstra (1959) The preparation and properties of uranium (IV) silicate. Am. Mineral., 44, 1057-1063.

Garrels, R. M. and C. L. Christ (1959) Behavior of uranium minerals during oxidation, part 6, of *Geochemistry and Mineralogy of the Colorado Plateau Uranium Ores.* U. S. Geol. Survey Prof. Pap. 320, 81-89.

Ghouse, K. M. (1965) A note on the refinement of the crystal structure of Indian monazite. Naturwiss., 52, 32-33.

Griffith, W. P. (1969) Raman studies on rock-forming minerals. Part 1. Orthosilicates and cyclosilicates. J. Chem. Soc. A, 9, 1372-1377.

Haines, E. L., A. J. Gancarz, A. L. Albee and G. J. Wasserburg (1972) The uranium distribution in lunar soils and rocks 12013 and 14310. Lunar Sci. III, Abstr., 350-351.

Heinrich, E. W. (1963) Xenotime and thorite from Nigeria. Am. Mineral., 48, 206-208.

Hubin, R. (1972) Application of ir absorption spectrometry to the study of metamict thorite recrystallization. Bull. Soc. Fr. Mineral. Cristallogr., 94, 471-476 (in French).

_____ (1974) La thorite métamicte: application de la résonance paramagnétique électronique (R.P.E.) à l'étude de sa recristallisation. Bull. Soc. Fr. Mineral. Cristallogr., 97, 417-421.

_____ and P. Tarte (1971) Etude infrarouge des orthosilicates et des orthogermanates - IV structures scheelite et zircon. Spectrochim. Acta, 27A, 683-690.

Karkhanavala, M.D. (1956) The synthesis of huttonite and monazite. Curr. Sci. (India), 25, 166-167.

Keller, C. (1963) Investigations of germanates and silicates of the type ABO_4 with quadrivalent elements thorium to americium. Nukleonik, 5, 41-48.

Khalezov, A. B. and A. S. Avdonin (1974) Coffinite from red beds. Soviet Geol., 3, 134-138.

Kim, Soo Jin (1978) Chemical composition of the coffinite from the Woodrow Mine, New Mexico, U.S.A. Mining Geol. (Korea) 11, 183-186.

Kostern, A. V. and V. N. Zuev (1962) Hydrothermal huttonite. Zap. Vses. Mineral. Obshchestva, 91, 99-102. Chem. Abstr., 57, 6904c.

Krol, O.F. (1960) Thorite containing manganese. Tr. Kazakhsk. Nauchn-Issled. Inst. Mineral. Syr'ya, No. 3, 162-167. Chem. Abstr., 57, 1880c.

_____ (1962) Thorite containing arsenic. Tr. Kazakhsk. Nauchn-Issled. Inst. Mineral. Syr'ya, 7, 136-144, Chem. Abstr., 61, 2829h.

Langmuir, D. and K. Applin (1977) Refinement of the thermodynamic properties of uranium minerals and dissolved species, with application to the chemistry of ground waters in sandstone-type uranium deposits. U.S. Geol. Survey Circular 753, 57-60.

Lunga, S. (1966) Etude des Courbes de liquidus et des properties thermodynamiques des Systems SiO_2-ThO_2 et SiO_2-ThO_2-UO_2. J. Nucl. Mat., 19, 157-159.

Moench, R. H. (1962) Properties and paragenesis of coffinite from the Woodrow Mine, New Mexico. Am. Mineral., 47, 26-33.

Mooney, R. C. L. (1948) Crystal structures of a series of rare earth phosphates. J. Chem. Phys., 16, 1003.

Mooney-Slater, R. C. L. (1962) Polymorphic forms of bismuth phosphate. Z. Kristallogr., 117, 371-385.

Mumpton, F. A. and R. Roy (1961) Hydrothermal stability studies of the zircon-thorite group. Geochim. Cosmochim. Acta, 21, 217-238.

Neumann, H. and B. Nilssen (1968) Tombarthite, a new mineral from Högetveit, Evje, South Norway. Lithos, 1, 113-123.

Nord, G. L. (1977) Characterization of fine-grained black uranium ores by transmission electron microscopy. U. S. Geol. Survey Circular 753, 29-31.

Pabst. A. (1951a) X-ray examination of uranothorite. Am. Mineral., 36, 557-562.

_____ (1951b) Huttonite, a new monoclinic thorium silicate. Am. Mineral., 36, 60-69.

_____ (1952) The metamict state. Am. Mineral., 37, 137-157.

Pavlenko, A. S., L. P. Orlava and M. V. Akhmanova (1965) Cerphosphorhuttonite, a monazite-group mineral. Tr. Mineral. Muzeya, Akad. Nauk SSSR, 16, 166-174. Chem. Abstr., 63, 4020b.

Pavshukov, V. V., L. V. Komlev, Ye. B. Anderson and I. G. Smyslova (1975) X-ray microprobe data on the state of the U-Pb system in uranium ores. Geochem. Internat'l, 2, 251-261.

Pellas, P. (1951) The spontaneous destruction of the crystal lattice of radioactive minerals. Comptes Rend., 233, 1369-1371.

Phair, G. and K. O. Shimamoto (1952) Hydrothermal uranothorite in fluorite breccias from the Blue Jay mine, Jamestown, Boulder County, Colorado. Am. Mineral., 37, 659-666.

Ramdohr, P. (1961) Das Vorkommen von Coffinit in hydrothermalen Uranerzgangen, besonders vom Co-Ni-Bi-Typ. N. Jb. Miner. Abh., 95, 313-324.

Renard, J. P. (1974) Etude pétrographie et géochimique des granites du district uranifère de Vendée. Sci. de la Terre, Mémoire, 30, 216 p.

Reynolds, R. W., L. A. Boatner, C. B. Finch, A. Chatelain and M. M. Abraham (1972) EPR investigations of Er^{3+}, Yb^{3+} and Cd^{3+} in zircon-silicate structure. J. Chem. Phys., 56, 5607.

Robie, R. A., B. S. Hemingway and J. R. Fisher (1978) Thermodynamic properties of minerals and related substances at 298.15 K and 1 bar (10^5 Pascals) pressure and at higher temperatures. U. S. Geol. Survey Bull. 1452, 456 p.

Robinson, S. C. and S. Abbey (1957) Uranothorite from eastern Ontario. Canad. Mineral., 6, 1-15.

Rose, D. (1980) Brabantite, CaTh $[PO_4]_2$, a new mineral of the monazite group. Neues Jahrb. Mineral. Monatsh., no. 6, 247-257.

Schuiling, R. D., L. Vergouwen and H. van der Rijst (1976) Gibbs energies of formation of zircon ($ZrSiO_4$), thorite ($ThSiO_4$), and phenacite (Be_2SiO_4). Am. Mineral., 61, 166-168.

Shannon, R. D. (1976) Revised effective ionic radii and systematic studies of interatomic distances in halides and chalcogenides. Acta Crystallogr., A32, 751-767.

Silver, L. T. and S. Deutsch (1963) Uranium-lead isotopic variations in zircons: a case study. J. Geol., 71, 721-758.

Sinha, D. P. and R. Prasad (1973) On the synthetic preparation and lattice structure of thorite. J. Inorg. Nucl. Chem., 35, 2612-2614.

Speer, J. A., S. W. Becker and S. S. Farrar (1980) Field relations and petrology of the post-metamorphic, coarse-grained granitoids and associated rocks of the Southern Appalachian Piedmont. Proceedings "The Caledonides in the USA, IGCP project 27: Caledonide Orogen," VPI & SU Dept. of Geological Sciences Memoir, 2, 137-148.

Staatz, M. H., J. W. Adams and J. S. Wahlberg (1976) Brown, yellow, orange, and greenish-black thorites from the Seerie Pegmatite, Colorado. J. Res. U. S. Geol. Survey, 4, 575-582.

Stieff, L. R., T. W. Stern and A. M. Sherwood (1956) Coffinite, a uranous silicate with hydroxyl substitution: a new mineral. Am. Mineral., 41, 675-688.

Sveshnikova, E. V., D. N. Knyazeva and M. T. Dmitrieva (1964) Metamict thorites from nepheline-syenite complex in the Enesei Ridge. Tr. Mineral. Muzeya, Akad, Nauk SSSR No. 15, 239-246. Chem. Abstr., 61, 13049a.

Syme, R. W. G., D. J. Lockwood and H. J. Kerr (1977) Raman spectrum of synthetic zircon $(ZrSiO_4)$ and thorite $(ThSiO_4)$. J. Phys. C: Sol. St. Phys., 10, 1335-1348.

Taylor, M. and R. C. Ewing (1978) The crystal structures of the $ThSiO_4$ polymorphs: huttonite and thorite. Acta Crystallogr., B34, 1074-1079.

Traill, R. J. (1969) A catalogue of Canadian minerals. Geol. Survey Canada Paper 69-45, 649 p.

Ueda, T. (1953) The crystal structure of monazite $(CePO_4)$. Mem. Col. Sci., Univ. Kyoto Ser. B, 20, 227-246.

Willgallis, A. (1970) Zur Mikrosondenanalyse der U-Th-Minerale in Malsburger Granit. N. Jb. Miner. Abh., 114, 48-60.

Xianjue, Wang (1978) A new mineral - lingaitukuang. Kexue Tongbao 23, 743-745 (in Chinese). New mineral names. Am. Mineral. 66, 878-879.

Zavarzin, A. V. (1977) Disseminated coffinite-brannerite ores. R. P. Petrov (ed) Tekstury Strukt. Uranovykh Rud Endog. Mestorozhd., pp. 163-167, 195-202, Izd. Atomizdat: Moscow, USSR. Chem. Abstr., 89, 166366m.

Chapter 5

TITANITE (SPHENE)   P. H. Ribbe

INTRODUCTION

Titanite has a fairly large stability field (see *e.g.* Hunt and Kerrick, 1977), and thus it is found as an accessory mineral in a wide range of igneous and metamorphic rocks. However, because of its softness (H = 5) it is rarely observed as detrital grains in sedimentary rocks. The major element chemistry of this orthosilicate is relatively simple: $Ca(Ti>>Al,Fe^{3+})(O>>OH,F)[SiO_4]$, but ubiquitous rare earth elements substitute for Ca in 7-fold coordination and there are many substituents other than Al and Fe (notably Ta and Nb) for Ti in octahedral coordination. Malayaite, $CaSnO[SiO_4]$, is a tin analog which occurs mostly in skarns and exhibits a limited solid solution with titanite (Takenouchi, 1971). Both synthetic $CaTiO[SiO_4]$ and its germanium analog[*], $CaTiO[GeO_4]$, crystallize in space group $P2_1/a$ (Robbins, 1968), but malayaite and titanite above 220°C have higher symmetry, $A2/a$. Impure titanites have linear anti-phase domains which increase in frequency with increasing (Al,Fe) + (OH,F) for Ti+O, leading to diffuseness and eventual extinction of $k+l$ odd reflections.

The structure of synthetic titanite will be discussed first because it leads to an understanding of the wide variety of chemical substitutions that are observed as well as to an understanding of the domain texture common in titanites containing Al,Fe-OH,F.

CRYSTAL STRUCTURES

The structure of titanite was solved by Zachariasen (1930), who studied a natural crystal with space group $C2/c$ and reported a wide range of Si-O distances (1.54 - 1.74 A). Mongiorgi and Riva di Sanseverino (1968) refined the structure of another titanite in the transformed space group[1] $A2/a$ and reported much more reasonable bond lengths for the $[SiO_4]$ tetrahedron.

- - - - - - - - - - - - - - - - - - -

[*] A $CaGeO[GeO_4]$ analog has space group $P1$ (Nevskii *et al.*, 1981).

[1] Transformation matrices for the variety of space groups reported for titanite to the right-handed coordinate system proposed by Donnay and Ondik (1973) and used herein are given by Speer and Gibbs (1976, p. 238).

The Structure of $P2_1/a$ Synthetic CaTiO[SiO$_4$]

Recently, Speer and Gibbs (1976) examined a synthetic titanite which has space group $P2_1/a$, Z = 4, lattice parameters a = 7.069, b = 8.722, c = 6.566 A, β = 113.86°, volume = 370.2 A^3, and calculated density

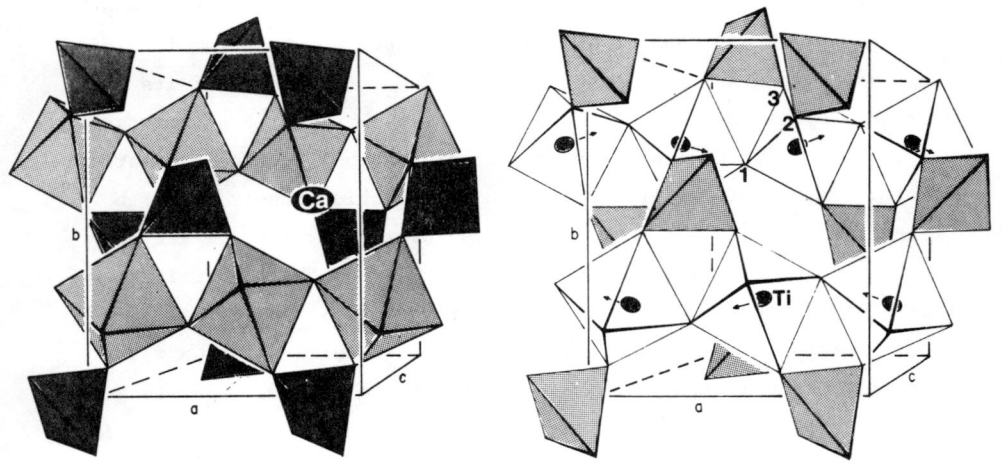

Figure 1. (A) Perspective drawing of the titanite structure showing the kinked chain of Ti-octahedra crosslinked by isolated Si-tetrahedra. One of the Ca-cations is shown in its seven-coordinated site between the chains. (B) Perspective drawing of the titanite structure showing the direction of movement of Ti-cations accompanying the phase transition $P2_1/a \rightarrow A2/a$. After Taylor and Brown (1976, p. 440, Fig. 3).

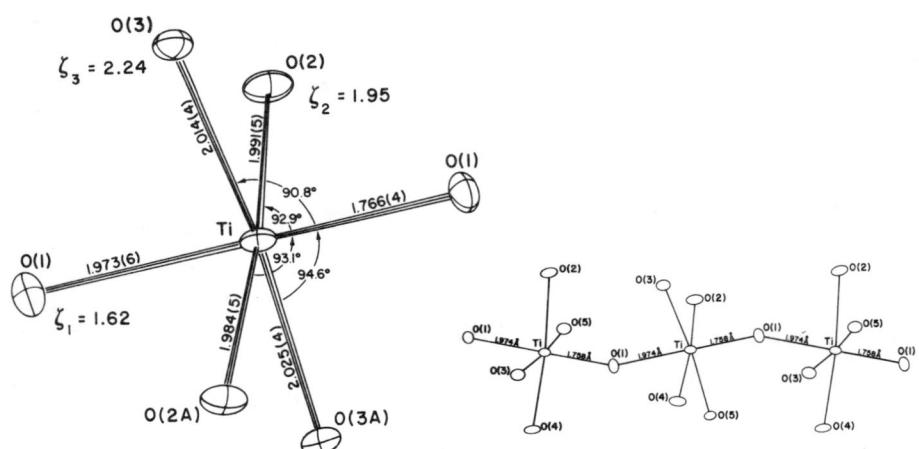

Figure 2. The titanium octahedron in synthetic $P2_1/a$ titanite (left). After Speer and Gibbs (1976, p. 243, Fig. 2). Compare with Figures 1 and 3 to determine ligancy of oxygens 1,2,3. The linkage of the corner-sharing octahedral chain is shown to the right.

138

3.52 g/cc. A slightly edited version of their description of the structure follows:

"The dominant structural units in titanite are kinked chains of corner-sharing TiO_6 octahedra running parallel to the a cell edge [Fig. 1]. The repeat unit in the chain is defined by two tilted octahedra which share a common oxygen atom designated O(1). While the oxygen atoms surrounding the titanium atoms form nearly regular octahedra, the titanium atoms are displaced from the geometrical centers of the octahedra, resulting in alternating long (1.974 A) and short (1.766 A) Ti-O(1) bonds. The four remaining Ti-O bond distances are longer than either of the Ti-O(1) bonds and range between 1.984 A and 2.025 A. The off-center displacement of the Ti is also reflected in the O-Ti-O(1) angles with the average of the four angles involving the shorter bond being wider than those involving the larger bond (Fig. 2). Two chains per unit cell are related by a center of symmetry, one chain having the octahedral titanium displaced in the +a direction from the geometrical center of the octahedra and the other chain having the titanium displaced in the −a direction. These chains are cross-linked by silicate tetrahedra sharing the remaining four oxygens [3-coordinated O(2) and O(2A) and 4-coordinated O(3) and O(3A)]. The silicate tetrahedra share oxygen atoms with four separate TiO_6 octahedral groups in three separate chains. The oxygen atoms O(3) and O(3A) belong to the same chain while O(2) and O(2A) are shared by two chains. This produces a $[TiOSiO_4]^{-2}$ framework with large cavities enclosing calcium atoms in irregular 7-coordination polyhedra. This coordination is based on a limiting Ca-O bond length of 3.0 A, which seems reasonable considering that the next closest cation, Si, is at a distance of 3.066 A.

"The Si-O bond length variations in the silicate ion are small, ranging from 1.641 A to 1.647 A, despite a range of $<O-Si-O>_3$ from 108.3° to 110.8° and a range of $\zeta(O)$ [formal Pauling bond strengths to oxygen] from 1.96 to 2.24. Moreover, neither of these parameters may be used to satisfactorily rationalize the small Si-O bond length variations that do occur in titanite."

The $P2_1/a \rightarrow A2/a$ Transformation at 220°C

In x-ray diffraction photographs of synthetic $P2_1/a$ titanite there is a class of reflections $k + \ell$ odd (those violating the A-centering condition) which are systematically weak, but sharp. Taylor and Brown (1976; see their Fig. 2) found that these reflections became weaker, but not streaked or diffuse, as the temperature of a single crystal was raised, gradually becoming unobservable at $\sim$220°C. They ascribe this nonquenchable $P2_1/a \rightarrow A2/a$ inversion to what is probably a second order "distortional transformation" in which the Ti atom is displaced by $\sim$0.01 A to the center of the $[TiO_6]$ octahedron (Fig. 1B) and the pseudo-symmetrically related members of the oxygen pairs O(2), O(2A) and O(3), O(3A) are slightly displaced and become related by a center of symmetry at the Ti site.

The $A2/a$ Structure of Malayaite at 25°C

By contrast with $P2_1/a$ CaTiO$[SiO_4]$ at room temperature, with its alternately long and short M-O(1) bond lengths along a, CaSnO$[SiO_4]$ has space group $A2/a$ and is in fact simply an expanded version of the structure of synthetic titanite above 220°C with Sn atoms centered in the octahedra. Other than expected differences between Ti-O and Sn-O bond lengths, most steric details of the two structures are very similar.[2] Figure 3 gives selected interatomic distances and angles. As judged by the M-O(1)-M angle, the corner-sharing octahedral chain is somewhat less kinked (133°) in malayaite than in titanite (141°).

Thus the point can be made that the effects both of heating titanite above $\sim$220°C and of substituting cations larger than Ti in the $[MO_6]$ octahedra are the same. Centrosymmetric octahedra and structures with $A2/a$ symmetry are produced with no evidence at the M site of positional disorder. Malayaite and high-temperature titanite are thus truly iso-structural, but they in turn are probably *not* isostructural with chemically impure titanites whose $A2/a$ symmetry arises by a different mechanism. Effects of the substitution of $(Al,Fe)^{3+} + (OH,F)^-$ for $Ti^{4+} + O(1)^{2-}$ can best be understood in terms of linear anti-phase domains [see section on Domain Texture below].

- - - - - - - - - - - - - - - -

[2]Crystal data for a malayaite CaTi$_{.05}$Sn$_{.95}$O$[SiO_4]$, from Devonshire, England: $Z = 4$, $a = 7.149$, $b = 8.906$, $c = 6.667$ A, $\beta = 113.3°$, volume = 389.9 A^3, calculated density = 4.485 g/cc. (Higgins and Ribbe, 1977a; *cf.* Eppler's (1976) values for synthetic CaSnO$[SiO_4]$.) Filippova and Bystrikov (1976) have synthezied Cr- and Fe-bearing malayaites for use as pigments in ceramics.

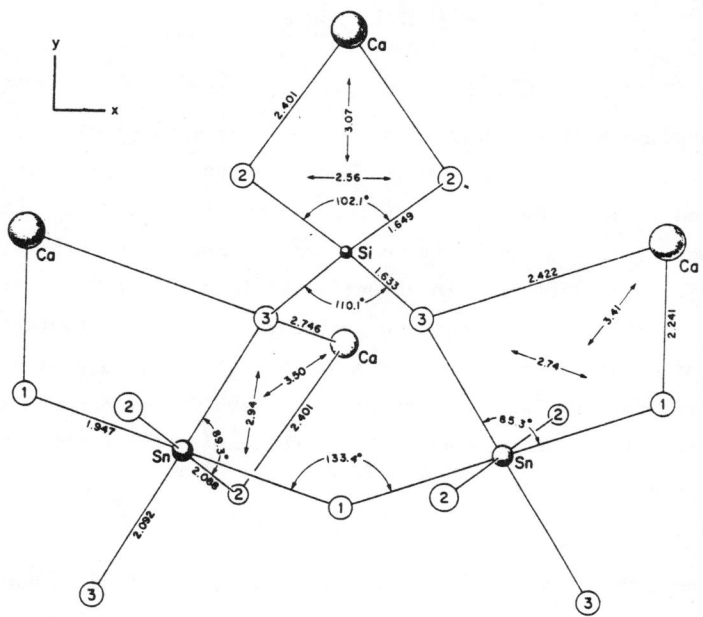

Figure 3. Projection of a portion of the malayaite structure onto (001) showing the linkage of two [SnO6] octahedra along [100]. The bridging [SiO4] tetrahedron and the edges shared between [CaO7] polyhedra and the [SiO4] tetrahedron and the [SnO6] octahedron are also illustrated. The oxygen atoms are shown as unshaded, labelled circles. All bond lengths and interatomic distances are in Angströms. After Higgins and Ribbe (1977, p. 804, Fig. 1).

Predicted Structures at High Pressures and High Temperatures

Dempsey and Strens (1976), using DLS (distance least-squares) methods, have calculated the structures of titanite at 25 kbar and at 1200°K, predicting decreases in mean Si^{IV}-O, Ti^{VI}-O and Ca^{VII}-O bond lengths of 0.003, 0.008 and 0.015 A, respectively, at 25 kbar and increases of 0.004, 0.017 and 0.027 A, respectively, at 1200°K. Remarkably, all three of the predictions at 1200°K are just 0.002 A less than the values extrapolated from the high-temperature refinements by Taylor and Brown (1976).

Nonsilicates Isostructural with Titanite and Malayaite

Isokite, $CaMgFPO_4$ (Povarennykh, 1972) and tilasite, $CaMgFAsO_4$ (Bladh et al., 1972), are isostructural with titanite and malayaite. In addition there are many nonsilicate "Isomer Type I" compounds which have 6.9 to 7.9 A repeats along the vertex-shared $[MO_5]^{\infty}$ octahedral chain and are structurally related to titanite, according to the classification scheme of Moore (1970). Speer and Gibbs (1976) and Higgins and Ribbe (1977b) enumerated these; Mayer and Völlenkle (1972) suggested that all are "stuffed derivatives" of the aristotype, $Ge(OH)PO_4$.

Immiscibility in the System $CaTiO[SiO_4]$ – $CaSnO[SiO_4]$

The only naturally occurring silicate isostructural with titanite is malayaite, and Takenouchi (1971) found complete Ti ⇄ Sn solid solution at 700°C and 7 kbar. He also found that below 615 ± 15°C at 1 kbar there is an asymmetric solvus with its peak near $Ti_{75}Sn_{25}$. The exsolution is attributed to the difference in effective ionic radii of Ti (r = 0.605 A) and Sn (r = 0.69 A). Takenouchi suggested that Sn-rich titanite and Ti-rich malayaite rarely occur in nature because the temperatures of formation of tin-bearing deposits are probably <500°C, but Ramdohr (1935) has reported a titanite in association with cassiterite containing ∿10 mole percent Sn.

Stoichiometry

Higgins and Ribbe (1976) surveyed the available chemical data and by comparing them to more modern analyses concluded that there is no significant evidence for non-stoichiometry in titanite. Most likely the analytical problems faced by Sahama (1946) were parallel to those he (1953) and others faced with F and OH in the humite minerals. The x-ray fluorescence analyses of Lee *et al.* (1969) are highly questionable (*cf.* Fig. 1 in Higgins and Ribbe, 1976).

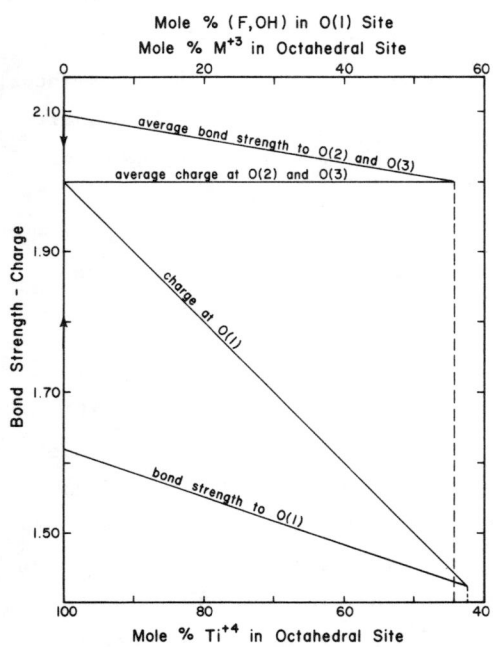

Figure 4. Change in Pauling bond strength and formal charge at the O(1), O(2), and O(3) anion sites with substitution of M^{3+} + (F,OH)⁻ for Ti^{4+} + O^{2-}. The average bond strength to the O(2) and O(3) sites reaches a value of 2.00 at almost the same composition at which the charge at the O(1) site is equal to the bond strength received at that site. Modified after Higgins and Ribbe (1977, p. 884, Fig. 2). Tips of arrows represent bond strengths for hypothetical $Na^+Nb^{5+}O[SiO_4]$.

Electrostatic Charge Balance and Chemical Substituents

As illustrated in Figure 2, the oxygen atoms in pure $CaTiO[SiO_4]$ are either underbonded ($\zeta_1 = 1.62$; $\zeta_2 = 1.95$) or overbonded ($\zeta_3 = 2.24$): O(1) is bonded to (2Ti + Ca), O(2) to (Si + Ti + Ca), and O(3) to (Si + Ti + 2Ca). The substitution of trivalent cations for Ti^{4+} and, concomitantly, monovalent OH and/or F for O(1) serves to decrease local electrostatic charge imbalance, as graphically presented in Figure 4. This may very well account for the extensive substitution (up to 30 mol %) of $(Al,Fe)^{3+} + (OH,F)$ in natural titanites.

The substitution $Na^+ + (Nb,Ta)^{5+} \rightleftarrows Ca^{2+} + Ti^{4+}$ is not an important one (Sahama, 1946), but it would have a similar charge-normalizing effect on the structure (see Fig. 4). In fact, wherever significant levels of Nb and Ta appear in titanite, the charge-balancing is accomplished by trivalent cations in the octahedral site: $2Ti^{4+} \rightleftarrows (Nb,Ta)^{5+} + (Al,Fe)^{3+}$ (Clark, 1974; Paul et al., 1981); where rare earths are dominant, the substitution is $2REE^{3+} + Fe^{2+} \rightleftarrows 2Ca^{2+} + Ti^{4+}$ (Exley, 1980).

Summary of Substituents in Titanite

Sahama (1946) thoroughly reviewed the chemistry and classification of titanite. Much of the subsequent discussion is based on his work, that of Zabavnikova (1957), who found no less than 40 elements in titanite, and that of Staatz et al. (1977), who reviewed much of the literature on rare earth elements (REE) and thorium in titanite.

There are two variety names that require comment: *keilhauite* is yttrotitanite (up to 12% $(Y,Ce)_2O_3$) and *grothite* has high $(Al + Fe)^{3+}$ in which REE's are comparatively low (Deer et al., 1962). These names should be abandoned in favor of chemically descriptive prefixes.

The 7-coordinated Ca polyhedron. Substituents for 7-coordinated Ca include alkaline earths (Sr usually exceeds Ba) and alkalis (Na < 0.03 mol %; K is questionable, although occasionally reported [Smith, 1970]). Mn ranges up to 5 mol % (Roy, 1974). The REE's rarely exceed 5%, but Exley (1980) found levels as high as 87% replacement of Ca in unusual titanites of the Skye granites: Ce is dominant by a factor of two over La and 3 to 4 over Nd; Sm/Nd = 0.10 - 0.13; Ce/Y = 40 - 104. Indeed, in most specimens REE's are predominantly Ce and Nd, but Y may range as high as the total lanthanide concentration.[3] The lanthanide distributions in titanites usually do not mimic the crustal abundance profile. In ion microprobe studies of titanites from many localities, Higgins and Ribbe (1976) found REE to be ubiquitous, though often at trace concentrations. They also reported Th, U and radiogenic Pb in all but two specimens (both were from pegmatites). These elements undoubtedly occupy the ^{VII}Ca site in titanite: thorium is always predominant, though at concentrations less than 3000 ppm (Hurley and Fairbairn, 1957), even in a totally metamict titanite (Černý and Povondra, 1972). Staatz *et al.* (1977, p. 623) made the point that "sphene in rocks containing either allanite or zircon has a lower thorium content than in rocks that do not contain allanite or zircon;" the same could be said for uranium. Because of its radiogenic element content, titanite may be used in fission-track age determinations (Naeser, 1967; Naeser and Faul, 1969).

The underbonded O(1) site. As mentioned earlier, substitution of OH, Fe and Cl (when present) undoubtedly takes place at the underbonded O(1) site[4] (*cf.* Fig. 3). Jaffe (1947) found F, at least in trace amounts, to be present in most titanites he analyzed, although OH is more abundant. Beran (1970) located the hydrogen atom in an OH-bearing titanite from Minas Gerais, Brazil, using infrared spectroscopy. The H-O(1) bond is in the (010) plane nearly normal to [100] and parallel to the principal optical vibration axis X.

The Si site. No substituents for Si have been reported in natural titanite other than Al, which is commonly assigned to tetrahedral coordination in order to balance the structural formula. This practice is for the most part unjustified. Careful analyses, especially using well-calibrated microprobe techniques, result in values within a percent or two of the expected stoichiometry for Si. On the basis of changes in d_{033}-spacings, Rosenberg (1974) suggested that synthetic "sphene tends

- - - - - - - - -

[3] Fleischer (1978) surveyed the relative proportions of lanthanide and yttrium by rock type and concluded that geologic environment has an important effect on these, although analyses of individual titanites are *not* sufficient to identify the host rocks.

[4] The distribution of F between coexisting titanite and biotite was determined by Ekström (1972) in three apatite-bearing iron ores.

to hydroxylate progressively with decreasing temperature by the substitutions $\square + 4H^+ = Si^{4+}$ and $Al^{3+} + H^+ = Ti^{4+}$, the latter being
favored and more extensive."

The Ti octahedron.   The REE^{3+}, Th^{4+} and U^{4+} cations are presumed to be
located in the ^{VII}Ca site, but they must be charge-balanced elsewhere;
Na^+ concentrations are too low. The $[TiO_6]$ octahedron is the only other
site in which such charge-balancing can take place. Chromium is an occasional minor substituent there (Sahama, 1946; Jaffe, 1947). Only small
amounts of Fe^{2+} are reported in most chemical analyses, but stoichiometric
balance requires that in the REE-rich titanites of the Skye granites the
substitution is $Ca_2^{2+} + Ti_n^{4+} \rightleftarrows REE_{2n}^{2+} + Fe_n^{2+}$ with n as high as 0.46 (Exley,
1980). Divalent Cu and Mg are usually of negligible concentration but
Exley found up to 0.035 Mg in REE, Fe^{2+}-rich titanites. As mentioned
earlier, Nb^{5+} (up to 0.1 atoms), Ta^{5+} (up to 0.15 atoms), and traces of
V substitute for Ti^{4+} and are not charge-balanced by Na^+ in the Ca site
but by $(Al,Fe)^{3+}$ in the Ti site.

In summary, the most common coupled substitution in "normal" titanite
is $(Al,Fe)^{3+} + O^{2-} \rightleftarrows Ti^{4+} + (OH,F)^-$, in which Al is usually predominant
and (Al + Fe) may approach 30 % of the octahedral cations. This has
structural implications which are manifested in ways discussed below.

Lattice Parameters

The ranges in lattice parameters observed for natural titanites are
a = 7.039 - 7.088 A; b = 8.643 - 8.740 A; c = 6.527 - 6.584 A; β =
113.74 - 114.15°. These small variations are consistent with the predominant, partially size-compensating substitution, (Al + Fe) → Ti (where
$r_{Al} < r_{Ti} < r_{Fe^{3+}}$ = 0.53 < 0.605 < 0.645 A). Higgins and Ribbe (1976)
found that cell dimensions increase linearly with increasing effective
octahedral cation radius (EOCR), even though other substituents are
ignored (see Fig. 5). EOCR = 0.605 - (0.075Al/Si) + (0.04Fe/Si) where
0.605 is the radius of Ti, Al/Si and Fe/Si are the atomic proportions of Al
and Fe normalized to one Si, and 0.075 and 0.04 are the respective differences between the radii of Al and Ti and Fe and Ti (radii from Shannon
and Prewitt, 1970).

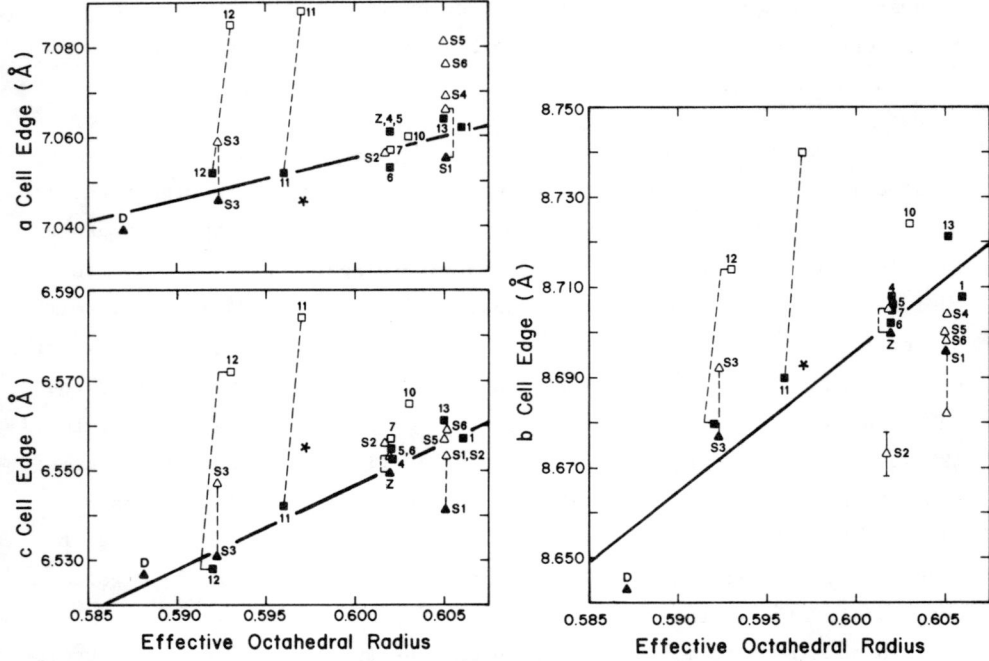

Figure 5. Variation of effective octahedral cation radius with a, b, and c cell edge. Triangles represent data from Cerny and Sanseverino (1972); squares are data from Higgins and Ribbe (1976). Open symbols indicate unheated specimens and filled symbols represent heated specimens. The regression lines were calculated using only data from heated specimens. After Higgins and Ribbe (1976, p. 884, Fig. 3). * is data from Paul et al. (1981).

Metamict Titanites

Metamictization of zircon is evidenced by decreased density and increased cell dimensions, and in severely radiation-damaged specimens x-ray diffraction peaks are diffuse and sometimes unmeasurable (see chapter on zircon in this volume). The same is true for titanite which commonly contains Th and U and radiogenic Pb. Cerny and Povondra (1972) described a completely metamict Al,F-rich titanite which, when heated, recrystallized and gave lattice parameters indicated by point D in Figure 5. Specimens 11 and 12 (Higgins and Ribbe, 1976) had expanded cells and significant Th-U levels, and when these were heated for 3 hours at 1100°C their lattice parameters helped define the curves of Figure 5. Careful microprobe analyses of one of these partially metamict specimens and its heated equivalents showed small (3 – 5%) increases in the weight percent of each oxide species after heating. It is possible that the higher density and oxide total was due to loss of nonstructural H_2O.

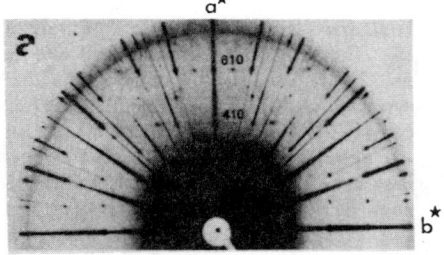

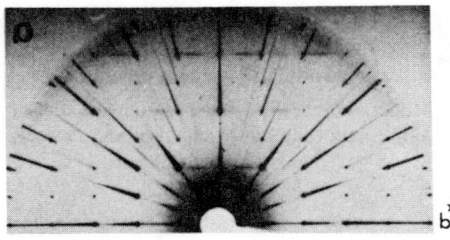

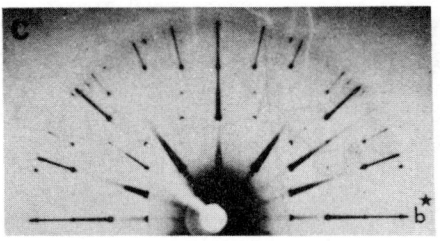

Figure 6. Zero-level $a*b*$ precession photographs of (1) specimen containing 3.1 mol % (Fe + Al) with sharp $k + l$ odd reflections (610,410) indicating a primitive lattice (space group $P2_1/a$), (2) specimen with 6.5 mol % (Fe + Al) showing diffuse $k + l$ odd reflections, and (3) specimen with 25.9 mol % (Fe + Al) showing no $k + l$ odd reflections or diffuse streaks indicating a centered lattice (space group $A2/a$). After Higgins and Ribbe (1976, p. 886, Fig. 5).

Domain Textures

Observations regarding domains in titanite began with Robbins (1968), but Speer and Gibbs (1976), Taylor and Brown (1976) and Higgins and Ribbe (1976) ultimately provided the theory, evidence, and explanation for the linear domains observed in natural titanites. In synthetic $P2_1/a$ titanite below $\sim$220°C the class of reflections $k + l$ odd, which violate the A-centering condition, are systematically weak but sharp. As the concentration of (Al + Fe) + (OH,F) for Ti + O increases (inset), it becomes obvious in x-ray precession photographs (Fig. 6) that the $k + l$ odd reflections become progressively

Mole percent		
Fe	Al	$\underline{k} + \underline{l}$ odd
0.0	0.0	sharp
2.8	0.3	sharp
2.0	4.5	diffuse
2.5	5.6	diffuse
3.6	2.8	diffuse
3.6	2.5	diffuse
7.1	18.8	absent
7.0	14.6	absent

more diffuse, eventually becoming unobservable above $\sim$ 20 mol % (Al + Fe). The diffuseness is

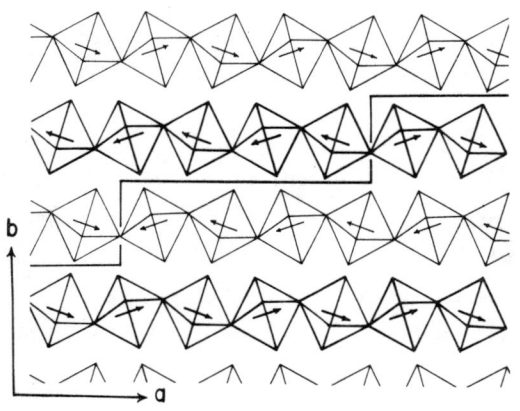

Figure 7. Domain model of titanite showing chains of Ti-octahedra parallel to a. Chains with dark and light outlines represent those above and below the plane of the drawing. Arrows represent directions of displacement of the Ti-cations. On either side of the heavy line through the figure are domains of opposite orientation ("anti-phase"). Two types of O(1) sites are located at the vertical domain boundaries: those with both Ti-cations displaced away from O(1) (upper right) and those with both displaced towards O(1) (center). After Taylor and Brown (1976, p. 446, Fig. 4).

147

localized with nodes on (100) planes in reciprocal space (cf. Fig. 6 and electron diffraction patterns in Higgins and Ribbe, 1976, Fig. 6), indicating that the antiphase domains which form are disordered, one-dimensional, and parallel to [100].

A domain model is schematically presented in Figure 7. Linear disorder in the [100] chains of vertex-sharing octahedra is introduced by displacement of the octahedral cation from its normal off-center position (*cf.* Figs. 1 and 2) to a similar position on the opposite side of the geometric center of the octahedron. If the normal sequence of M-O bonds parallel to [100] is long-short-long-short (L-S-L-S), an antiphase domain is introduced by a mistake: ...L-S-L-S|S-L-SL... The presumption is that the substitution of Al ± Fe for Ti in the chain and the accompanying substitution of OH,F at O(1) (which is on the domain boundary) cause the shift in M position. Domains on op-posite sides of a vertical boundary in Figure 7 are related by two-fold axes parallel to [010] which, when added to the $P2_1/a$ space group gives the apparent $A2/a$ space group of impure natural titanites. [See Speer and Gibbs (1976, p. 242) for full discussion.]

PHYSICAL PROPERTIES AND TWINNING

Electric Properties

"The off-center displacement of Ti in an octahedron is character-istic of oxygen-octahedral ferroelectrics. The antiparallel displace-ments of the titanium atoms in adjacent chains [which are related by 2_1 or 2-fold axes in titanite or malayaite]...would compensate the electric moments produced by the ionic displacements, resulting in a net zero electric polarization for the unit cell. This state for a dielectric crystal is called antiferroelectric" (Speer and Gibbs, 1976, p. 242).

Optical Properties

Titanite ranges from colorless to yellow, green, red-brown or black, depending mainly on iron and REE content. It is usually yellow or brown in thin section and moderately to weakly pleochroic in these colors, with absorption $\alpha < \beta < \gamma$, unless REE concentration is high (Exley, 1980). Refractive indices lie in the ranges $\alpha = 1.843 - 1.950$, $\beta = 1.870 - 2.034$, $\gamma = 1.943 - 2.110$ with birefringence $\delta = 0.100 - 0.192$, $2V_\gamma = 17° - 40°$, and optic axis dispersion $r > v$, strong (all data from Deer *et al.*, 1962, pp. 69, 74).

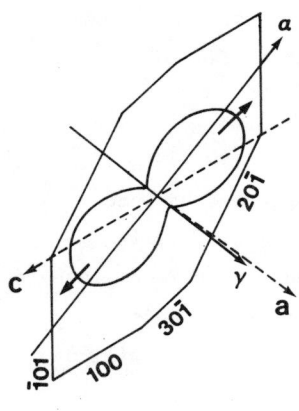

The orientation of the optical indicatrix in titanite is shown in the inset. The γ refractive index is nearly parallel to [100], the octahedral chain axis (Ribbe, 1976). In OH-bearing specimens α is nearly parallel to the H-O(1) bond directions as indicated by the opposite-pointing arrows intersecting the absorption figure for the OH-stretching frequency at λ = 2.90μm (from Beran, 1970 -- axial orientation and crystal forms relabeled to conform to Speer and Gibbs (1976) choice of unit cell). The complexity of the compositions of natural titanites precludes using optical properties in a chemically determinative way.

Cleavage, Hardness, Density

Titanite has good {011} cleavage which, not surprisingly, is parallel to the octahedral chains. It is relatively soft (H = 5 to 5.5), possibly due to the fact that the octahedra do not share edges, and the structure is relatively open (volume per anion = 18.5 A^3, *vs* 17.3 A^3 for grossular). The calculated density is 3.52 g/cc for pure CaTiO[SiO_4], ranging to higher or lower values depending on chemical composition and metamict state.

Thermal Expansion

The lattice parameters of synthetic titanite increase with temperature except for a slight, but barely significant decrease in *a* and *c* between 165° and 270°C; unit cell volume is the same at these temperatures on either side of the transition from $P2_1/a \rightarrow A2/a$ at $\sim$220°C. Taylor and Brown (1976) make no comment on these their observations. They do, however, note that "changes in polyhedral volumes over the temperature range 25°C to 740°C are -0.31, 1.52, and 2.05 percent, respectively for the SiO_4 tetrahedron, the TiO_6 octahedron, and the CaO_7 polyhedron. The negligible changes in tetrahedral volume and mean Si-O distance with temperature are consistent with those found in other silicates studied at high temperature. The relative volume and bond length changes among the three polyhedra support the commonly held notion that bond strengths decrease in the series Si^{IV}-O, Ti^{VI}-O, and Ca^{VII}-O" (p. 441).

Twinning

In order to preserve the axial orientation suggested by Donnay and Ondik (1973) for titanite and used exclusively herein, it is necessary to transform $\{hkl\}$ notation for twins indexed on Zachariasen's (1930) $C2/c$ cell by the matrix $[\bar{1}0\bar{1}/010/100]$. However, because the literature on twinning refers only to the Zachariasen cell, I will use untransformed indices for twin planes and axes with a subscript z to help avoid confusion, *e.g.*, $\{221\}_z$.

Growth twins. Single twins with twin plane $\{100\}_z$ are fairly commonly observed in titanite. $\{100\}_z$ is also a dominant crystal face, which along with $\{001\}_z$, $\{\bar{1}01\}_z$ and $\{\bar{1}02\}_z$ give titanite its familiar wedge-shaped habit and the name *sphene* (from the Greek for *wedge*). See inset under Optical Properties for morphology and axial orientation relabelled according to the Donnay and Ondik cell.

Mechanical twins. Mügge (1889) considered the possibility that the irrational compositional planes of polysynthetic twins near $(221)_z$ and $(2\bar{1}1)_z$ were due to mechanical twin gliding. In an investigation of polysynthetically twinned titanite from granite shocked in nuclear explosions at the Nevada Test Site, Borg (1970) found two sets of composition planes at ∿55° to each other developed within individual grains. She deduced the twin glide system to be that predicted by Mügge (1889): K_1 = irrational slip plane, near $\{221\}$; $K_2 = \{\bar{1}31\}_z$ (the other plane of no distortion); $N_1 = <110>_z$, the "slip line" or twin axis about which twin and host are related by 180° rotation; and the magnitude of shear, $s = 0.60$. She reports (p. 1887) that "the cation network can be almost perfectly restored by homogeneous shear, but additional movements or shuffles are necessary in order to restore the oxygen atoms to their proper positions about the cations. Either Ti or Si polyhedra apparently rotate as discreet units during the twin process."

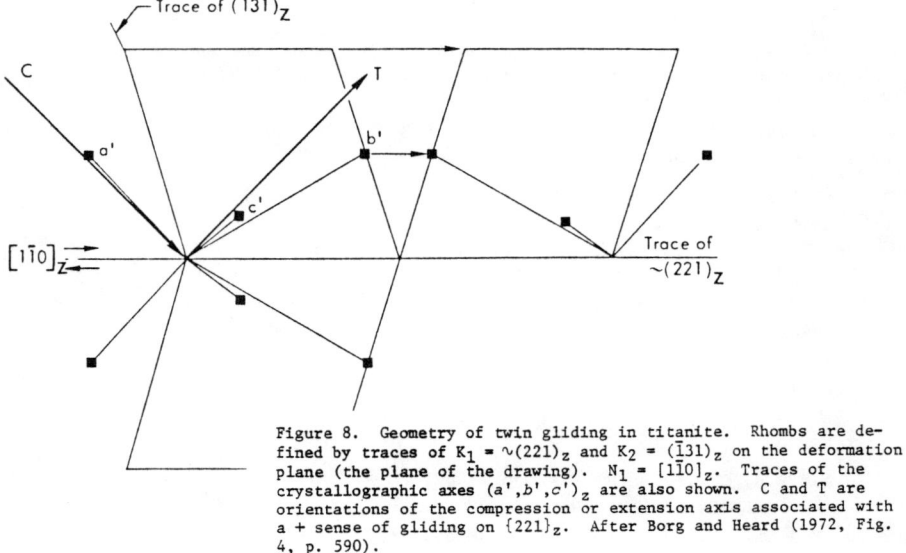

Figure 8. Geometry of twin gliding in titanite. Rhombs are defined by traces of $K_1 = \sim(221)_z$ and $K_2 = (\bar{1}31)_z$ on the deformation plane (the plane of the drawing). $N_1 = [1\bar{1}0]_z$. Traces of the crystallographic axes $(a',b',c')_z$ are also shown. C and T are orientations of the compression or extension axis associated with a + sense of gliding on $\{221\}_z$. After Borg and Heard (1972, Fig. 4, p. 590).

In a later laboratory study of mechanical twinning at 8 kbar and temperatures ranging from 25°-500°C, Borg and Heard (1972) found that the same twin glide system operates, and they learned that increasing the temperature of deformation experiment by 500°C lowers the critical resolved shear stress for twinning by $\sim$50 percent to 1.3 kbar. The fact that $<110>_z$ mechanical twinning can be induced in only one direction makes primary titanite a potentially valuable tool for deducing principal stress directions in granitic rocks during deformation. The geometric elements of twin gliding in titanite are shown in Figure 8.

The morphologies of titanite grown from aqueous supercritical solutions at 2 kbar are illustrated below (from Franke and Ghobarkar, 1980). At temperatures below 600°C, twinning on (100) was commonly found. See Goldschmidt (1922) for a compilation of drawings of titanite in its multitudinous habits.

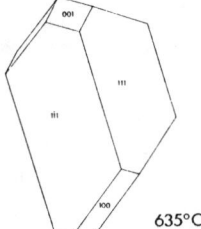

635°C

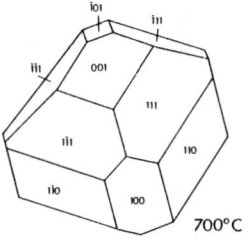

700°C

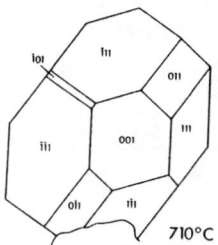

710°C

TITANITE: REFERENCES

Beran, A. (1970) Measurement of infra-red pleochroism in minerals IX. The pleochorism of OH-streaking in titanite. *Tschermaks Mineral. Petrogr. Mitt., 14*, 1-5.

Bladh, K.W., R.K. Corbett, W.J. McLean and R.B. Laughon (1972) The crystal structure of tilasite. *Am. Mineral., 57*, 1880-1884.

Borg, I.Y. (1970) Mechanical <110> twinning in shocked sphene. *Am. Mineral., 55*, 1876-1888.

────── and H.C. Heard (1972) Mechanical twinning in sphene at 8 kbar, 25° to 500°C. *Geol. Soc. Am. Memoir 132*, 585-591.

Cerny, P. and P. Povondra (1972) An Al,F-rich metamict titanite from Czechoslovakia. *Neues Jahrb. Mineral. Monatsh.*, 400-406.

────── and L.R. DiSanseverino (1972) Comments on the crystal chemistry of titanite. *Neues Jahrb. Mineral. Monatsh.*, 97-103.

Clark, A.M. (1974) A tantalum-rich variety of sphene. *Mineral. Mag., 39*, 605-607.

Deer, W.A., R.A. Howie and J. Zussman (1962) *Rock-Forming Minerals,* Vol. 1. Longmans, Green and Co., London, pp. 69-76.

Dempsey, M.J. and R.G.J. Strens (1976) Modelling crystal structures. In, R.G.J. Strens, ed., *The Physics and Chemistry of Minerals and Rocks,* John Wiley and Sons, New York, pp. 441-458.

Donnay, J.D.H. and H.M. Ondik (1973) *Crystal Data, Determinative Tables Volume 2.*

Ekström, T.K. (1972) The distribution of fluorine among some coexisting minerals. *Contrib. Mineral. Petrol., 34*, 192-200.

Eppler, R.A. (1976) Lattice parameters of tin sphene. *J. Am. Ceram. Soc., 59*, 455.

Exley, R.A. (198) Microprobe studies of REE-rich accessory minerals: implication for Skye granite petrogenesis and REE mobility in hydrothermal systems. *Earth Planet. Sci. Letters, 48*, 97-110.

Filippova, E.A. and A.S. Bystrikov (1976) Synthesis of ceramic iron-containing pigments based on tin sphene. *Izv. Akad. Nauk SSSR., 12*, 2241-2244.

Fleischer, M. (1978) Relation of the relative concentrations of lanthanides in titanite to type of host rocks. *Am. Mineral., 63*, 869-873.

Higgins, J.B. and P.H. Ribbe (1976) The crystal chemistry and space groups of natural and synthetic titanites. *Am. Mineral., 61*, 878-888.

────── and ────── (1977a) The structure of malayaite, $CaSnOSiO_4$, a tin analog of titanite. *Am. Mineral., 62*, 801-806.

────── and ────── (1977b) The classification of vertex-sharing octahedral chain structures. *Neues Jahrb. Mineral. Monatsh.*, 310-319.

Hunt, J.A. and D.M. Kerrick (1977) The stability of sphene; experimental redetermination and geologic implications. *Geochim. Cosmochim. Acta, 41*, 279-288.

Hurley, P.M. and H.W. Fairbairn (1957) Abundance and distribution of uranium and thorium in zircon, sphene, apatite, epidote, and monazite in granitic rocks. *Am. Geophys. Union Trans., 38*, 939-944.

Jaffe, H.W. (1947) Reexamination of sphene (titanite). *Am. Mineral., 32*, 637-642.

Lee, D.E., R.E. Mays, R.E. Van Loenen and H.J. Rose, Jr. (1969) Accessory sphene from hybrid rocks of the Mount Wheeler mine area, Nevada. *U.S. Geol. Surv. Prof. Paper 650-B*, B41-B46.

Mayer, H. and H. Völlenkle (1972) Die Kristallstruktur und Fehlordnung von Ge(OH)PO$_4$. *Z. Kristallogr., 136*, 387-401.

Mongiorgi, R. and L.R. Di Sanseverino (1968) A reconsideration of the structure of titanite, CaTiOSiO$_4$. *Mineral. Petrogr. Acta, 14*, 123-141.

Moore, P.B. (1970) Structural hierarchies among minerals containing octahedrally coordinating oxygen. I. Stereoisomerism among corner-sharing octahedral and tetrahedral chains. *Neues Jahrb. Mineral. Montash.*, 163-173.

Mügge, O. (1889) Ueber durch Druck enstandene Zwillinge von Titanit nach den Kanten [110] und [1$\bar{1}$0]. *Neues Jahrb. Mineral. Geol. Palaontol., 2*, 98-115.

Naeser, C.W. (1967) The use of apatite and sphene for fission track age determinations. *Geol. Soc. Am. Bull., 78*, 1523-1526.

————— and H. Faul (1969) Fission track annealing in apatite and sphene. *J. Geophys. Res., 74*, 705-710.

Nevskii, N.N., V.V. Ilyukhin and N.V. Belov (1979) The crystal structure of CaGe$_2$O$_5$ - a germanium analog of sphene. *Dokl. Akad. Nauk SSSR, 246*, 1123-1126.

Paul, B.J., P. Cerný, R. Chapman and J.R. Hinthorne (1981) Niobian titanite from the Huron Claim pegmatite, southeastern Manitoba. *Canadian Mineral., 19*, 549-552.

Povarennykh, A.S. (1972) *Crystal Chemical Classification of Minerals*, 2 vols. Plenum Press, New York, 766 p.

Ramdohr, P. (1935) Ein Zinnvorkommen im Marmor bei Arandis, Deutsch Südwestafrika. *Neues Jahrb. Min., Abt. A., 70*, p. 1.

Ribbe, P.H. (1976) Polyhedral chains in the structures of gem orthosilicates: relationships to physical properties (abstr.). 25th Ann. I. Geol. Congr. *Abstracts, 2*, 593-594.

Robbins, C.R. (1968) Synthetic CaTiSiO$_5$ and its germanium analogue (CaTiGeO5). *Mat. Res. Bull., 3*, 693-698.

Rosenberg, P.E. (1974) Compositional variations in synthetic sphene (abstr.). Geol. Soc. Am. *Abstr. with Progr., 6*, 1060.

Roy, S. (1974) Manganoan sphene from Garra Balaghat District, Madhya Pradesh, India. *Acta Minera.-Petrogr., 21*, 275-276.

Sahama, T.G. (1946) On the chemistry of the mineral titanite. *Bull. Comm. Geol. Finlande, 24*, 88-118.

————— (1953) Mineralogy of the humite group. *Ann. Acad. Sci. Fennicae, III. Geol. Geogr., 31*, 1-50.

Shannon, R.D. (1976) Revised effective ionic radii and systematic studies of interatomic distances in halides and chalcogenides. *Acta Crystallogr., A32*, 751.

————— and C.T. Prewitt (1970) Revised values of effective ionic radii. *Acta Crystallogr., B26*, 1046-1048.

Smith, A.L. (1970) Sphene, perovskite and coexisting Fe-Ti oxide minerals. *Am. Mineral., 55*, 264-269.

Speer, J.A. and G.V. Gibbs (1976) The crystal structure of synthetic titanite, CaTiOSiO4, and the domain textures of natural titanites. *Am. Mineral., 61*, 238-247.

Staatz, M.H., N.M. Conklin and I.K. Brownfield (1977) Rare earths, thorium, and other minor elements in sphene from some plutonic rocks in West-Central Alaska. *J. Res.*, *5*, 623-628.

Takenouchi, S. (1971) Hydrothermal synthesis and consideration of the genesis of malayaite. *Mineral. Deposita*, *6*, 335-347.

Taylor, M. and G.E. Brown (1976) High-temperature structural study of the $P2_1/a \rightleftarrows A2/a$ phase transition in synthetic titanite, $CaTiSiO_5$. *Am. Mineral.*, *61*, 435-447.

Zabavnikova, I.I. (1957) Diadochic substitutions in sphene. *Geochem.*, 271-278.

Zachariasen, W.H. (1930) The crystal structure of titanite. *Z. Kristallogr.*, *73*, 7-16.

Chapter 6

CHLORITOID   P. H. Ribbe

INTRODUCTION

Chloritoid is found most commonly in regionally metamorphosed rocks, but it occurs in contact metamorphic aureoles, hydrothermal deposits and emery deposits as well. Halferdahl (1961) lists over a hundred localities. His detailed review and evaluation of the literature on chloritoid is a valuable historical document, and in it he gave evidence that most chemical analyses of chloritoid have been biased by a wide variety of inclusions, accounting for most cations other than Fe, Mg, Mn, Al and Si. He concludes that its ideal composition is $FeO \cdot Al_2O_3 \cdot SiO_2 \cdot H_2O$. The structure of chloritoid consists of two types of octahedral layers (Fig. 1a,b), one trioctahedral whose formula is $[Fe_2^{2+}AlO_2(OH)_4]^{-1}$ with Mg and Mn substituting for Fe^{2+} and Fe^{3+} for Al;[1] the other has composition $[Al_3O_8]^{-7}$ with three-fourths of the octahedral sites filled apparently only with Al (cf. Hålenius et al., 1981). These layers are bound together by isolated $[SiO_4]$ tetrahedra whose basal planes are located over the empty octahedral sites in the $[Al_3O_8]^{-7}$ layer and whose apical oxygens are part of the trioctahedral layer (Fig. 1c,d). The two tetrahedral layers are related by centers of symmetry. Although it does not convey the structural details completely, the formula could be written

$$[(Fe^{2+},Mg,Mn)_2(Al,Fe^{3+})O_2(OH)_4][Al_3O_8][Si]_2.$$

The layer-like nature of chloritoid produces a perfect, mica-like {001} cleavage. It is pseudo-hexagonal with good {110} cleavage, and it exhibits the same sort of polytypism that is characteristic of micas.

The variety of optic orientations of intergrown triclinic and monoclinic chloritoid (Hietanen, 1951) contained the clues to its polytypism, but full disclosure awaited the advent of high resolution transmission electron microscopy. Recently Jefferson and Thomas (1977, 1978) have observed three polytypes in "non-random disorder" in a number of chloritoid crystals, and they suggest that crystals free of stacking

- - - - - - - - - - - - - - - - - -

[1] Halferdahl (1961) suggested that the name *magnesium chloritoid* be used for the Mg-end member, but since Fe is always greater than Mg, the prefix *magnesian* should suffice. *Ottrelite* is the suggested name for the Mn-end member. A specimen from Ottré, Belgium contains $Mn_{.88}Fe_{.58}^{2+}Mg_{.54}$ in the two *M(1B)* sites (Fransolet, 1978).

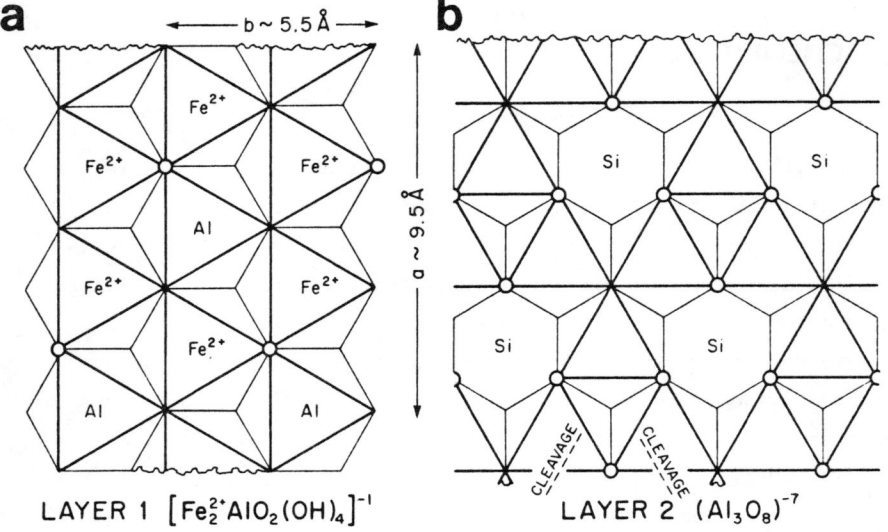

LAYER 1 $[Fe_2^{2+}AlO_2(OH)_4]^{-1}$ LAYER 2 $(Al_3O_8)^{-7}$

Al is in $M(1A)$ sites, Fe^{2+} in $M(1B)$ sites.

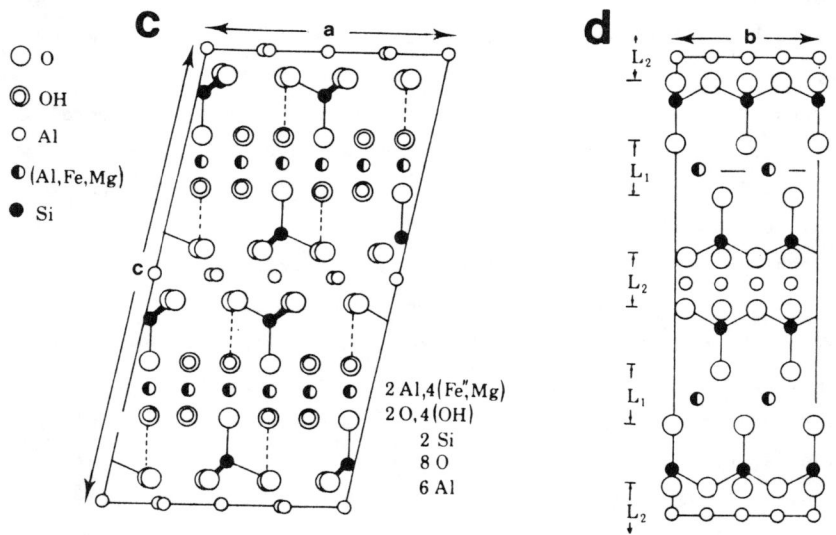

○ O
◎ OH
○ Al
◗ (Al, Fe, Mg)
● Si

2 Al, 4 (Fe″, Mg)
2 O, 4 (OH)
2 Si
8 O
6 Al

Figure 1. Schematic views of the crystal structure of chloritoid. (a) Idealized projection on (001) of the trioctahedral layer, L_1, showing the ordered Fe^{2+}/Al^{3+} cation distribution. Mg and Mn presumably substitute for Fe^{2+} in this layer. (b) Idealized projection on (001) of the closest-packed layer, L_2, in which 3/4ths of the octahedral sites are occupied by Al. The (110) and (1$\bar{1}$0) cleavage traces are shown; presumably (100) is equally probable. The three open circles at alternate vertices of the open hexagons labeled "Si" are the basal oxygens of the Si tetrahedra which bond layer 2 to layer 1. The vertices of these tetrahedra are the open circles shown in (a). (c) The $2M_2$ structure projected along [010] and (d) along [100] (after Harrison and Brindley, 1957). The (001) cleavage is perfect.

faults may be rare indeed. As the basis for our discussion of poly-typism, crystal chemistry and physical properties of chloritoid, we must examine its crystal structures in detail.

CRYSTAL STRUCTURES

Brindley and Harrison (1952) solved the structure of a monoclinic chloritoid by one-dimensional Fourier synthesis, and Harrison and Brindley (1957) published a refinement which was described in general terms in the INTRODUCTION and is pictured schematically in Figures 1c and 1d. This turned out to be the $2M_2$ polytype, and although the 1Tc structure is the simplest, we discuss $2M_2$ first because it is the most common and the most precisely characterized.

Monoclinic Chloritoid

Recently Hanscom (1975) reported full-matrix least-squares refinements of two very similar $C2/c$ chloritoids in which he was able to prove by site-refinement techniques that Mg substitutes for Fe^{2+} and Fe^{3+} for Al^{3+} only in Layer 1 (L_1 in Fig. 1). The latter observation contradicts that of Faye et al. (1968) who found no evidence of electron transfer between the two oxidation states of iron in polarized optical absorption spectra and concluded that Fe^{2+} and Fe^{3+} must occur in octa-hedra that do not share edges, i.e., Fe^{3+} must be in the $[Al_3O_8]^{-7}$ layer, L_2. But in Hanscom's specimens only Al occupies the edge-sharing octa-hedra in L_2, and he argues that the very restricted size of the octa-hedra ($<M-O> < 1.9$ A) prevents the substitution of larger cations. Tricker et al. (1978) confirmed by Mössbauer spectroscopy that there is only one type of octahedral site (in L_1) occupied by Fe^{2+}, but they state (p. 174) that "the Fe^{3+} ions are present to such a small extent that it did not prove feasible to evaluate their preferred siting."

In L_2 each $[AlO_6]$ octahedron shares four edges with other octa-hedra, and unpolymerized $[SiO_4]$ tetrahedra sit above the unoccupied octahedral sites, sharing corners with octahedra in L_1 as well as L_2. In Figures 1a and 1b open circles mark the apical oxygens in L_1 and those in the basal planes of the $[SiO_4]$ groups in L_2. Note the pseudo-hexagonal geometry of the closest-packed anion and cation arrays. Layer 1 is like the "brucite" layer in trioctahedral micas, but with

Table 1. Simple possible polytypic variants of chloritoid based on combinations of the six principal displacement vectors shown in the inset (after Jefferson and Thomas, 1978) and lattice parameters for three polytypes, 1Tc, $2M_1$ and $2M_2$.

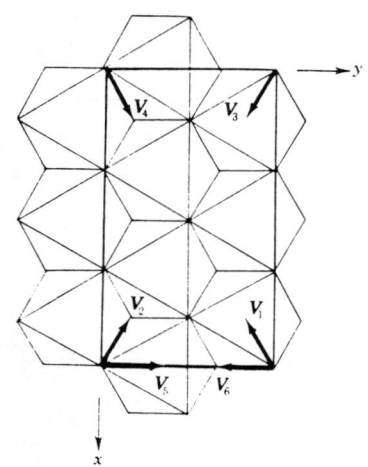

[001] projection of L_1 octahedral layer showing the six displacement vectors.

CONTINUOUS VECTOR SEQUENCE	NOTATION
Any one vector, *e.g.*, $V_1V_1V_1$...	1Tc*
Any combination V_1V_2, V_5V_2, V_3V_4..., *etc.*	$2M_2$
Any combination V_1V_4, V_2V_3, or V_5V_6...	$2M_2^{\frac{1}{}}$**
Any combination V_1V_3, V_2V_4, V_1V_5..., *etc.*	$2M_1$
Either $V_1V_3V_5$ or $V_2V_4V_6$...	3T
$V_1V_2V_3V_4V_5V_6$...	6H

- - - - - - - - -

* Comparable to 1*M* mica.

** Comparable to 2*O* mica.

LATTICES PARAMETERS* for three Fe-rich polytypes and ottrelite ([$Mn_{.88}Fe^{2+}_{.58}Mg_{.54}$] in *M(1B)*).

Chloritoid polytype (Ref.)	Lattice parameters						Z	Space group
	a(A)	b(A)	c(A)	α(°)	β(°)	γ(°)		
1Tc (Hanscom, 1980)	9.46	5.50	9.15	97.1	101.6	90.1	4	$C\bar{1}$
†$2M_2$ (Hanscom, 1973)	9.482	5.485	18.180	90	101.8	90	8	$C2/c$
$2M_1$ (Jefferson & Thomas, 1978	5.47	9.47	17.90	90	97.4	90	8	$C2/c$
- - - - - - - - - - -								
M&T Ottrelite (Fransolet, 1978)	9.505	5.484	18.214	90	101.8	90	8	[mix]

*See Hålenius *et al.* (1981, Table 1) for a compilation of analyses and lattice parameters for six chloritoids, including Fransolet's (1978) ottrelite.

†Averages for four structures -- all four observed values were within one part in one thousand for a, b, c and β.

composition $[Fe_2^{2+}AlO_2(OH)_4]^{-1}$. The anion monolayers of L_1 are nearly ideally closest packed (as in L_2), but octahedra are rotated 30° relative to those in L_2, and the anion density is only two-thirds that in L_2. Thus the tetrahedrally coordinated Si atoms are not sandwiched between closest packed layers in register. Perhaps this explains the fact that although the $[SiO_4]$ tetrahedron is quite regular, the mean Si-O distance (1.642 A) is slightly larger than that observed in sheet silicates or most other orthosilicates. Weak hydrogen bonds are formed between OH in L_1 and the oxygens not bonded to Si in L_2 (Hanscom, 1973; 1975); the O-H$\cdots$O angle is 163.5° and the shortest OH$\cdots$O interlayer distance is 2.66 A.

As mentioned earlier, the structures described by Hanscom (1973, 1975) are $2M_2$ polytypes, but there are others.

Possible Polytypes of Chloritoid

By virtue of the pseudo-hexagonal geometry of the octahedral and tetrahedral layers, stacking faults similar to those described for micas (Ross *et al.*, 1966) are common also in chloritoid. There are six possible principal displacement vectors which relate adjacent layers of ∿9.1 A thickness along [001]. They are illustrated on the sketch of layer L_1 in Table 1. Six of the many possible polytypic variants that may be derived by different combinations of these vectors are also listed there (after Jefferson and Thomas, 1978) along with lattice parameters of the three polytypes isolated thus far, namely $2M_2$ (discussed above), $2M_1$ (structure not refined), and 1Tc (the only one-layer structure). The 4-layer polytype described from x-ray precession photographs (Hanscom, 1973) has not yet been deciphered.

The question of whether these several chloritoids are true polytypes or homologues of slightly differing composition will be addressed later. A description of the 1Tc structure follows.

Triclinic Chloritoid, 1Tc

Triclinic chloritoids were found by a number of mineralogists in the course of optical investigations (see detailed reviews by Hietanen, 1951, and Halferdahl, 1961), but van der Plas *et al.* (1958) were the first to confirm these observations using x-ray diffractometry.

Single-crystal x-ray studies led Halferdahl (1961) to choose a non-primitive $C\bar{1}$ unit cell for triclinic chloritoid consistent in axial orientation with $C2/c$ monoclinic chloritoid ($a_t \sim a_m$, $b_t \sim b_m$, $c_t \sim 1/2c_m$, $\beta_t \sim \beta_m$; see Table 1).[2] He calculated fractional coordinates for atoms in triclinic material by matrix transformation of those in a monoclinic polytype, assuming that the monoclinic structure resulted from successive unit cell twinning (composition plane (001)) of the triclinic structure. Hanscom (1973 - unpublished, and 1980) tested this model using structure factors from another triclinic crystal and failed to obtain a satisfactory refinement (R = 0.46!). He finally solved the structure by symbolic addition and in retrospect saw that Halferdahl's predicted coordinates (adjusted for a different choice of origin of the unit cell) were correct except for the M2B atom (Al in L_2), which turned out to be on the center of symmetry and not in a general position.

All structural details for the 1Tc polytype are similar to those described above for $2M_2$ (idealized drawings in Figure 1); only the c dimension is halved (to ~ 9.15 A) and slight coordinate shifts are observed. Mean M-O distances for the [AlO_6] octahedra in L_2 are ~ 0.01 A larger in 1Tc, but those for the [Fe^{2+},MgO_6] and [$Al,Fe^{3+}O_6$] octahedra differ only by small amounts consistent with their slightly different iron contents.

A high degree of "mosaic spread" was observed in x-ray reflections from Hanscom's (1973) $C\bar{1}$ chloritoid, and omega scans (his Fig. 16) showed severe "splitting of intensity." No doubt this arose from "intergrowths" of monoclinic chloritoid in the predominantly triclinic matrix, as described below.

STACKING DISORDER IN CHLORITOID: POLYTYPISM OR HOMOLOGY?

Hanscom (1973) studied other chloritoids by x-ray precession methods and reported that "diffuse streaking parallel to c^* is common in all

- - - - - - - - - - - - - - - - - - -

[2] The Delaunay primitive cell (Donnay et al., 1954) has lattice parameters $a = 5.48$, $b = 9.16$, $c = 5.48$ A; $\alpha = 96.7$, $\beta = 120.0$, $\gamma = 96.9°$ (Halferdahl, 1961, p. 68, Table 10).

‾single crystal photographs of chloritoid. It is usually more pronounced in triclinic structures and those where mechanical distortion
is present [including one with $c \sim 36$ A]. This diffuse streaking
parallel to c^* results from random stacking faults..." parallel to
(001). Intergrowths of monoclinic and triclinic chloritoid are indeed
common, and early optical microscopic investigation (Hietanen, 1951)
led to the identification of two distinct monoclinic varieties, one
with optic vibration direction X = b and one with Y = b. It is likely
that these are, respectively, the $2M_1$ and $2M_2$ polytypes, but the first
positive x-ray identification of these polytypes was made by Jefferson
and Thomas (1977). They describe the $2M_2$ polytype as derived from 1Tc
by "twinning" about [010] with an (001) composition plane at every
layer.[3] The $2M_1$ polytype is described as related to 1Tc by "twinning"
about [001] rather than [010]. Lattice parameters are given in Table
1: note the reversal of a and b from $2M_2$ to $2M_1$.

High resolution transmission electron microscopy was used by
Jefferson and Thomas (1977, 1978) to obtain (00l) lattice images of
chloritoid with the electron beam parallel either to [100], as referred
to the $2M_2$ polytype, or to [010]. The former condition gave one-dimensional lattice resolution, the latter a complete two-dimensional image.
These are shown in Figures 2 and 3, respectively; and the following
description is quoted freely from their 1977 abstract.

"Two types of planar defect are observed, that at A, which shows
8.9 Å fringes interspersed within the normal 17.8 Å fringes of the
monoclinic form, and can readily be interpreted as an intergrowth of
the 1Tc variant, and apparently larger periodicities such as at B,
where the spacing between strong fringes increases to 35.6 Å. The
structure of the second type of defect is somewhat ambiguous. It could
possibly correspond to a true four-layer structure [Hanscom, 1973, observed such a periodicity in one crystal], but the occurrence of defects
of this type with a width which is invariably a multiple of 17.8 Å suggests that such intergrowths are of small blocks of the $2M_1$ form. Although the majority of defects appear to extend completely across the

- - - - - - - - - - - - - - - - - - -

[3] Hanscom (1973) showed that the operation was not strictly twinning.
See discussion in the previous section.

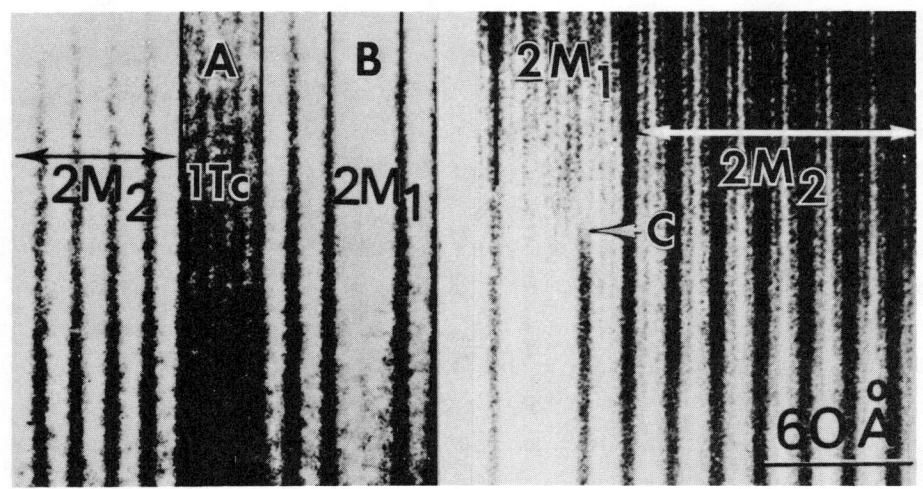

Figure 3. (a) One-dimensional lattice image showing two types of planar defect in chloritoid. (b) Similar image showing a terminating defect at c. After Jefferson and Thomas (1977, their Fig. 1).

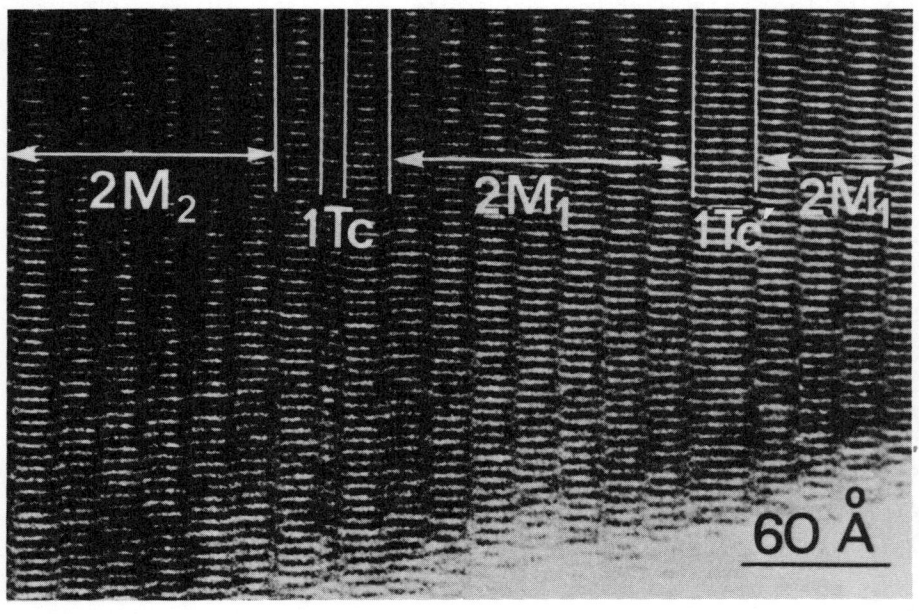

Figure 4. Two-dimensional lattice image, showing the two types of planar defect resolved as intergrowths of alternating monoclinic and triclinic polytypes. After Jefferson and Thomas (1977, their Fig. 2). See their more elaborate treatment (1978).

crystals, some cases were noted, such as at point C in Figure 2b, where the fault terminated in a linear defect. Considerable structural re-arrangement must occur at such a feature, where an apparently four-layer defect abruptly changes to a six-layer one, but the interfacial energy appears to be slight, as evidenced by the lack of strain contrast in that region [see Jefferson and Thomas, 1978, Figs. 10 and 11 for structural details].

"The possibility of the second type of defect discussed above representing an intergrowth of the $2M_1$ variant is confirmed by two-dimensional images, one of which is shown in Figure 3. In this case the crystal is predominantly made up of the $2M_1$ structure, but a region of intergrown $2M_2$ polytype can clearly be observed. Also evident are two strips of 1Tc forms, one of which is only one unit cell wide, and the disposition of fringes indicates that these two regions are not identical. This, however, is quite possible in the chloritoid structure, there being four possible triclinic variants, in different orientations, but all crystallographically equivalent. If the micrograph of Figure 3 is examined at a point further away from the crystal edge than that illustrated, the $(h0l)$-type fringes are observed to disappear, and the $(00l)$ fringes assume the nature of those in Figure 2a, the $2M_2$ intergrowth appearing as a defect of 106.8 Å width, and the triclinic regions as 8.9 Å fringes.

"No true longer-period polytypes of chloritoid have been discovered in the course of this examination...nor is there any evidence as yet for a relationship between chemical composition and polytypism, as found in micas, but a microanalytical study is in progress, which may reveal possible changes in stoichiometry on an ultrastructural scale." This ultimately will determine whether the structural variants of chloritoid are true polytypes or homologous structures of slightly varying composition. The latter seems unlikely.

CRYSTAL CHEMISTRY, PHYSICAL PROPERTIES AND TWINNING

Chemical Variation

Halferdahl (1961) listed more than a hundred chemical and partial

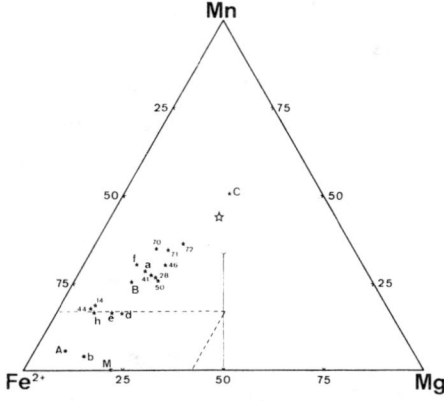

Figure 5. Triangular diagram showing divalent (Fe,Mg,Mn) occupancies of the *M(1B)* octahedral sites from analyses plotted by Fransolet (1978; his Fig. 3) from Béthune (1977), Beugnies (1976) and Kramm (1973). Fransolet's ottrelite is the large star. The broken lines delineate the substitution of Fe^{2+} by Mg and Mn given by Halferdahl (1961).

chemical analyses, including those he considered "unreliable." From these he concluded that only a few substituents are of importance in chloritoid. They are $Mg \rightarrow Fe^{2+}$, $Mn \rightarrow Fe^{2+}$, $Fe^{3+} \rightarrow Al$, and $F \rightarrow OH$. "There appears to be no convincing evidence that chloritoid actually contains more than traces of CaO, Na_2O or K_2O," and titanium probably does not exceed 0.01 atoms per formula unit. Hanscom (1973, his Fig. 8) plotted $Mg/(Fe + Mg)$ versus unit cell volume for four $2M_2$ chloritoids and by extrapolation concluded that "a pure ferrous iron chloritoid ought to have a cell volume of about 927 $Å^3$."

The best summary of the compositional range of chloritoid was assembled by Fransolet (1978),[3] who plotted on a triangular diagram the Fe^{2+}, Mg and Mn contents of the *M(1B)* octahedral sites (Fig. 5). Note that Fe is by far the predominant divalent cation, and in this plot Mg exceeds 25% only in the two Mn-rich specimens (ottrelites). Halferdahl (1961), however, found that in the Mn-poor chloritoids Mg replaced as much as 42% of the Fe^{2+} (his data would plot in the parallelogram outlined in Fig. 5). He also concluded that Fe^{3+} replaced up to 14% of the Al; only minor amounts of F were found, replacing up to 0.25% of the OH. Mössbauer studies of six natural chloritoids by Hålenius *et al.* (1981) indicated that Fe^{2+} is restricted to the relatively large *M(1B)* octahedron and Fe^{3+} to the smaller *M(1A)* octahedron, both in the "brucite-type layer" of chloritoid (layer L_1 -- see Fig. 1).

Sector zoning ("hourglass structure") is characteristic of chloritoids from many localities but no microprobe or careful structural study of it has been done.

Specific Gravity, Hardness

Both observed and calculated specific gravities are given for

- - - - - - - - - - - -

[3] Fransolet's work was inadvertently overlooked in the first edition of this volume.

chloritoids by Halferdahl (1961), Hanscom (1973) and Fransolet (1978), who found them to be in reasonable agreement, ranging between 3.48 and 3.61 depending primarily on Fe and Mn content. Higher values (to 3.80) reported by Deer *et al.* (1962, p. 161) presumably result from inclusions such as rutile, ilmenite and iron oxides. Deer *et al.* give the hardness of chloritoid as 6.5; it is likely to be much softer on the {110} cleavage than on {001}.

Cleavage and Curved Basal Plates

The perfect, mica-like {001} cleavage of chloritoid is an obvious product of weak Fe-OH,O bonds in layer L_1 (Fig. 1). The {110} cleavage is less well developed, but rhomboidal, hexagonal and triangular plates have been described in the literature, as have pseudo-hexagonal etch figures (summary in Halferdahl, 1961, p. 65 *ff.*). This cleavage can best be understood with reference to Figure 1b which shows partial traces of the (110) and ($1\bar{1}0$) planes on L_2, indicating that Al-O bonds to shared octahedral edges (but not Si-O bonds) are broken. A {100} cleavage is equally probable.

There is a tendency for the basal plates of chloritoid to curve, and no explanation has been offered for this phenomenon. One possible cause is misfit between layers L_1 and L_2 (analogous to antigorite). In a chemically zoned crystal L_2 might contain successively more Mg along [001], causing shrinkage of L_2 relative to chemically unsubstituted L_1 and warping of the (001) cleavage plates. Another possibility is partial or differential dehydration [see below].

Optical Properties

Halferdahl accumulated all the optical data available for chloritoid through 1961 (his Tables 19 and 20) and Fransolet (1978, Table 4) added two more. In summary: α ranges from 1.709 to 1.730, β from 1.712 to 1.734 and γ from 1.716 to 1.740. Understandably, the lowest of each of these three values is for the most Mn-rich specimen (Fransolet, 1978). The change of refractivity with composition has not been investigated sufficiently well to be used in a determinative manner. Actually, $2V_\gamma$ ranges from 36° to 72° from grain to grain and specimen to specimen in optically positive monoclinic chloritoid and from 92° to 125° in optically negative triclinic chloritoid. Because of ubiquitous (001) stacking faults which frequently are not optically

visible, mixtures of the two or three polytypes with their different
optical orientations can confuse the systematics. Obviously single-
crystal x-ray photographs should be used routinely in specifying the
nature of any chloritoid observed in rocks. Conventional optical data
are of relatively little value.

Chloritoid is commonly dark green, but ranges from colorless to
green in thin section. The general pleochroic formula is α = green,
β = blue to indigo, γ = colorless to yellow. Dispersion is $r > v$
strong.

In a careful polarized optical absorption study, Hålenius et $al.$
(1981; $contra$ Faye et $al.$, 1968) assigned the broad absorption band at
16,300 cm^{-1} to a $Fe^{2+} + Fe^{3+} \rightarrow Fe^{3+} + Fe^{2+}$ charge-transfer transition
in layer L_1. Two bands at 10,900 cm^{-1} and 8,000 cm^{-1} were assigned
unequivocally to spin-allowed d-d transitions in octahedrally coordinated
Fe^{2+}, and the intensity of the band at ~28,000 cm^{-1} was related to
Fe^{3+} concentration. They calculated eight highly correlated ($R > 99\%$)
regression equations for band intensities as functions of the concentrations of Fe^{2+} and Fe^{3+} and the concentration product $[Fe^{2+}][Fe^{3+}]$. A non-
destructive 'microscope-spectrophotometric method' related to this work was
reported by Hålenius and Langer (1980). It permits quantitative determination of Fe^{2+} (to $\pm$ 0.15 g-atom/l) and Fe^{3+} (to $\pm$ 0.05 g-atom/l) on
chloritoid grains in thin section with areal resolution of ~10μm.

Twinning

"The twinning observed in chloritoid is always lamellar with the
composition plane parallel to the basal cleavage.

"Theoretically all twinning laws are possible which maintain an
arrangement in the immediate neighbourhood of any atom close to the
composition plane essentially similar to that which would occur if
normal growth took place. Thus the composition plane in chloritoid is
limited to a position where in the untwinned structure the sequence of
atomic sheets on either side of the composition plane is the same,
$i.e.$ it must lie at the level of the cations..." in L_1 or L_2
(Harrison and Brindley, 1957, p. 81).

The darker lines in Figure la trace the directions of permissible
twin axes in L_1: [100], [130], [1$\bar{3}$0]; in L_2 (Fig. 1b) they trace
[010], [110], [1$\bar{1}$0]. As Harrison and Brindley pointed out, [130]|[1$\bar{3}$0]

and $[110]|[1\bar{1}0]$ are symmetry-related and $[010]$ is the two-fold axis in
$C2/c$ chloritoids, leaving only $[100]$, $[130]$ and $[110]$, all of which
have been observed experimentally. Possibilities are wider in $C\bar{1}$
chloritoids. Hietanen (1951) listed three other twin axes: $[120]$,
$[210]$ and $[310]$. From Figure 1 it is clear that these and others are
possible because of the pseudo-hexagonal closest-packed nature of this
structure.

DEHYDRATION AND STABILITY

Topotactic Dehydration-Oxidation

Bachmann (1956) proposed that dehydration in air of chloritoid
takes place according to the reaction [using structural formulas]:

$$[Fe_2^{2+}AlO_2(OH)_4]^{-1}[Al_3O_8]^{-7}[Si_2]^{+8} + \tfrac{1}{2} O_2 \rightarrow$$
$$[Fe_2^{3+}AlO_4]^{-1}[Al_3O_8]^{-7}[Si_2]^{+8} + 2H_2O.$$

He observed that the spacing normal to (001) increased from 8.9 A per
layer in two-layer chloritoid (possibly $2M_1$?) to 9.36 A in what he
described as a single-layer defect structure. Halferdahl (1961,
p. 90) questioned his experiment, and Jefferson and Thomas (1979)
suggested that his proposed structure (Fig. 37, p. 164 in Deer et $al.$,
1962) was correct in essence but not in detail. The latter repeated
his experiment under more controlled conditions and with careful
characterization of their starting product.

In $vacuo$ for several hours at 700°C, $2M_2$ chloritoid dehydrated
with no change in morphology to a completely amorphous, nearly opaque
(black-brown) mass with no evidence of crystallinity visible even at
the highest resolution of the electron microscope. In air at the same
temperature the same color and morphology were observed, but the dehydrated, topotactically oxidized product is a distinctly crystalline,
three-layer, rhombohedral structure. Designated 3R by Jefferson and
Thomas (1979), it has cell dimensions $a = 5.8$, $c = 28$ A. The oxygens
and Al atoms of layer L_2, the Si atoms and the iron atoms of L_1 are
essentially unchanged in position, but the anions of L_1 are rearranged
to form a 3-fold array of symmetrically equivalent $[Fe^{3+}O_5]$ polyhedra
(their Fig. 7b).

At relatively rapid heating rates (> 2.5°C/min), van der Plas

167

et al. (1958) reported an endothermic reaction in chloritoid at 770°C, as seen in TGA and DTA plots. All the water was not driven off until 1050°C. Fransolet (1978) reviewed the data for ottrelites: at 10° C/min his specimen showed an endothermic reaction between 670 and 770°C and complete dehydration by 820°C.

Stability

Because it is an important marker mineral in the low and middle grades of metamorphism (Winkler, 1965), chloritoid has recently been the subject of numerous experimental investigations, particularly since the development of oxygen fugacity buffers. Halferdahl reviews all the work, including his own, up to 1961. The next major work was that of Ganguly and Newton (1968) on the thermal stability of chloritoid at high pressure (10 - 25 kbar) and relatively high oxygen fugacity. This was followed by an investigation of its stability limits relative to the coexisting minerals of regionally metamorphosed pelitic rocks (Ganguly, 1969), and by a study of the reaction relations of chloritoid and staurolite (Albee, 1972). Grieve and Fawcett's (1970, 1974) investigation of the stability of chloritoid below 10 kbar water pressure yielded, among other important data, conclusive evidence that chloritoid is not a "stress mineral," as had been suggested by Harker (1932).

The papers on occurrences and parageneses of chloritoid are too numerous to mention and are in any case beyond the scope of this work. See Halferdahl (1961) for an important review and the aforementioned works on stability and reaction relations for later important references. Liou and Chen's (1978) recent work contains references to many of the important petrologic studies. See also Atherton (1980), Baltatzis (1980), Holdaway (1978), and Cruickshank and Ghent (1978); Fransolet (1978) has summarized studies of the Mn-rich chloritoids.

Albee, A.L. (1972) Metamorphism of pelitic schists: reaction relations of chloritoid and staurolite. Geol. Soc. Am. Bull., 83,
3249-3268.

Atherton, M.P. (1980) The occurrence and implications of chloritoid
in a contact aureole and alusite schist from Ardara, County Donegal.
J. Earth Sci. R. Dublin Soc., 3, 101-109.

Bachmann, H.G. von (1956) Dehydration von Chloritoiden. Z. Kristallogr., 108, 145-156.

Baltatzis, E. (1980) Chloritoid-forming reaction in the eastern Scot-
tish Dalradian: a possibility. N. Jahrb. Mineral. Mh., 1980,
306-313.

Bethune, P. (de) (1977) La composition chimique des chloritoides belges.
Bull. Soc. belge Géol. 86, 9-11.

Beugnies, A. (1976) Structure et métamorphisme du paléozoique de la
région de Muno, un secteurclef du domaine hercyniende l'Ardenne.
Ann. Mines Belgique, 6e livraison, 481-509.

Brindley, G.W. and F.W. Harrison (1952) The structure of chloritoid.
Acta Crystallogr., 5, 698-699.

Cruickshank, R.D. and E.D. Ghent (1978) Chloritoid-bearing pelitic rocks
of the Horsethief Creek Group, southeastern British Columbia.
Contrib. Mineral. Petrol., 65, 333-339.

Deer, W.A., R.A. Howie and J. Zussman (1962) *Rock-Forming Minerals*,
Vol. 1. Longmans, London. pp. 161-170.

Donnay, J.D.H., W. Nowacki and G. Donnay (1954) *Crystal Data*. Mem.
Geol. Soc. Am., 60, 138.

Faye, G.H., P.G. Manning and E.H. Nickel (1968) The polarized optical
absorption spectra of tourmaline, cordierite, chloritoid and
vivianite: ferrous-ferric electronic interaction as a source of
pleochroism. Am. Mineral., 53, 1174-1201.

Fransolet, A.M. (1978) Donnés nouvelles sur l'ottrélite d'Ottré,
Belgique. Bull Minéral., 101, 548-557.

Ganguly, J. (1969) Chloritoid stability and related parageneses:
Theory, experiments, and applications. Am. J. Sci., 267, 910-944.

——— and R.C. Newton (1968) Thermal stability of chloritoid at high
pressure and relatively high oxygen fugacity. J. Petrol., 9, 444-
446.

Grieve, R.A.F. and J.J. Fawcett (1970) The synthesis of chloritoid at
low pressures. Am. Mineral., 55, 49-135.

——— and ——— (1974) The stability of chloritoid below 10 kb P_{H_2O}.
J. Petrol., 15, 113-139.

Hålenius, U., H. Annersten and K. Langer (1981) Spectroscopic studies
on natural chloritoids. Phys. Chem. Minerals, 7, 117-123.

——— and K. Langer (1980) Microscope-photometric methods for non-destruc-
tive Fe^{2+}-Fe^{3+} determinations in chloritoids $(Fe^{2+},Mn^{2+},Mg)_2(Al,Fe^{3+})_4$
$Si_2O_{10}(OH)_4$. Lithos, 13, 291-294.

Hanscom, R.H. (1973) *The Crystal Chemistry and Polymorphism of Chloritoid.* Ph.D. Dissertation, Harvard Univ., Cambridge Massachusetts.

—— (1975) Refinement of the crystal structure of monoclinic chloritoid. Acta Crystallogr., B31, 780-784.

—— (1980) The structure of triclinic chloritoid and chloritoid polymorphism. Am. Mineral., 65, 534-539.

Harker, A. (1932) *Metamorphism.* Methuen, London.

Harrison, G.W. and G.W. Brindley (1957) The crystal structure of chloritoid. Acta Crystallogr., 10, 77-82.

Hietanen, A. (1951) Choritoid from Rawlinsville, Lancaster County, Pennsylvania. Am. Mineral., 36, 859-868.

Holdaway, M.J. (1978) Significance of chloritoid-bearing and staurolite-bearing rocks in the Picuris Range, New Mexico. Geol. Soc. Am. Bull., 89, 1404-1414.

Jefferson, D.A. and J.M. Thomas (1977) Structural variation in chloritoid. E.C.M.-4 Proc. (Oxford), 626-628.

—— and —— (1978) High resolution electron microscopic and X-ray studies of non-random disordered in an unusual layered silicate (chloritoid). Proc. Roy. Soc. Lond. A., 361, 399-411.

Jefferson, D.A. and J.M. Thomas (1979) Topotactical dehydration of chloritoid. Acta Crystallogr., A35, 416-421.

Kramm, U. (1973) Chloritoid stability in manganese rich low-grade metamorphic rocks, Venn-Stavelot Massif, Andennes. Contrib. Mineral. Petrol., 41, 179-196.

Liou, J.G. and P.-Y. Chen (1978) Chemistry and origin of chloritoid rocks from eastern Taiwan. Lithos, 11, 175-187.

Plas, L. van der, T. Hügi, M.H. Mladeck and E. Niggli (1958) Chloritoid vom Hennensädel südlich Vals (nördliche Aduladecke). Schweiz. min. petrogr. Mitt., 38, 237-246.

Ross, M., H. Takeda and D.R. Wones (1966) Mica polytypes: Systematic description and identification. Science, 151, 191-193.

Tricker, M.J., D.A. Jefferson, J.M. Thomas, P.G. Manning and C.J. Elliott (1978) Mössbauer and analytical electron microscopic studies of an unusual orthosilicate: chloritoid. J. Chem. Soc., Faraday Trans. II, 74, 174-181.

Winkler, H.G.F. (1965) *Die Genese der metamorphen Gesteine.* Springer-Verlag, Heidelberg.

Chapter 7

STAUROLITE   P. H. Ribbe

INTRODUCTION

Other than a recently discovered occurrence "of undoubted igneous origin" (Gibson, 1978), staurolite is almost exclusively found in medium-grade regionally metamorphosed rocks and in sediments derived therefrom. Inasmuch as this chapter does not address the subject of staurolite paragenesis, the interested reader is referred to footnote 1 for a selected list of references relating to experimental investigations of its stability and major recent field-related petrologic studies.[1]

Uncertainties about the exact chemical composition of staurolite — in particular the H_2O content and the valence state(s) of iron — and confusion about the space group and thus the crystal structure have persisted until very recently. For example, Náray-Szabó (1929) determined that the structure of staurolite was based on a closest packed array of (O,OH) anions, but refining in space group $Ccmm$, he overlooked four partially occupied octahedral sites and suggested a formula of $H_8Fe_4Al_{16}Si_8O_{48}$. On the basis of six wet chemical analyses Juurinen (1956) postulated an incorrectly balanced formula $H_4Fe_4Al_{18}Si_8O_{48}$, but Ganguly (1972) gave $Fe_4^{2+}Al_{18}Si_8O_{46}(OH)_2$ as "the idealized stoichiometry of staurolite, at least as a limiting composition." This formula had been suggested in 1915 by Hörner in his unpublished inaugural dissertation and, except for the exact number of cations and hydrogens, was essentially affirmed in the more nearly correct refinement of the structure by Náray-Szabó and Sasvári (1958; discussed below).

With the advent of more precise chemical analyses, least-squares site refinements of the crystal structure using both x-ray and neutron

---

[1]*Experimental investigations:*

Hoschek (1967,1968), Richardson (1968), Ganguly and Newton (1965), Schreyer and Seifert (1969), Ganguly (1972), Hellman and Green (1979), *Field-related petrologic studies:*   Yardley (1981).

Chinner (1967), Schreyer and Chinner (1966), Hietanen (1969), Hollister (1969), Guidotti (1970), Albee (1972), Kwak (1974), Smellie (1974), Ashworth (1975), Gibson (1978).

A book entitled "Staurolite" by V.V. Fed'kin (1975, Moscow: Academia Nauk USSR, 272 pp.) contains much crystal chemical data and detailed discussions of staurolite parageneses in context of occurrences in the USSR.

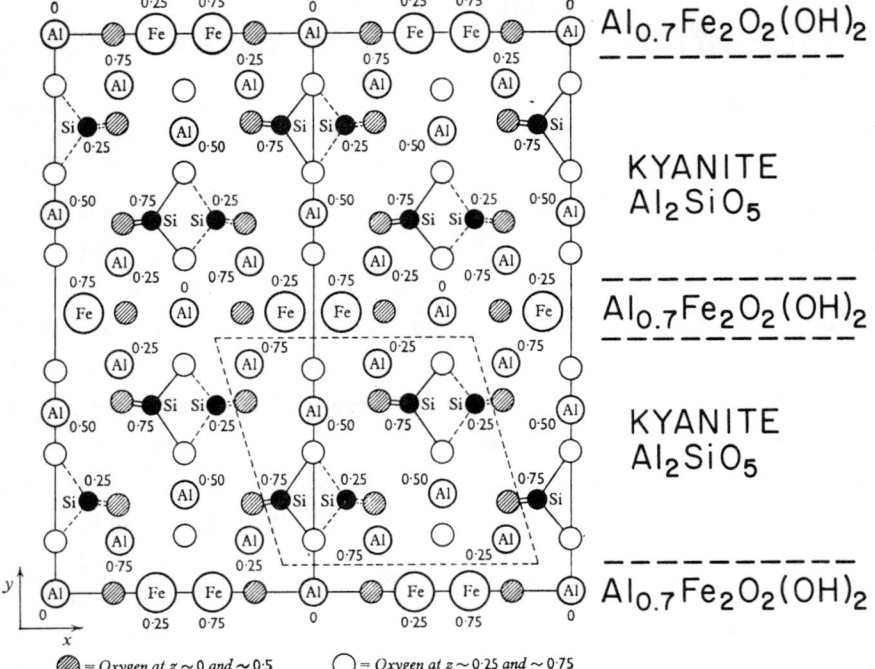

Figure 1. The structure of staurolite projected on (001). Two unit cells are illustrated and the kyanite portion of the structure is outlined by dashed lines (modified from Náray-Szabó and Sasvári (1958, Fig. 3, p. 863)). Wenk (1980) studied epitaxial intergrowths of staurolite and kyanite in specimens from Alpe Sponda, Switzerland and found the interface to be coherent with a few dislocations (see his Figs. 1 and 2).

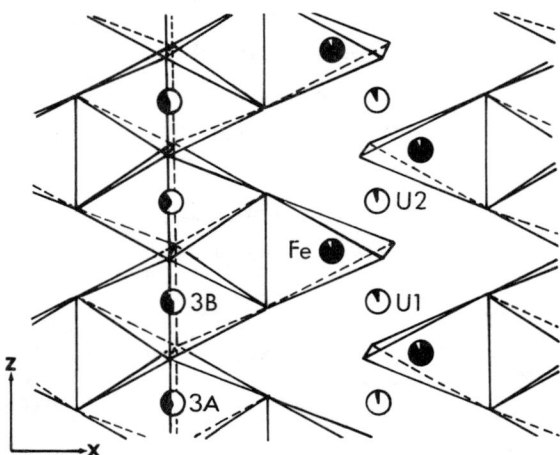

Figure 2. Schematic drawing of the $Al_{0.7}Fe_2O_2(OH)_2$ layer of staurolite showing arrangement of Fe tetrahedra and Al octahedra. Approximate site occupancies as determined by Smith (1968) in a specimen from St. Gotthard are indicated by filled portions of circles: see text for details. Modified from Dickson and Smith (1976, Fig. 7, p. 214).

172

diffraction data, Mössbauer spectroscopic studies of iron distribution and valence, and nuclear magnetic resonance studies of dipole-dipole interactions between H^+ and the electron spins of paramagnetic Fe^{2+}, both the structure and the chemistry of staurolite are now fairly well understood. The generalized structural formula suggested by Smith (1968) is:

[7 octahedral sites, principally Al]$_{\sim 18}$

[1 tetrahedral site, principally Fe]$_4$ or <4

[1 tetrahedral site, principally Si]$_8$O$_{48}$H$_4$ or <4.

The generalized chemical formula determined by Griffen et al. (1982) is:

$(Fe,Mg,Zn)_{25.6-1.25x}Al_{1.5x-8.2}Si_{16.2-0.5x}O_{48}H_{\sim 4}$ $[16.6 \leqslant x \leqslant 18.6]$
(see later discussion).

CRYSTAL STRUCTURE

The initial model of Náray-Szabó (1929) for the structure of staurolite based on octahedral Al and tetrahedral Fe and Si in a cubic closest packed array of (O,OH) anions was essentially correct. However, most staurolites are in fact monoclinic with β = 90.00±0.05°, as discovered by Hurst et al. (1956) in morphological, optical and x-ray studies of twinning. Náray-Szabó and Sasvári (1958) then re-refined the structure in space group C2/m and for the first time located partially-filled Al octahedra which, in addition to the filled Al octahedra and Fe and Si tetrahedra, led to their drawing as modified in Figure 1.

Hanisch (1966) reported the structure of a zincian staurolite that he considered to be truly orthorhombic (Ccmm) with octahedral sites Al(3A) and Al(3B) [Fig. 2] exactly half occupied and therefore symmetrically equivalent. By contrast Smith (1968) used the very weak 0kℓ (ℓ-odd) reflections in a ferroan staurolite from St. Gotthard to determine slight but significant deviation from orthorhombic symmetry in both atomic positional parameters and site occupancies, confirming space group C2/m.

Regardless of exact symmetry, the structure of staurolite may be considered to be made up of slabs of the kyanite structure alternating with monolayers of approximate composition $Al_{0.7}^{IV}Fe_2^{IV}O_2(OH)_2$ along [010], as indicated in Figure 1. Details of this layer and the partial occupancies of octahedral Al(3A,3B) and U(1,2) sites and tetrahedral Fe^{2+} sites in a C2/m staurolite are shown in Figure 2. Because these sites

173

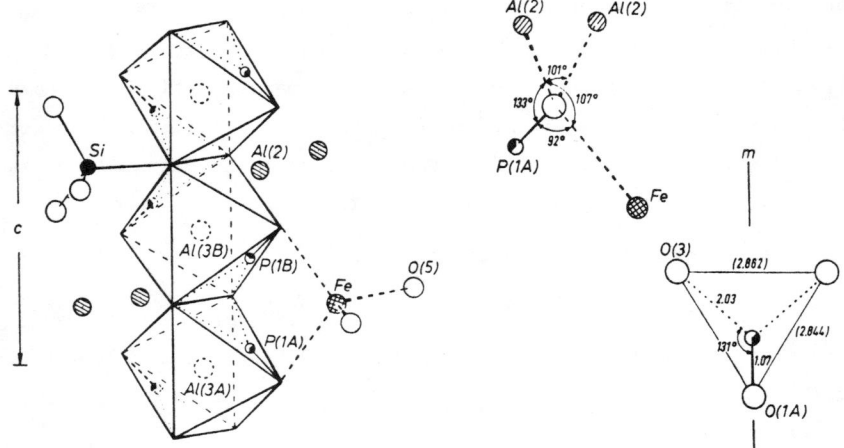

Figure 3. Location of the partially filled proton sites P(1A) and P(1B) in staurolite: (left) nearly in the face of the Al(3A,3B) octahedra, (center) relative to other cations, and (right) relative to oxygens O(3) and O(1A), to the latter of which it is bonded at 1.07A to form hydroxyl. After Takéuchi, Aikawa and Yamamoto (1972, Figs. 3 and 4, p. 13).

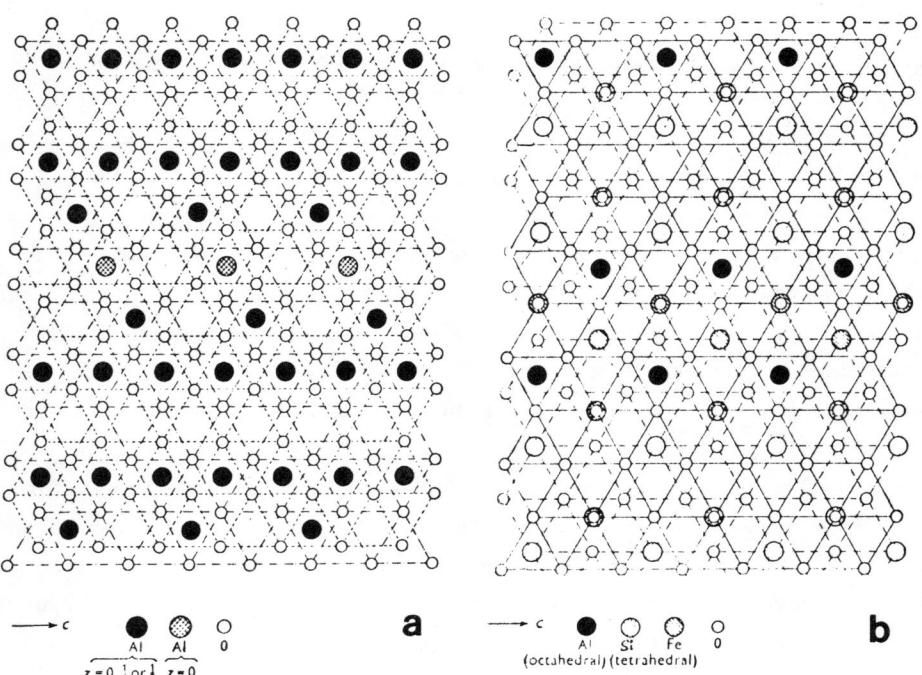

Figure 4. Layers of the cubic closest packed staurolite structure parallel to (130). After Náray-Szabó and Sasvári (1958, Figs. 1 and 2, p. 863).

are only partially occupied, Smith (1968) found it impossible in his least-squares refinement to make unequivocal assignments of atomic species, but he suggested the following: $\sim$0.06(Fe+Mn) in the U(1) and U(2) octahedra which share faces with the Fe tetrahedron, itself only 94% occupied by $0.6Fe^{2+}$ + 0.3(Mg and/or Al) + 0.04Ti; 0.4 and 0.3 (Al+ Fe), respectively, in the Al(3A) and Al(3B) octahedra; and — in the "kyanite" portion of the structure — Al plus minor (Mg+Fe) in three octahedral sites and Si plus sufficient Al to fill it in the Si tetrahedron.

Other important structural details were discovered in a combined neutron diffraction and nuclear magnetic resonance spectral study by Takéuchi *et al.* (1972). Three hydrogen atoms (protons) were found to be disordered on two sets of sites, each of which is fourfold in the structure. Individual proton sites are located approximately on the faces of the partially occupied Al(3A,3B) octahedra (Fig. 3) on a line between these and the Fe^{2+} sites, and it is most likely that when a proton site is occupied, the nearby Al(3) site is vacant. The partial occupancy of the Fe tetrahedron (Smith, 1968) may also be related to the presence of protons. If the limit of four H per unit cell is reached (as Griffen and Ribbe (1973) suggest is commonly the case), the average occupancy of the proton sites would be 0.5 H.

STACKING FAULTS, TOPOTAXY, AND ANTI-PHASE DOMAINS

Using transmission electron microscopy, Fitzpatrick (1976) observed what she called "cation stacking faults" on the closest-packed planes (010) and (130) in staurolite. Their displacement vectors are 1/6[010] and 1/6[130], respectively. The (010) faults occur in pairs bounding minute regions of kyanite (*cf.* Fig. 1) and are probably due to depletion of Fe, because the incidence of (010) faults (in a specimen from Death Valley) were found to be higher in Fe-poor sectors than in Fe-rich sectors of the same crystal. The cause of (130) faults is a more complex matter, since depletion of Fe and/or Al or the addition of Si can effect them (*cf.* idealized drawings of the (130) layers in Figure 4).

Given the possibility of stacking faults which produce regions of kyanite within staurolite, it is not surprising to find parallel growths

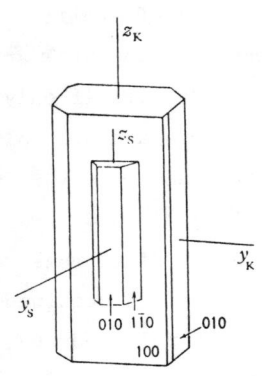

of the sort illustrated in the inset (after Deer *et al*., 1962, Fig. 35). Wenk (1980) examined the staurolite-kyanite interface (see comments in Fig. 1 legend). Upon heating at 1000°C, staurolite transforms topotactically to kyanite with $[100]_{st} \parallel [010]_{ky}$ and $[001]_{st} \parallel [001]_{ky}$ (Dasgupta, 1974).

Antiphase domain boundaries (APB's) with R = 1/2[001] were imaged in a staurolite from Kwoiek, British Columbia by Fitzpatrick (1976), using one of the more intense $0k\ell$, ℓ-odd reflections. These support Smith's (1968) hypothesis that ordering of cations is mainly responsible for the intensity of the $0k\ell$, ℓ-odd reflections: the greater the intensity, the greater the difference in occupancy between Al(3A)[$\pm$U(1)] and Al(3B)[$\pm$U(2)] (see Fig. 2), as also suggested by Dollase and Hollister (1969). Of course the presence of APB's suggests that ordering must have occurred sometime after crystallization (*contra* Hollister, 1970, p. 763), and that this staurolite probably grew initially with orthorhombic symmetry and then inverted to monoclinic *C2/m* (Smith, 1968, p. 1152).

MORPHOLOGY, CLEAVAGE AND SECTOR ZONING

By calculating "the fraction of the total bond energy" required to attach a new layer to its substrate on each potential crystal growth face in staurolite, Dowty (1976a, p. 457) found that the forms should be ranked in the order {010}, {110}, {001}, in agreement with Niggli (1926). The {101} form often seen was not well prognosticated by Dowty's model, but the indistinct (010) cleavage was correctly predicted by his "cleavage energy calculations." Griffen and Ribbe (1973) suggested that the cleavage result from weak bonding in the $Al_{0.7}Fe_2O_2(OH)_2$ layer parallel to (010) because the cation sites are only partially occupied and weak metal-OH bonds are concentrated there (*cf*. Figs. 1 and 2).

The origin, mechanism and nature of sector zoning in staurolite were addressed in great detail by Hollister (1970), based on descriptions by Hollister and Bence (1967). A sketch showing the geometry of the {001}, {010} and {001} sectors is compared in Figure 5 with a projection of the staurolite structure down [001]. The latter was drawn by Dowty (1976b)

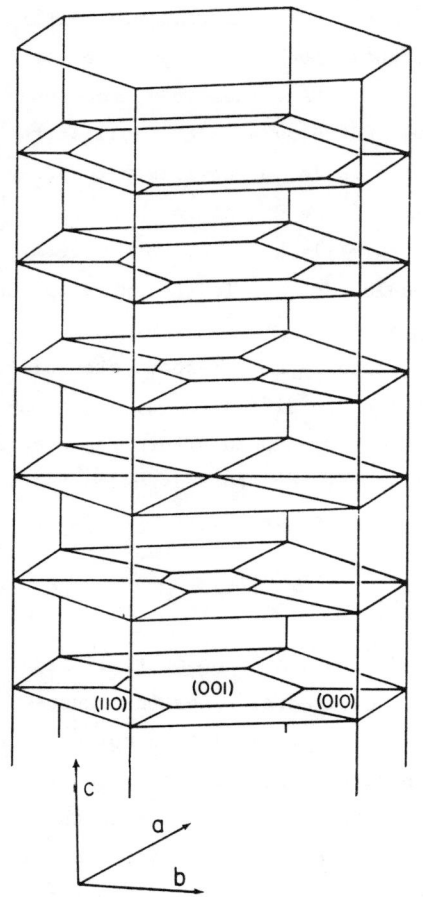

STAUROLITE [00ĪĪ] PROJECTION

Si - ●   Al(3)- □
Al(1)- ▽   Fe - ▼
Al(2)- ◇   O - ○

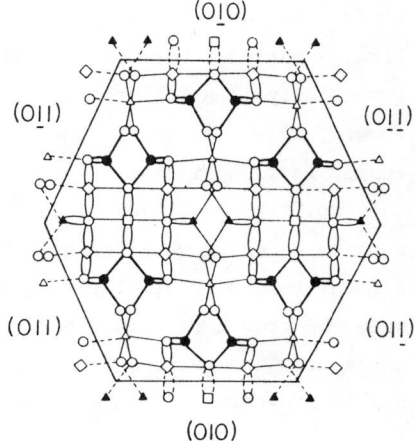

Figure 5. Left: an idealized sketch showing the {001}, {010} and {110} sectors of staurolite (after Harker, 1939, p. 44). Above: A projection of the staurolite structure down [001] after Dowty (1976b, Fig. 4, p. 464). The "exposed sites" which govern the chemistry of the growing sectors are shown outside the hexagon.

See further discussion by Tracy, p. 369, Volume 10 of this series.

who comments further on the zoning phenomena in light of his model for predicting surfaces of least bonding (Dowty, 1976a).

In summary it can be said that, relative to other sectors, (1) the {001} sector has excess Al with somewhat depleted Si, due in large part to the exposure on {001} of the partially filled Al(3A,3B) sites and the fact that Si sites are not exposed (Dowty, 1976b, Fig. 5, *contra* Hollister, 1970, Fig. 7). These sectors appear to have less intense $0k\ell$, ℓ-odd reflections, indicating more nearly disordered Al distributions in the two Al(3) sites than the {010} sectors (see earlier discussion of anti-phase domains and Dollase and Hollister, 1969). (2) The {010} sectors have $Ti^{4+} + Mg^{2+} \rightarrow 2Al^{3+}$, indicating that "to account for the difference in Ti and Mg content between {010} and {001}, we apparently must call on a rejection of Ti from the exposed half-sites on {001}. Perhaps the Al(1) and Al(3) sites exposed on {001} are inherently much less receptive to Ti than the Al(2) sites exposed on {110} [Fig. 5]. Alternatively, the abundant excess Al adsorbed onto Al(3) on {001} may play a role. Hollister (1970) considered that Al(1) and Al(2) sites would not be likely to accept foreign cations, because they are in a part of the structure which is essentially identical to that of kyanite, normally a rather pure mineral" (Dowty, 1976b, p. 465). Smellie (1974) would add Mn^{2+} to the substitution couple involved in this zoning.

Hollister's (1970) conclusions that "strongly sector-zoned staurolite appears to occur when the growth rate exceeds the diffusion rates perpendicular to the growth layers" and that weakly or unzoned staurolite occurs when the growth rate is about the same or slower than the diffusion rate, emphasize the obvious fact that sector zoning is the product of metastable growth.

COMPOSITION

Given the complexity of its crystal structure and the multiplicity of partially occupied Al-octahedral and Fe-tetrahedral sites in staurolite, it is not surprising to find considerable variability in composition. At issue are not only the ranges of elemental concentration but also the occupancies of the coordination polyhedra by the various substituents. Griffen and Ribbe (1973), in the most recent discussion of compositional variability *per se*, gave the following ranges for elemental concentrations in number of atoms per 48 (O,OH):

Mg	0.38 to 1.13 [1.28]		Zn	0 to 1.41 <1.54>
Al	17.07 to 17.60 [18.12]		Mn	0 to 0.09
Si	7.64 to 7.91 [7.45]		Co	no data (1.94)
Ti	0.09 to 0.14 [0.16]		Ni	no data (0.20)
Fe	2.19 to 3.44 (0.82)			

The numbers in square brackets are from an exceptionally low Si, high (Al, Mg) staurolite from a rock "of undoubted igneous origin" (Gibson, 1978); an even more aluminous specimen was reported by von Knorring *et al.* (1979) -- 60.5 wt % Al_2O_3 ($\approx$ 19.9 Al atoms based on a cation sum of 30). The numbers in parentheses are from *lukasite*, a cobaltan staurolite from Zambia (Skerl and Bannister, 1934)[2], and the high zinc value is from a partial analysis of the *Ccmm* staurolite refined by Hanisch

- - - - - - - - -

[2] Compare the study of Co-staurolite by Čech *et al.* (1981).

(1966; *cf.* Nemec, 1978). Other elements present in trace amounts are
vanadium and chromium. Calcium is not uncommonly reported in bulk analyses (up to 0.1 atoms), but much of it may be associated with impurities.
Griffen and Ribbe (1973) did not detect calcium at the 0.005 level in
microprobe analyses of specimens from many localities, but Foster (1977)
reports 0.05 Ca in a specimen from a pelitic schist in Maine. It is thought
unlikely that sodium and potassium are incorporated into the staurolite
structure. No implication is intended that these cation ranges specify
the limits of staurolite composition if for no other reason than Fe-, Mg-,
and Zn-rich specimens have been synthesized well outside these ranges
(Ganguly and Newton, 1965; Richardson, 1967; Schreyer and Seifert, 1969;
Griffen, 1981).

It is generally agreed that all the silicon is in tetrahedral coordination together with sufficient aluminum to fill the site, and it is presumed that the generally minute amounts of titanium and manganese are octahedral. But apart from the site-occupancy refinement by Smith (1968) and
numerous Mössbauer studies of iron distribution, the only clue to the distribution of cations amongst the seven octahedral sites (principally Al)
and the one tetrahedral site (principally Fe^{2+}) are principal component
analyses (PCA) by Griffen and Ribbe (1973). Their statistical studies do
not *prove* any substitutional relationships, but the results are of interest
nonetheless: Fe and Mg were deduced to be the primary substituents in the
^{VI}Al sites with Zn and Al in the ^{IV}Fe site. The generally accepted view
that Mg is the primary substituent for Fe in the ^{IV}Fe site was not supported
by PCA, but this conclusion may be in error -- synthetic Fe-Mg staurolites
certainly contain Mg in tetrahedral coordination.

A synthesis of data for 82 chemically-analyzed staurolites (Griffen
et al., 1982) indicated a strong correlation between Al' (the Al remaining
"after any Si deficiency below 8.0 atoms is made up by Al") and the sum
total of divalent cations: $Al' = 20.5 - 0.80 (Fe+Mg+Zn)$; $R^2 = -0.96$. An
"average chemical formula" was calculated, $(Fe,Mg,Zn)_{4.0}Al_{17.3}(Si_{7.6}Al_{0.4})$
$O_{48}H_{\sim4}$, but the generalized formula is

$$(Fe,Mg,Zn)_{25.6-1.25x}Al_{1.5x-8.2}Si_{16.2-0.5x}O_{48}H_{\sim4}$$

where $16.58 \leqslant x \leqslant 18.61$ for their data set. When $Si = 8.0$, x cannot be
smaller than 16.4, but the upper limit for x is "elusive."

Iron

Mössbauer spectroscopy has been used extensively to study iron valence and distribution in staurolite. Ferric iron has been reported by Dowty (1972) and Regnard (1976), but it is considered to be less than 3% of total iron by these investigators and is assumed to be absent by most. But interpretations of the Fe^{2+} spectra vary significantly. Consistent with Smith's (1968) site refinement and Griffen and Ribbe's (1973) principal component analyses are the interpretations of Mössbauer spectra by Bancroft *et al.* (1967), Smith (1968), Dickson and Smith (1976), Scorzelli *et al.* (1976) and Regnard (1976) that Fe^{2+} is both tetrahedrally and octahedrally coordinated. By contrast Dowty (1972), following a suggestion of possible positional disorder at the ^{IV}Fe site by Smith (1968), propounded the idea that there may be two distinct tetrahedral sites and (possibly) no octahedral iron at all. Dzhemats and Nikitina (1977) appear to support Dowty's idea of two non-equivalent tetrahedral sites but profess to observe octahedral Fe^{2+} as well. They suggest that "the distribution of the Fe^{2+} among tetrahedral and octahedral sites could relate to the conditions of formation of staurolite."

Yet another interpretation is advocated by Regnard (1976) who assigns doublets from the spectra of oriented single crystals of staurolite to one tetrahedral and two different octahedral sites (he prefers Al(3A) and Al(3B)), a reasonable idea in light of Smith's (1968) results and those of Griffen and Ribbe (1973, p. 491) who suggest that Fe substitutes in the $Al_{0.7}Fe_2O_2(OH)_2$ layer, which contains Al(3A,B), and not in the kyanite portion of the structure -- see Figures 1 and 2.

Dickson and Smith's (1976) low-temperature Mössbauer study of staurolite did not give a clear indication of octahedral Fe^{2+} until they discovered antiferromagnetic ordering and peak intensity changes at 4.2°K that suggested the possibility. Regnard (1976) gives the magnetic ordering temperature as 6±1°K (*cf.* the work of Borg and Borg (1975) on other silicates). The suggestion of Scorzelli *et al.* (1976) that an electron exchange mechanism may be responsible for one of the doublets in the staurolite spectrum bears further investigation.

Vaughn *et al.* (1974), using SCF Xα scattered wave molecular orbital calculations, were able to determine spectral energies in good agreement with the observed optical absorption spectra of ^{IV}Fe in staurolite.

Zinc

Guidotti (1970) observed an increase in Zn content with decreasing modal staurolite, suggesting that, within the limits as yet undefined, staurolite accommodates most of the Zn available in the rocks in which it occurs at the expense of other elements. This fact is undoubtedly related to the availability of a large tetrahedral site into which Zn^{2+} (r = 0.57A -- radius from Griffen (1981)) easily substitutes for Fe^{2+} (r = 0.63A -- radius from Shannon (1976)). Ashworth (1975) suggests that staurolite formed in equilibrium at "anomalously high grade" in muscovite-free migmatized semipelites was stabilized by its high Zn content. In a study of zinc in staurolite, Nemec (1978; see also 1980) states that "staurolites of the staurolite-almandite subfacies display up to 1 percent ZnO, those of the upper metamorphic subfacies of the almandite-amphibolite facies usually contain 1 to 3 percent ZnO ...", wisely adding the caveat that "individual areas differ in ZnO content of staurolites in rocks of identical metamorphic grade."

Griffen (1981) synthesized zinc and iron end-member staurolites, as well as intermediate compositions at $\sim$30 kbar and 750°C, and he suggests, based on molar volume data, that Zn^{2+} and Fe^{2+} are disordered, but he is not certain whether the $Zn \rightleftarrows Fe$ substitution is entirely in the tetrahedral site (cf. Griffen and Ribbe, 1973).

Hydroxyl

There is no question but that there needs to be developed a more reliable analytical method for this petrologically important constituent of staurolite (Ashworth, 1975, finds it critical). Conventional analyses have too large a margin of error, as does the technique based on assumptions of charge balance propounded by Griffen and Ribbe (1973). The neutron diffraction technique is excellent but costly and unduly time-consuming (Takéuchi et al., 1972). Most workers would agree that there are 3±1 hydroxyl ions and 45∓1 oxygens in the closest packed anion array. It seems likely that the range may be narrower than that, possibly 3 < OH ≤ 4, although Lonker (private comm., 1980) devised an electrolytic technique to determine water content and found OH ranges between 1.79 and 3.57 in New England occurrences.

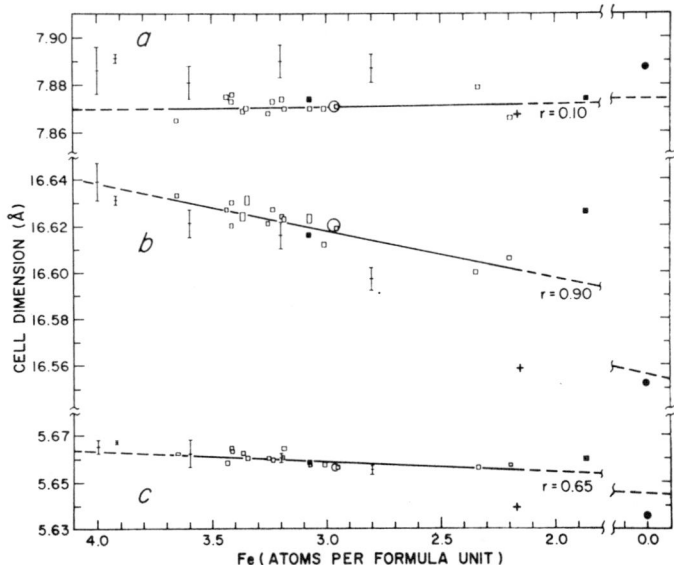

Figure 6. Variations of cell edges with unit cell Fe content. Data from Richardson (1967) (vertical bars) and Schreyer and Seifert (1969) (solid circles) are for
synthetic staurolites. Other data are from Juurinen (1956) recalculated from 2θ
values by least-squares methods (solid squares); von Knorring et al. (1979) (crosses);
Smith (1968) (open circles); and Griffen and Ribbe (1973) (open rectangles). All
symbols are 2 e.s.d.'s in height. Regression lines were calculated using only data
from Griffen and Ribbe (Fig. 2, p. 490).

LATTICE PARAMETERS

Lattice parameters for the pseudo-orthorhombic cell of natural and
synthetic staurolites are compiled in Figure 6 as a function of the number
of Fe atoms per unit cell. The angle β was found to be within a few
minutes of 90°, regardless of composition (Hurst et al., 1956; Griffen and
Ribbe, 1973). The following structural rationalization of cell edge
variation is paraphrased from Griffen and Ribbe (pp. 491-492).

Parallel to the a cell edge, which is independent of Fe content, the
$Al_{0.7}Fe_2O_2(OH)_2$ layer consists of partially-filled Al-octahedra alternating
with two Fe-tetrahedra (see Figs. 1 and 2). With an increase in Fe, the
largest major cation, the Fe-tetrahedra become larger on the average;
however, a does not increase because the Al(3) octahedra, two-thirds of
which are unoccupied, are easily distorted. If the Fe $\rightleftarrows$ Al substitution
were to occur in the filled Al-octahedra of the kyanite layer, that layer
would presumably expand and a would increase. Since such is not the case,
it is proposed that this substitution occurs primarily in the $Al_{0.7}Fe_2O_2(OH)_2$
layer, as also suggested by Regnard (1976). This substantiates Hollister's

182

(1970) assumption (based on microprobe analyses of kyanites by Chinner *et al.* (1969) and Albee and Chodos (1969)) that the kyanite layer in stauro-lite is nearly pure Al$_2$SiO$_5$.[3]

The alternating kyanite and Fe-Al monolayers are perpendicular to *b*, the cell edge along which the greatest percent variation is observed. An increase in Fe increases the thickness of the monlayer at both the Fe-tetrahedra and the partially-filled Al-octahedra. Parallel to *b* the Fe-tetrahedra share corners with filled Al-octahedra of the kyanite layer, while the partially-filled octahedra of the Fe-Al monolayer share corners with filled Si-tetrahedra. Thus an increase in the thickness of this layer causes an increase in *b* (Fig. 6).

In the kyanite layer the key structural units along *c* are chains of edge-sharing Al-octahedra cross-linked by Si-tetrahedra. In the Fe-Al monolayer, they are slightly kinked chains of partially-filled Al-octa-hedra with corner-sharing Fe-tetrahedra alternating on either side of the chain (see Fig. 2). Since parallel kyanite layers are linked by these Fe-Al polyhedral chains, any increase in the size of the latter causes expansion of the former in the direction of *c*. The amount of expansion is restricted by the geometrical integrity of the nearly pure kyanite layer, thus the increase in *c* is slight.

These rationalizations were tested by Griffen *et al.* (1982) and found to be inadequate. As might be expected, their plots of cell dimensions as functions of the weighted mean radii of cations in the monolayer between kyanite slabs (see Fig. 1) has improved correlations over those in Figure 6 in which only Fe content is plotted on the abscissa. Griffen (1981, Table 2) lists lattice parameters of all synthetic staurolites in the [IV](Zn-Fe) and [IV](Fe-Mg) series. See Gibbons *et al.* (1981) for high-temperature data on [IV]Fe staurolites. Thermal expansion of synthetic Fe-staurolite was examined by Gibbons *et al.* (1981) under very difficult conditions, because dehydroxylation occurs during heating. But using distance least-squares (DLS) modelling, they obtained values for "hydroxy-lated staurolite" of 7.87, 7.17 and 7.57 × 10^{-6}°C^{-1} for *a*, *b*, and *c*, re-spectively, over the range 20-800°C.

- - - - - - - - - - - - - - -

[3]If substitution does occur in the kyanite layer, it will likely be divalent cations for Al in the Al(2) octahedral site, because the ligancy of this site consists of three charge balanced oxygens (ζO(2A) = ζO(2B) = ζO(5) = 2.0), one slightly underbonded oxygen (ζO(3) = 1.85), and the two O(1) atoms that are presumed to be half oxygen, half hydroxyl (ζO(1) = 1.2) (*cf.* Smith, 1968).

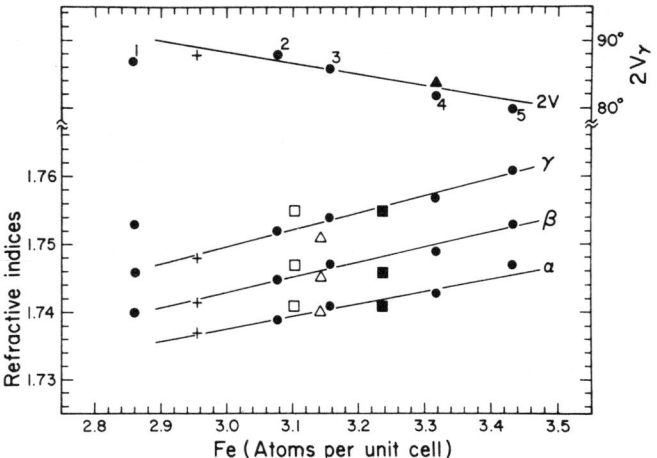

Figure 7. Variation of optical properties with Fe content. Data are taken from Juurinen (1956) (solid circles); Leake (1958) (solid triangle); Chinner (ms) (open squares); Snelling (1957) (solid squares); Hietanen (1969) (open triangles); and Griffen and Ribbe (1973, Fig. 1, p. 485) (crosses). Juurinen's specimen #1 was omitted in computing the regression lines.

OPTICAL AND OTHER PHYSICAL PROPERTIES

Staurolite's optically positive character is consistent with the complex edge-sharing octahedral chains which trend in the [001] direction and result as well in its pronounced prismatic habit parallel to [001] (Ribbe, 1976). Variation of refractive indices and $2V$ with the number of Fe atoms per unit cell is shown in Figure 7. It is evident from the scatter of data points, particularly Juurinen's specimen #1, that there may be chemical constituents other than iron that need to be considered in accounting for the optical properties. For example, refractive indices for pure zinc staurolite are $\alpha = 1.683$, $\beta = 1.694$, $\gamma = 1.700$ (Griffen, 1981); OH-content is likely to have a secondary effect. The brownish colors of staurolite and its golden yellow color and pleochroism in thin section are attributed mainly to tetrahedrally coordinated Fe^{2+} (see Vaughn et al., 1974; Dickson and Smith, 1976). Dispersion is $r > v$.

Because its structure is based on a cubic closest packed anion array, staurolite is dense (ranging from $\sim$3.74 g/cc for 2.8 Fe atoms per formula unit to $\sim$3.78 g/cc for 3.5 Fe) and hard (7.5 on the Moh's scale); it is used as an abrasive (Fulton, 1975). Staurolite resists weathering and is not uncommonly found as a "heavy mineral" in sediments derived from medium-grade regionally metamorphosed rocks.

The mineral staurolite got its name from two Greek words, *stauros* = cross and *lithos* = stone, which are descriptive of its well-known penetration twins, the 60°-cross and the less common 90°-cross. Hurst *et al.* (1956) studied these twin types optically and by x-ray precession methods and observed two of five possible twin laws for the 60°-crosses -- $[313]_{180°}$ and $[102]_{120°}$ -- and both possible twin laws for the 90°-crosses -- $[100]_{90°}$ and $[013]_{180°}$. The staurolite twin with (230) or ($\bar{1}30$) as twin plane reported by Dana (1876) was not found, and it may have been a random intergrowth. Smith (1968, p. 1139) suggested that "submicroscopic twinning on (010) [which he did not in fact observe] may have reduced the observed deviation [of his *C2/m* staurolite] from orthorhombic symmetry." Stacking faults may have the same effect.

Diversity in the appearance of staurolite twins is attributable to five factors: (1) unequal development of forms (Figs. 10,11,13); (2) multiple twinning (Fig. 12); (3) repeated twinning (Fig. 17); (4) combination of twinning with parallel intergrowths (Figs. 15 and 16); and (5) differences in the relative sizes of the twinned crystals (Hurst *et al.*, 1956).

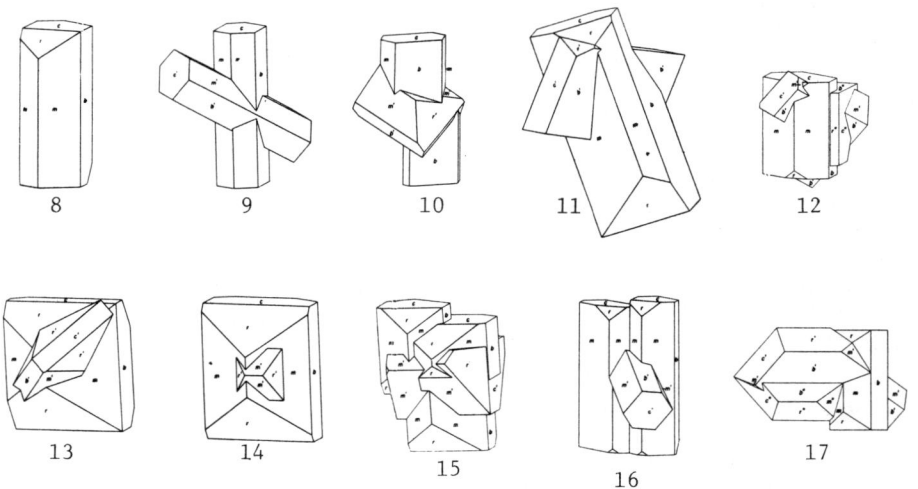

Figures 8-17. Portrait-drawings of staurolite twins, selected from 2100 specimens from Fannin County, Georgia: 8, Typical habit of staurolite crystal; 9, Typical 60°-cross; 10, 11, Malformed twins (60°-crosses); 12, Combined 60°-cross and 90°-cross; 13, Malformed 60°-cross; 14, Typical 90°-cross; 15, Parallel intergrowth and 90°-cross; 16, Parallel intergrowth and 60°-cross; 17, Trilling (repeated 60°-twin). From Hurst *et al.* (1956, Figs. 3-12, p. 151).

Albee, A. L. (1972) Metamorphism of pelitic schists: reaction relations of chloritoid and staurolite. Geol. Soc. Am. Bull., 83, 3249-3268.

_____ and A. A. Chodos (1969) Minor element content of coexistent Al_2SiO_5 polymorphs. Am. J. Sci., 267, 310-316.

Ashworth, J. R. (1975) Staurolite at anomalously high grade. Contrib. Mineral. Petrol., 53, 281-291.

Bancroft, G. M., A. G. Maddock and R. G. Burns (1967) Applications of the Mössbauer effect to silicate mineralogy. I. Iron silicates of known crystal structures. Geochim. Cosmochim. Acta, 31, 2219-2246.

Borg, R. J. and I. Y. Borg (1975) Proc. Int. Conf. Mössbauer Spectroscopy. Cracow, Poland, p. 167.

Čech, F., P. Povondra and S. Vrana (1981) Cobaltan staurolite from Zambia. Bull. Minéral., 104, 526-529.

Chinner, G. A. (1967) Chloritoid and the isochemical character of Barrow's zones. J. Petrology, 8, 268-282.

_____, J. V. Smith and C. R. Knowles (1969) Transition-metal contents of Al_2SiO_5 polymorphs. Am. J. Sci., 267-A, 96-113.

Dana, E. S. (1876) On new twins of staurolite and pyrrhotite. Am. J. Sci., Ser. 3, v. 11, 384-387.

Dasgupta, D. R. (1974) Topotactic transformations. Indian J. Earth Sci., 1, 60-72.

Deer, W. A., R. A. Howie and J. Zussman (1962) *Rock-forming Minerals, Ortho- and Ring Silicates*, Vol. 1. London: Longmans, p. 151-160.

Dickson, B. L. and G. Smith (1976) Low-temperature optical absorption and Mössbauer spectra of staurolite and spinel. Canadian Mineral., 14, 206-215.

Dollase, W. A. and L. S. Hollister (1969) X-ray evidence for ordering differences between sectors of a single staurolite crystal. Geol. Soc. Am. Progr. with Abstr., 268-270.

Dowty, E. (1972) Site distribution of iron in staurolite. Earth and Planet. Sci. Lett., 15, 72-74.

_____ (1976a) Crystal structure and crystal growth: I. The influence of internal structure on morphology. Am. Mineral., 61, 448-459.

_____ (1976b) Crystal structure and crystal growth: II. Sector zoning in minerals. Am. Mineral., 61, 460-469.

Dzhemats, A. C. and L. P. Nikitina (1977) Mössbauer spectroscopy of staurolite. J. Leningrad Univ., 24, 42-46 (in Russian).

Fitzpatrick, J. J. (1976) *Studies in the Microstructure and Crystal Chemistry of Minerals*. Ph.D. Diss., Univ. of California, Berkeley.

Foster, C. T., Jr. (1977) Mass transfer in sillimanite-bearing pelitic schists near Rangeley, Maine. Am. Mineral., 62, 727-746.

Fulton, R. B., III (1975) Indus. Minerals Rocks, 4th Ed. New York, AIME, p. 1095-1097.

Ganguly, J. (1972) Staurolite stability and related paragneiss: Theory, experiments, and applications. J. Petrol., 13, 335-365.

_____ and R. C. Newton (1965) Synthesis and stability of staurolite (abstr.). Geol. Soc. Am. Spec. Pap. 87, 63.

_____ and _____ (1968) Thermal stability of chloritoid at high pressures and relatively high oxygen fugacities. J. Petrology, 9, 444-466.

Gibbons, K., M. J. Dempsey and C. M. B. Henderson (1981) The thermal expansion of staurolite, $Fe_4Al_{18}Si_8O_{44}(OH)_4$. Mineral. Mag., 44, 69-72.

Gibson, G. M. (1978) Staurolite in amphibolite and hornblendite schists from the Upper Seaforth River, central Fiordland, New Zealand. Mineral. Mag., 42, 153-154.

Griffen, D. T. (1981) Synthetic Fe/Zn staurolites and the ionic radius of $^{IV}Zn^{2+}$. Am. Mineral., 66, 932-937.

_____ and P. H. Ribbe (1973) The crystal chemistry of staurolite. Am. J. Sci., 273-A, 479-495.

_____ T. C. Gosney and W. R. Phillips (1982) The chemical formula of natural staurolite. Am. Mineral., 67, 292-297.

Guidotti, C. V. (1970) The mineralogy and petrology of the transition from the lower to upper sillimanite zone in the Oquossoc Area, Maine. J. Petrology, 11, 277-236.

Hanisch, K. (1966) Zur Kenntnis der Kristallstruktur von Staurolith. N. Jahrb. Mineral. Monatsh., 362-366.

Harker, A. (1939) *Metamorphism*. London: Methuen, 2nd ed., 362 p.

Hellman, P. L. and T. H. Green (1979) The high pressure experimental crystallization of staurolite in hydrous mafic compositions. Contrib. Mineral. Petrol., 68, 369-372.

Hietanen, A. M. (1969) Distribution of Fe and Mg between garnet, staurolite, and biotite in aluminum-rich schist in various metamorphic zones north of the Idaho batholith. Am. J. Sci., 267, 422-456.

Hollister, L. S. (1969) Contact metamorphism in the Kwoieck area of British Columbia: An end member of the metamorphic process. Geol. Soc. Am. Bull., 80, 2465-2494.

_____ (1970) Origin, mechanism, and consequences of compositional sector-zoning in staurolite. Am. Mineral., 55, 742-766.

_____ and A. E. Bence (1967) Staurolite: sectorial compositional variations. Science, 158, 1053-1056.

Hörner, F. (1915) *Beiträge zur Kenntnis des Stauroliths*. Inaugural-Dissertation, Heidelberg.

Hoschek, G. (1967) Zur unteren Stabilitätsgrenze von Staurolith. Naturwiss., 8, 200.

_____ (1968) Zur oberen Stabilitätsgrenze von Staurolith. Naturwiss., 5, 226-227.

Hurst, V. J., J. D. H. Donnay and G. Donnay (1956) Staurolite twinning. Mineral. Mag., 31, 145-163.

Juurinen, A. (1956) Composition and properties of staurolite. Ann. Acad. Sci. Fenn., Ser A, III Geol., 47, 1-53.

Kwak, T. A. P. (1974) Natural staurolite breakdown reactions at moderate to high pressures. Contr. Mineral. Petrol., 44, 57-80.

Leake, B. E. (1958) Composition of pelites from Connemara, Co. Galway, Ireland. Geol. Mag., 95, 281-296.

Náray-Szabó, I. (1929) The structure of staurolite. Z. Kristallogr., 71, 103-116.

_____ and K. Sasvári (1958) On the structure of staurolite, $HFe_2Al_9Si_4O_{24}$. Acta Crystallogr., 11, 862-865.

Nemec, D. (1978) Zink in Staurolith. Chem. Erde, 37, 307-314.

_____ (1980) Zinkhaltiger Staurolith aus den Leptyniten der Blanicer Furche und aus dem übrigen Moldanubikum der Böhmisch - Märischen Höhe. Chem. Erde, 39, 311-320.

Niggli, P. (1926) Baugesetze kristalliner Materie. Z. Kristallogr., 63, 49-121.

Regnard, J. R. (1976) Mössbauer study of natural crystals of staurolite. J. Phys., 37, C6-797 to C6-800.

Ribbe, P. H. (1976) Polyhedral chains in the structures of gem orthosilicates: relationships to physical properties (abstr.). 25th Intern. Geol. Congr. Abstracts, 2, 593-594.

Richardson, S. W. (1967) The composition of synthetic Fe-staurolite. Carnegie Inst. Washington Year Book, 66, 397-398.

_____ (1968) Staurolite stability in a part of the system Fe-Al-Si-O-H. J. Petrol., 9, 467-488.

Schreyer, W. and G. A. Chinner (1966) Staurolite-quartzite bands in kyanite-quartzite at Big Rock, Rio Arriba County, New Mexico. Contr. Mineral. Petrol., 12, 223-244.

_____ and F. Seifert (1969) High-pressure phases in the system $MgO-Al_2O_3-SiO_2-H_2O$. Am. J. Sci., 267-A, 407-443.

Scorzelli, R. B., E. Baggio-Saiovitch and J. Danon (1976) Mössbauer spectra and electron exchange in tourmaline and staurolite. J. Phys., 37, C6-801 to C6-805.

Skerl, A. C. and F. A. Bannister (1934) Lusakite, a cobalt-bearing silicate from Northern Rhodesia. Mineral. Mag., 23, 598-606.

Smellie, J. A. T. (1974) Compositional variation within staurolite crystals from the Ardara aureole, Co. Donegal, Ireland. Mineral. Mag., 39, 672-683.

Smith, J. V. (1968) The crystal structure of staurolite. Am. Mineral., 53, 1139-1155.

Snelling, N. J. (1957) A note on the composition of staurolite from the Caenlochan schists. Mineral. Mag., 31, 603-604.

Takéuchi, Y., N. Aikawa and T. Yamamoto (1972) The hydrogen locations and chemical composition of staurolite. Z. Kristallogr., 136, 1-22.

Vaughan, D. J., J. A. Tossell and K. H. Johnson (1974) The bonding of ferrous iron to sulfur and oxygen in tetrahedral coordination: A comparative study using SCF Xα scattered wave molecular orbital calculations. Geochim. Cosmochim. Acta, 38, 993-1005.

von Knorring, O., Th. G. Sahama and J. Siivola (1979) Zincian staurolite from Uganda. Mineral. Mag., 43, 446.

Wenk, H. R. (1980) Defects along kyanite-staurolite interfaces. Am. Mineral., 65, 766-769.

Yardley, B. W. D. (1981) A note on the composition and stability of Fe-staurolite. N. Jahrb. Mineral. Mh., 1981, 127-132.

Chapter 8
KYANITE, ANDALUSITE and Other Aluminum Silicates   P.H. Ribbe

INTRODUCTION

It is hoped that a separate volume in this monograph series will be devoted to the aluminum silicate polymorphs, but in the meantime it is expedient to present brief descriptions of the crystal structures, chemical variations and physical properties of kyanite, andalusite, and sillimanite.[1] Using the Zoltai (1960) classification scheme, only kyanite and andalusite would be considered true *orthosilicates*, i.e., having "isolated" tetrahedra with sharing coefficients of 1.00, but sillimanite with its alternating, corner-sharing $[AlO_4]$ and $[SiO_4]$ tetrahedra would be considered a double chain aluminosilicate with sharing coefficient 1.75, a classification eminently suited to its physical properties. On the other hand Liebau (Chapter 1 , this volume) would consider all three polymorphs to be *monosilicates*, but sillimanite would only be included (along with pure anorthite) so long as there is never any demonstrable disorder among the corner-shared $[AlO_4]$ and $[SiO_4]$ tetrahedra. In the following discussion I will not be concerned with a solution to this taxonomic problem.

The importance of the aluminum silicate polymorphs to the metamorphic petrologist cannot be overstated, and these minerals have been the subjects of extensive petrographic, chemical, structural and experimental investigations. Although it is not strictly considered one of the aluminum silicate polymorphs, mullite has a structure in many ways similar to sillimanite; it is included in this chapter for the sake of completeness. Mullite is rare in nature, but it is very important as a ceramic material.

Phase Equilibria

There have been numerous attempts to ascertain the triple point in the stability diagram of andalusite-sillimanite-kyanite, whose free energies differ by less than 100 calories per mole.[2] Only two PT

---

[1] A selected bibliography of these minerals was compiled in 1959 by A. Grametbaur (Bull. of the U.S. Geological Survey, *1019-N*, 973-1046).

[2] Holdaway's (1971) opening statement is quotable: "The Al_2SiO_5 phase diagram is perhaps the most studied and least well defined silicate phase diagram. A glance at Figure 1 of Zen (1969) serves to illustrate this fact."

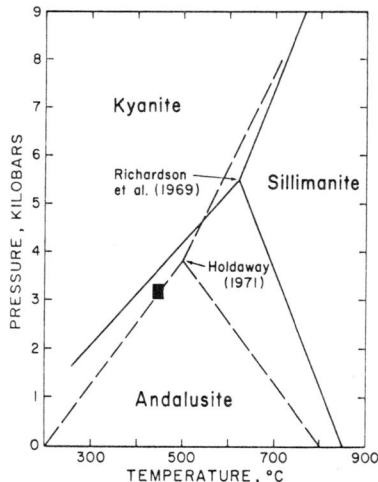

Figure 1. Experimental phase diagrams for
the aluminum silicate polymorphs determined
by Richardson et al. (1969) [solid lines]
and Holdaway (1971) [dashed lines]. The
rectangle represents the triple point at
P = 3.0 - 3.2 kbar, T = 420 - 440°C determined from lattice vibrations (phonon spectra)
by Salje and Werneke (1982)..

diagrams are reproduced here. One
is from Richardson *et al.* (1969)
with the triple point near 5.5 kbar
and 622°C, the other is from Holdaway (1971) with the triple point at
3.76 kbar and 510°C. The "polygon
of experimental uncertainty" of the
former does not overlap the triple
point determined by the latter, nor
that of Salje and Werneke (1982).
Holdaway (1971) listed some of the
factors which may contribute to discrepancies among studies of aluminum
silicate stability: (1) grain size
differences -- intense grinding produces strain in kyanite leading to
errors of 2 kbar in the kyanite-sil-
limanite boundary, and the presence
of fibrolite in sillimanite may raise the andalusite-sillimanite boundary
by 200°C in runs of short duration; (2) minor elements in solid solution
-- notably trivalent Fe and Mn (with occasional Cr); (3) disorder of
tetrahedral Al and Si in sillimanite. Salje and Werneke (1982) claim to
have superceded the problems arising from impurities, lattice distortions
and intergrowths by calculating specific heats from measured phonon
spectra associated with thermal lattice vibrations. Their triple point
has been added to Figure 1; they found that lattice faults in sillimanite
increase its T and P "considerably", but they did not address the problem
of $(Al,Si)^{IV}$ disorder.

$(Al,Si)^{IV}$ order/disorder in sillimanite

In a statistical thermodynamic evaluation of the Gibbs' free energy
of sillimanite as a function of Al,Si disorder at various temperatures,
Greenwood (1972, p. 570) concludes:

"Assumptions of the possibility of Al^{IV}- Si^{IV} disorder in sillimanite leads to the conclusions that sillimanite should coexist
metastably with kyanite and andalusite over a range of temperature, that use of sillimanite formed at one temperature to test
equilibrium at another temperature may be misleading, that dis-

190

crepant experimental results may be due in part to different
degrees of order in the sillimanite used, that $Al^{IV}-Si^{IV}$ disorder is as important as solid solution in displacing Al_2SiO_5
equilibria, and that in generaly any solid-solid reaction having a small entropy change will be highly sensitive to small
perturbations in free energy whatever the source, be it disorder, solid solution, or even grain size and surface energy."

However, Saxena (1974) warns that Al,Si disorder in sillimanite cannot be considered as ideal mixing on tetrahedral sites, because very
slight disordering results in impossibly great shifts of the ky-sil
phase equilibrium boundary. He advises using a non-ideal mixing model
which produces much more reasonable results, in line with observed variations in experimental data. Discussion of aluminum silicate equilibria
continues unabated. Uncertainties remain, but by contrast the crystal
structures, chemical variations, and physical properties are now rather
well characterized.

CRYSTAL STRUCTURES AND RELATED PHYSICAL PROPERTIES

Taylor (1928, 1929) determined the structures of andalusite and
sillimanite, and he was part of the team who deduced the structure of
kyanite from that of staurolite (Naray-Szabo *et al.*, 1929). Several
refinements appeared (Burnham and Buerger, 1961; Burnham, 1963a,b),
including neutron diffraction studies of sillimanite and andalusite
(Finger, 1972), but the latest work by Winter and Ghose (1979) is
the most precise and most useful, because it includes refinements at
25, 400, 600, 800 and 1000°C for sillimanite and andalusite and at
25, 400 and 600°C for kyanite, thus providing insight to the thermal
expansion behavior of these polymorphs at high temperatures.

All these structures have silicon in tetrahedral coordination
only, and each has one aluminum atom in octahedral coordination. The
major difference among the three is that the remaining aluminum is
4-coordinated in sillimanite, 5-coordinated in andalusite, and 6-
coordinated in kyanite. The kyanite structure is the simplest of the
three, being based on a cubic close-packed array of oxygen anions,
and we will examine it first.

Table 1. Summary of the physical properties of the Al_2SiO_5 polymorphs (see also Tables 2, 3 and 4).

SILLIMANITE	ANDALUSITE	KYANITE
$Al^{VI}[Al^{IV}Si^{IV}O_5]$	$Al^{VI}Al^{V}O[SiO_4]$	$Al^{VI}Al^{VI}O[Si^{IV}O_4]$

Space group (all polymorphs contain 4 formula units per unit cell)

Pbnm	*Pnnm*	$P\bar{1}$

Optical properties (after Deer *et al.* (1962) Vol. 1 pp. 121, 129, 137)*

	SILLIMANITE	ANDALUSITE	KYANITE
α	1.654-1.661	1.629-1.640	1.712-1.718
β	1.658-1.662	1.633-1.644	1.721-1.723
γ	1.673-1.683	1.638-1.650	1.727-1.734
	$2V_\gamma = 21\text{-}30°$	$2V_\alpha = 73\text{-}86°$	$2V_\alpha = 82\text{-}83°$
	$r > v$ strong	$r < v$	$r > v$ weak

[Figures by permission of Longman Group Ltd.]

Density of end-members (after Langer, 1976, p. 389)

	SILLIMANITE	ANDALUSITE	KYANITE
	3.144 g cm^{-3}	3.247 g cm^{-3}	3.674 g cm^{-3}
Hardness	6.5 - 7.5	6.5 - 7.5	4-5 ∥ z on {100}
			6-7 ∥ y on {100}
			∿7.5 on {1$\bar{1}$0}
Cleavage	{010} good	{110} good	{100} perfect
		{100} poor	{010} good
			{001} parting
Twinning		(101), rare	(100): t.a. ⊥
			(100), [010],
			[001]

Isothermal compressibility at 1 bar, 25°C: 0.77 ±0.04 $Mbar^{-1}$, all polytypes (Brace *et al.*, 1969)

---

*There has been considerable confusion in the literature about the optical properties of andalusite. Tröger (see Bambauer *et al.*, 1979, p. 50), Deer *et al.* (1961, p. 129), and Winchell and Winchell (1951, p. 521) all give refractive indices that would indicate that andalusite is positive but list 2Vα values of less that 90°, indicating a negative sign. Only Phillips and Griffen (1981) have them correctly described. Gunter (1982) determined that the sign *and* the index designations change with composition (see inset to the left below). He also measured (per. comm., 1981) optical properties for sillimanite (0.021 Fe per formula unit) and kyanite (0.003 Fe) whose structures were refined by Peterson (1980). They are recorded below together with refractive indices and 2V for "pure" andalusite whose values were determined by extrapolation in the figure to the left. *See also p. 207 & 208.*

	Sillimanite	Andalusite	Kyanite
Sign	(+)	(−)	(−)
$α_D$	1.6595	1.6325	1.7140
$β_D$	1.6609	1.6385	1.7236
$γ_D$	1.6812	1.6434	1.7296
$2V_{400}$	19.5°	$2V_{589}$	83.7°
$2V_{666}$	27.4°	≈ 84°	85.6°
$2V_{900}$	29.5°		82.3°

Table 2. Sillimanite, andalusite, and kyanite: cell dimensions as a function of temperature (with standard deviations in parentheses)

	25°C	400°C	800°C
$(\delta V/\delta T)_P$	7.230×10^{-4}		
Sillimanite			
a (Å)	7.4883(7)	7.4932(9)	7.4998(8)
b (Å)	7.6808(7)	7.7035(9)	7.7255(8)
c (Å)	5.7774(5)	5.7872(5)	5.7978(6)
V (Å³)	332.29(5)	334.06(6)	335.92(6)
V (cm³/mol)	50.049(7)	50.315(9)	50.595(9)
Andalusite			
a (Å)	7.7980(7)	7.8355(13)	7.8759(14)
b (Å)	7.9031(10)	7.9289(18)	7.9567(18)
c (Å)	5.5566(5)	5.5611(10)	5.5664(10)
V (Å³)	342.45(6)	345.50(11)	348.82(11)
V (cm³/mol)	51.58(1)	52.04(2)	52.54(2)
$(\delta V/\delta T)_P$	11.774×10^{-4}		
Kyanite			
a (Å)	7.1262(12)	7.1423(8)	7.1687(9)
b (Å)	7.8520(10)	7.8724(10)	7.8917(11)
c (Å)	5.5724(10)	5.5968(6)	5.6182(6)
α (°)	89.99(2)	89.94(1)	89.89(1)
β (°)	101.11(2)	101.18(1)	101.20(1)
γ (°)	106.03(1)	105.99(1)	105.98(1)
V (Å³)	293.60(9)	296.31(7)	299.29(7)
V (cm³/mol)	44.22(1)	44.63(1)	45.08(1)

Table 3. Sillimanite, andalusite, and kyanite: principal strain components of thermal expansion (with standard deviations in parentheses)

Variable Range	Principal Strain Components per 1°C $(\times10^{-5})$	Orientation of Principal Axes Angle (degrees) with respect to: +a	+b	+c
Kyanite 25→800°C	1.10(3)	109(5)	76(3)	15(3)
	0.81(3)	40(6)	68(6)	91(5)
	0.60(2)	123(5)	27(5)	105(3)
Sillimanite 25→800°C	0.66	90	0	90
	0.40	90	90	0
	0.16	0	90	90
Andalusite 25→800°C	1.29	0	90	90
	0.87	90	0	90
	0.23	90	90	0

Table 4. Calculated and calorimetric reaction enthalpies at atmospheric pressure and equilibrium temperature for the aluminum silicate polymorphic inversion reactions

Reaction	T_{eq}	Slope (bars deg.$^{-1}$)	ΔH at T_{eq} (cals/mol.) Skinner et al.	Present Study	Calorimetric
And.-Sill.	770°C	-13.97	800	670	670±230
Ky.-And.	220°C	12.49	1051	1043	956±200
Sill.-Ky.	326°C	20.19	-1647	-1617	-1720±170

a.

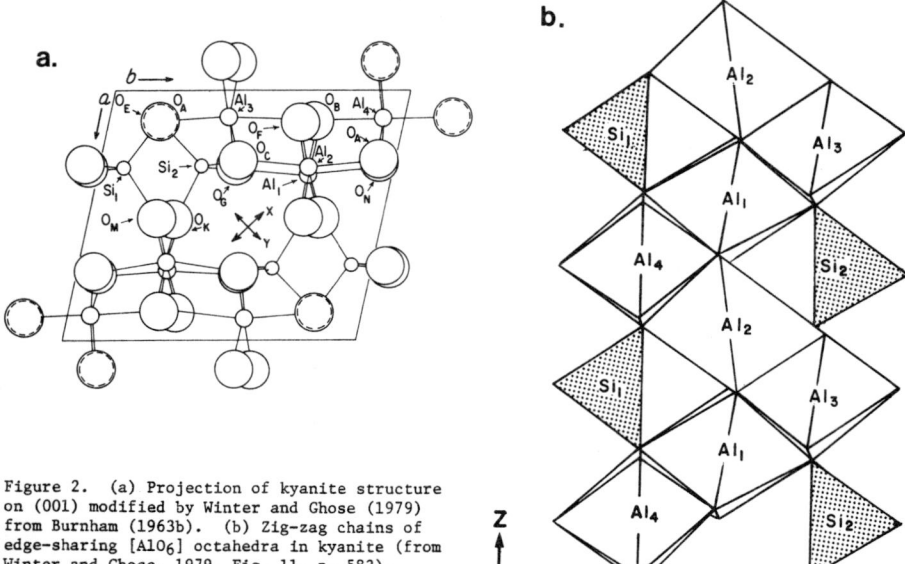

b.

Figure 2. (a) Projection of kyanite structure on (001) modified by Winter and Ghose (1979) from Burnham (1963b). (b) Zig-zag chains of edge-sharing [AlO$_6$] octahedra in kyanite (from Winter and Ghose, 1979, Fig. 11, p. 583).

For general reference purposes Table 1 contains a summary of physical properties of the nearly pure end-member aluminum silicates selected from the literature. The lattice parameters of all three polytypes vary linearly with temperature, and Table 2 lists the cell dimensions at 25, 400 and 800°C. Table 3 gives the principal strain components of thermal expansion and Table 4 the calculated and calorimetric reaction enthalpies at atmospheric pressure and equilibrium temperature for the inversion reactions; Tables 2-4 are from Winter and Ghose (1979).

Kyanite, $Al_2^{VI}O[SiO_4]$

Kyanite, the high-pressure polymorph, is ten percent or more dense than sillimanite and andalusite (Table 1). This is due in part to the fact that its structure is based on cubic close-packed oxygens, with 10% of the tetrahedral sites filled with Si and 40% of the octahedral sites filled with Al. All the Al is in zig-zag edge-sharing octahedral chains parallel to [001] (see Fig. 2 and compare Fig. 2a to Fig. 1 in Chapter 7 on staurolite). Burnham (1963b) noted that the unit cell dimensions of the triclinic kyanite cell were very similar indeed to those calculated for an ideal cubic close-packed oxygen arrangement.

The equations below are from Burnham (1963b),
the angles from Langer (1976), and the figure from Kostov (1968).

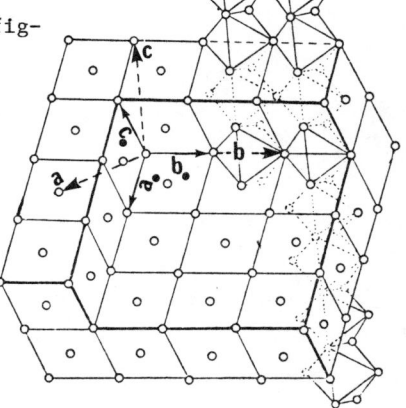

$$a_{ky} = \tfrac{3}{2} a_\bullet - \tfrac{1}{2} b_\bullet + c_\bullet$$
$$b_{ky} = 2b_\bullet$$
$$c_{ky} = - a_\bullet + c_\bullet$$

Angle	(a) calculated for ideal fcc oxygen	(b) observed in synth. Al-kyanite
α	90.0°	89.97±0.03°
β	100.9°	101.25±0.03°
γ	105.5°	106.03±0.03°

Differential hardness and cleavage.   The pronounced differential hardness of
kyanite is easily measured because of its excellent cleavage faces, but it is
widely variable (Winchell, 1945; see the comment under *Thermal expansion* below). The Moh's hardness on {100} is 4-5 parallel to [001] and 6-7 parallel
to [010]; kyanite is hardest ($\sim$7.5) on {1$\bar{1}$0} (Table 1). Cleavages and partings are not obviously related to the oxygen packing, but rather to the zig-
zag octahedral chain in the [001] direction. The perfect {100} cleavage is
parallel to the flat aspect of the chain (the plane of Fig. 2b); it is also
parallel to the only cleavage plane in staurolite ({100}$_{ky}$ || {010}$_{st}$ -- *cf.*
Fig. 1, Chapter 7). The {010} cleavage of kyanite is also parallel to the
octahedral chains, and it involves breakage of the next fewest number of Al-O
bonds (*cf.* Figs. 2a and 2b).

Twinning, deformation mechanisms, and antiphase domain boundaries.
In kyanite one type of antiphase domain boundary occurs on the close-packed
planes (220) with a fault vector $\tfrac{1}{2}[1\bar{1}1]$. Boland *et al.* (1977) further determined that kyanite deforms predominantly by the movement of superdislocations
(Burgers vector [001], fault vector $\tfrac{1}{2}[0\bar{0}1]$) associated with a second type of
antiphase domain boundary on (100). Deformation twinning also occurs on (100)
with twin vectors $\tfrac{1}{2}[001]$, $\tfrac{1}{2}[010]$ or $\tfrac{1}{2}[011]$. These mechanisms explain the ex-
istence of fine-scale (100) lamellae ("striae") described by Mügge (1883) and
ascribed by Raleigh (1965) to gliding deformation with a slip system (100)[001]
Menard *et al.* (1979) found the same slip system in a study of kyanite under
uniaxial compression: they report an elastic limit of < 80 bars.

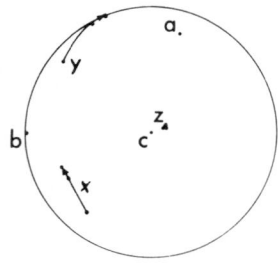

Thermal expansion. "The orientation of the maximum thermal expansion direction z for kyanite is within 12° of c" (Winter and Ghose, 1979; see Tables 2 and 3 and the inset, their Fig. 5). It is thus nearly parallel to the octahedral chain and nearly normal to the (001) parting. Of further interest is the fact that Buttgenbach (1923) showed that hardnesses determined parallel to [001] on both (100) and (010) were less by $\sim$1 unit on the Moh's scale than those measured normal to [001]. Highest thermal expansion and lowest hardness indicate that the weakest bonds are parallel to [001].

Since the $[SiO_4]$ tetrahedra do not expand at high temperatures, it is the $[AlO_6]$ octahedra, two of which (Al_1, Al_3) share five edges and two of which (Al_2, Al_4) share four edges with other polyhedra, that are responsible for the expansion. Only a careful study of a three-dimensional model will shed much light on the expansion of these complexly interlinked octahedra. The expansion directions $y > x$ are indicated by arrows in Figure 2a.

Andalusite, $Al^{VI}Al^{V}O[SiO_4]$

The structure of andalusite differs from that of kyanite in several respects, as illustrated in Figure 3. The chain of edge-sharing octahedra parallel to [001] contains only half of the Al atoms, and it is straight. Adjacent chains are cross-linked by pairs of edge-sharing distorted trigonal bipyramids which contain the other half of the Al atoms in irregular 5-fold coordination. This accounts in part for the fact that andalusite, $Al^{VI}Al^{V}O[SiO_4]$, is considerably less dense than kyanite, $Al_2^{VI}O[SiO_4]$; its mean refractive index is correspondingly lower -- 1.638 *vs* 1.720. The predominance of the [001] chains in andalusite account for its {110} prismatic habit and the {110} and {100} cleavages which are parallel to the chains (Table 1). The andalusite structure is in many ways more like sillimanite than kyanite, as is evidenced by a comparison of Figures 3 and 4, although reconstructive transformations of significant magnitude

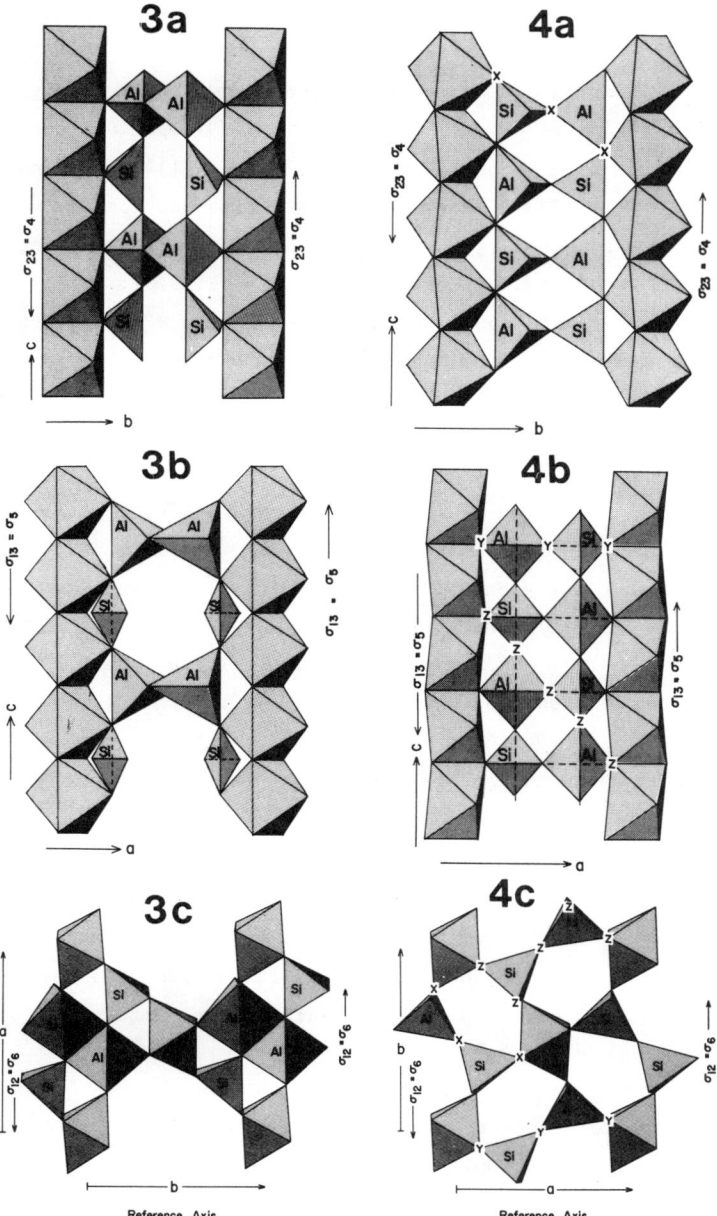

Fig. 3a – c. Crystal structure of andalusite. **a** Projection down a. Note chains of octahedra parallel to c, and additional chains of alternating SiO_4 tetrahedra and AlO_5 hexahedra also extending parallel to c. **b** Projection down b. SiO_4 tetrahedra are artificially shortened in the a direction to make octahedral chains more visable in this view. **c** Projection down c. The pair of arrows in each view indicated by σ_x represents the sense of shear stress and corresponding shear strain measured by the elastic constant with the repeated index x (cxx or sxx)

Fig. 4a – c. Crystal structure of sillimanite. **a** Projection down a. **b** Projection down b. **c** Projection down c. The pair of arrows in each view indicated by σ_x represents the sense of shear stress and corresponding shear strain measured by the elastic constant with repeated index x (cxx or sxx)

[copied directly from Figs. 4 and 5 of Vaughn & Weidner (1978, p. 141).]

197

are required to produce any one of the aluminum silicates from any other of the polymorphs.

There is an Mn^{3+} isotype of andalusite called *kanonaite*, recently discovered in Zambia (Vrána *et al.*, 1978). *Viridine* as a species name has been replaced by the term *manganoan andalusite* $(Mn < Al)^{VI}$; *aluminous kanonaite* has $(Mn > Al)^{VI}$. Chemical variations are discussed below, but it should be noted here that Mn^{3+}, Fe^{3+} and minor Fe^{2+} substitution occur in the octahedral *M1* site (Strens, 1968; Hålenius, 1978, 1979); trivalent Mn and Fe in lesser amounts are found in the 5-coordinated *M2* site (Abs-Wurmbach *et al.*, 1981; Weiss *et al.*, 1981).

Thermal expansion and compressibility. Schneider (1979a) determined the thermal expansion of andalusite at temperatures up to 1000°C, reporting a discontinuous expansion of *a* between 600 and 800°C, but the work of Winter and Ghose (1979) indicated linear expansions (*contra* Skinner *et al.*, 1961) for all axes and the volume, with the expansion of *a* > *b* >> *c* (Tables 2 and 3). Details of individual Al–O bond length variation with heating, which provide a rationale for the observed expansions, are discussed by Winter and Ghose, who refined the andalusite structure at 25, 400, 600, 800 and 1000°C. They report that the four short bonds in the $[AlO_5]$ polyhedron remain relatively unchanged, whereas the longest bond "expands considerably."

Vaughn and Weidner's (1978) determinations of the elastic constants (by the technique of Brillouin scattering) indicate that the *a*-axis in andalusite is the most compressible (c_{11} = 2.33 megabars) and the *c*-axis, parallel to the octahedral chains, is predictably the least compressible (c_{33} = 3.80 Mbars). These data are in perfect agreement with the observed thermal expansions.

Deformation. According to Schneider (1979b), shock-loaded andalusite distorts by gliding parallel to (001) and/or (100), and the common direction [010] of these planes is "probably the main direction of shock induced lattice deformation," although he suggests that "the discrepancy of [his] experiment [relative to structural considerations and the elastic properties] may be explained by the high shock energy and disequilibrium of dynamic deformation."

Sillimanite, $Al^{VI}[Al^{IV}Si^{IV}O_5]$

The crystal structure of sillimanite is characterized by straight chains parallel to [001] of edge-sharing [AlO$_6$] octahedra which are very similar to those in andalusite (*cf.* Fig. 3a with 4b and 3b with 4a). The primary difference is the fact that half of the Al in sillimanite is in 4-fold coordination rather than 5-fold, as in andalusite. Each [AlO$_4$] tetrahedron shares three corners with [SiO$_4$] tetrahedra, and vice versa, forming a unique sort of double chain which Liebau (Ch. 1, Fig. 8a, this volume) calls an "unbranched *einer* double chain." The thermodynamic significance of whether or not the Al,Si distribution is always, at all temperatures, fully ordered was discussed in the introduction to this chapter: there is no evidence to the contrary.

The chain-like polymerization of both tetrahedra and octahedra parallel to [001] is responsible for sillimanite's {010}, {110} prismatic habit, its common fibrous character (to which the name *fibrolite* has been given), its good {010} cleavage, and its strong optically positive character (partial birefringence (γ-β) $\sim$ 0.02).

Thermal expansion and compressibility. Winter and Ghose (1979; see Tables 2 and 3) give the axial expansion of sillimanite $b > c > a$, and this is structurally explicable in terms of opposite rotations, parallel to [001], of the XXX and YYY ($\equiv$ ZZZ) [SiAlO$_7$] groups that have a bow-tie appearance in Figure 4c. These structural units in the [001] projection are in fact the symmetrically non-equivalent Al,Si tetrahedral chains which rotate with increasing temperature into the surrounding open "tunnels" with pentagonal cross-section (*e.g.*, the upper-left polygon labelled XZZXX). The compressibilities are c_{22} = 2.32 Mbars parallel to b, c_{11} = 2.82 Mbars parallel to a, which are consistent with thermal expansions $b > a$. Although the axis of least compressibility is the [001] chain direction (c_{33} = 3.88 Mbars), as in andalusite, this is not consistent with the $b > c > a$ expansions with temperature. For a detailed discussion of the elastic constants and an excellent comparison of the structural deformation under shear stress of both sillimanite and andalusite, see Vaughn and Weidner (1978).

Fibrolite. This name is given to a fine-grained acicular variety of sillimanite which has been thought to rapidly crystallize as a meta-stable phase, possibly containing excess silica (Holdaway, 1971) or to differ in Al,Si order from sillimanite. Using electron diffraction and microprobe analysis, Cameron and Ashworth (1972) have shown that natural fibrolite does not differ in ordering scheme or composition from coarser grained sillimanites, and they suggest (p. 136) that "One must conclude that the 'fibrolite effect,' involving an overstepping of the univariant line andalusite - sillimanite in *P-T* space as discussed at length by Holdaway, if it exists, is more likely to be caused by an increase in surface free energy due to grain size" [see *Introduction*]. Chinner *et al.* (1969) warn that many petrographically described fibro-lites have been proven by microprobe analysis to be muscovite pseudo-morphs.

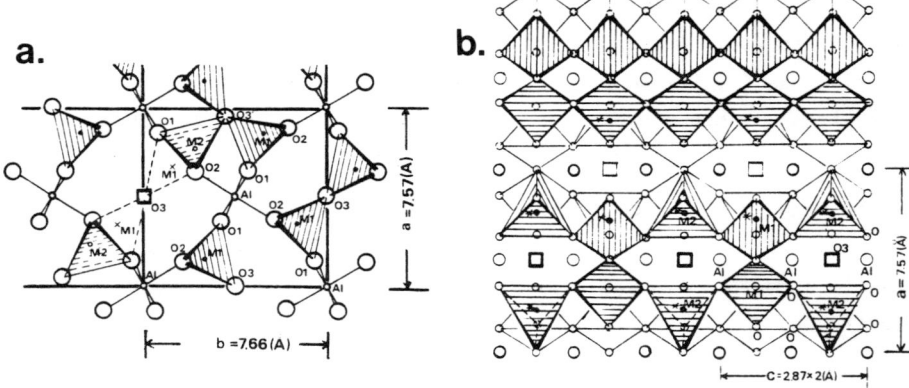

Figure 5. Projections of the 2:1 mullite structure after Nakajima *et al.* (1975, Fig. 1, p. 173). Al- and Si-tetrahedra are shaded. Oxygen vacancies at *O3* (from the sillimanite structure) are shown by squares. In Figure 5a Al has shifted from the *M*1 into the *M*2 site.

Mullite, $Al^{VI}[Al_{1+2x}Si_{1-2x})^{IV}O_{5-x}]$

Mullite is rare in nature, but it is an important ceramic material, melting at $\sim 1850°C$. According to Agrell and Smith (1960), the composi-tion of mullite ranges between the extremes $3Al_2O_3 \cdot 2SiO_2$, $2Al_2O_3 \cdot SiO_2$ and $3(Al_{0.9}Fe_{0.1})_2O_3 \cdot 2SiO_2$. Sadanaga *et al.* (1962) and Burnham (1963c) found the average structure of mullite to be very similar to that of sillimanite with straight chains of edge-sharing $[AlO_6]$ octahedra, but with oxygen vacancies and with excess Al in a new tetrahedral site co-ordinated to the displaced O_c (= O3) atom and the three other oxygens of the structure of stoichiometric sillimanite (see Figs. 5,6,7).

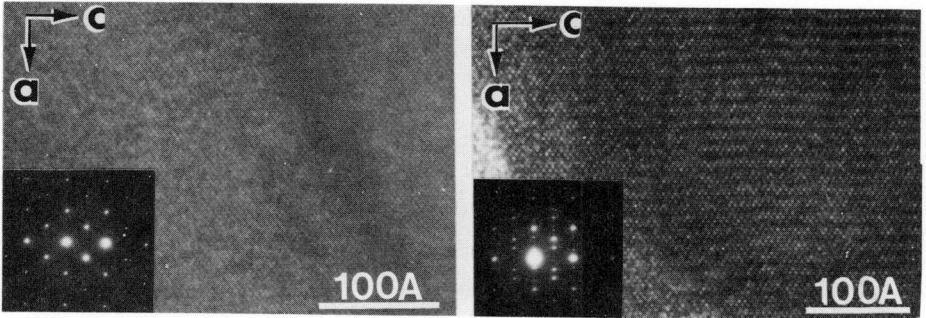

Figure 6. High resolution TEM lattice images from sillimanite (left) and mullite (right), both taken with the electron beam parallel to [010]. White spots in the mullite image represent oxygen vacancies. After Nakajima *et al.* (1975, Figs. 4 and 2, pp. 177 and 175).

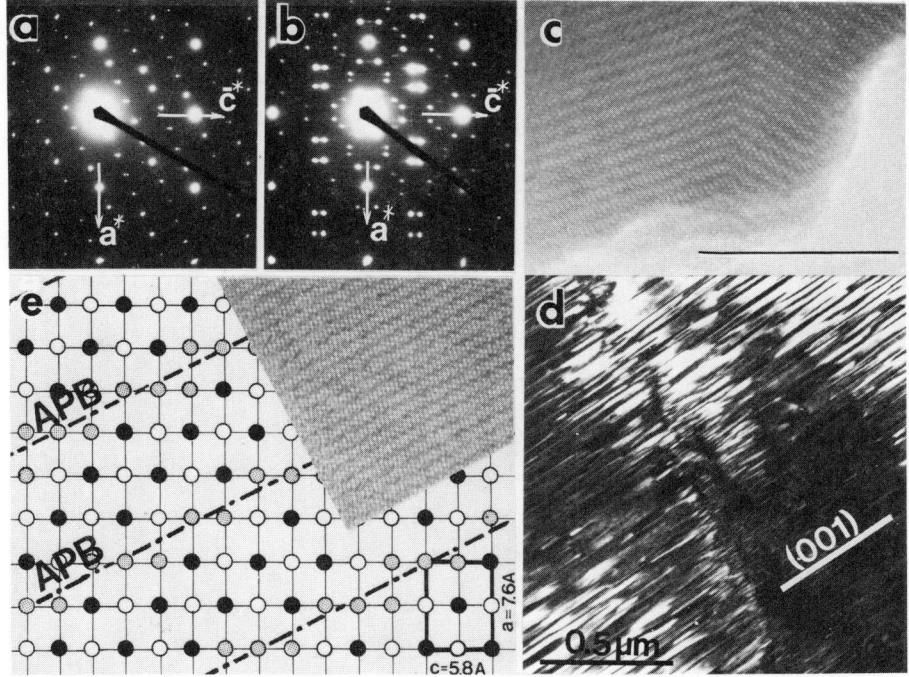

Figure 7. Electron diffraction patterns of Al-rich mullite: (a) untwinned and (b) twinned with 'e' and 'f' superlattice spots doubled in number. (c) High resolution TEM image of a twin boundary (vertical). Scale bar equals 150Å. (d) Lower magnification image of multiple twins. (e) Superstructure model of nonrational antiphase boundaries on a drawing of the O3 sites: open circles are vacant sites, filled circles are oxygens, shaded circles are partially occupied sites. Inset shows lattice image of $x = 0.55$ mullite; same scale as (c).

The charge-balancing substitution is $2Al^{3+} + \square = 2Si^{4+} + O^{2-}$, and the general formula is $Al_{2+2x}Si_{1-2x}O_{5-x}$, with $0.085 < x < 0.295$ (Cameron, 1977a,b). Burnham (1963c,d) considered mullite to have a disordered Al,Si distribution, but this is true only of the *average* unit cell with

$c \simeq 2.9A$. Correctly interpreted, the complex diffraction patterns of Al-rich mullites[3] indicate that mullite has an incommensurate antiphase domain structure with a nonintegral periodicity ranging from 9 to ~15A which changes orientation from parallel to [100] for $x < 0.5$ to a range of orientations with increasing [001] component for $x > 0.5$ (Nakajima and Ribbe, 1981). Tokonami et al. (1980) proposed and successfully tested an elaborate model for mullite based on tricline superstructures involving ordered oxygen vacancies and primitive "orthorhombic" sub-cells with $a = 7.6A$, $b = 7.7A$ and $c = 2.9A$ (see Fig. 7a). Their model supercedes that of Guse and Saalfeld (1976) and adequately explains the intensities and positions of subsidiary non-Bragg diffraction maxima observed in most mullites (cf. Cameron, 1977b, Fig. 3a-h, p. 752). Twinning is the cause of the doubling of satellite reflections in a^*c^* diffraction patterns (which were misinterpreted by Cameron, 1977b; see his Fig. 3). The electron micrograph in Figure 7d illustrates the fine scale of twinning common in Al-rich mullites (Nakajima and Ribbe, 1981).

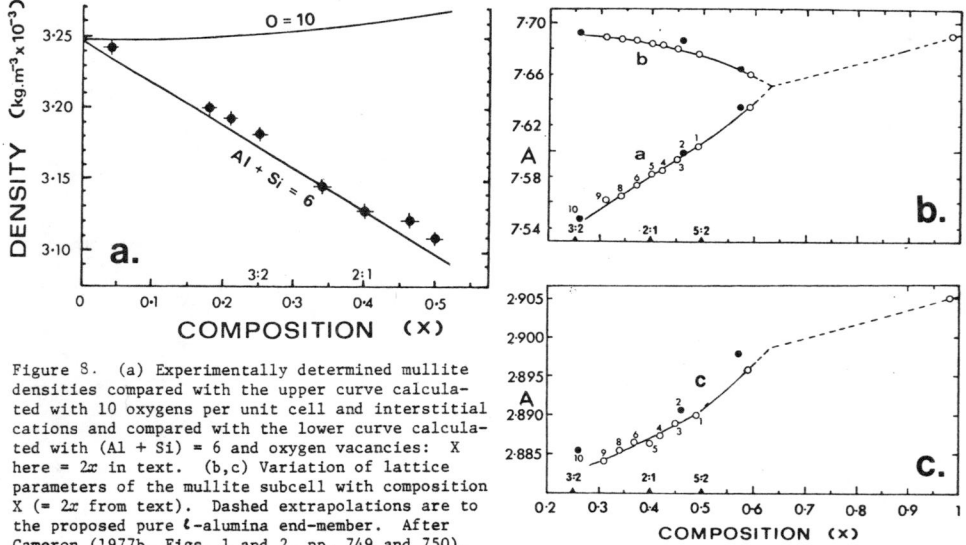

Figure 8. (a) Experimentally determined mullite densities compared with the upper curve calculated with 10 oxygens per unit cell and interstitial cations and compared with the lower curve calculated with (Al + Si) = 6 and oxygen vacancies: X here = $2x$ in text. (b,c) Variation of lattice parameters of the mullite subcell with composition X (= $2x$ from text). Dashed extrapolations are to the proposed pure ℓ-alumina end-member. After Cameron (1977b, Figs. 1 and 2, pp. 749 and 750).

[3]   The a^*c^* diffraction patterns of mullite bear strong resemblance to the b^*c^* nets of intermediate plagioclase with 1st, 3rd... order 'e' reflections and 2nd, 4th... order 'f' reflections. Compare Figures 1 and 3 of Nakajima and Ribbe (1981) with Figure 1 of Grove (1976), and see Figures 7 and 8.

Cameron (1977a, p. 265) stated that "complete solid solution between sillimanite and mullite can be shown to occur when Fe^{3+}, in the
virtual absence of Ti, replaces Al^{VI}", but by contrast, in a "possible
phase diagram" for the system $Al_2O_3 \cdot SiO_2 - SiO_2$ (Cameron, 1977b, his Fig.
6), he suggests that exsolution between Fe-free sillimanite and mullite
occurs below $\sim 900°C$. He found that the density and cell dimensions of
mullite vary regularly with x, where x = number of oxygen vacancies per
formula unit as in the formula above (see Fig. 8 and cf. the results of
Hariya et al., 1969, which show the effects of temperature, pressure
and Al_2O_3 content on cell parameters of natural and synthetic compounds
in the range 2:1 mullite $\rightarrow$ sillimanite).[4] The vector $2s$ joining pairs
of prominent satellite peaks in a^*c^* electron diffraction patterns also
varies almost linearly with x but may be affected by quench rates
(Nakajima and Ribbe, 1981, their Fig. 6). Cameron considers mullite to
be a substituted iota alumina ($\iota - Al_2O_3$).

The reader is referred to the literature for further details of
the structure and chemistry of this important (non-orthsilicate!) refractory material. Of particular value is a paper by Davis and Park
(1971) in which are reviewed phase relations in the system $Al_2O_3 - SiO_2$,
the details of mullite synthesis, its thermodynamic and chemical properties, and its mechanical and electrical properties (cf. Shaffer, 1964).

A "new polymorph of Al_2SiO_5" was reported by Aramaki and Roy (1963)
in their experimental studies in the system $Al_2O_3 - SiO_2 - H_2O$. It crystallizes in space group $Pbam$ and appears to be structurally similar to
andalusite. However, it may not have the 1:1 $Al_2O_3 \cdot SiO_2$ composition
(see also Turco, 1964).

COMPOSITIONAL VARIATONS AND RELATED PHYSICAL PROPERTIES

Naturally-occurring Aluminum Silicates

As indicated in the introduction to this chapter, minor elements
in solid solution -- notably Fe^{3+} and Mn^{3+} and possibly Cr^{3+} -- may
have significant effects on the aluminum silicate phase equilibria in
nature, and for that reason they deserve special consideration.

---

[4]Earlier studies of cell dimensions, "solid solution" and polymorphism
of mullite and sillimanite worthy of note are those by Agrell and Smith
(1960), Aramaki and Roy (1963), Hariya et al. (1969), and Cameron (1976a,
b,c).

The quantitative spectrographic study of 12 kyanites, 3 sillimanites and 7 andalusites by Pearson and Shaw (1960, p. 813) concluded that there "were no indisputable differences in trace-elements between [sic] the three minerals, but there is a possible concentration of B, Be, and Ba in sillimanite and a probable paucity of Cr in andalusite. These results suggest that minor and trace-elements cannot be considered as factors influencing the polymorphic behavior of the minerals." Ga, Cr, Mg, Ti, Li, Cu, V, Zr and Mn were measured in most minerals, B, Be and Ba in several, and Ni, Co, Mo, Sn, Ag, Y, Sc, Sr and Pb were usually below detection limits. Alkali metals were found to be present in small amounts (in "lattice voids"?) but are not believed to be essential constituents of the structure (*contra* Jakob, 1937; 1940, 1941; *cf.* Henriques, 1957; Eigenfeld and Machatski, 1957). Grew and Hinthorne (pers. comm. 1982) found up to 0.45 wt % B_2O_3 and 0.23 wt % MgO in sillimantites in a kornerupine-bearing rock from Paderu, India.

Chinner *et al.* (1969) claim that neither Na, K, Ca nor P occur in the structures of aluminum silicates, at least at their detection level of 0.01 wt %. Evidence is given that inclusions are responsible for these elements wherever they are detected; *e.g.*, Ca and P vary simultaneously and are presumed to be in apatite. Mg, Sc and Mn were also undetected at the 0.01 wt % level in the 10 kyanites, 6 sillimanites, 3 andalusites (viridine excepted), and 6 mullites they analyzed. Ti, V, Cr and Fe were found, with Fe the only metal occurring regularly in amounts greater than 0.2 wt %, although Cr is known to reach 18 mol % Cr_2SiO_5 in a kyanite from a Cr-rich grospydite nodule (Sobolev *et al.*, 1968). The highest iron content reported in natural specimens is 2.5 mol % Fe_2SiO_5 in a kyanite associated with hematite and quartz.

Although the Fe^{3+} content may change abruptly within zoned grains, indicating "lack of equilibrium probably caused by diffusion control of chemical supply during growth" (Chinner *et al.*, 1969, p. 96), it is partitioned almost equally among the phases with but a slight increase through the essentially Mn-free sequence kyanite < sillimanite < andalusite. Based on the assumption that coexisting phases have equilibrated with respect to Fe^{3+}, Holdaway (1971) gives partition coefficients K_D of 1.3 for ky-and, 1.1 for ky-sil and 0.8 for and-sil, where K_D (at low iron concentrations) may be approximated by Fe_{ky}/Fe_{and}, Fe_{ky}/Fe_{sil}, etc.

Albee and Chodos (1969), Chinner *et al.* (1969), and Okrush and Evans (1970) would all agree that in regional metamorphosed rocks, the effects of solid solution of Fe^{3+} or any other minor element on the stability fields of the aluminum silicate polymorphs is probably very much overshadowed by metastable persistence of one phase in the field of another during a multiphase metamorphism. The calculations of Strens (1968) and Holdaway (1971) also show that the effect of iron content on stabilities is negligible, particularly since most aluminum silicates contain < 1 mol % of the Fe end member.

Andalusite, "Viridine" and Kanonaite (the $Mn^{3+}AlSiO_5$ end member)

Manganoan andalusite (often called *viridine*) is most commonly found in "low- to high-grade regional or contact metamorphic manganiferous sediments such as some quartzites or schists of specific sedimentary origins and relatively high oxidation states" (Abraham and Schreyer, 1975, p. 1) and not uncommonly coexists with nearly Mn-free kyanite or sillimanite. Andalusite is strongly favored by high-spin Mn^{3+} with its $3d^4$ configuration (Strens, 1968). Crystal field theory predicts that it will have three spin-allowed *d-d* transitions when found in a distorted octahedron with D_{4h} symmetry. The Al octahedron of andalusite is both the most appropriately distorted site and the largest of all Al^{VI} sites among the aluminum silicate polymorphs (see Fig. 9a).[5] The energy level diagram for Mn^{3+} in viridine is shown in Figure 9b, along with the

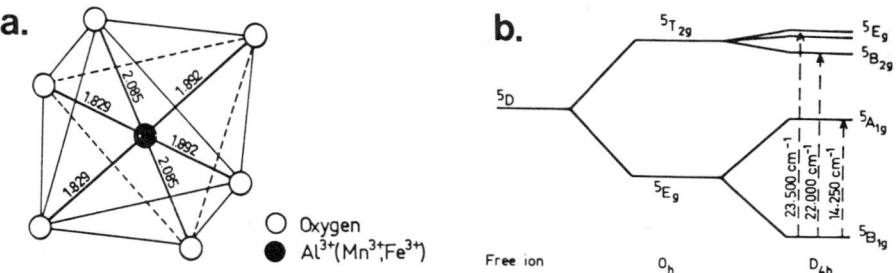

Figure 9. (a) The Al_1 octahedron in andalusite. (b) Proposed energy-level diagram for Mn^{3+} in the Al_1 octahedron in Mn-andalusite, showing transitions observed in polarized absorption spectra. From Hålenius (1978, Figs. 2 and 6, pp. 568 and 574).

---

[5] Mean Al^{VI}-O distances in kyanite octahedra are Al_1 = 1.902, Al_2 = 1.913, Al_3 = 1.919 and Al_4 = 1.896 A, in sillimanite 1.912 A and in andalusite 1.935 A. The mean Al-O distance for the [AlO$_5$] site in andalusite is only 1.836 A. Data from room-temperature refinements by Winter and Ghose (1979).

transitions observed in polarized absorption spectra; its crystal field stabilization energy is 47.6 kcal (Hålenius, 1978). In their four crystal structure refinements on the andalusite-kanonaite join, Abs-Wurmbach *et al.* (1981) assumed that neither trivalent Mn nor Fe entered the very much smaller 5-coordinated $M2$ polyhedron, even though their Mössbauer results showed up to 15% of the iron present (i.e. ~ 0.01 Fe atoms) in $M2$ and the mean $M2$-O bond lengths increased from 1.841A to 1.849A with total Mn in the formula. In *kanonaite*, $(Mn^{3+}_{\sim.75}Al_{\sim.25})AlSiO_5$, Weiss *et al.* (1981) found that $\sim$15% of the total Mn was in $M2$, but they recognized certain problems with their x-ray site-refinement that made them less than dogmatic about their results.

Not surprisingly, Abs-Wurmbach *et al.* (1980) found experimentally that andalusite, the low pressure-low temperature phase, is stabilized at the expense of both kyanite and sillimanite by incorporation of Mn^{3+}. Their calculations indicate that the Jahn-Teller effect, not volume change, is responsible for the stabilization of manganoan andalusite up to rather high pressures.

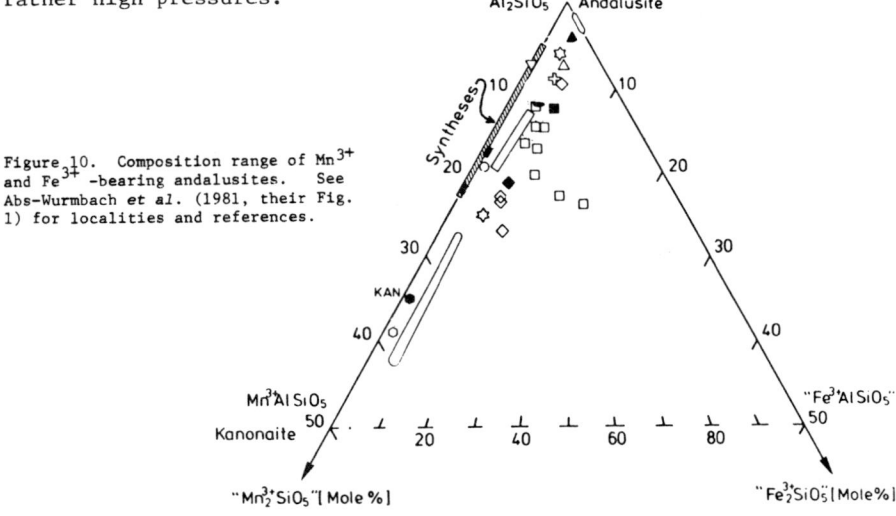

Figure 10. Composition range of Mn^{3+} and Fe^{3+}-bearing andalusites. See Abs-Wurmbach *et al.* (1981, their Fig. 1) for localities and references.

Electron spin resonance studies by Holuj *et al.* (1966) and Mössbauer parameters obtained by Hålenius (1978) show that Fe^{3+} is also concentrated in the octahedral site in andalusite, together with very minor Fe^{2+}. As mentioned above a small proportion of Fe^{3+} may occur in the 5-coordinated site, though apparently not in the tetrahedral site. Only Mn-rich andalusites contain significant amounts of Fe^{3+}; in rare specimens Fe^{3+} may exceed Mn^{3+} (Fig. 10).

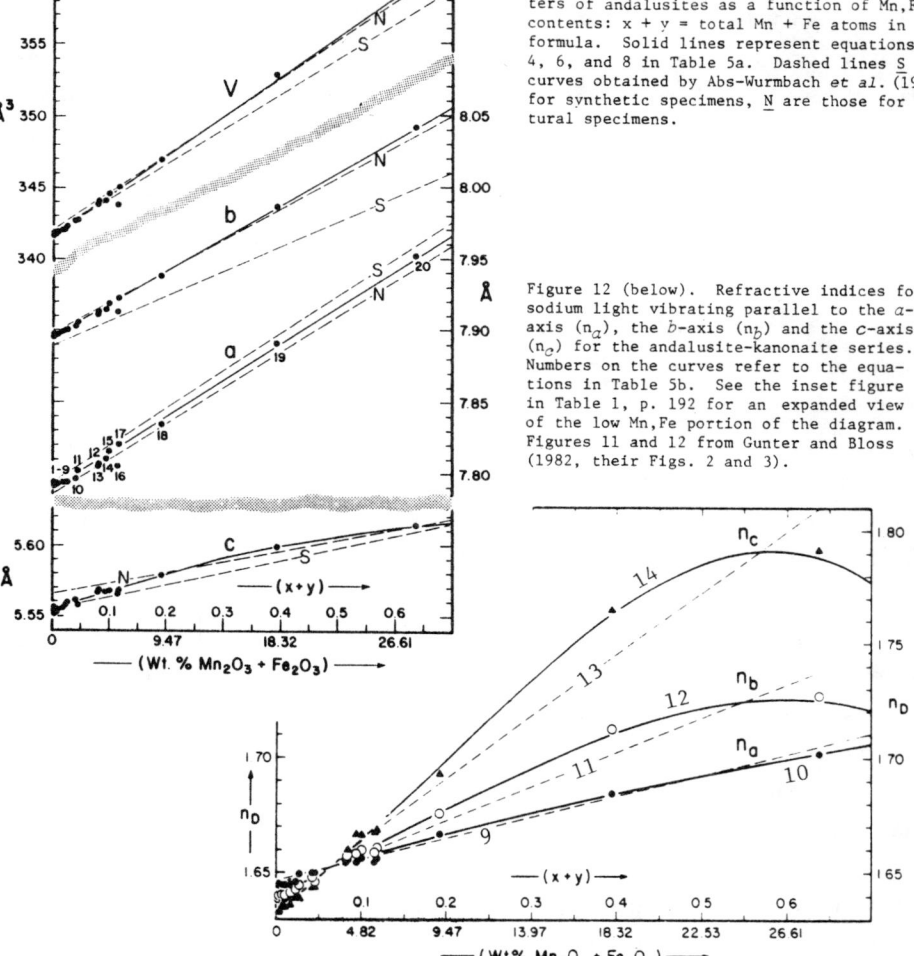

Figure 11 (to the left). Unit cell parameters of andalusites as a function of Mn,Fe contents: x + y = total Mn + Fe atoms in the formula. Solid lines represent equations 2, 4, 6, and 8 in Table 5a. Dashed lines S̲ are curves obtained by Abs-Wurmbach *et al.* (1981) for synthetic specimens, N̲ are those for natural specimens.

Figure 12 (below). Refractive indices for sodium light vibrating parallel to the *a*-axis (n_*a*), the *b*-axis (n_*b*) and the *c*-axis (n_*c*) for the andalusite-kanonaite series. Numbers on the curves refer to the equations in Table 5b. See the inset figure in Table 1, p. 192 for an expanded view of the low Mn,Fe portion of the diagram. Figures 11 and 12 from Gunter and Bloss (1982, their Figs. 2 and 3).

Table 5. Least-squares regression equations for (a) lattice parameters and (b) refractive indices for sodium light vibrating parallel to the a-axis (n_a), the b-axis (n_b), and the c-axis (n_c) in the system andalusite-kanonaite. x = Mn and y = Fe, in number of atoms per formula unit. From Gunter and Bloss (1982). See curves and data plotted in Figures 11 and 12.

Eqn.	(a) LATTICE PARAMETERS	Eqn.	(b) REFRACTIVE INDICES
1	$a = 7.7908(13) + 0.258(6)x + 0.162(39)y$	9	$n_a = 1.6436(4) + 0.091(2)x + 0.158(11)y$
2	$a = 7.7887(10) + 0.253(6)[x+y]$	10	$n_a = 1.6440(3) + 0.120(5)[x+y] - 0.046(8)[x+y]^2$
3	$b = 7.8956(8) + 0.232(4)x + 0.191(25)y$		
4	$b = 7.8948(6) + 0.230(4)[x+y]$	11	$n_b = 1.6371(8) + 0.141(3)x + 0.331(24)y$
5	$c = 5.5529(4) + 0.091(2)x + 0.230(12)y$	12	$n_b = 1.6389(4) + 0.18(1)[x+y] + 0.12(5)[x+y]^2$
6	$c = 5.5538(6) + 0.148(9)[x+y] - 0.086(14)[x+y]^2$		$\qquad\qquad - 0.30(6)[x+y]^3$
7	$V = 341.58(11) + 27.5(5)\ x + 29.2(32)y$	13	$n_c = 1.6289(19) + 0.260(8)x + 0.524(57)y$
8	$V = 341.61(8) + 27.6(4)[x+y]$	14	$n_c = 1.6324(8) + 0.27(2)[x+y] + 0.47(9)[x+y]^2$
			$\qquad\qquad - 0.8(1)\ [x+y]^3$

The variation of lattice parameters of the andalusite-kanonaite series was examined by Abs-Wurmbach *et al.* (1981) and Gunter and Bloss (1982). The latter have produced regression equations for a, b, c and volume as functions of total Mn + Fe (= x + y) -- see Table 5a and Figure 11 -- and multiple regression equations with x and y as separate variables (Table 5a, equations 1, 3, 5 and 7).

Refractive indices and optic signs of minerals in this solid solution series have been the source of considerable confusion, as discussed in the footnote of Table 1 (p. 192). Gunter and Bloss have resolved these problems with careful optical studies on individual grains whose compositions were subsequently determined by electron microprobe analysis. They found (Table 5b) that "substitution of Fe^{3+} increases n_a, n_b and n_c significantly more than does Mn^{3+}. Thus the standard plots of index *versus* [x + y] can only approximately conform to the data." The curves are in fact non-linear and intersect near x + y = 0.066, where andalusite will appear to be isotropic (compare Fig. 12 with the very different curves of Vrána *et al.*, 1978, their Fig. 4, p. 331).

Beran and Zemann (1969) reported that (110) plates of andalusite show strong pleochroism in absorption bands related to OH groups, but Wilkins and Sabine (1973) show a maximum of 0.05 wt % H_2O in andalusite. The spectacular pink-greenish-yellow optical pleochroism in andalusites was attributed to $O^{2-} \rightarrow Fe^{3+}$ and $Ti^{3+} \rightarrow Ti^{4+}$ charge transfer by Faye and Harris (1969), but Mn^{3+} is clearly responsible for the green color and pleochroism in viridines. Kanonaite is greenish-black and strongly pleochroic with n_a = green, n_b = deep emerald green, and n_c = deep golden yellow (see Gunter and Bloss, 1982, Table 6 for pleochroic and absorption formulae for twenty andalusite-kanonaites).

Sillimanite

Except for the possibility of Al,Si disorder [discussed in the *Introduction* to this chapter], the chemistry of sillimanite is relatively simple. Minor amounts of iron, chromium and titanium are detectable by microprobe analysis: Mn^{3+} is notably absent. Iron usually predominates, and there is some controversy as to its distribution between the Al^{vi} and Al^{iv} sites. LeMarshall *et al.* (1971) assigned Fe^{3+} to both sites in an interpretation of their EPR spectra, as did Grew and Rossman (1976) and Rossman *et al.* (1982), who relied on optical absorption and Mössbauer spectra for their arguments. By contrast, Hålenius (1979) interpreted his Mössbauer spectra in terms of Fe^{3+} and small amounts of Fe^{2+} in octahedral coordination, with no Fe^{3+} in the Al^{IV} site. His optical absorption study confirmed these assignments, but he conceded that tetrahedral Fe^{3+} may exist in some specimens in minute amounts. Cameron's (1977a) Mössbauer investigation of mullite showed Fe^{3+} only in octahedral coordination. The colors of sillimanite are summarized in the table below (from Rossman et al., 1982).

Color	Pleochroic formula	Contributing causes
YELLOW	X = yellow; Y = green-yellow Z = colorless	Fe^{3+} in 4- and 6-coord'n; Cr^{3+}
BROWN	X,Y = colorless to pale yellow Z = violet brown	Fe^{3+} in 6-coord'n and Fe^{2+}-Ti^{4+}, Fe^{2+} -Fe3+ charge transfer
BLUE	X,Y = colorless to pale yellow Z = blue	Same as blue with possible Fe^{2+}

Natural Kyanites

As with sillimanite, only iron, chromium and titanium are of any significance in natural kyanites. Parkin et al. (1977) summarize the history of spectroscopic investigations of transition metals in kyanite. Their conclusions are that Fe^{3+} predominates in green kyanite, as well as in blue kyanite, but that in the latter there is always a significant proportion of Fe^{2+}. Both $Fe^{2+} \rightarrow Fe^{3+}$ charge-transfer and Fe^{2+} spin-allowed crystal-field transitions thus contribute to the electronic spectra of blue kyanites. Smith and Strens (1976) correlated color intensity with Ti concentration. They assumed that charge-balancing coupled substitution $Fe^{2+} + Ti^{4+} \rightleftarrows 2Al^{3+}$ takes place in adjacent octahedral sites in the structure, giving rise to $Fe^{2+} \rightarrow Ti^{4+}$ charge-transfer, which produces a prominent band at 16,500 cm^{-1}. The assignment of this band to a spin-allowed crystal-field transition in Ti^{3+} (White and White, 1967) is not viable (see Parkin et al., 1977, p. 303).

Pleochroism is weak in kyanite, with α colorless, β violet-blue, γ cobalt-blue sometimes visible in thick sections (Deer *et al.*, 1961).

"The relation between chemical substitution and petrological growth environment of kyanite is more clearly marked than for andalusite [and sillimanite]. Cr contents above 0.10 percent are found only in those metamorphic facies in which kyanite can appear in rocks of basaltic composition, namely glaucophane schists and eclogites. [E.g., Sobolev *et al.* (1968) reported one Cr-rich kyanite with 18 mol % 'Cr_2SiO_5' from a grospydite inclusion in a Siberian kimberlite.] Fe contents show a strong correlation with the Fe^{3+}/Fe^{2+} ratio of the environment [and] one may conclude...that substitution of Fe for Al in kyanite is controlled by the oxygen partial pressure of the environment" (Chinner *et al.*, 1969, p. 103).

Water content of natural kyanites is given as 0.009 - 0.062 wt % H_2O by Wilkins and Sabine (1973).

Synthetic Transition Metal-Bearing Kyanites, $(Al_{2-x}M^{3+}_x)^{VI}O[SiO_4]$

Although, as we have seen, transition metal concentrations are negligibly low in most natural kyanites, it is obvious that solid solution of these elements in the aluminum silicate polymorphs will significantly affect stability relations among them. Following the work of Strens (1968) on the effects of Fe^{3+} and Mn^{3+} (*cf.* later calculations by Holdaway, 1971), Langer and co-workers began a systematic study of the influence of transition metal substitutions on the polymorphic transformations of kyanite, investigating the pseudobinary systems $Al_2SiO_5-M^{3+}_2SiO_5$ (where M^{3+} = Ti, V, Cr, Mn and Fe) at high temperatures and at pressures up to 20 kilobars. The studies of Seifert and Langer (1970) and Langer and Seifert (1971) on Cr^{3+}, Langer and Fentrup (1973) on Fe^{3+} and V^{3+}, and Abs-Wurmbach and Langer (1975) on Mn^{3+} were summarized by Langer (1976).

The maximum solubility of M^{3+} in kyanite at 20 kbar and $\sim$1000°C is shown in Figure 13 to decrease linearly with increasing effective ionic radius of the substituent, ranging from 24 mol % M_2SiO_5 with Cr^{3+} (r = 0.615 A) to zero with Ti^{3+} (r = 0.67 A).

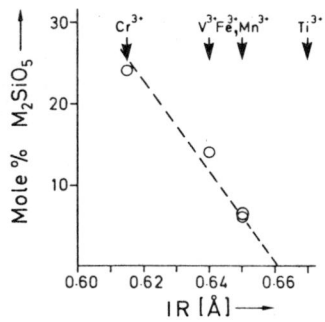

Figure 13. Maximum solid solubility of trivalent transition metals in kyanite at 20 kilobars and ∿1000°C, plotted as a function of the effective ionic radii (IR) of the substituent cations. Note that using the revised radii from Shannon (1976) [table below] the Fe^{3+} and Mn^{3+} values superpose. Modified from Langer (1976, Fig. 3, p. 394).

Lattice parameters increase with $M^{3+} \rightarrow Al^{3+}$, because the metal cations are all larger than Al ($r = 0.535$ A). The table below (modified from Langer, 1976, Table 4) summarizes the increases from pure Al_2SiO_5 with 10 mol % M_2SiO_5. Ionic radii are from Shannon (1976).

M^{3+}	Δa (A)	Δb (A)	Δc (A)	ΔVol (A^3)	Radius (A)
V	0.024	0.028	0.026	2.2	0.640
Cr	0.016	0.024	0.019	1.2	0.615
Mn*	0.029	0.015	0.026	2.2	0.645
Fe*	0.029	0.029	0.022	2.2	0.645

* extrapolated values

Langer (1976) concluded that the sizes of kyanite octahedra provide no reason for an Al/M^{3+} ordering in the substituted mixed crystals, but he did not have access to the data of Winter and Ghose (1979) [tabulated in footnote 3 above] which indicate $Al_3 > Al_2 > Al_1 > Al_4$ with a range of 0.023 A. Comparing the intensity ratio of the 300 and 200 x-ray reflections in powder patterns of $(Al_{1.5}Cr_{0.5})SiO_5$ with those calculated for different models of Al,Cr distribution, Langer and Seifert (1971) found a good numerical match with $Al_{0.5}Cr_{0.5}$ in the Al_1 and Al_2 octahedra, which have approximate symmetry D_{4h}, and only Al in the other sites. Crystal-field effects apparently predominate over size effects, as substantiated by the interpretation of diffuse reflectance spectra of Cr-bearing synthetic kyanites (Langer, 1976, Fig. 8 and p. 401).

The color effects of transition metals in the kyanites $(Al_{2-x}M^{3+}_x)SiO_5$ are tabulated below.

V^{3+} light greyish-green Cr^{3+} deep emerald green
Mn^{3+} light orange-yellow Fe^{3+} slightly yellowish-green

211

Abraham, K. and W. Schreyer (1975) Minerals of the viridine hornfels from Darmstadt, Germany. Contrib. Mineral. Petrol., 49, 1-20.

Abs-Wurmbach and K. Langer (1975) Synthetic Mn^{3+}-kyanite and viridine, $(Al_{2-x}Mn^{3+})SiO_5$ in the system $Al_2O_3-MnO-MnO_2-SiO_2$. Contrib. Mineral Petrol., 49, 21-38.

_____, _____, and W. Schreyer (1980) Viridine, $(Al_{1-x}Mn_x^{3+})_2(O|SiO_4)$: its stability relations as a function of P, T, and fO_2. Proc. 11th Gen'l Meeting. IMA, Novosibirsk 1978, Vol. 9: *Experimental Mineralogy*.

_____, _____, F. Seifert and E. Tillmanns (1981) The crystal chemistry of (Mn^{3+}, Fe^{3+})- substituted andalusites (viridines and kanonaite), $(Al_{1-x-y}Mn_x^{3+}Fe_y^{3+})_2(O/SiO_4)$: crystal structure refinements, Mössbauer, and polarized optical absorption spectra. Z. Kristallogr., 155, 81-113.

Agrell, S.O. and J.V. Smith (1960) Cell dimensions, solid solution, polymorphism, and identification of mullite and sillimanite. J. Am. Ceramic Soc., 43, 69-76.

Albee, A.L. and A.A. Chodos (1969) Minor element content of coexistent Al_2SiO_5 polymorphs. Am. J. Sci., 267, 310-316.

Anderson, P.A.M., R.C. Newton, and O.J. Kleppa (1977) The enthalpy change of the andalusite-sillimanite reaction and the Al_2SiO_5 diagram. Am. J. Sci., 277, 585-598.

Aramaki, S. and R. Roy (1963) A new polymorph of Al_2SiO_5 and further studies in the system $Al_2O_3-SiO_2-H_2O$. Am. Mineral., 48, 1322-1347.

Bambauer, H.U., F. Taborszky, and H.D. Trochim (1979) *Tröger's Optical Determination of Rock-forming Minerals*. Stuttgart: E. Schweizerbart, 188 p.

Beran, A. and J. Zemann (1969) Messung des Ultrarot-Pleochroismus von Mineralien, VIII Der Pleochroismus der OH-Streckfrequenz in Andalusit. Tschermaks Mineral. Petrogr. Mitt., 13, 285-292.

Boland, J.N., B.E. Hobbs, and A.C. McLaren (1977) The defect structure in natural and experimentally deformed kyanite. Phys. Stat. Sol. (a) 39, 631-641.

Brace, W.F., C.H. Scholz, and P.N. La Mori (1969) Isothermal compressibility of kyanite, andalusite, and sillimanite from synthetic aggregates. J. Geophys. Res., 74, 2089-2098.

Burnham, C. W. (1963a) Refinement of the crystal structure of sillimanite. Z. Kristallogr., 118, 127-148.

_____ (1963b) Refinement of the crystal structure of kyanite. Z. Kristallogr., 118, 337-360.

_____ (1963c) Crystal structure of mullite. Carnegie Inst. Wash. Year Book, 62, 223-227.

_____ (1963d) Composition limits of mullite, and the sillimanite-mullite solid solution problem. Carnegie Inst. Wash. Year Book, 62, 227-228.

_____ and M.J. Buerger (1961) Refinement of the crystal structure of andalusite. Z. Kristallogr., 115, 269-290.

Buttgenbach, H. (1923) Description des minéraux du Congo belge (Sixiéme memoire). Mem. Acad. R. Belgique, Cl. de Sci., 2, 7, 6 [Mineral. Abstr. 2-265].

Cameron, W.E. (1976a) Coexisting sillimanite and mullite. Geol. Mag., 113, 497-514.

_____ (1976b) Exsolution in 'stoichiometric' mullite. Nature, 264, 736-738.

_____ (1976c) A mineral phase intermediate in composition between sillimanite and mullite. Am. Mineral., 61, 1025-1026.

_____ (1977a) Composition and cell dimensions of mullite. Am. Ceram. Soc. Bull., 56, 1003-1007.

_____ (1977b) Mullite: a substituted alumina. Am. Mineral., 62, 747-755.

_____ and J.R. Ashworth (1972) Fibrolite and its relationship to sillimanite. Nature, 235, 134-136.

Chinner, G.A., J.V. Smith and C.R. Knowles (1969) Transition-metal contents of Al_2SiO_5 polymorphs. Am. J. Sci., 267-A, 96-113.

Davis, R.F. and J.A. Pask (1971) Mullite. *In* A.M. Alper, Ed., *High Temperature Oxides*, New York: Academic Press.

Deer, W.A., R.A. Howie, and J. Zussman (1961) *Rock-forming Minerals, vol. 1, Ortho- and Ring Silicates*, London: Longmans, 119-144.

Eigenfeld, I. and F. Machatski (1957) The supposed alkali content of kyanite. Osterr. Akad. Wiss. Math.-naturv. Kl. Anz., 94, 151-152 [C.A., 5984d (1959)].

Faye, G.H. and D.C. Harris (1969) On the origin of colour and pleochroism in andalusite from Brazil. Canadian Mineral., 10, 47-56.

Finger, L.W. and E. Prince (1972) Neutron diffraction studies: andalusite and sillimanite. Carnegie Inst. Wash. Yearbook, 71, 496-500.

Greenwood, H.J. (1972) Al^{IV}-Si^{IV} disorder in sillimanite and its effect on phase relations of the aluminum silicate minerals. Geol. Soc. Amer. Mem., 132, 553-571.

Grew, E.S. and G.R. Rossman (1976) Color and iron in sillimanite. Int'l Geol. Congr., Sydney, Abstr., 2, 564-565.

Grove, T.L. (1976) Exsolution in metamorphic bytownite. In H.-R. Wenk et al., eds., *Electron Microscopy in Mineralogy*. Berlin: Springer-Verlag, p. 265-270.

Gunter, M. (1982) *Relationship Between Chemcial Composition, Lattice Parameters, and Optical Properties of Andalusite and its Isostructural Analogs*. M.S. Thesis, Virginia Polytechnic Institute and State University, Blacksburg, Virginia.

_____ and F.D. Bloss (1982) Andalusite-kanonaite series: lattice and optical parameters. Am. Mineral. 67, in press.

Guse, W. and H. Saalfeld (1976) Das diffuse Beugungsbild von Mullit, $2Al_2O_3 \cdot SiO_2$. Z. Kristallogr., 143, 177-187.

Hålenius, U. (1978) A spectroscopic investigation of manganian andalusite. Canadian Mineral., 16, 567-575.

_____ (1979) State and location of iron in sillimanite. N. Jahrb. Mineral. Mh., 4, 165-174.

Hariya, Y., W.A. Dollase and G.C. Kennedy (1969) An experimental investigation of the relationship of mullite to sillimanite. Am. Mineral., 54, 1419-1441.

Henriques, Å. (1957) The alkali content of kyanite. Arkiv. Mineral. Geol., 2, 271.

Holdaway, M.J. (1971) Stability of andalusite and the aluminum silicate phase diagram. Am. J. Sci., 271, 97-131.

Holuj, F.J., J.R. Thyer, and N.E. Hedgecock (1966) ESR spectra of Fe^{3+} in single crystals of andalusite. Canadian J. Phys., 44, 509-523.

Jakob, J. (1937) Über den alkaligehalt der disthene. Schweiz. Mineral. Petr. Mitt., 17, 214-219.

_____ (1940) Über den Chemismus des andalusits. Schweiz. Mineral. Petr. Mitt., 20, 8-10.

_____ (1941) Chemische and strukturelle untersuchungen am disthene. Schweiz. Mineral. Petr. Mitt., 21, 131-135.

Kostov, I. (1968) *Mineralogy*. Edinburgh: Oliver and Boyd, 587 pp.

Langer, K. (1976) Synthetic $3d^{3+}$-transition metal bearing kyanites, $(Al_{2-x}M_x^{3+})SiO_5$. In R.G.J. Strens, (ed.), *The Physics and Chemistry of Minerals and Rocks*, John Wiley & Sons, London, 389-402.

_____ and K.R. Frentrup (1973) Synthesis and some properties of iron- and vanadium-bearing kyanites, $(Al,Fe^{3+})SiO_5$ and $(Al,V^{3+})_2SiO_5$. Contrib. Mineral. Petrol., 47, 47-46.

_____ and F. Seifert (1971) High pressure-high temperature synthesis and properties of chromium kyanite, $(Al,Cr)_2SiO_5$. Z. Anorg. Allgem. Chem., 383, 29-39.

Le Marshall, J., D.R. Hutton, G.J. Troup and J.R.W. Thyer (1971) A paramagnetic resonance study of Cr^{3+} in sillimanite. Phys. Stat. Sol., 5, 769-773.

Menard, D., J.C. Doukhan and J. Paquet (1979) Uniaxial compression of kyanite $Al_2O_3SiO_2$. Bull. Minéral., 102, 159-162.

Mügge, O. (1883) N. Jahrb. Geol. Palaeontol., Abh 1, 71.

Nakajima, Y., N. Morimoto, and E. Watanabe (1975) Direct observation of oxygen vacancy in mullite, $1.89Al_2O_3 \cdot SiO_2$ by high resolution electron microscopy. Proc. Japan Acad., 51, 173-178.

_____ and P. H. Ribbe (1981) Twinning and superstructure of Al-rich mullite. Am. Mineral., 66, 142-147.

Náray-Szabo, St., W.H. Taylor and W.W. Jackson (1929) The structure of cyanite. Kristallogr., 71, 117.

Okrush, M. and B.W. Evans (1970) Minor element relationships in coexisting andalusite and sillimanite. Lithos, 3, 261-268.

Parkin, K.M., B.M. Loeffler, and R.G. Burns (1977) Mössbauer spectra of kyanite, aquarmarine, and cordierite showing intervalence charge transfer. Phys. Chem. Minerals, 1, 301-311.

Pearson, G.R. and D.M. Shaw (1960) Trace elements in kyanite, sillimanite and andalusite. Am. Mineral., 45, 808-817.

Peterson, R.C. (1980) *Bonding in Minerals: I. Charge Density of the Aluminosilicate Polymorphs*. Ph.D. Dissertation, Virginia Polytechnic Institute and State University, Blacksburg, Virginia.

_____ and R.K. McMullen (1980) Neutron structure refinements of the Al_2SiO_5 polymorphs. Trans. Am. Geophys. Union, EOS 61, 409.

Petreus, Ion (1974) The divided structure of crystals. N. Jahrb. Mineral. Abh., 122, 314-338.

Phillips, W.R. and D.T. Griffen (1981) *Optical Mineralogy: The Nonopaque Minerals.* San Francisco: W.H. Freeman and Co., 677p.

Raleigh, C.B. (1965) Glide mechanisms in experimentally deformed minerals. Science, 150, 739-741.

Richardson, S.W., M.C. Gilbert, and P.M. Bell (1969) Experimental determination of kyanite-andalusite and andalusite-sillimanite equilibria: the aluminum silicate triple point. Am. J. Sci., 267, 259-272.

Rossman, G.R., E.S. Grew and W.A. Dollase (1982) The colors of sillimanite. Am. Mineral. 67, 749-761.

Saalfeld, H. (1979) The domain structure of 2:1-mullite ($2Al_2O_3 \cdot 1SiO_2$). N. Jahrb. Mineral. Abh., 134, 305-316.

Sadanaga, R., M. Tokonami, and Y. Takéuchi (1962) Structure of mullite $2Al_2O_3 SiO_2$, and relationship with the structures of sillimanite and andalusite. Acta Crystallogr., 15, 65-68.

Salje, E. and C. Werneke (1982) The phase equilibrium between sillimanite and andalusite as determined from lattice vibrations. Contrib. Mineral. Petrol., 79, 56-67.

Saxena, S.K. (1974) Order-disorder in sillimanite. Contrib. Mineral. Petrol., 45, 161-167.

Schneider, H. (1979a) Thermal expansion of andalusite. J. Am. Ceram. Soc., 62, 5, 6.

_____ (1979b) Deformation of shock loaded andalusite studied with x-ray diffraction techniques. Phys. Chem. Minerals, 4, 245-252.

Seifert, F. and K. Langer (1970) Stability relations of chromium kyanite at high pressures and temperatures. Contrib. Mineral. Petrol., 28, 9-18.

Shaffer, T.B. (1964) *Materials Index* I. New York: Plenum Press, 407-408.

Shannon, R.D. (1976) Revised effective ionic radii and systematic studies of interatomic distances in halide and chalcogenides. Acta Crystallogr., A32, 751-767.

Skinner, B.J., S.P. Clark, Jr. and D.E. Appleman (1961) Molar volumes and thermal expansions of andalusite, kyanite, and sillimanite. Am. J. Sci., 259, 651-668.

Smith, D.G.W. and J.D.C. McConnell (1966) A comparative electron-diffraction study of sillimanite and some natural and artificial mullites. Mineral. Mag., 35, 810-814.

Smith, G. and R.G.J. Strens (1976) Intervalence-transfer absorption in some silicate, oxide and phosphate minerals. *In* R.G.J. Strens (Ed.), *The Physics and Chemistry of Minerals and Rocks,* John Wiley & Sons, London, 583-612.

Sobolev, N.V., Jr., I.K. Kuznetsova, and N.I. Zyuzin (1968) The petrology of grospydite xenoliths from the Zagodochnaya kimberlite pipe in Yakutia. J. Petrol., 9, 353-380.

Strens, R.G.J. (1968) Stability of Al_2SiO_5 solid solutions. Mineral. Mag., 36, 839-849.

Taylor, W.H. (1928) The structure of sillimanite and mullite. Z. Kristallogr., 68, 503-521.

_____ (1929) The structure of andalusite, Al_2SiO_5. Z. Kristallogr., 71, 205-218.

Tokonami, M., Y. Nakajima, and N. Morimoto (1980) Diffraction aspect and structural model of mullite, $Al(Al_{1-2x}Si_{1-2x})O_{5-x}$. Acta Crystallogr., A36, 270-276.

Turco, G. (1964) Formation d'un nouveau silicate d'aluminum de formule Al_2SiO_5 au cours de le synthèse hydrothermale de la zunyite. C. R. Acad. Sci. Paris, 258, 3331-3334.

Vaughn, M.T. and D.J. Weidner (1978) The relationship of elasticity and crystal structure in andalusite and sillimanite. Phys. Chem. Minerals, 3, 133-144.

Vrána, S.M., M. Rieder, and J. Podlaha (1978) Kanonaite, $(Mn^{3+}_{0.76}Al_{0.23}Fe^{3+}_{0.22})^{[6]}Al^{[5]}[O|SiO_4]$, a new mineral isotopic with andalusite. Contrib. Mineral. Petrol., 66, 325-332.

Weiss, Z., S.W. Bailey and M. Rieder (1981) Refinement of the crystal structure of kanonaite, $(Mn^{3+},Al)^{[6]}(Al,Mn^{3+})^{[5]}O[SiO_4]$. Am. Mineral., 66, 561-567.

White, E.W. and W.B. White (1967) Electron microprobe and optical absorption study of colored kyanites. Science, 15B, 915-917.

Wilkins, R.W.T. and W. Sabine (1973) Water content of some nominally anhydrous silicates. Am. Mineral., 58, 508-516.

Winchell, H. (1945) The Knoop microhardness tester as a mineralogical tool. Am. Mineral., 30, 583.

Winchell, A.N. and H. Winchell (1951) *Elements of Optical Mineralogy, Part II.* New York: John Wiley and Sons, 551 p.

Winter, J.K. and S. Ghose (1979) Thermal expansion and high-temperature crystal chemistry of the Al_2SiO_5 polymorphs. Am. Mineral., 64, 573-586.

Zen, E-an (1969) The stability relations of the polymorphs of aluminum silicate: a survey and some comments. Am. J. Sci. 267, 297-309.

Zoltai, T. (1960) Classification of silicates and other minerals with tetrahedral structures. Am. Mineral., 45, 960-973.

Chapter 9

TOPAZ   **P. H. Ribbe**

INTRODUCTION

Topaz, $Al_2(F,OH)_2[SiO_4]$, occurs primarily in acid igneous rocks, often in late-stage hydrothermal veins in pegmatites or in greisen. It may be found in sediments as a detrital heavy mineral ($\rho \sim 3.5$) or as the product of pneumatolytic action in sediments near an igneous intrusion. Chemically, only very small concentrations (<0.04 wt %) of Fe and Cr for Al have been confirmed in topaz by microprobe analysis (Ribbe and Rosenberg, 1971), and no substituents for Si are known. Up to ~ 30 mol % OH may substitute for F, and this measurably affects optical properties (Penfield and Minor, 1894) and lattice parameters (Rosenberg, 1967).

The structure of the orthorhombic *Pbnm* polymorph was determined independently by Pauling (1928) and Alston and West (1928). When OH is present, it is disordered (Zemann *et al.*, 1979). Recently the structure of a triclinic *P*1 polymorph was refined (Parise *et al.*, 1980) using neutron diffraction techniques. In *P*1 topaz hydrogen associated with the hydroxyl ion is ordered on only one of eight sites which are otherwise symmetrically equivalent in *Pbnm* topaz. This ordering is apparently responsible for what have been called "anomalous" optical and electrical properties in both monoclinic and triclinic topazes (Akizuki *et al.*, 1979). These crystal structures will be discussed first to provide a context for a consideration of the physical properties of topaz.

CRYSTAL STRUCTURE

The Structure of Orthorhombic Topaz

Most topaz crystallizes in space group *Pbnm* with cell parameters $a \sim 4.65$, $b = 8.79 - 8.83$ (increasing with OH $\rightarrow$ F), $c \sim 8.39$ A. A very precise set of atomic coordinates for a specimen "from Brazil" containing < 5 mol % OH for F was refined by Ladell (1965), and based on those data Ribbe and Gibbs (1971, quoted freely below) gave a detailed account of the steric details of the crystal structure and their relation to physical properties. A compilation of mean bond lengths are shown in Figure 1, together with drawings of the orthorhombic

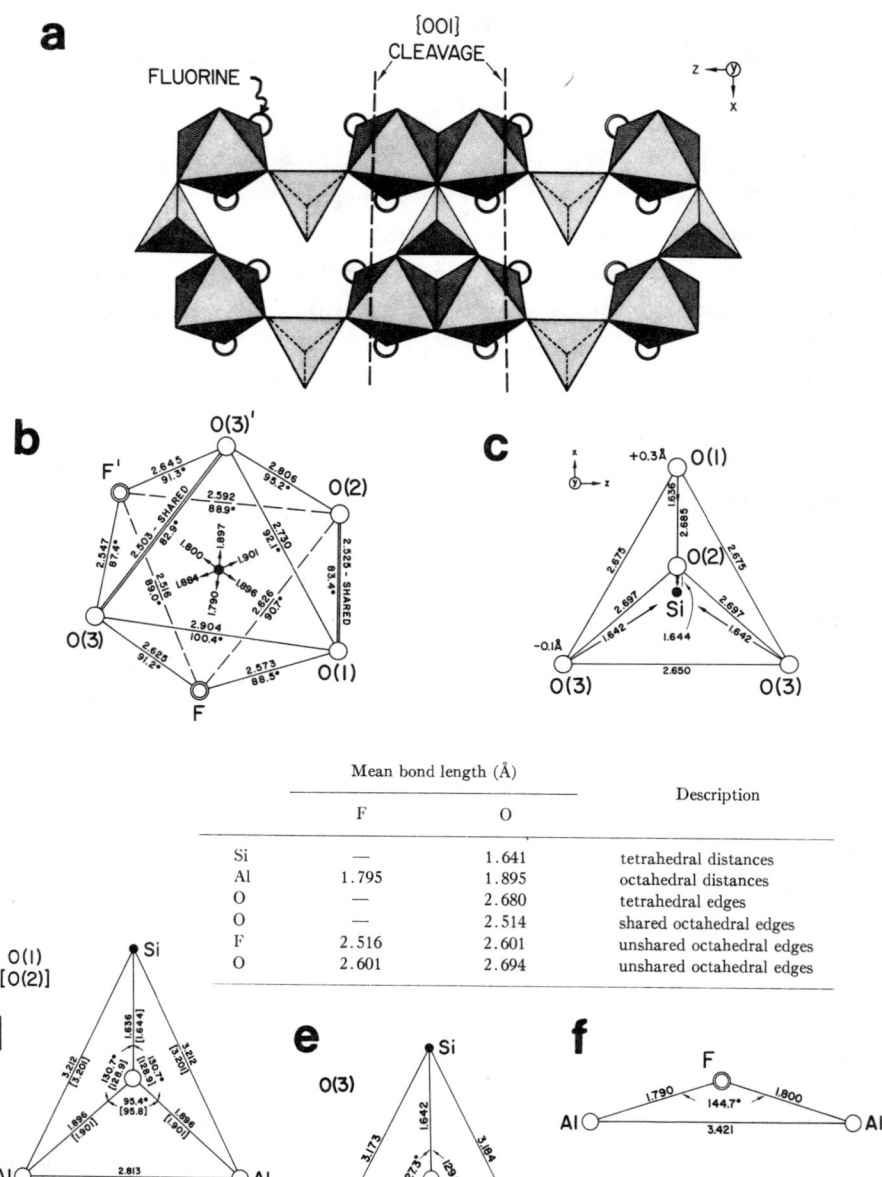

Figure 1. (a) Chains of edge-sharing [AlO₄F₂] octahedra and negative corner-sharing [SiO₄] tetrahedra cross-linked in the *x* direction by positive SiO₄ tetrahedra. The trace of the {001} cleavage is shown by dashed lines. Note that only Al-O and Al-F bonds are broken. (b & c) Cation coordination polyhedra: the [AlO₄F₂] octahedron and [SiO₄] tetrahedron. (d,e,f) Anion coordination diagrams for O(1) and O(2) [in brackets], for O(3), and for (F,OH). Interatomic distances (in Ångstroms) and angles are labeled. After Ribbe and Gibbs (1971, Figs. 1-3).

structure. The structure is based on closest-packed monolayers of oxygen anions alternating with monolayers of $[(F,OH)_2 O_1]_\infty$ in the mixed hcp-ccp sequence ABAC. One-third of the octahedral sites are filled with Al and one-twelfth of the tetrahedral sites with Si. Oxygen atoms are coordinated by 1 Si + 2 Al atoms, F,OH by two Al atoms (Fig. 1d,e,f). The key structural units are crankshaft-like chains of edge-sharing $[AlO_6]$ octahedra parallel to c cross-linked by corner-sharing $[SiO_4]$ tetrahedra (Fig. 1a). These chains are related to similar ones in the next lower layer by the n-glide parallel to (010) at 1/4 along b. The $[SiO_4]$ tetrahedron, like that in chloritoid (Chapter 6), is quite regular; the Si-O bonds are within 0.005 A of the mean, 1.641 A for the x-ray refinement (Fig. 1c), and within 0.008 A of the mean, 1.643 A for neutron diffraction refinement (Zemann et al., 1979). The O$\cdots$O shared edges and O-Al-O angles are the smallest in the octahedron (Fig. 1b). All distortions of both cation and anion polyhedra from the ideal geometry expected in closest-packed structures are explicable in terms of simple electrostatic interactions (Ribbe and Gibbs, 1971).

The fluorine ligancy is illustrated in Figure 1f. In orthorhombic topaz containing OH in the monovalent anion sites H is disordered over eight equivalent sites. A neutron refinement of $Al_2 F_{1.44}(OH)_{0.56}[SiO_4]$ by Zemann et al. (1979) located the proton at a distance of 0.98 A from the oxygen nucleus (at $x = 0.0027$, $y = 0.7508$, $z = 0.1619$ and equivalent positions -- compare Isetti and Penco, 1967).

"Anomalous" Topazes

Rinne (1926) made the first systematic study of "sectoral texture" in topaz. He found that 2V, the direction of the acute bisectrix, and the orientation of the optic plane varied from sector to sector, all of which had identical crystallographic orientations. He reported that these "anomalous" optical properties disappeared after heating the crystals to 950°C for five hours.

Recently Akizuki et al. (1979) examined uniterminal prismatic topaz crystals from Ouro Preto, Brazil which had pronounced sectoral textures (Fig. 2). They found no measurable differences in F content or crystallographic orientation from sector to sector, and sector boun-

Figure 2. Sectoral textures in optically "anomalous" topaz from Ouro Preto, Brazil, as viewed in (001) thin sections seen between crossed polars (× 10). Courtesy of Dr. M.S. Hampar, University of Manchester.

daries were not distinguishable in transmission electron microscopic studies of ion-thinned samples. They proposed that the sectors are differentiated primarily on the basis of (F,OH) ordering. They state (p. 240) that "the immediate surroundings of the fluorine-hydroxyl sites are different with respect to the growth plane [in a given sector] for each of the sites and therefore, at the growth surface, the four sites [in a unit cell projected on (001) -- see Figure 3 below] are not equivalent. If this non-equivalence results in different relative preferences for fluorine and hydroxyl on the sites then ordering will occur and the symmetry of the crystal will be reduced. By con-

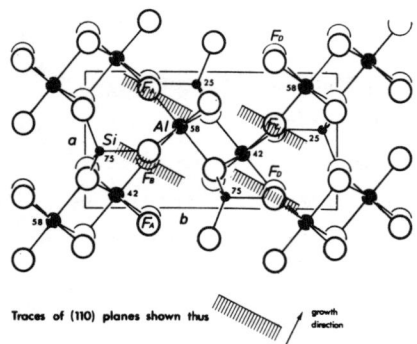

Traces of (110) planes shown thus /////// / growth direction

Figure 3. The crystal structure of topaz projected onto (001), with (110) growth planes indicated. After Akizuki *et al.* (1979; their Fig. 4).

sidering the symmetry of the crystal relative to different growth planes it is possible to predict the symmetry that would be expected to occur by ordering for each type of plane. For a general $\{hkl\}$ growth plane the symmetry would be triclinic, for $\{0kl\}$, $\{h0l\}$, and $\{hk0\}$ growth planes the symmetry would be monoclinic

(with the mirror plane perpendicular to the a-, b-, and c-axes respectively) and for {100}, {010}, and {001} growth planes the symmetry
would remain orthorhombic. The symmetries predicted in this way for
the various sectors in the topaz are consistent with the optical effects
observed. The degree of ordering, and hence the actual values of $2V_\gamma$
and the extinction angle would be expected to vary for different growth
planes producing the same crystal symmetry, as the environment of a
particular fluorine/hydroxyl site will not be the same in relation to
the different planes."

Akizuki *et al.* (1979, p. 239) found that "the {010} growth sectors
in the crystal rim behave as if they had retained their orthorhombic
symmetry, the {$hk0$} sectors in the rim behave as if they had monoclinic
symmetry with the diad parallel to the c-axis and the {hkl} sectors
in the core of the crystal behave as if their symmetry were triclinic."
Heating at 950°C for four hours apparently caused F,OH to disorder. As
a result, the symmetry of all sectors inverted to orthorhombic and 2V
values became uniform, although the specimen from North Queensland
investigated by Parise (1980; discussed below) did not invert with
heating at 950°C for 18 hours (see Fig. 4)

The Structure of Triclinic Topaz

Because the nucleus of a hydrogen atom has a high negative neutron
scattering power which is independent
of scattering angle, neutron diffraction is preferred over x-ray diffraction for studying crystal structures
in which the location of H (or OH or
H_2O) is diagnostically critical.
Parise (1980) used neutron methods to
unambiguously assign the H atom to a
single site in the structure of a $P\bar{1}$
triclinic topaz $Al_2F_{1.8}(OH)_{0.2}[SiO_4]$.

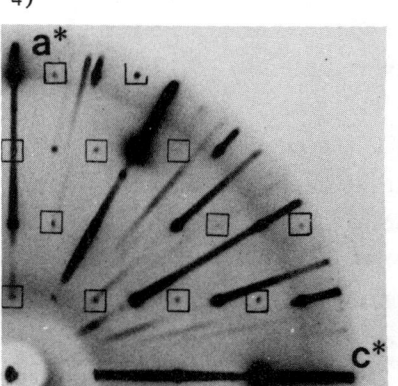

Figure 4. A quadrant of the *hOl* reciprocal
net from a precession photograph of topaz
from North Queensland which was heated at
950°C for 18 hours. *t* reflections with
h + l odd are outlined. Courtesy of J.B.
Parise.

Intensities of "t" reflections --
those which violate space group *Pbnm* (hOl, $h + l$ odd; Okl, k odd; see
Fig. 4) -- were used to locate hydrogen at $x = 0.498$, $y = 0.254$, $z =$
0.161, which is only one of the eight symmetrically equivalent sites
each of which were found by Zemann *et al.* (1979) to contain 0.07 H in

disordered orthorhombic topaz. Figure 5 contains two perspectives of
the hydrogen environment. It is bonded to oxygen in the O/F(8) site
with H - O = 0.97 A. This ordering produces a small but significant
distortion of the orthorhombic structure (*cf*. Figs. 1 and 5), which is
not evident in the cell dimensions of the triclinic crystal. In this
$P1$ topaz there are, however, four non-equivalent [SiO_4] tetrahedra with
the following ranges in individual bond lengths (in Angstroms) and their
respective mean values [in brackets]: 1.632 - 1.676 [1.649]; 1.616 -
1.646 [1.631]; 1.636 - 1.657 [1.647]; 1.618 - 1.643 [1.634]. The eight
[$AlO_4(F,OH)_2$] octahedra have mean bond lengths ranging from 1.853 -
1.872 A.

The hydrogen atom occupies a cavity surrounded by a "cage-like"
arrangement of octahedra and tetrahedra which is best envisioned with
a ball-and-spoke model (but see Fig. 5). Parise concludes that in addition to the strong hydroxyl bond to O/F(8) at 0.97 A, H is weakly
bonded to O(3) at 2.28 A, O(7) at 2.29 A and F(3) at 2.39 A in this particular polymorph. He contends that it is these hydrogen bonds that
"play an important part in determining the ordering of hydroxide in
the structure." Except for very slight distortions, the atomic environments around each of the monovalent anion sites in triclinic topaz are
very similar indeed, and one might expect a random substitution of OH
for F. But as Akizuki *et al.* (1979) suggested, even in orthorhombic
topaz the environment of each site along a particular growth surface
(e.g., the (110) plane illustrated in Fig. 3) is different, and depending on the direction of growth there will be different affinities
for OH based on possible hydrogen bonding arrangements and thus a
variety of ordering schemes. In other words, the configuration of the
anions surrounding potential sites for hydrogen will, at the time of
incorporation of OH onto the growth surface, largely control which of
the eight sites is (are) preferred and which way the O-H bond(s) is
(are) oriented.

Akizuki *et al.* (1979) found that heating (950°C, 4 hrs.) caused
triclinic topaz to invert to orthorhombic, and they suggested that this
was due to the introduction of vacancies into the OH sites as water
was lost, permitting diffusion and disordering of these anions. The
fact that Parise was unable to alter his specimen after 18 hours at

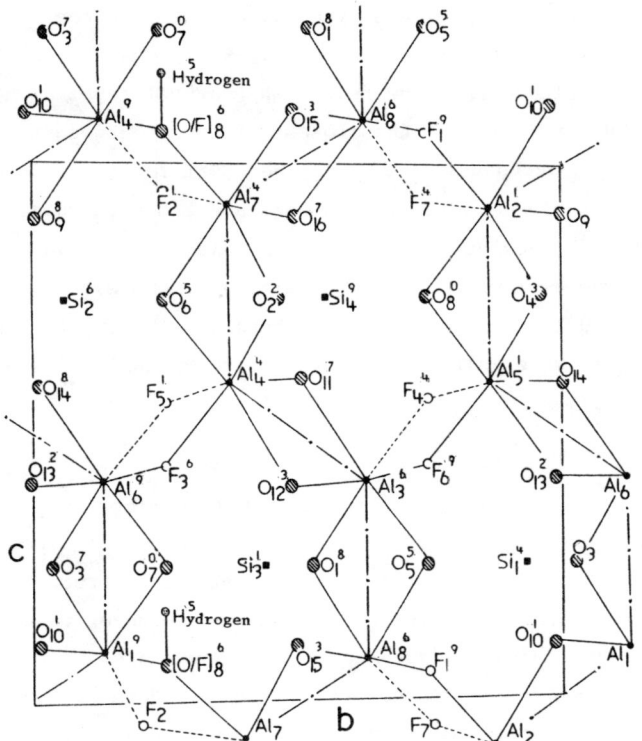

Figure 5a. The structure of $P\overline{1}$ topaz projected onto (100). Parallel to c are zig-zag chains of octahedra (dot-dash lines). Compare Figure 1a. Corner-shared F atoms link octahedra to one another (dotted lines). Superscripted small numbers are the decimal-fractional x co-ordinates ($\times$ 10) of the labelled atoms. Si-O bonds are omitted for clarity. After Parise (1980, his Fig. 3.3, p. 73).

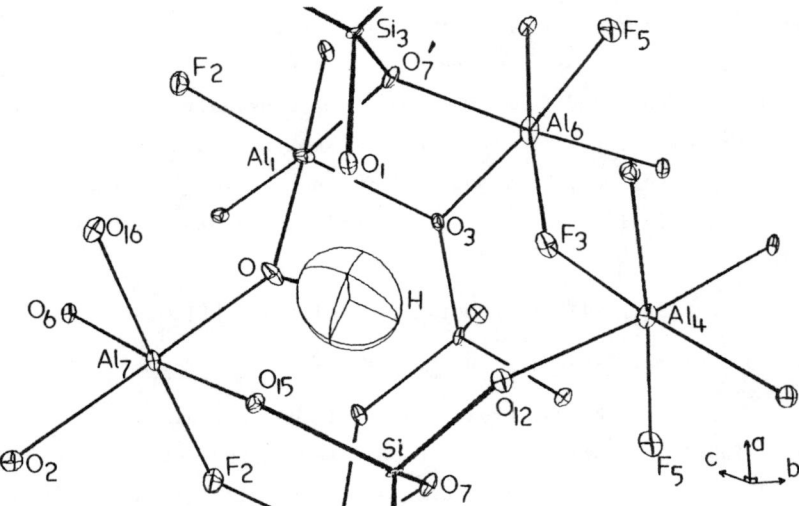

Figure 5b. The environment around the hydrogen atom (large triaxial ellipsoid). The O_8-H bond is 0.97 A. Nearest-neighbor anions are O_3 and O_7 at 2.3 A and F_3 and O_{15} at 2.4 A. After Parise (1980).

221

950°C may imply that there are certain types of ordering schemes for OH that are significantly more stable than others.

Parise (1980) concluded that the reason why natural topaz always has less than 50% OH for F is that the closest distance between two hydrogen atoms (1.5 Å) at occupancies >50% is too short for two adjacent sites to be simultaneously occupied (*cf*. Gebert and Zemann, 1965). Barton (1982) found that his calculations based on this "proton-avoidance [activity] model agree well with natural assemblages, whereas those based on the ideal model [100% OH for F] do not." Needless to say, the amount and degree of order of OH significantly affect both optical and electrical properties (see below).

Twinning, Intergrowths and Dislocations

Rare (010) twins were observed in topaz by Goldschmidt (1910) and confirmed by Gliszczynski (1949), whose structural study indicated that they must be "contact" twins. The latter also reported a structural explanation for rarely observed (010) coaxial intergrowths of topaz.

Phakey and Horney (1976), using x-ray topography and scanning and transmission electron microscopy, have examined growth defects and dislocation densities in relation to impurities in topaz, and Epelborn *et al*. (1973) presented calculations which relate minimum dislocation energy (parallel to [001]) to the observed orientations of dislocations which occur during growth of topaz (*cf*. Giacovazzo, 1973). The first authors showed a direct correlation between etch pits on (001) and dislocations: Burgers vectors of the edge dislocations are [100], [001], [01$\bar{1}$] and [$\bar{1}$11]. Isogami and Sunagawa (1975) painstakingly reconstructed the growth history of a topaz crystal from a pegmatite druse using x-ray topographic techniques; they and Phakey and Horney (1976) examined the boundaries of optically distinct sectors which were discussed in detail above.

PHYSICAL PROPERTIES; FLUORINE DETERMINATIVE METHODS

Habit, Cleavage, Thermal Expansion and Density

The edge-sharing octahedral chains parallel to z explain the characteristic [001] prismatic habit of topaz (see morphological study by Russell, 1924). The perfect {001} cleavage is parallel to the only planes which can be passed through the structure without breaking Si-O bonds (see Fig. 1a). Al-O and Al-F,OH bonds are broken in equal numbers.

Kôzu and Ueda (1929) measured the thermal expansion of topaz and found it to be greatest parallel to the chain axis and normal to the {001} cleavage, which of course is the plane of the weakest bonds. Density decreases systematically with increasing OH content (Deer *et al.*, 1962, p. 148, Fig. 31) because OH is both lighter in weight and larger in effective anionic volume than F. [See the discussion of OH → F substitution in the humites (Chapter 10).]

Optical and Electrical Properties

Color. The generally pale colors of topaz, which in addition to high refractivity and hardness make it valued as a gemstone, are variable and due primarily to radiation-induced color centers and/or trace levels of transition metals (Petrov and Berdesinski, 1975; Lehman, 1975; Petrov, 1977). The radiation-induced colors (yellow - orange - brown and blue, which have been simulated by gamma-ray bombardment -- Nassau and Prescott, 1975) are unstable and may be destroyed or altered by relatively gentle heating (< 500°C) or even by exposure to sunlight (Webster, 1975; Petrov, 1978). The absorption spectra of irradiated topazes were examined by Petrov *et al.* (1977). The transition metal most commonly identified with pink topaz is chromium; octahedral ferric iron may also be important (*cf.* Thyer *et al.*, 1967). Other transition elements identified spectroscopically at less than 300 ppm are Co, Ni, V, Mn and Cu (El-Hinnawi and Hofmann, 1966). Combinations of transition metals and radiation-induced color centers account for the variability of color in topaz (see excellent study by Petrov, 1977), and gemmologists commonly heat-treat crystals to obtain more desirable colors.

Many colored topazes are weakly pleochroic; dispersion is $r > v$. In a study of the infrared pleochroism of topaz, Gebert and Zemann (1965) determined the three-dimensional absorption figure for the OH-stretching-frequency (λ = 2.77μ) and almost exactly predicted the coordinates of the hydrogen atoms which have since been determined by Zemann *et al.* (1979) using neutron diffraction. Gervais *et al.* (1973) discussed the infrared active modes of topaz based on reflectivity measurements.

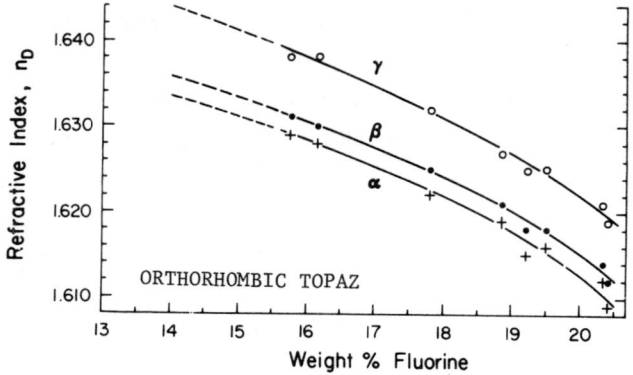

Figure 6. Refractive indices of topaz (±0.001) *vs* F-content, from Ribbe and Rosenberg (1971, Fig. 2). Improved regression equations for these data (from Carmen, 1981) are given to the right.

Weight % flourine =

$-14091.671 + 17669.224\alpha - 5530.429\alpha^2$

$-14882.367 + 18626.868\beta - 5820.051\beta^2$

$- 9572.430 + 12019.085\gamma - 3763.960\gamma^2$

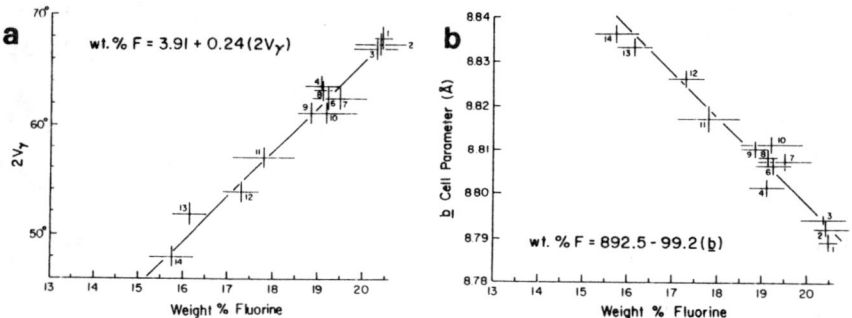

Figure 7. (a) Optic axial angle $2V_\gamma$ *vs* wt % F to be used only for orthorhombic topaz. $2V_\gamma$ from Rosenberg (1967). (b) *b* cell edge *vs* wt % F. Size of data points represent estimated standard errors of the measurements. Specimen numbers, figures and regression equations from Ribbe and Rosenberg (1971, Figs. 1b and 3a).

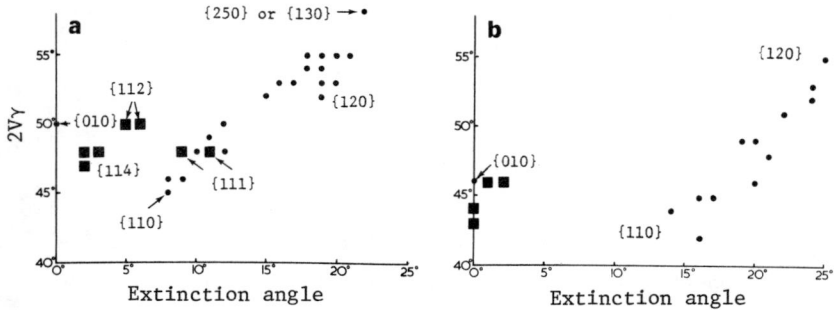

Figure 8. The relationship of $2V_\gamma$ and extinction angles $\beta \hat{} b$ for (a) the topaz specimen shown in Figure 2a and (b) the crystal shown in Figure 1b of Akizuki *et al.* (1979, p. 238). Points from core sectors are indicated by crosses and those from rim sectors by dots. The trend of dots upward from the left near the label {110} to the right near the label {120} presumably are data that represent a range of intergrowths of {110} and {120} sectors. The fluorine content of the crystal in 8b may be somewhat less than that in 8a. After Akizuki *et al.* (1979, their Fig. 3, p. 239).

Refractive indices and 2V: orthorhombic topaz. Both α and γ are parallel to the closest-packed (010) layers in orthorhombic topaz; γ is also parallel to the crankshaft-like chains of edge-sharing $[AlO_4F_2]$ octahedra, and topaz is another example of an optically positive orthosilicate (Ribbe, 1976).

Because of higher polarizability of OH, refractive indices increase with OH $\rightarrow$ F. Based on optical data for a suite of orthorhombic topazes (Rosenberg, 1967) whose F-contents were carefully measured by microprobe methods (Ribbe and Rosenberg, 1971), polynomial regression equations were calculated for the refractive indices *vs* wt % F curves (Fig. 6) and the linear regression equation for $2V_\gamma$ *vs* wt % F (Fig. 7a). Since errors in refractive index and 2V measurements are likely to be small, and because α and β and $2V_\gamma$ are all easily accessible from {001} cleavage fragments, it is a simple matter to use any or all of these parameters in determining F concentration in orthorhombic topaz.

Optic orientation and 2V: monoclinic and triclinic topaz. As mentioned earlier, Rinne (1926) was the first to systematically investigate "anomalous" topazes, finding that 2V, the direction of the acute bisectrix (Bx_a), and the orientation of the optic plane (OAP) varied from sector to sector, although orientations of the crystallographic axes were the same in all of them. The optic orientation of "normal" orthorhombic topaz is $\alpha = a$, $\beta = b$, $\gamma = c$, OAP = (010) and $Bx_a = [001]$. After heating at 950°C for 3-5 hours, the "anomalous" topazes studied by Rinne (and those investigated later by Akizuki *et al.*, 1979) assumed this orientation, although Parise's (1980) specimen did not.

Although many details remain to be explained, the hypothesis of Akizuki *et al.* (1979) which related ordering to the variation of optical properties in different growth sectors of topaz has been confirmed by the work of Parise (1980) on at least one triclinic specimen. Figure 8 contains a collection of their data from two crystals, both of which had essentially constant F content across all their various sectors. The data in Figure 8a came from the (001) section pictured in Figure 2a: The outer rim of sectors corresponds to prism growth planes and the core corresponds to pyramidal growth planes. Akizuki *et al.* found that the OAP in the rim sectors {110}, {120} and {010} to be parallel to the

c axis. The indicatrix rotates about the Bx_a which is normal to (001).
"The prism faces are believed to be alternations of {110} and {120}
faces. The optical axial angle varies from 45° for {110} growth sectors
to 58° for {120} sectors with intermediate values corresponding to intergrowths [of the two sectors]. The extinction angles ($\beta \wedge b$) vary
from 9° for {110} sectors to 20° for {120} sectors" (p. 238). Extinction
angles and 2V values are given for various other identifiable sectors
in Figure 8.

The range of 2V values approaches 15° in these two specimens whose
sectors are variously orthorhombic, monoclinic, and triclinic in symmetry.
Thus we may conclude that at least for non-orthorhombic (F,OH)
ordered topaz, the determinative curves of Figures 6 and 7a must not
be used. Although Akizuki *et al.* (1979) found that the extinction
angles became 0° and 2V for all sectors converged to values within 2°
of one another with heating at 950°C for 4 hours, some caution must
still be exercised in drawing conclusions regarding F/(F + OH) ratios
in the heated material from optical data. Disordering of (F,OH) is
probably accompanied by loss of water, and Parise's (1980) experience
indicates that at least one triclinic specimen retains a high degree
of order after heating at 950°C for 18 hours.

Piezoelectricity and pyroelectricity. The (F,OH) ordering observed in topaz accounts for its non-centric character and thus the
piezo- and pyroelectric properties that have been observed in certain
topazes since the early work of Friedel and Curie (1885). In an extension of work begun nearly a century ago (Mack, 1886), Akizuki *et al.*
point out that "the pyroelectric properties appear to mirror the
anomalous optical properties" of topaz.

Lattice Parameters

Rosenberg (1967) discussed the significance of the variation in
lattice parameters in terms of OH → F: the c cell dimension *(Pbnm*
space group setting) varies over a range of only 0.006 A and a is
poorly correlated with F/(F + OH). But the b cell edge is normal to
the closest-packed anion layers of alternating $[(F,OH)_2O_1]_\infty$ and O_∞
composition, and the regression equation (Fig. 7b) has a correlation

coefficient of -0.97 (Ribbe and Rosenberg, 1971)[1]. Because of the cummulative errors of cell edge measurements and the low correlation of a and c with F content, volume is less highly correlated with F content than b (-0.88): wt % F = 465.5 - 1.3 × Vol.

Since (F,OH) ordering does not affect unit cell edges or volume (Parise, 1980), it is advisable to use Figure 7b rather than Figure 7a to determine F content of topaz, regardless of its symmetry.

Fluorine Analysis

A relatively precise and simple method of determining the fluorine content of topax involves use of a specific ion electrode. See Van Loon (1968) for details.

TOPAZ-VAPOR EQUILIBRIA and THERMODYNAMIC PROPERTIES

In his study of compositional variations in F-rich assemblages, Rosenberg (1972) found that Al and F substitute in topaz for Si and O, respectively and where Al is "not readily available," anion substitutions appear to be compensated by tetrahedral vacancies. These observations are of academic interest but have no application to known natural topaz-bearing assemblages. He also reported (1978) that topaz hydrolyzes with decreasing temperature according to the equation $Al_2SiO_4F_2$ + $xH_2O = Al_2SiO_4F_{2-x}(OH)_x + xHF$. Thus F/OH ratios in topaz are dependent on the fugacities of HF and H_2O which are buffered by assemblages of four phases, including vapor (see Fig. 9).

Barton (1982) predicted fugacity ratios for coexisting gaseous species (H_2O and HF) in relation to the composition of topaz that differ greatly from Rosenberg's. His results are based on new thermodynamic data for HF, H_2O and OH- and F-topaz. The free energies of formation at 298°K and 1 bar, G°, are -2693251 and -2910660 J/mol for OH- and F-topaz; entropies, S°, are 112.04 and 105.40 J/K·mol. Heat capacities, C_p, are $504.413 - 0.08737T + 1358330T^{-2} - 5869.53T^{-0.5}$ J/K·mol for OH-topaz and $471.414 - 0.08165T - 1269470T^{-2} -$

- - - - - - - - - -

[1] The physical parameters recorded by Saito and Ushio (1968) for synthetic topazes are totally inconsistent with the recent and more carefully determined data presented herein. The cell parameters given by Chaudhry and Howie (1970) appear to be systematically in error.

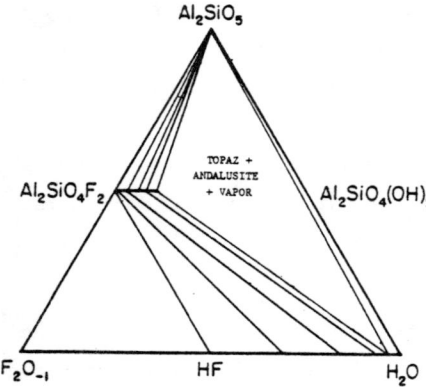

Figure 9. Schematic composition plot showing the limiting buffering assemblage for topaz solid solutions on the join Al_2SiO_5-H_2O-F_2O_{-1} in the Al_2O_3 - SiO_2-H_2O-F_2O_{-1} system. Note that compositions to the F_2O_{-1} side of the line between HF and the composition Al_2SiO_5 + $5F_2O_{-1}$ are physically unattainable. From Barton (1982, Fig. 1).

$5485.54T^{-0.5}$ J/K·mol for F-topaz; activities, a, are $X_{OH-T}^2/(1-2X_{OH-T})$ and $2X_{F-T}-1$, respectively (cf. Barton et al., 1982). Barton's schematic composition plot showing the limiting buffering assemblage for topaz solid solutions in the system Al_2O_3-SiO_2-H_2O-F_2O_{-1} is presented in Figure 9.

PETROLOGIC APPLICATIONS

Although petrologic applications are beyond the scope of this volume, the reader is referred to an excellent treatment of the subject by Barton (1982). The following is quoted from his abstract: "In the presence of topaz, muscovite may be stable on the solidi of some granites to pressures as low as 1-2 kbar. Calcic plagioclase is incompatible with topaz, especially in the presence of potassium feldspar. Hydroxyl-topaz contents of topaz coexisting with other aluminous phases increase with increasing pressure and decreasing temperature. With decreasing temperature at constant composition topaz removes HF from a water-rich fluid." This last conclusion is in apparent disagreement with Rosenberg (1978, p. 1218), who stated that "topaz is much more effective in removing fluoride from the fluid phase at high temperatures than at low temperatures."

228

TOPAZ: REFERENCES

Akizuki, M., M.S. Hampar and J. Zussman (1979) An explanation of anomalous optical properties of topaz. *Mineral. Mag., 43,* 237-241.

Alston, N.A. and J. West (1928) The structure of topaz. *Proc. Roy. Soc. A, 121,* 358-367.

Barton, M.D. (1982) The thermodynamic properties of topaz solid solutions and some petrologic applications. *Am. Mineral., 67,* in press.

_____, H.T. Haselton, Jr., B.S. Hemingway, O.J. Kleppa and R.A. Robie (1982) The thermodynamic properties of fluor-topaz. *Am. Mineral., 67,* 350-355.

Carman, M.F., Jr. (1981) A spindle stage study of the optical properties of a topaz. *Bull Minéral., 104,* 742-749.

Chaudhry, M.N. and R.A. Howie (1970) Topaz from the Meldon aplite, Devonshire. *Mineral. Mag., 37,* 717-720.

Deer, W.A., R.A. Howie and J. Zussman (1962) *Rock-forming Minerals,* vol. 1. John Wiley and Sons, Inc., New York. p. 145-150.

El-Hinnawi, E.E. and R. Hofmann (1966) Distribution of trace elements in topaz. *Chem. Erde, 25,* 230-236.

Epelborn, Y. and A. Zarka (1973) Étude théorique et expérimentale d'énergie de dislocations. *J. Crystal Growth, 20,* 103-108.

Friedel, C. and J. Curie (1885) *Bull. Soc. Minéral. France, 8,* 16-27.

Gebert, W. and J. Zemann (1965) Messung des Ultrarot-Pleochroismus von Mineralien. III. Der Pleochroismus der OH-Streckfrequenz in Topas. *N. Jb. Mineralogie, Mh., 1965,* 380-384.

Gervais, F., B. Piriou and J.-L. Servoin (1973) Étude par réflexion infraroque des modes internes et externes de quelques silicates. *Bu.. Soc. franc. Mineral. Cristallogr., 96,* 81-90.

Giacovazzo, C. (1973) Cenni di teoria dinamica nei cristalli imperfetti: studio topografico di campioni di topazio e di GaSe. *Rend. Soc. Italiana Min. Petr., 29,* 195-218.

Gliszczynski, S. von (1949) Über gesetzmässige Verwachsungen von Topasen I. Teil. Zwillinge. II. Teil. Koaxiale Verwachsungen. *Neues Jahrb. Mineral. Geol. Abt. A., Monatsh.,* 1-23.

Goldschmidt, V. (1910) Topaszwillinge aus Brasilien. *Z. Kristallogr., 47,* 639-644.

Isetti, G. and A.M. Penco (1967) La determinazione della posizione dell' idrogeno nell'ossidril - topazio mediante la spettrofotometria infrarossa in luce polarizzata. *Per. Mineral. (Roma), 36,* 995-1010.

Isogami, M. and I. Sunagawa (1975) X-ray topographic study of a topaz crystal. *Am. Mineral., 60,* 889-897.

Kôzu, S. and J. Ueda (1929) Optical and thermal studies of topaz from Naegi, Japan. *Sci. Rep. Tohoku Imperial Univ., Ser. 3, 3,* 162-170.

Ladell, J. (1965) Redetermination of the crystal structure of topaz: a preliminary account. *Norelco Rep., 12,* 34-39.

Lehman, G. (1975) Bemerkung zu der Arbeit 'Thermolumineszenz als Untersuchungsmethode der Farbursachen von Topasen.' *Zeits. deutsch. gemmolog. Gesell., 24,* 73.

Mack, K. (1886) *Ann. Phys. Chem. (Weidemann), 28,* 153-167.

Nassau, K. and B.E. Prescott (1975) Blue and brown topaz produced by gamma radiation. *Am. Mineral., 60,* 705-709.

Parise, J.B. (1980) *Order-Disorder Phenomena in Minerals: Structural Studies*. Ph.D. Dissertation, James Cook Univ. of N. Queensland.

_____, C. Cuff, F. H. Moore (1980) A neutron diffraction study of topaz: evidence for lower symmetry. *Mineral. Mag.*, *43*, 943-944.

Pauling, L. (1928) The crystal structure of topaz. *Proc. Nat. Acad. Sci. U.S.A.*, 14, 603-606.

Penfield, S.L. and J.C. Minor, Jr. (1894) On the chemical composition and related physical properties of topaz. *Am. J. Sci. Ser. 3*, *47*, 387-396.

Petrov, I. (1977) Farbuntersuchungen an Topas. *Neues Jahrb. Mineral., Abhdl. 130*, 288-302.

────── (1978) Farbe, Farbusachen und Farbveränderungen bei Topasen. *Zeits. deutsch. gemmolog. Gesell.*, *27*, 3-11.

────── and W. Berdesinski (1975) Thermolumineszenz als Untersuchungsmethode der Farbursachen von Topasen. *Zeits. deutsch. gemmolog. Gesell.*, *24*, 73-80.

──────, ────── and H. Bank (1977) Bestrahlte gelbe und rotbraune Topase und ihre Erkennung. *Zeits. deutsch. gemmolog. Gesell.*, *26*, 148-151.

Phakey, P.P. and R.B. Horney (1976) On the nature of grown-in defects in topaz. *Acta Crystallogr.*, *A32*, 177-182.

Ribbe, P.H. (1976) Polyhedral chains in the structures of gem orthosilicates: Relationships to physical properties (abstr.). 25th Ann. I. Geol. Congr. *Abstracts*, *2*, 593-594.

────── and G.V. Gibbs (1971) The crystal structure of topaz and its relation to physical properties. *Am. Mineral.*, *56*, 24-30.

────── and P.E. Rosenberg (1971) Optical and x-ray determinative methods for fluorine in topaz. *Am. Mineral.*, *56*, 1812-1821.

Rinne, F. (1926) Bemerkungen über optische Anomalien, insbesondere des brasilianer Topas. *Z. Kristallogr.*, *63*, 236-246.

Rosenberg, P.E. (1967) Variations in the unit-cell dimensions of topaz and their significance. *Am. Mineral.*, *52*, 1890-1895.

────── (1969) Topaz-vapor equilibria in the presence of excess silica and water (abstr.). *Geol. Soc. Am. Ann. Mtg., Atlantic City*, 192-193.

────── (1972) Compositional variations in synthetic topaz. *Am. Mineral.*, *57*, 169-187.

────── (1978) Fluorine-hydroxyl exchange in topaz (abstr.). Trans. Am. Geophys. Union *EOS*, *59*, 1218.

Russell, A. (1924) Topaz from Cornwall, with an account of its localities. *Mineral. Mag.*, *20*, 221.

Saito, H. and M. Ushio (1968) Relationship between fluorine content and physical properties of synthetic topaz $[Al_2SiO_4F_x(OH)_{2-x}]$. *J. Ceramic. Assoc. Japan*, *76*, 412-420.

Thyer, J.R., S.M. Quick and F. Holuj (1967) The E.S.R. spectrum of Fe^{3+} in topaz. *Can. J. Phys.*, *45*, 3597-3610.

Van Loon, J.C. (1968) Rapid determination of fluoride in mineral fluorides using a specific ion electrode. *Anal. Lett.*, *1*, 393-398.

Webster, R. (1975) *Gems*, 3rd edition. Butterworth & Co., London, p. 118-125.

Zemann, J., E. Zobetz, G. Heger and H. Völlenkle (1979) Strukturbestimmung eines OH-reichen Topases. *Oesterr. Akad. Wiss., Math. - naturwiss. Kl., Anzeiger, 116*, 145-147.

Chapter 10

The HUMITE SERIES and Mn-ANALOGS   P. H. Ribbe

INTRODUCTION

"The minerals of the humite group have a very limited paragenesis, and their occurrence is, with rare exceptions, restricted to metamorphosed and metasomatized limestones and dolomites, and to skarns associated with ore deposits at contacts with acid and, less frequently, alkaline plutonic rocks" (Deer *et al.*, 1962, pp. 47 & 57). In recent years it has been the exceptional occurrences that have generated considerable interest in the humites. Nielsen and Johnsen (1978) reported an unusual occurrence of titaniferous clinohumite [symbol: Ti-Cl] as a major constituent in late magmatic veins in a Tertiary ultramafic, strongly alkaline intrusion. But even more intriguing is the subject of humites as possible mineralogical sites for water in the earth's upper mantle. Möckel (1939), in a long-overlooked report, recorded Ti-Cl [which has OH rather than F as its predominant monovalent anion] in serpentinized peridotites in ophiolites of the Swiss Alps. Sun (1954) found Ti-Cl in a kimberlitic tuff plug, Voskresenskays *et al.* (1965) in Siberian kimberlites, and McGetchin *et al.* (1970) in a kimberlite in the Moses Rock (Utah) dike, noting its similarity to Ti-Cl in a kimberlite in the Buell Park diatreme 160 km distant (Balk, 1954). Since then titanian *chondrodite* has also been found in the Buell Park kimberlite (Aoki *et al.*, 1976).

In the meantime, Merrill *et al.* (1972) studied the dehydration of Ti-Cl up to 1170°C at 30 kbar and concluded that it can exist as a hydrous mineral in the upper mantle. They felt it was unlikely that it "supplied water for emplacement and eruption of the Moses Rock kimberlite through dehydration above 150 km" (p. 259) but did not exclude the possibility at greater depths. More recently Yamamoto and Akimoto (1974; 1977) investigated the system $MgO-SiO_2-H_2O$ at high pressures and synthesized hydroxyl-chondrodite $2Mg_2SiO_4 \cdot Mg(OH)_2$ and hydroxyl-clinohumite $4Mg_2SiO_4 \cdot Mg(OH)_2$ for the first time. Aoki *et al.* (1976, p. 243) suggested that "it is probable that these two minerals are stable in the upper mantle at a depth from 70 to 120 km." On the other hand Mitchell (1978), who found Ti-Cl occurring as metasomatic reaction rims on forsterite inclusions in the Jacupiranga (Brazil) carbonatite, states (in agreement with Smith, 1977) that "the rare reaction formation of titanian clinohumite occurring upon the immersion of forsterite in carbonated ultrabasic liquids appears to place in doubt the suggestion that Ti-Cl is ever a liquidus phase in kimberlites or is important in the mantle." Evans and Tromsdorff (1978) observed the breakdown of Ti-Cl to olivine + ilmenite + vapor in metamorphosed ultramafics at middle amphibolite conditions, and Engi and Lindsley (1979) reported similar breakdown products (+ brucite?) as low as 550°C at P_{H_2O} = 4 kbar and 600°C at 5 kbar. The former authors (1979) suggest that F is required for Ti-Cl to survive in crustal ultramafics outside the stability field of antigorite.

Against this background of revived interest in these minerals, we present a review of humite crystal chemistry.

CHEMISTRY

Strictly speaking, the humite minerals are a homologous series of *magnesium* orthosilicates whose structures are based on hexagonal closest-packed arrays of anions (O,F,OH) with octahedral and tetrahedral cation distributions related to that in forsterite (see Ch. 11 and Fig. 1). The general chemical formula is $n[M_2SiO_4] \cdot M_{1-x}Ti_x(F,OH)_{2-2x}O_{2x}$ where M is Mg >> Fe > (Mn,Ca,Ni,Zn), $0 < x < 0.5$, and n = 1 for norbergite, 2 for chondrodite, 3 for humite, and 4 for clinohumite. Manganese analogs of three of the four Mg-homologues are known (n = 2 alleghanyite, 3 manganhumite, 4 sonolite) and an analog of norbergite has been synthesized. Unpublished microprobe analyses (Richardson, pers. comm.) suggest extensive, if not complete solid solution between Mg and Mn end-members. The Mn-rich isotypes contain OH rather than both (F+OH) as is usual in Mg-rich humite minerals. In nature only Ti-rich chondrodite and clinohumite ($0.25 < x < 0.5$) are fluorine free among humite group minerals, although Ti-free hydroxyl chondrodite and clinohumite have been synthesized at high pressures and temperatures (Yamamoto and Akimoto, 1974, 1977).

Phase relations in the system $MgO-MgF_2-SiO_2$ are shown in Figure 2.

Table 1 contains a summary of crystal data and references for end-members of the humite minerals and their isotypes. The appendix contains 55 previously unpublished microprobe analyses of Mg-rich humites from the dissertation of Jones (1968); these analyses were used in many of the crystal chemical arguments made by Ribbe *et al.* (1968), Jones *et al.* (1969), and others since then. Fluorine values

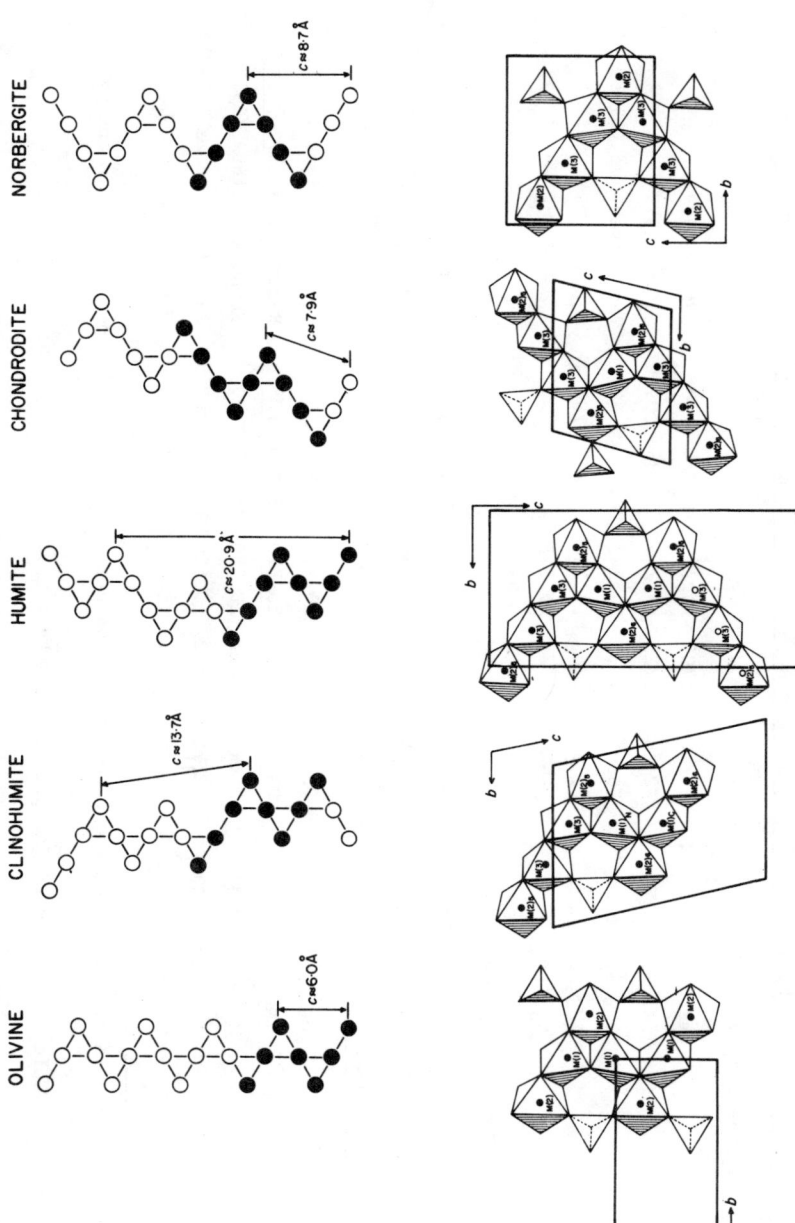

Figure 1. Crystal structures of olivine and the humite minerals projected down a. The upper part of each figure schematically shows the serrated chains of edge-sharing octahedra of the various structures. Solid circles indicate those octahedra which are detailed in the polyhedral drawing of the structure directly below which correlates in both the schematic and polyhedral drawings. After Papike and Cameron (1976, *Rev. Geophys. Space Phys. 14*, Fig. 4, p. 40).

Table 1. Crystal data for selected humite group minerals and their isotypes. Chemically related olivine end-members are included for comparison. Synthetic end-members are indicated by an asterisk *. Jones et al. (1969, p. 401, Table 3) list lattice parameters for eight natural specimens, Duffy (1977, appendix) for eight synthetic compounds of intermediate F,OH compositions, and Fujino and Takéuchi (1978) for inter-grown Ti-chondrodite and Ti-clinohumite.

Name	Formula	Space group	a(Å)	b(Å)	c(Å)	α(°)	Reference
Magnesium-fluorine end members							
*Norbergite	$Mg_2SiO_4 \cdot MgF_2$	Pbnm	4.707	10.265	8.724	90	Duffy (1977)
*Chondrodite	$2Mg_2SiO_4 \cdot MgF_2$	$P2_1/b$	4.725	10.249	7.788	109.2	Duffy (1977)
*Humite	$3Mg_2SiO_4 \cdot MgF_2$	Pbnm	4.735	10.243	20.72	90	Van Valkenburg (1961)
*Clinohumite	$4Mg_2SiO_4 \cdot MgF_2$	$P2_1/b$	4.740	10.226	13.582	100.9	Duffy (1977)
*Forsterite	Mg_2SiO_4	Pbnm	4.756	10.195	5.981	90	Yoder & Sahama (1957)
Magnesium-hydroxyl end members							
*Hydroxyl-chondrodite	$2Mg_2SiO_4 \cdot Mg(OH)_2$	$P2_1/b$	4.752	10.350	7.914	108.7	Yamamoto & Akimoto (1974)
*Hydroxyl-clinohumite	$4Mg_2SiO_4 \cdot Mg(OH)_2$	$P2_1/b$	4.747	10.284	13.695	100.6	Yamamoto & Akimoto (1977)
Manganese-hydroxyl isotypes							
*MAN[1]	$Mn_2SiO_4 \cdot Mn(OH)_2$	Pbnm	4.869	10.796	9.179	90	Francis & Ribbe (1978)
Alleghanyite	$2Mn_2SiO_4 \cdot Mn(OH)_2$	$P2_1/b$	4.850	10.720	8.275	108.6	Rentzeperis (1970)
Manganhumite	(see Table 4)	Pbnm	4.815	10.580	21.448	90	Francis & Ribbe (1978)
Sonolite[2]	(see Table 4)	$P2_1/b$	4.872	10.669	14.287	100.3	Kato (pers. comm., 1978)
*Tephroite	Mn_2SiO_4	Pbnm	4.875	10.521	6.238	90	Finch et al. (1975)

[1] Manganese Analog of Norbergite; not known to occur in nature. This powdered material was not analyzed subsequent to synthesis.

[2] See footnote to Table 4; unbalanced formula reported by Kato.

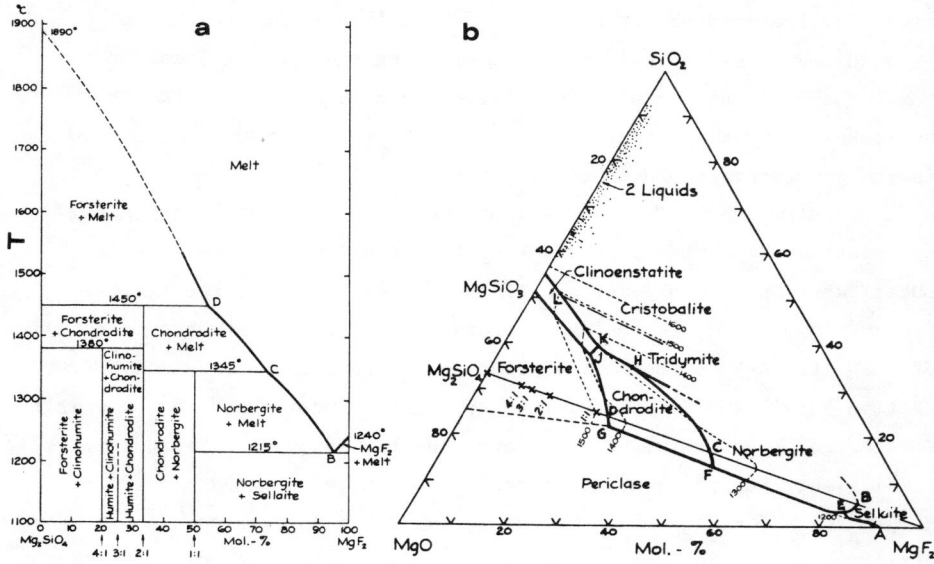

Figure 2. Phase equilibrium diagrams of the systems (a) Mg_2SiO_4-MgF_2 and (b) MgO-MgF_2-SiO_2. **From** Hinz and Kunth (1960, Figs. 4 and 5).

reported there have since been confirmed by Hinthorne and Andersen (1975) using an ion microprobe and $^{19}F^+/^{28}Si^+$ ratio analysis.

Substituents in Octahedral Sites

In the Mg-homologues, calcium substitutes in the *M* sites in amounts up to 0.2 weight percent CaO, and the ZnO content also may range as high, especially in humites from Franklin and Sterling Hill, New Jersey and the Kafveltorp area of Sweden.[1] Zinc is apparently more abundant in the Mn-isotypes. NiO has been detected at the 0.2 weight percent level in both titanian chondrodite and clinohumite from the Buell Park (Arizona) kimberlite (Fujino and Takéuchi, 1978).

Although listed in most bulk chemical analyses, aluminum has not been detected by electron microprobe analysis (Jones *et al.*, 1969), suggesting that it occurs in Al-rich inclusions such as spinel, which may also contain some of the ferric iron commonly reported in humite

- - - - - - - - - - - - - - - - - -

[1] A calcium analog of chondrodite, $Ca_5[SiO_4]_2(OH)_2$, was prepared hydrothermally (Buckle and Taylor, 1958; structure by Kuznetsov *et al.*, 1980), as was $Cd_5[SiO_4]_2(OH)_2$ (Egorov-Tismenko *et al.*, 1971).

minerals. Borneman-Starynkevitch and Myasnikov (1950) found that Fe^{2+} is oxidized on mechanical crushing, implying that Fe^{3+} values for humites are alomst certainly much lower than reported. In the Mg-humites Fe^{+2} usually exceeeds Mn^{+2}, but of course $Mn>Mg\geqslant Fe>(Zn,Ca,Ni)$ is the order for the Mn-isotypes.

Titanium, the only quadrivalent ion that is found in octahedral coordination in humites, receives special consideration because it substitutes only in one portion of the structure, *i.e.*, the $M(F,OH)O$ "layer" in which $M3$ is the only octahderal site. $M3$ is the smallest of the octahedra in the humite group structures (see *CRYSTAL CHEMISTRY* section below) and for that reason as well as local charge balance it is "preferred" by titanium. The $M3$ atom has two monovalent ligands if it is divalent, but the general "formula" for the $M3$ octahedron is $M^{2+}_{1-x}Ti^{4+}_{x}O_{4+2x}(OH,F)_{2-2x}$, and since Ti may range as high as $x \simeq 0.5$ atoms per formula unit in humites, as many as five of the ligands to $M3$ may be oxygens. Jones *et al.* (1969) noted that when x exceeds 0.25, $F \simeq 0.0$ and (OH) becomes the only monovalent ligand. There is no apparent crystal chemical reason for this, but it is not an unexpected observation when paragenesis is considered: most, if not all occurrences of titanian clinohumite and the one occurrence of titanian chondrodite (Aoki *et al.*, 1976) are in ultrabasic rocks.

Using the data from Tables A-3 and A-4 in the appendix and from recent structure analyses, Figure 3 shows that Mg constitutes more than 90 percent of all octahedral cations except where Ti is present in amounts greater than ~ 0.25 atoms per formula unit (*i.e.*, $x > 0.25$). In titanian chondrodite and clinohumites, Mg totals 80 percent or more of the octahedral cations, and Fe is generally found in higher concentrations than in Ti-poor specimens. In the quadrilateral shown here Fe always predominates over Mn, but of course Mn is the major cation in the isotypes, alleghanyite, manganhumite, and sonolite, as shown on the triangular inset in Figure 3.

Substituents in Tetrahedral Sites

Boron may substitute for silicon in tetrahedral coordination. Using the ion microprobe, Hinthorne and Ribbe (1974) measured ~ 5 atomic percent B for Si in two chondrodites, and P.B. Moore (pers. comm.) has

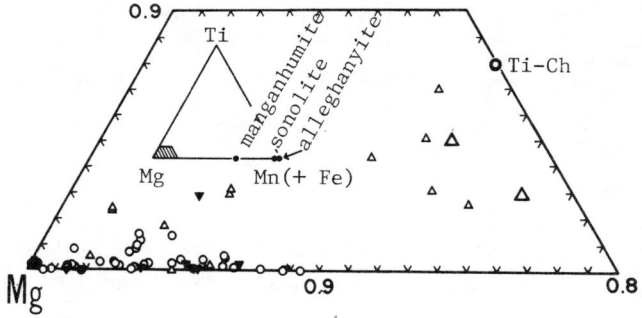

Figure 3. The quadrilateral is an enlargement of the hachured area
on the Mg-Ti-Mn(+Fe) triangular diagram (inset), showing substitution for Mg in the humite minerals based on ratios of atomic proportions of Mg, Ti and Mn(+Fe) from Tables A-3 and A-4 (Jones, 1968) in
the appendix. Solid circles, norbergite; open circles, chondrodite;
solid triangles, humite; open triangles, clinohumite. *Ti-Ch* is the
titanian chondrodite and the upper heavy triangle is the Ti-clinohumite whose structures were refined by Fujino and Takéuchi (1978).
The lower heavy triangle is the Ti-clinohumite refined by Kocman and
Rucklidge (1973). Three Mn-isotypes are shown in the inset. See
Table 4 for analyses. Modified from Jones *et al.* (1969, Fig. 1).

discovered a series of boron-rich humites at Franklin, New Jersey.[2]
Schäfer (1896) reported 1.04 and 1.68 weight percent BeO in two clinohumites, and Fleischer and Cameron (1955) and Ross (1964) suggested that
beryllium may be a relatively important constituent of the humite minerals,
although this has not yet been confirmed.

Stoichiometry

Because most wet chemical analyses of the humite group minerals
showed rather substantial stoichiometric imbalances in the ratio
$Si:(F + OH + O_{Ti})$, as summarized at the top of the next page,

- - - - - - - - - -

[2] Synthetic analogs of humite minerals have been crystallized in the
system $MgO-GeO_2-H_2O$ (Lyons and Ehlers, 1970) and $MgO-GeO_2-MgF_2$ (Van
Valkenburg, 1961; McCormack, 1966). Capponi and Marezio (1975)
synthesized borate analogs of chondrodite: $M^{3+}_{4-x}M^{2+}_{1+x}[BO_4]_2(OH)_xO_{2-x}$,
where M^{3+} = Al,Ga,Fe and M^{2+} = Ni,Co,Mg,Fe and the structure of
$Al_4Co[BO_4]_2O_2$ was reported. Fe_3BO_6 is isostructural with norbergite,
$Mg_3SiO_4F_2$ (White *et al.*, 1965).

Mineral	Number of chemical analyses	Si:(F+OH+O_{Ti}) Sahama (1953)		
		Theoretical	Range	Average
Norbergite	3	4:8	4:7.14–8.78	4:8.23
Chondrodite	20	4:4	4:3.37–4.80	4:3.93
Humite	10	12:8	12:5.98–9.19	12:8.07
Clinohumite	13	8:4	8:2.90–5.30	8:4.25

Sahama (1953) concluded that deviations from ideal compostitions were mainly due to analytical errors, particularly for F and H_2O. Jones *et al.* (1969, pp. 405–407) confirmed his assessment: their microprobe data showed relatively insignificant (±2%) departure from stoichiometric balance (Fig. 4). They did mention the possibility that "very fine

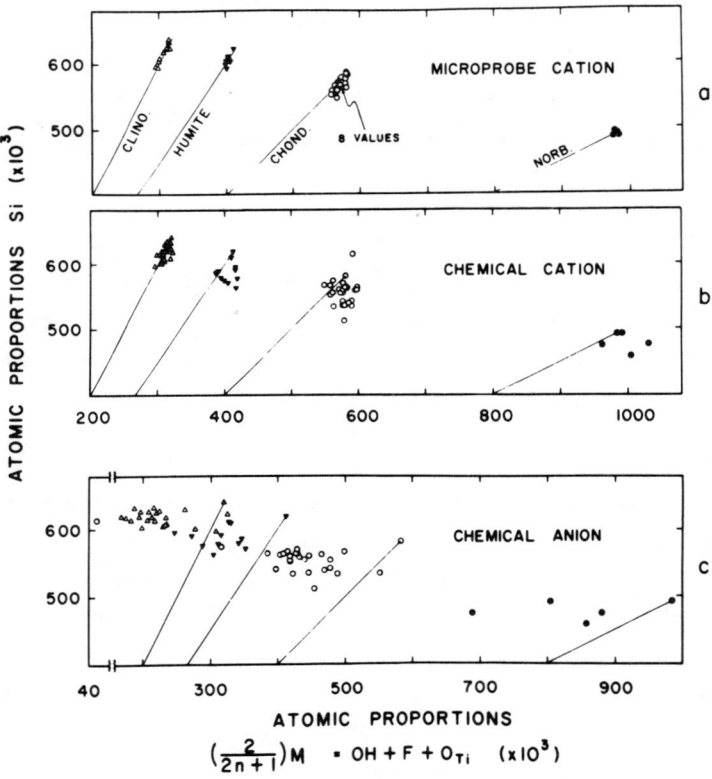

Figure 4. A summary of stoichiometry in the humite minerals. Straight lines represent lines along which analyses should fall for stoichiometric clinohumite (open triangles), humite (solid triangles), chondrodite (open circles) and norbergite (solid circles). a. Si *vs* (2/[2n+1])*M* for microprobe analyses. b. Si *vs* (2/[2n+1])*M* for bulk chemical analyses, where $O_{Ti} = O_{2x}$, the number of oxygens required to charge-balance Ti substituting for M^{2+} in octahedral coordination. The lack of agreement between b and c implies a substantial electrostatic charge imbalance.

scale epitaxial intergrowths of more than one humite mineral" might
account for part of the nonstoichiometry but saw no evidence of such
intergrowths in single-crystal x-ray photographs. Müller and Wenk(1978)
found a few such "compositional faults", but White and Hyde (1982a,b)
reported extensive faulting and intergrowths in both natural and synthetic Mg- and Mn-humites and clinohumites [as discussed at the end of
this chapter]. Some microprobe analyses reported by White (1982a)
showed large (>10%) deviations from ideal stoichiometry, but because
there was no correlation between "nonstoichiometry" and the specimens
that were found by electron microscopy to be the most structurally
inhomogeneous, the analyses are clearly suspect.

CRYSTAL STRUCTURES

Structural Homology; Octahedral Chains

Olivine and the four humite minerals constitute a polysomatic series of orthosilicates based on hexagonal closest-packed anion arrays
in which one-half the octahedral sites are filled. The structural
homology is derived from a replacement of four oxygens by four (F,OH)
which is charge-balanced by the replacement of one tetrahedrally coordinated silicon atom by a tetrahedral void according to the formula
$M_{2y}Si_{y-1}O_{4y-4}(F,OH)_4$, where y = 3, 5, 7, 9 for norbergite, chondrodite,
humite, and clinohumite, respectively (Table 2). [Note that y = 2n+1.]

Given consistent choices of space groups and axial orientation
[see below], the a and b cell edges are similar for all five minerals
(Table 1). The key structural unit in each is a serrated chain of
edge-sharing M-octahedra which is parallel to the c-axis. Figure 1
(from Papike and Cameron, 1976) illustrates the chains, showing the
types of octahedra present in the unit cell of each structure as projected down the a axis (cf. Table 2). Adjacent chains in the (100)
plane are related by the translation vector b and are cross-linked by
sharing corners with up- and down-pointing (+ and -) tetrahedra (cf.
Figs. 1 and 5). Each (+) tetrahedron shares its three basal edges
(i.e., those approximately parallel to (100) with octahedra in the
layer below, each (-) tetrahedron with octahedra in the layer above
(Fig. 5). Octahedral chains in adjacent (100) layers are related by
b-glide planes, but this is a consequence of an intentional but unconventional choice of space group for the monoclinic species.

239

Table 2. Summary of some structural data for the humite homologs, $M_{2y}Si_{y-1}O_{4y-4}(F,OH)_4$, and olivine for comparison. The parameter q is the number by which d_{001} must be divided to put normalized unit cell volumes on the same scale. The number of symmetrically non-equivalent Si tetrahedra and the fraction of the tetrahedral sites which are filled in the hcp anion array are also given. The names of manganese isotypes are shown in parentheses.

Name, *space group*	y	q	Fraction of tet'l sites	No. of Si tetrahedra	$M1_6$	$M2_6$	$M2_5$	$M2_4$	$M3_4$
Norbergite, *Pbnm* (?)	3	6	1/12	1				*m*	*n*
Chondrodite, *P2₁/b* (Alleghanyite)	5	5	1/10	1	*c*		*n*		*n*
Humite, *Pbnm* (Manganhumite)	7	14	3/28	2	*n*	*m*	*n*		*n*
Clinohumite, *P2₁/b* (Sonolite)	9	9	1/9	2	*c,n*	*n*	*n*		*n*
Forsterite, *Pbnm* (Tephroite)	–	4	1/8	1	*c*	*m*			

* The subscript indicates the number of oxygens coordinated to the M atom (Mg, Fe, Mn, Ca, Ni, Zn in $M1$ and $M2$; these ±Ti in $M3$). For example, $M2_4 = MO_4(F,OH)_2$. The symbol m indicates that the octahedron has mirror symmetry, c indicates that M is on a center of symmetry, and n that there is no symmetry.

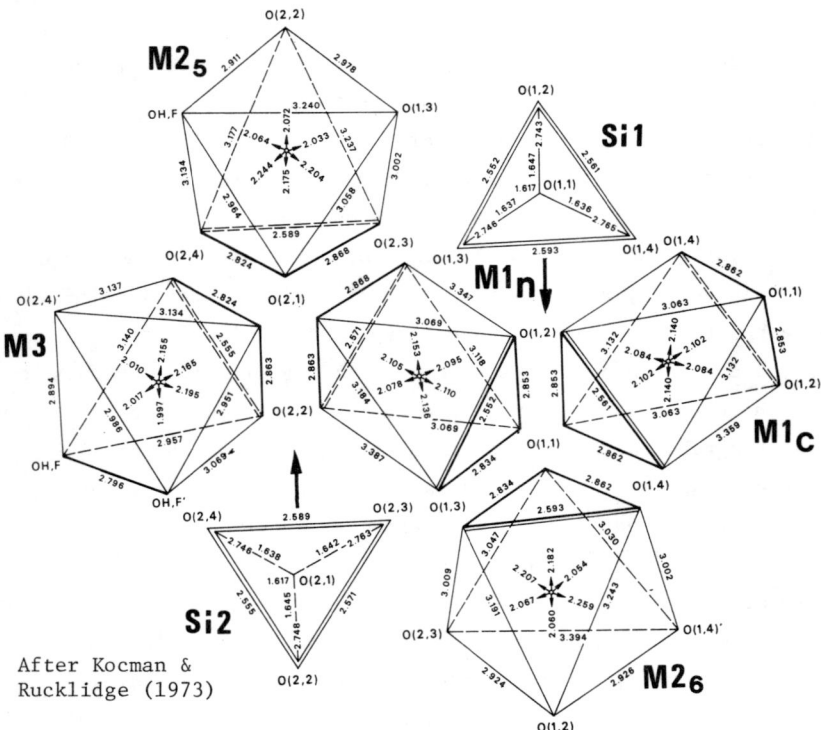

After Kocman &
Rucklidge (1973)

Figure 5. Exploded (100) projection of the steric features of the cation co-ordination polyhedra in clinohumite. Heavy lines indicate polyhedral edges shared between two octahedra; double lines are edges shared between an octahedron and a tetrahedron. In the real structure the Si(1) tetrahedron is translated in the direction of the arrow and sits above the level of the octahedra, likewise Si(2), but it projects into the layer below. If these were drawn with the Si atoms in the same interlayer as the M atoms, Si(1) would be a (–) tetrahedron sharing the O(1,3) corner with M(2)₅ and Si(2) and (+) tetrahedron sharing the O(2,3) corner with M(2)₆ (*cf.* Fig. 2 of Robinson *et al.* (1973a, reprinted 1973b)).

240

Space Groups

The space groups for chondrodite and clinohumite and their Mn-isotypes are designated $P2_1/b$. The unique axis a is chosen parallel to the 6_3 axis of an ideal hexagonal closest-packed anion array for all the humite minerals and olivine; c is the translation vector in an octahedral chain, and b relates adjacent chains in the (100) plane. Thus the orthorhombic minerals have space group $Pbnm$, a choice consistent with that adopted by Bragg and Brown (1926) for olivine and by Taylor and West (1928; 1929) for humite and norbergite (see also Bragg and Claringbull, 1965). Jones (1969) unscrambled what can only be described as a confusing mess in the standard reference literature (Deer *et al.*, 1962; Strunz, 1957,; *etc.*) in which authors have listed space group symbols inconsistent with one another and with axial labels in their sketches of the structures or crystal habits of the humite minerals. The confusion has been compounded in subsequent discussions of twinning, cleavage and orientations of the optical indicatrix. It is strongly suggested that the conventions used by Jones (1969) and listed in Table 1 be adopted permanently to avoid further misorientation of unsuspecting readers. [The choice of axes for alleghanyite by Rentzeperis (1970) has been corrected in Table 1 but is at variance with his choices for all humites (see the review article by Rentzeperis, 1972).]

Individual Structures

Bragg and West (1927) and Taylor and West (1928; 1929) solved the structures of the humite minerals by analogy with that of olivine, assuming from charge-balance considerations that the (F,OH) anions are bonded to three Mg and the oxygens are bonded to one tetrahedrally co-ordinated Si and three octahedrally coordinated Mg, as in olivine. This requires ordering of the (F,OH) anions within the structure. By writing the formula for the humites $nMg_2SiO_4 \cdot Mg(F,OH)_2$, Taylor and West were misled in describing their structures as stacking sequences of "unit blocks" of composition $Mg(F,OH)_2$ and nMg_2SiO_4. Although their drawings of the idealized structures are correct (*cf.* Bragg and Claringbull, 1965, Fig. 122), they should have designated compositions of the "unit blocks" as $Mg(F,OH)O$ and $Mg_2SiO_3(F,OH) \cdot (n-1)Mg_2SiO_4$. This misunderstanding was compounded by later authors who, in thinking

about F ⇄ OH substitution in humites, assigned structural names and
crystal-chemical significance to the "unit blocks," *viz.*, "...Mg(OH)$_2$
has a lattice of the cadmium iodide type, while MgF$_2$ has the symmetri-
cally·coordinated rutile structure" (Rankama, 1947). Terms such as
"brucite structure" and "sellaite" and "brucite-sellaite" were also
inappropriately used.

To avoid repetition in subsequent discussions of individual structures of the humite minerals, a tabulation of polyhedral edge-sharing
is provided below (Table 3). Octahedral ligancy is specified, and the
polyhedra are listed in order of decreasing mean M-(O,F,OH) distance.
Site symmetry is listed in Table 2 for each structure in which a particular octahedron occurs.

Table 3.

Octahedron	Number of edges shared with	
	Si-tetrahedra	M-octahedra
$M(2)O_6$	1	2 ($M1$)
$M(2)O_5(F,OH)$	1	2 ($M1+M3$)
$M(2)O_4(F,OH)$	1	2 ($M3$)
$M(1)O_6$	2	2 ($M2$)
		2 ($M1,M3$)*
$M(3)O_4(F,OH)_2$	1	2 ($M2+M3$)
		1 ($M1,M3$)**

* 2$M1$ in olivine, $M1+M3$ in clinohumite and
  humite, 2$M3$ in chondrodite.
** $M3$ in norbergite only.

Norbergite. Like olivine and all the humites, *Pbnm* norbergite,
Mg$_3$[SiO$_4$](F,OH)$_2$, has one-half of its octahedral sites filled, but
only one-twelfth of its tetrahedral sites are filled (see Table 2).
The inset (from Figure 1) shows polyhedral linkages in the (100) projection. The cations occupying both the $M3$ and the $M2_4$ octahedral
sites have four oxygen and two (F,OH) ligands,

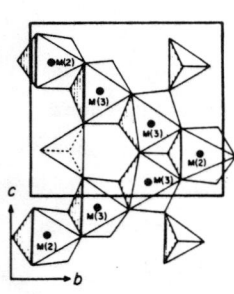

the latter are in *cis* arrangement in both octahedra, although the (F,OH)···(F,OH) edge is
shared between $M3$ octahedra and is unshared --
and ∿0.23 A longer -- in the $M2_4$ octahedron. By
virtue of its larger number of shared edges
(Table 3) $M3$ is an inherently smaller octahedron
than $M2_4$, which is unique to F-rich norbergite.

$M2_4$, Si and two of the four anions are on mirror planes, just as in olivine. All atoms are formally charge-balanced.

Gibbs and Ribbe (1969) discuss the norbergite structure in terms of cation and anion ligancy and strains in bond angles as measured by their deviation from values expected in a structure with ideal hcp geometry, *i.e.* with perfectly regular tetrahedra and octahedra. Since the conclusions of this type of analysis are the same for all the humite structures, a summary of results will be presented later.

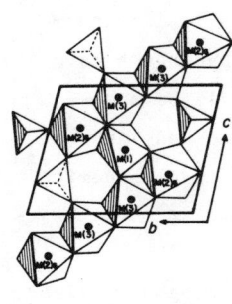

Chondrodite. Chondrodite, $Mg_5[SiO_4]_2(F,OH)_2$, has one-tenth of its tetrahedral sites filled. As in norbergite, the monovalent anions are ordered in the hcp anion array and are nearly coplanar with the three M cations to which they are bonded; each oxygen is bonded to one Si and 3 M atoms. As shown in the inset, the $M2_5$ octahedron is distorted from the mirror symmetry of the $M2_6$ polyhedra in olivine and humite and the $M2_4$ in norbergite by virtue of having a single monovalent ligand. Gibbs *et al.* (1970) suggested that the Fe atoms in the structure they refined (Table 4) preferred the $M1$ octahedron, the only site with no (F,OH) ligands, although this preference is not evident in the refinement by Fujino and Takéuchi (1978) of a titanian chondrodite (Table 4) in which Ti is nearly all concentrated in the $M3$ site. The preference of Mn for $M2_5 > M1 >> M3$ in magnesian alleghanyite was noted by Francis (1980). Cation ordering is further discussed in the section on *CRYSTAL CHEMISTRY.*

Table 4. Compositions of the humite group minerals and their isotypes whose structures have been determined.

Mineral or Compound*	Mg	Fe	Mn	Ni	Ca	SiO₄	Ti	F	OH	O_Ti	Reference
Norbergite	2.98 ·	<.01	0	--	<.01	1	0.01	1.81	0.17	0.02	Gibbs & Ribbe (1969)
Chondrodite	4.95	0.03	0.01	--	0.01	2	<.01	1.27	0.73	<.01	Gibbs *et al.* (1970)
*Hydroxyl-chondrodite	5	0	0	0	0	2	0	0	2	0	Yamamoto (1977)
Titanian chondrodite	3.99	0.57	0.01	0.01	0	2	0.42	0	1.15	0.85	Fujino & Takéuchi (1978)
Alleghanyite	--	--	5	--	--	2	--	--	2	--	Rentzeperis (1970)
Magnesian alleghanyite	1.95	0.02	2.84	*Zn .18*	<.01	2	--	*F > OH*		--	Francis (1980)
Humite	6.60	0.35	0.05	--	<.01	3	0.01	1.06	0.93	0.01	Ribbe & Gibbs (1971)
Magnesian manganhumite	2.10	0.07	4.76	--	0.04	3	--	0	2	--	Francis & Ribbe (1978)**
Clinohumite	8.42	0.50	0.06	--	<.01	4	0.02	1.04	0.93	0.03	Robinson, Gibbs &
Titanian clinohumite	7.33	1.04	0.05	--	<.01	4	0.47	0	1.06	0.94	Ribbe (1973)
Titanian clinohumite	7.44	1.09	0.02	0.02	--	4	0.43	0	1.14	0.86	Fujino & Takéuchi (1978)
Titanian clinohumite	7.34	1.36	0.08	--	--	4	0.26	0.40	1.08	0.52	Kocman & Rucklidge (1973)
Sonolite	1.01	0.07	7.93	--	0.37	4	0.01	0.33	0.87	0.80	Kato (pers. comm., 1978)***

Formula, normalized to n SiO₄

* Hydroxyl chondrodite was synthesized from a mixture of 13Mg(OH)₂ + 4SiO₂ at 77 kbar and 1125°C (Yamamoto and Akimoto, 1974; 1977).

** Composition reported by Moore (1978).

*** Composition reported exactly as communicated. It is likely that there is an excess of 0.38 M^{2+} cations and that F,OH,O_Ti should be 0.33, 1.65, 0.02.

The structure of synthetic hydroxyl-chondrodite (OH-Ch) was refined by Yamamoto (1977) who found it to be a somewhat expanded version of a F-rich natural specimen refined by Gibbs et al. (1970). The hydroxyls constitute a shared edge between $M3$ octahedra, and Yamamoto postulated positional disorder for the protons of the two OH ions. They are thought to be equally distributed either both to the left of their respective oxygens or both to the right along the shared edge, i.e., either H·O···H·O or O·H···O·H. This arrangement is different from that proposed for the hydrogens in alleghanyite, $Mn_5[SiO_4]_2(OH)_2$, which is isostructural with OH-Ch (Rentzeperis, 1970; 1972).[3] Since the hydrogens have eluded x-rays, perhaps neutron diffraction experiments will be necessary to locate them more precisely in chondrodite. Fujino and Takéuchi (1978) have located the half-occupied H sites in a titanian clinohumite, and Kato (pers. comm.) has found the partly-occupied sites in sonolite.

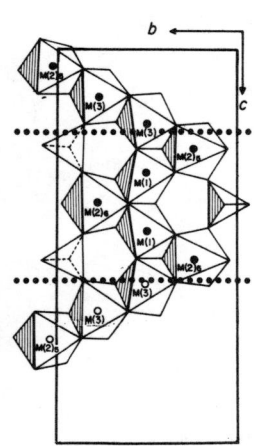

Humite. The $n = 3$ member of the homologous series, $Mg_7[SiO_4](F,OH)_2$, has orthorhombic symmetry (*Pbnm*) and the longest octahedral chain repeat of all ($\sim$ 21 A; cf. Figure 1). Humite has an easily recognizable bit of the olivine structure in it (between the horizontal dotted lines in the inset), the only difference being that the $M1$ site has point symmetry $\bar{1}$ in olivine and 1 in humite (Table 2). Three-twenty-eighths of the tetrahedral sites are filled, and there are two types of Si tetrahedron, Si(1) on the mirror plane (with $M2_6$ and two oxygen atoms) and Si(2) which is only slightly distorted from m symmetry (see Fig. 4 in Ribbe and Gibbs, 1971); both have similar mean bond lengths to oxygen, 1.629 and 1.628 A, respectively. In the humite whose composition is recorded in Table 4 site refinement of the Fe/Mg distribution led to the conclusion that Fe^{2+} prefers the more distorted octahedral sites and those with the more polarizable ligands, i.e. $M2_6$ and $M1 > M2_5 > M3$. In manganhumite

---

[3] Unfortunately Rentzeperis (1970, 1972) has labeled the $M3$ octahedron Mn(2), the $M2_5$ octahedron Mn(1) and $M1$ octahedron Mn(3), but we will persist with conventional nomenclature.

244

(analysis in Table 4) the large $M2_6$ and $M2_5$ sites contain only Mn, smaller $M1$ contains 0.7 Mn and smallest site, $M3$, contains just 0.25 Mn/(Mn + Mg).

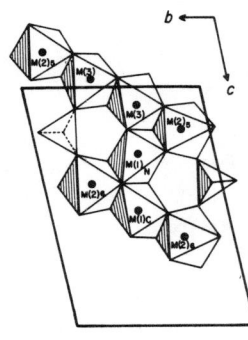

Clinohumite. The last member of the humite mineral group is monoclinic $P2_1/b$, like chondrodite, and it contains an even larger bit of olivine-like structure than humite (*cf.* Figure 1), including one $M1$ octahedron with $\bar{1}$ symmetry. $M1$ is the only atom on a special position. In the asymmetric unit there are two Si atoms, eight oxygens and one (F,OH,O) site occupied by up to 1-x hydroxyls in titanian clinohumite (Ti-Cl). Along with titanian chondrodite (Ti-Ch), Ti-Cl has received considerable attention in the recent literature as a possible site for water in the Earth's upper mantle (see *INTRODUCTION*). Crystal structures of no less than three Ti-Cl's have been refined, partly due to an unexpected and probably incorrect result on the first one (Table 4) which reported a disordered distribution of Ti among the octahedral sites. It is in fact the restricted Ti content (x apparently does not exceed 0.5 in nature) and its ordered (disordered?) arrangement that have been of primary crystal chemical interest [see below].

Fujino and Takéuchi have located the centrosymmetrically related, partially-occupied hydrogen sites in a Ti-Cl whose 3-coordinated (F,OH,O) site has the average composition $(OH)_{.57}O_{.43}$. They are illustrated in Figure 6a. Kato (pers. comm.) has found that the hydrogen positions in sonolite, the Mn-isotype of clinohumite which contains only OH in the (F,OH,O) site, are nearly the same as those in Ti-Cl. This suggests that the fully occupied H sites in hydroxyl-chondrodite (Yamamoto, 1977) and alleghanyite (Rentzeperis, 1970), will be the same in these structures as in Ti-Cl and sonolite. Proton-proton repulsion is quite likely the reason for the instability of hydroxyl end members of the Mg-humites in nature (Van Valkenburg, 1961). OH-Ch and OH-Cl were found to be stable under nearly the same P-T conditions, but only between ∿700° and 1000°C and 29-77 kbar (Yamamoto and Akimoto, 1977). It may be that the OH-Mn-isotypes have a lower P-T stability range only because the larger Mn^{2+} cations expand the structure more than Mg^{2+}, thus separating the hydroxyl ions.

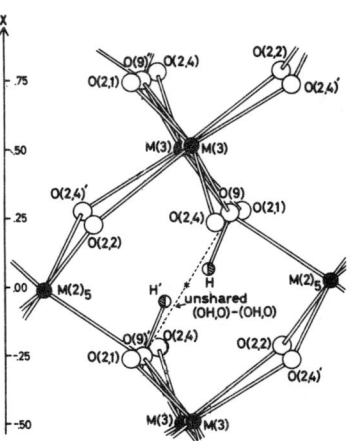

Figure 6a

Atomic configuration in Ti-Cl
viewed along an axis normal to
the plane defined by [100] and
the (OH,O)···(OH,O) atoms (*not*
the two involved in the edge
shared between *M*3 octahedra).
The partially-occupied H positions are indicated by half-
filled circles. After Fujino
and Takéuchi (1978, p. 542,
Fig. 3).

Bonding in the humite minerals

It is not an oversimplification to state that the steric details
of the structures of the humite homologues can be explained by a
straightforward application of Pauling's rules for ionic structures,
although another vocabulary could be invoked to describe the bonding.
An analysis of bond-angle strain at the anions was initiated by Gibbs
and Ribbe (1969) for norbergite and summarized for No, Ch and Hu by
Ribbe and Gibbs (1971); it is generally applicable also to Fo. This
analysis is based on comparisons of real structures with idealized
hexagonal closest-packed equivalents in which the anions are in perfect,
undistorted hcp arrays and the M and Si cations exactly in the centers
of their respective octahedral and tetrahedral holes. There are three
unique anions in this scheme, $(F,OH,O)^{III}$ with ideal $M2$-(F,OH,O)-$M3$
angles = 131.8°, $M3$-(F,OH,O)-$M3$ = 90°, and two sorts of 4-coordinated
oxygens whose cation arrays are sketched in Figures 6b and 6c. The
ideal angles are M_A-O-Si$_B$ = M_B-O-Si$_A$ = 125.3°, M_A-O-M_B = 131.8°,
M_B-O-M_B = 90° and M_B-O-Si$_B$ = 79.5°. Figure 6b contains composite O^{IV}
coordination diagrams constructed using average interatomic distances
and bond-angle strains (observed minus ideal angles) for No, Ch and Hu.
The numbers would be changed only slightly by addition of data from
more recent structure analyses. Electrostatic interactions between
cations in adjacent polyhedra qualitatively explain all bond-angle distortions (Figs. 6b,c): "both the M_A-O and Si$_A$-O bonds are somewhat

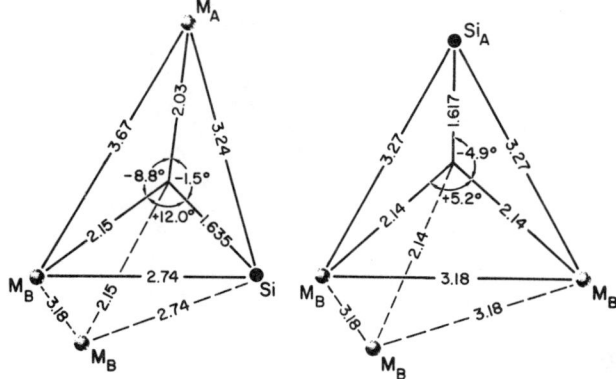

Figure 6b

Composite oxygen coordination diagrams using means of inter- atomic distances in No, Ch and Hu from 10 oxygens for the diagram to the left, from 4 oxygens to the right. After Ribbe and Gibbs (1971, Fig. 2).

The ball and spoke figure on the left below shows cation coordination about oxygens 1,2,3,4 in Ch. A (apical) and B (basal) apply to the relative position of the cation in the anion coordination array (see above). On the right are drawings of faces of the distorted "cube" whose corners are the four oxygens, the three M_B cations and Si_B. The heavy lines between anions signify shared edges. Bond angle strains (°) are labeled at vertices. After Gibbs et $al.$ (1970, Fig. 2).

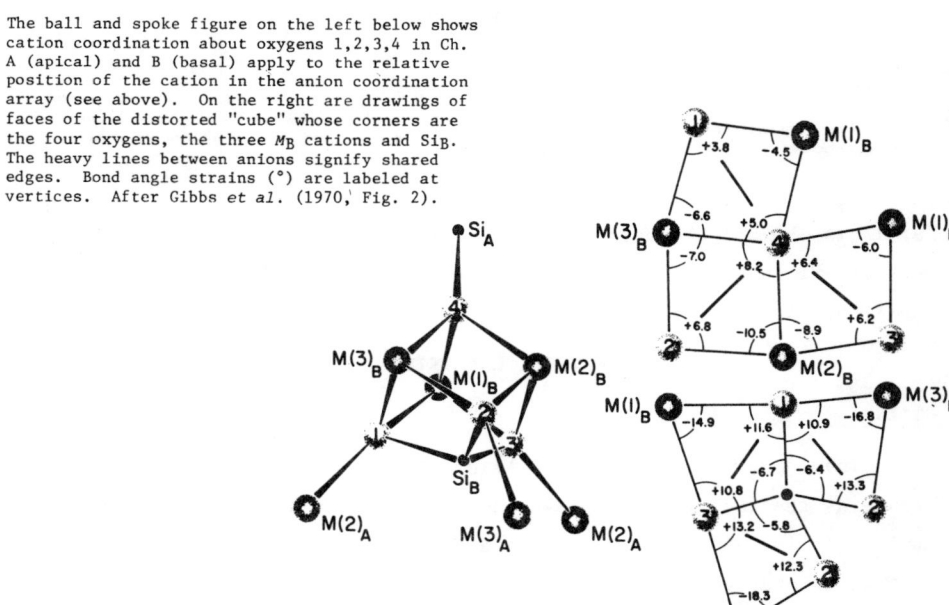

shorter than the respective M_B-O and Si_B-O bonds. This arises from the fact that the apical-basal cation pairs are in adjacent $corner$-$sharing$ polyhedra whereas the basal-basal pairs are in $edge$-$sharing$ polyhedra. The angle strains at oxygen are related to the intercation distances, and because of repulsion across shared edges, the strains are always positive for the M_B-O-Si_B and M_B-O-M_B angles and negative for the M_A-O-M_B bond angles. The M_B-O-M_B bond angle strains are on the average one degree greater when Si is in the basal array (solid dots) than when

247

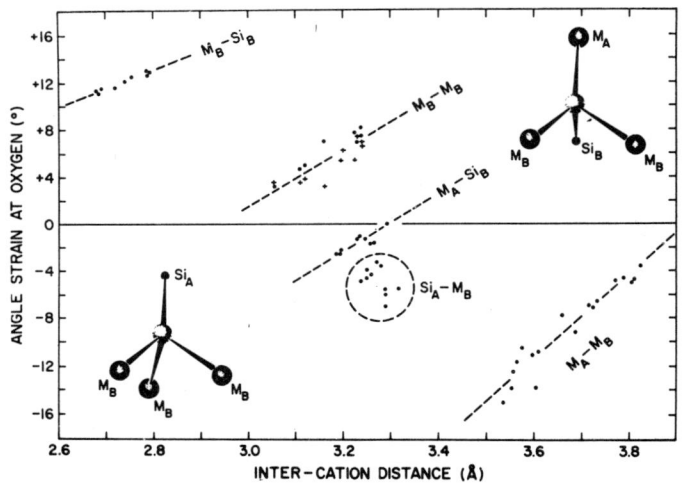

Figure 6c. The relation of angle strains (observed minus ideal for hcp) at oxygen as a function of inter-cation distance using data from No, Ch, and Hu (Table 4). The largest positive strains are associated with the large M_B-Si_B repulsion across shared polyhedral edges. After Ribbe and Gibbs (1971, Fig. 3).

Si is in the apical position (crosses), reflecting the greater repulsion between Si and the two M cations in the basal array" (Ribbe and Gibbs, 1971, p. 1160).

CRYSTAL CHEMISTRY

Titanium

There are two problems involving titanium in humites: the first concerns the fact that x appears never to exceed 0.5 Ti atoms per formula unit in natural specimens (see Fig. 2) and the second whether or not Ti is ordered into the $M3$ site, and if so, why?

Analyses of the most Ti-rich norbergite, chondrodite and clinohumite are listed in Table 4; the most Ti-rich humite (specimen 3 in appendix Table A-4) contains only 3.25 wt % TiO_2 (x = 0.20). Fujino and Takéuchi (1978) hinted that restriction of Ti levels in humites might be attributed to Pauling's electrostatic valence principle. That local charge balance might be important is evinced in Figure 7 where electrostatic imbalances are plotted as a function of x for each of the anions in the humite structures for two extreme cases: (a) Ti disordered into all octahedra and (b) Ti ordered into $M3$ sites.

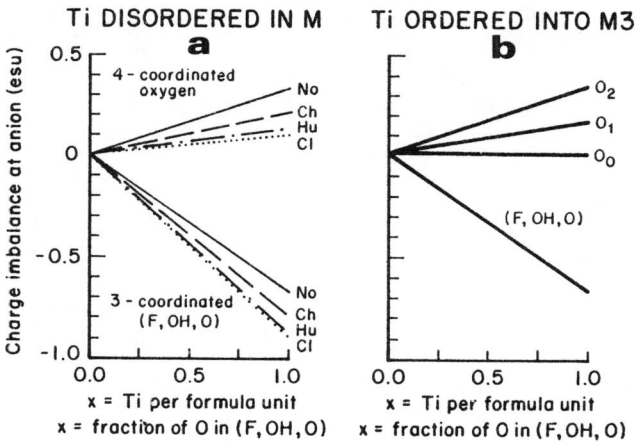

Figure 7. Plots of x versus charge imbalance at the anion for the humite minerals as-suming (a) Ti disordered in all the M octahedra and (b) Ti ordered into $M3$. Symbols are explained in the text and Table 5. The charge imbalance was calculated in the following manner for the (F,OH,O) site: $S - (1 + x)$, where S is the sum of the Pauling bond strengths to the anion and x is the fraction of O in that site. For the oxygens, charge imbalance is simply $S - 2$.

In case (a) there are only two types of anions as far as local charge balance is concerned, regardless of mineral species. Since Ti is disordered, the average charge on all M cations is the same: $<z> = 4[x/(2n + 1)] + 2[(2n + 1 - x)/(2n + 1)]$. The oxygens are 4-coordinated (to $3M + 1$ Si) and are all overbonded by $[1 + 3<z>/6] - 2.00$ esu. The (F,OH,O) anion is bonded to $3M$ atoms and is always underbonded by an amount $3<z>/6 - (1 + x)$ esu, where $x = O_{Ti}/(F + OH + O_{Ti})$. Figure 7a indicates the imbalances as a function of x for each of the four humites in which Ti is disordered.

In case (b) there are collectively among the humite homologues three types of 4-coordinated oxygens, O_0, O_1, O_2, the subscripts refer-ring to the number of $M3$ atoms to which they are bonded, plus the (F,OH,O) anion. From the multiplicities of M sites in the humites (Table 5) it can be seen that if all the Ti is ordered into the $M3$ sites, the maximum Ti occupancy of $M3$ is $x/2$, because $x/(2n + 1)$ is the fraction of Ti in the $(2n + 1)$ M sites and $2/(2n + 1)$ is the fraction of M sites which are $M3$ sites. The O_0 oxygens are charge-balanced because they are not bonded to $M3$, the O_1 oxygens are overbonded by

249

Table 5. Number and types of octahedral sites (per formula unit) and number and types of oxygen atoms and (F,OH,O) sites (per unit cell) for the humite minerals.

Mineral (Z)	Number of octahedral sites of type:						Number of oxygens of type:			Number of (F,OH,O)
	Ml_n	Ml_c	$M2_6$	$M2_5$	$M2_4$	$M3$	O_0	O_1	O_2	
Norbergite (4)					1	2			16	8
Chondrodite (2)		1		2		2	4	8	4	4
Humite (4)	2		1	2		2	30	12	6	8
Clinohumite (2)	2	1	2	2		2	20	8	4	4

x/6, the O_2's by 2x/6, and the (F,OH,O) anions are always underbonded by 4x/6. Figure 7b shows the imbalances for case (b); the proportions of O_0, O_1, O_2 and (F,OH,O) anions per formula unit are listed in Table 5 for all the humites.

It is clear from Figure 7 that the more Ti there is, the greater the local charge imbalances, regardless of whether Ti is ordered or disordered in any of the homologues. Quite likely these imbalances place a practical upper limit on Ti substitution. Comparing the two graphs, it is also evident that charge balance is somewhat improved when Ti is ordered into $M3$, except for norbergite in which all anions are bonded to 2 $M3$ sites. Thus one may argue that Ti prefers $M3$ for reasons of charge balance, but it should also be observed that $M3$ is always smaller than the other octahedra (even in alleghanyite, $Mn_5[SiO_4]_2(OH)_2$), although it has fewer shared edges than Ml (see Table 3). This is in part explained by the fact that $M3$ is bonded to two monovalent (F,OH) atoms when no Ti is present. Quadrivalent Ti ($r \sim 0.61$ Å) is smaller than Mg^{2+} ($r \sim 0.72$ Å) and the other M^{2+} cations found in humites and quite likely prefers the $M3$ site for that reason as well.

Thus the crystal chemical arguments all point to an ordered distribution of Ti^{4+} in $M3$, the somewhat questionable least-squares site refinements (Robinson et al., 1973) notwithstanding. Probably the most compelling argument for Ti^{4+} order is that made by Fujino and Takéuchi (1978, p. 537-9). In addition to least-squares site refinements of Ti-Ch and Ti-Cl, which both converged with relatively large estimated standard errors (± 0.08 Ti) to indicate all Ti in $M3$, they found that in these epitaxially intergrown specimens Ti-Ch had a Ti/Si ratio of 0.212, almost exactly twice as great as that in Ti-Cl (0.107). This is precisely the relationship expected if Ti is preferentially ordered into $M3$ in both structures: the $M3$/Si ratio is 2 in Ch, 1 in Cl.

Epitaxial intergrowths such as the one reported by Tilley (1951), but more particularly those containing substantial Ti, should be carefully analyzed in the future for additional confirmation of these conclusions regarding the ordering of Ti.

Iron

The compositions of all the structures discussed in this section are listed in Table 4. In a site refinement of the structure of a humite with 0.35 Fe^{2+}, Ribbe and Gibbs (1971) found that $3d^6$ Fe^{2+} preferred not the larger octahedra but the more distorted ones and those with the more polarizable ligands, i.e., $M(2)O_6$ 0.12 Fe, $M(1)O_6$ 0.09 Fe, $M(2)O_5(F,OH)$ 0.03 Fe, and $M(3)O_4(F,OH)_2$ 0.01 Fe. Their refinement of an iron-poor chondrodite gave the same indication with 0.05 Fe in $M1$ and 0.00 Fe in $M2_5$ and $M3$ and a refinement of Ti-poor clinohumite (Robinson et al., 1973) gave 0.08 – 0.12 Fe in $M2_6$, 0.06 – 0.10 Fe in $M1$, 0.03 – 0.06 Fe in $M2_5$ and 0.00 – 0.03 Fe in $M3$. By way of confirmation, Kocman and Rucklidge (1973) found a similar pattern in a titanian clinohumite in which they assigned all the Ti to $M3$ and then refined the Mg-Fe distribution.

Those least-squares site refinements of Ti-Cl discussed in the previous section gave equivocal results for Fe-Ti distribution because the problem of assigning three cations (Mg, Fe, and Ti) among five nonequivalent octahedral sites cannot possibly be solved without some unsatisfying assumptions (such as that made by Kocman and Rucklidge, 1973).

Based on the 55 microprobe analyses listed in the Appendix, Ribbe and Gibbs (1971, their Fig. 10) report that on the average the Fe/Mg ratio increases rapidly with decreasing (F + OH)/O in the humite mineral series. This would suggest that Fe/Mg partitioning may be controlled by the relative number of polarizable F,OH ligands in the individual structures. However, this suggestion is essentially negated by the fact that epitaxially intergrown Ti-Ch and Ti-Cl of Fujino and Takéuchi (1978; analyses in Table 4) show no such partitioning: Fe/Mg is nearly identical in both structures, but there are twice as many OH ligands and only one-fifth as many MO_6 octahedra in Ti-Ch as in Ti-Cl. As with titanium, the iron distributions among coexisting pairs or sequences of humite minerals should be more carefully examined.

Manganese and Calcium

Manganese is a minor substituent in the Mg-homologues (Tables A-3 and A-4), but it ranges up to ∿100% of the M atoms in the Mn-isotypes (see Winter et al., 1983, Table 2). It appears from site refinements of Mg-manganhumite, Mg-alleghanyite, and sonolite (details in Table 4) that size considerations alone are sufficient to explain the Mn-Mg distribution. Separate lines of best fit are drawn in Figure 8 for the several octahedra. Sonolite contains 0.37 Ca atoms per formula unit, and Kato's refinement (pers. comm.) indicates 0.92 and 0.93 (Mn+Ca)/(Mg+Mn+Ca+Zn) in the $M(2)_6$ and $M(2)_5$ octahedra, respectively. If calcium is apportioned equally into these $M2$ sites (as in olivine; see Ch. 11), then the mean (Mn,Mg)-(O,OH,F) distances would be 0.016A smaller as indicated by the arrows in Figure 8a, and the relationship of Mn-content versus mean M-(O,OH,F) would appear to be quite reasonable, given the inherent differences in octahedral sizes due to edge-sharing and ligancy (Table 3; cf. discussion of the Mg-humite analogs, p. 242f.)

Using data from four Mg,Mn-olivine refinements and eleven humite minerals in which Mg and Mn are the predominant M^{+2} cations, Francis (1980) calculated linear regression equations for mean M-O distances versus mean ionic radius of the M^{2+} cations for each type of octahedral site (Table 3), except $M(2)_4$ for which he had only one datum. Because the ligancy for the $M(1)$ sites always consists of six oxygens, the correlation is excellent (Fig. 8b); the regression equation is simply:

$$\langle M(1)\text{-O}\rangle = 1.445 + 0.9172\langle r_{M(1)}\rangle \quad (r = 0.995).$$

However, in order to combine $M(2)O_6$, $M(2)O_5(OH,F)$ and $M(2)O_4(OH,F)_2$ octahedra into a single regression and to treat all the $M(3)$ octahedra together, a second factor f had to be introduced to account for the fact that $F \not\simeq OH \not\simeq O$, as well as $\langle r_{M(2)}\rangle$, affect the mean octahedral M-anion bond lengths significantly (Figs. 8c,d). f was defined as one-half the number of fluorine atoms per formula unit times the number of monovalent ligands associated with the octahedron. The resulting equations are

$$\langle M(2)\text{-}(O,OH,F)\rangle = 1.541 + 0.8185\langle r_{M(2)}\rangle - 0.01649\,f \quad (r = 0.994)$$

$$\langle M(3)\text{-}(O,OH,F)\rangle = 1.502 + 0.8386\langle r_{M(3)}\rangle - 0.02122\,f \quad (r = 0.996).$$

The most comprehensive study of the Mn-humites is by Winter *et al.*
(1983), who found nearly pure (>96 mol % Mn) sonolite, manganhumite
and alleghanyite (all with OH:F ≃ 3:1) in association with tephroite,
rhondonite and Mn-carbonates at Bald Knob, North Carolina. They discuss
phase equilibria of these assemblages (*cf.* Momoi, 1980), listing cell
dimensions for these Mn-humites together with indexed powder diffraction
patterns of carefully selected material, and noting that Cook's (1969)
x-ray data for alleghanyite (as it appears in the ASTM Index, File No.
22-726) has non-indexable lines, probably due to contaminaton. See
White and Hyde (1982b, discussed below) for TEM studies of compositional
faulting and intergrowths of natural Mn-humites. White (1982b) gives
crystal data for the series nMn$_2$SiO$_4$·MnF$_2$ $(1 \leqslant n \leqslant 4)$.

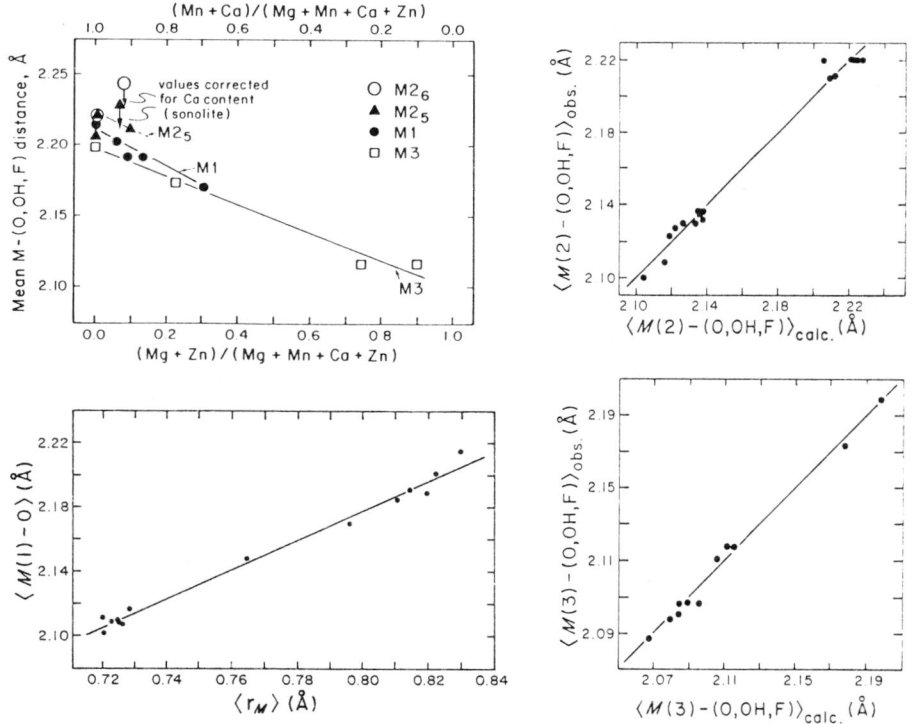

Figure 8

(a) Variation of mean M-(O,OH,F) distance with (Mg + Zn)/(Mg + Mn + Ca + Zn) in alleghanyite,
magnesian alleghanyite, magnesian manganhumite, and sonolite (references in Table 4).   Two
data points from the sonolite refinement are corrected for Ca content.
(b) Plot of observed mean M(1)-O bond lengths as a function of the mean radius of the cations
occupying M(1).
(c) Plot of observed mean M(2)-(O,OH,F) and
(d) M(3)-(O,OH,F) bond lengths versus values calculated from the regression equations in text.
Data from Francis (1980; see his Figs. 4.1, 5.1, 5.2, 5.3).

253

Table 6 Numbers and ratios of three-coordinated anions in synthetic and natural humites, forsterite, sellaite, and brucite together with normalized volumes, V, corrections to volume due to transition metal content, $\Delta V = [\langle r_M \rangle - r_M] V / r_{Mg}$, and calculated densities.

Name, q, reference*	Symbols, Fig. 3	No. of 3-coord. anions F	OH	O_{Ti}	$\dfrac{F+0.5\,O_{Ti}}{F+OH+O_{Ti}}$	$\dfrac{(F+OH)^{III}}{(F+OH)^{III}+O^{IV}}$	$V'(\text{Å}^3)$	$\Delta V'(\text{Å}^3)$	Calc'd density
Norbergite 1.0	●	2.00			1.00	0.33	70.25		3.199
$q=6$ 0.9	●	1.80	0.20		0.90		70.56		
0.7	●	1.40	0.60		0.70		71.19		
V	○	2.00			1.00		70.37		
1	○	1.81	0.17	0.02	0.91		70.54	0.00	
Chondrodite 1.0	⊙	2.00			1.00	0.20	71.23		3.205
$q=5$ 1.0	⊙	2.00			1.00		71.27		
0.8	⊙	1.60	0.40		0.80		71.68		
0.8	⊙	1.60	0.40		0.80		71.67		
0.7	⊙	1.40	0.60		0.70		71.82		
0.6	⊙	1.20	0.80		0.60		72.14		
2	⊕	1.27	0.73	0.00	0.63		71.86	-0.03	
9	⊕	1.86	0.11	0.03	0.94		71.50	-0.10	
24	⊕	1.06	0.90	0.04	0.54		72.24	-0.22	
(OH-Ch) Y	⊙		2.00		0.00	0.20	73.74		3.060
(Ti-Ch) F&T	φ	0.00	1.11	0.89	0.22		72.76	+0.17	
Humite V	□	2.00			1.00	0.14	71.78		
$q=14$ 3	□	0.84	0.80	0.36	0.51		72.40	0.00	
7	□	1.06	0.93	0.01	0.53		72.44	-0.22	
Clinohumite 1.0	△	2.00			1.00	0.11	71.83		3.211
$q=9$ 0.5	△	1.00	1.00		0.50		72.36		
0.4	△	0.80	1.20		0.40		72.44		
7	▲	1.04	0.92	0.04	0.53		72.52	-0.29	
(OH-Cl) Y&A	△		2.00		0.00	0.11	73.02		3.139
(Ti-Cl) K&R	▲	0.42	1.05	0.53	0.34		73.08	-0.45	
10	▲	0.00	1.05	0.95	0.24		72.98	-0.12	
F&T	▲	0.00	1.10	0.90	0.22		72.89	-0.16	
Forsterite, Mg_2SiO_4 (Yoder and Sahama, 1957) $q=4$					0.00		72.50		3.222
Sellaite, MgF_2 (Duffy, 1977) $q=1$					1.00		65.18		3.174
$MgOHF$ (Duffy, 1977) $q=2$					1.00		73.10		2.743
$Mg(OH)_2$ hypothetical $q=1$					1.00		81.03		2.312
Brucite, $Mg(OH)_2$					1.00		47.22		2.051

* Specimen numbers with decimals were synthesized by Duffy (1977): $0.9 \equiv n Mg_2SiO_4 \cdot Mg(F._{90}OH._1)_2$ Those with integral numbers are from Jones *et al.* (1969), V = Van Valkenburg (1961), Y = Yamamoto (1977), F&T = Fujino and Takéuchi (1978), Y&A = Yamamoto and Akimoto (1977), and K&R = Kocman and Rucklidge (1973).

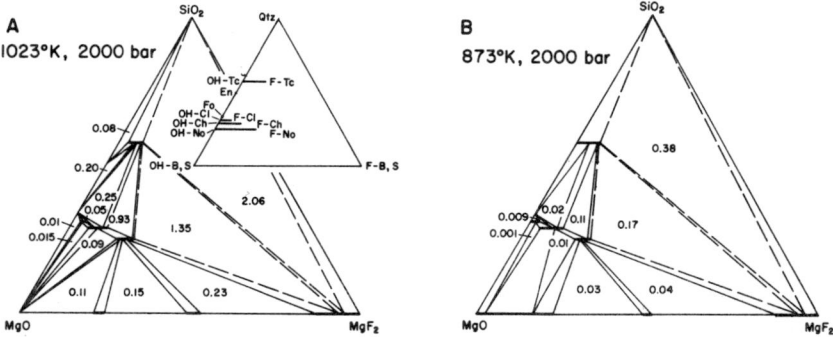

Figure 9. The system MgO–MgF_2–SiO_2–H_2O projected from H_2O onto MgO–MgF_2–SiO_2. Numbers shown are f_{HF} for each three phase field. New symbols: Tc = Talc, B = Brucite, S = Sellaite, En = Enstatite. Modified from Duffy and Greenwood (1979, Fig. 1).

Fluorine and Hydroxyl

Van Valkenburg (1961) synthesized F-end members of the humite series from melts and by solid state reactions, but OH-end members have been synthesized only at high temperatures and pressures -- OH-Ch and OH-Cl were found to be stable between 700-1000°C and 29-77 kbars (Yamamoto and Akimoto, 1977). Duffy (1977; see Duffy and Greenwood, 1979) synthesized F-No, F-Ch and F-Cl and many intermediate (F,OH)-humites as part of his investigation of the system $MgO-MgF_2-SiO_2-H_2O$. Projections of the stable assemblages from H_2O onto $MgO-MgF_2-SiO_2$ are shown in Figures 9a (1023°K, 2000 bars) and 9b (873°K, 2000 bars). Duffy (p. 29) said, "These provide insight into the limited compositional ranges of naturally occurring minerals of the humite group," and he discussed the analyses of Bourne (1974) and Jones *et al.* (1969 -- see appendix) in light of his elegant thermodynamic model for the system.

Duffy and Greenwood (1979) calculated expressions for molar volumes as a function of composition for the phases of variable F-OH content in this system and found that with the exception of brucite "all the volume functions are linear in composition. For sellaite $[MgF_2]$ and humites the difference in volume between the hydroxyl and fluoro-end members is 2.95 ± 0.19 cm^3 mol^{-1} of $(OH)_2$" (pp. 1159-1160; see their Table 4). They also give equations for variations of interplanar d-spacings with F $\rightleftarrows$ OH for sellaite (Sel),[3] No, Ch and Cl (their Table 5, p. 1160). Their data and some of those listed in Tables 1, 4, 6, and the appendices A-3, and A-4 are used in the following more detailed discussion of the relative effects of F $\rightleftarrows$ OH and other substitutions on the unit cell volumes of sellaite, the humites and forsterite. In all these structures the octahedral sites in the hcp anion array are half-filled with Mg, and to all intents and purposes the $[SiO_4]$ tetrahedra (where present) are the same size (mean Si-O $\sim$ 1.63 A) and each shares the same three edges with M octahedra. But the compositional variation from Sel $\rightarrow$ No $\rightarrow$ Ch $\rightarrow$ Hu $\rightarrow$ Cl $\rightarrow$ Fo is more complex than just (F,OH) for O. The charge

---

[3] Tetragonal sellaite, MgF_2, has the hcp rutile structure (Baur, 1976): $a = 4.6213$, $c = 3.0519$ A, $V' = a^2c$. "Intermediate sellaite," MgFOH, is orthorhombic and is similar in two dimensions and c/q to the humites and olivine (*cf.* Table 1). Its cell parameters (Duffy, 1977) are $a = 4.686$, $b = 10.123$, $c = 3.078$ A; $q = 2$.

balanced substitution is $\square^{IV} + 4(F,OH)^{III} \rightarrow Si^{IV} + 4O^{IV}$, where $\square^{IV}$ is a tetrahedral void. The fraction of tetrahedral sites occupied by Si in the hcp array decreases from 1/8 in Fo to 1/12 in No to zero in Sel (see Fig. 10 and Table 2).

For the sake of comparing these structurally related compounds, Ribbe *et al.* (1968) suggested that it would be convenient to normalize unit cell volumes to a reduced cell of volume $V' = [abc\sin\alpha] \div q$, where q is 4 for Fo, 6 for No, 5 for Ch, 14 for Hu, 9 for Cl. The parameter q was chosen to give a third dimension $d_{(001)}/q \sim 1.5$ A, *i.e.*, approximately equal to the radius of an average anion in the hcp array. Another perhaps more physically useful volume term is $V_A = V'/4$, the volume per anion, from which may be calculated a mean radius for the anion, $<r_A>_V = 1/2V_A^{1/3}$, on the assumption that V_A is a cube of dimensions $2<r_A>_V$.

Effective radii of F and OH. In Figure 10 there are three lines representing anion radii calculated in different ways (see Table 6); all are plotted against the ratio $(F + OH)^{III}/(F + OH)^{III} + O^{IV}$ which is of course related to the fraction of tetrahedral sites filled with Si (*cf.* upper and lower abcissas). The first is a line labelled "Expected $<r_A>$" which joins the 1.300 A value for the radius of F^{III} and the 1.378 A value for the radius of O^{IV} (Shannon and Prewitt, 1969) and assumes that a proportional mix of F^{III} and O^{IV} will produce the line $<r_A> = 1.378 - 0.078[F/(F + O)]$. Of course there are no structures to which such a line may apply, but it serves as reference.

An effective "radius" of F^{III} in Sel was calculated by subtracting $r_{Mg} = 0.72$ A from the grand mean Mn-F distance (1.988 A; Baur, 1976), and an effective "radius" of O^{IV} in Fo was determined similarly using $<<Mg-O>> = 2.114$ A (Smyth and Hazen, 1973). Since the $F^{III}/(F^{III} + O^{IV})$ ratio *decreases* and the fraction of filled tetrahedral sites *increases* linearly from left to right in Figure 10, one might expect that the dashed straight line, $<r_A>_M = 1.394 - 0.123[(F/(F + O)]$, joining Sel and Fo would contain points representing all the F-Mg-humite minerals. Unfortunately no structures of F-Mg end members have been refined, but if the grand mean M-O,F,OH distances corrected for the mean radii of the M cations, $<r_M>$, are plotted for each of the refined natural specimens (dotted tails of arrows), they all fall above the dashed line.

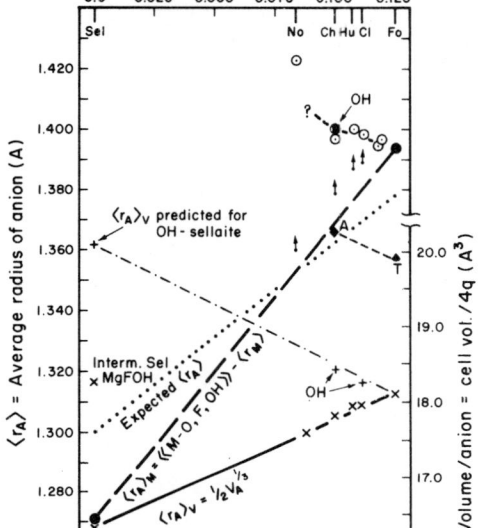

Fraction of tetrahedral sites filled

Figure 10

The average radii of anions $\langle r_A \rangle$ plotted against $(F + OH)^{III} / [(F + OH) + O^{IV}]$ and the fraction of tetrahedral sites filled for the series of compounds Sel → No → Ch → Hu → Cl → Fo. A = alleghanyite, T = tephroite. Dots at the tails of arrows represent $\langle r_A \rangle_M$ for natural Mg-equivalent homologues (Table 4) corrected for effects of M-site substituents: arrows point to open circles which are values of $\langle r_A \rangle_M$ predicted for OH-end members by simple proportionation.

Figure 10 and 11 from Ribbe (1979), Figs. 2 and 3, pp. 1030 and 1033.

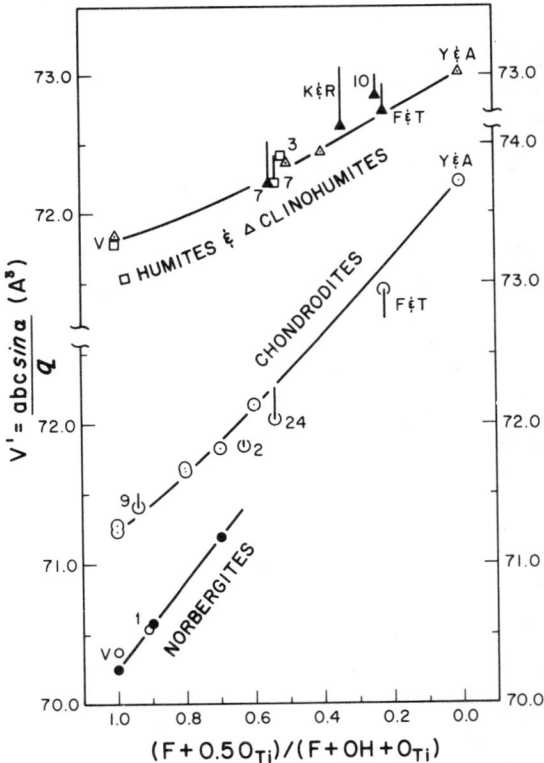

Figure 11

A plot of normalized cell volumes, $V' = Vol \div q$ versus $(F + 0.5\ O_{Ti}) / (F + OH + O_{Ti})$. Tails of vertical lines represent data points uncorrected for the deviation of $\langle r_M \rangle$ from 0.72 A, the radius of Mg^{2+}. Symbols and labels are listed in Table 6.

This is mainly due to OH^{III} substituting for F^{III} in the hcp anion array. On that assumption, values of $<r_A>_M$ (centered circles in Fig. 10) have been predicted for OH-end members by simple proportionation of the $OH/(F + OH)$ ratios observed (Table 6) and the difference between observed radii and those expected for pure F-end members. As Ribbe (1979, p. 1031) pointed out, the net effect of OH substitution for F (r = 1.300A) is to expand the structure as characterized by $<<M-0,F,OH>>$ rather more than expected based on the value of 1.358A suggested by Yamamoto (1977) for the radius of OH^{III}. In fact, some simple arithmetic using $<r_A>_M$ = 1.370A predicted for F-Ch and $<r_A>_M$ = 1.400A observed for OH-Ch shows that the effective "radius" of OH^{III} must be 0.150A larger than F^{III}.

In attempting to extract a more reasonable idea of the OH^{III} radius from the humite data summarized in Figure 10, Ribbe (1979) experienced the frustration that is common to such endeavors. One explanation is that "the replacement of F by OH causes an expansion of the whole anion array" (Yamamoto, 1977, p. 1483). This is true, but the reason for it evaded a careful study of the *unoccupied* polyhedra in the humites. Proton-proton interactions (see Fig. 6a) *do* serve to expand the structures of OH-rich humites considerably more than is indicated by comparing $M-(0,F,OH)$ bond lengths, and perhaps that explains why OH-No, which would have an abundance of both shared and unshared $OH\cdots OH$ octahedral edges, has not yet been synthesized and may be unstable except at very high pressure (> 150 kbar?). OH-No has a predicted normalized volume of 74.6 A^3. Fujino and Takéuchi (1978, p. 541-2) argue that $H\cdots H$ repulsion is the cause of structural instability and suggest that "under a F-free condition, OH-Ch and OH-Cl appear to be stabilized through incorporation of Ti in place of Mg." Obviously it is rather the concommitant substitution of O_{Ti} for OH that "stabilized" the structure, because Ti itself "destabilizes" the structure by increasing local electrostatic charge imbalance (Fig. 7).

Unit cell volumes and ionic radii. Using the normalized volumes, V' = unit cell volume $\div q$, corrected by an amount $\Delta V'$ to account for differences in the mean octahedral cation radius from one structure to the next, all the humite minerals may be systematized according to the average occupancy of the 3-coordinated (F,OH,O_{TI}) site. Pertinent data are compiled in Table 6 and plotted in Figure 11 where curves have been drawn through the points representing synthetic specimens. Because the substitution of Ti into $M3$ is accompanied by a charge-balancing substitution of O_{Ti}^{III} for $(F,OH)^{III}$, O_{Ti} had to be apportioned properly based on its effective ionic radius in order to model the volume relationships established in Figure 11. Although there is considerable uncertainty about anion radii, the abscissa was labelled $(F + 0.5\ O_{Ti})/(F + OH + O_{Ti})$, and this scheme is apparently successful (see Ribbe, 1979, for a full justification). In any case only five of the 27 specimens listed in Table 6 have sufficient Ti content to be significantly affected.

Several observations on Figure 11 are pertinent. (1) The humites and clinohumites are apparently indistinguishable, although there are no synthetic humites other than the F-end member to test this observation. (2) The slopes of the lines vary from -3.5 for No to -2.5 for Ch to -1.2 for Hu-Cl, suggesting that the relative effect of OH^{III} on the normalized volume increased with the proportion of OH in the total $(F,OH,O)^{III}O^{IV}$ anion array. This lends further support to the idea that increasing numbers of proton-proton interactions decrease the stability of the humite homologues, accounting for the extremely high pressures required to synthesize OH-Ch and OH-Cl and for the fact that OH-No has not yet been made.

PHYSICAL PROPERTIES

Optical Properties

Misplaced emphasis upon the isolated $[SiO_4]$ tetrahedron as the key structural unit in orthosilicates has diverted attention from the fact that most orthosilicates have prismatic habits and cleavage and are for the most part optically positive (Ribbe, 1976). The Mg-humites are optically positive because in all of them serrated edge-sharing octahedral chains are in fact the dominant structural units.

Sahama (1953) calculated mean refractive indices, $<R.I.> = (\alpha\beta\gamma)^{1/3}$ or $(\omega^2\varepsilon)^{1/3}$, for the F- and OH-members of the humites based on molar refractivity, plotting them against $1/(n + 1)$, with Fo $(n = \infty)$ and the sellaites $(n = 0)$, MgF_2 and $Mg(OH)_2$ (hypothetical), as end members. His curves $(cf.$ Deer $et\ al.,$ 1962, p. 55, Fig. 18) become essentially

259

straight lines when the data are plotted against (F,OH)/[(F,OH) + O], as in Figure 12. Refractive indices of the F-series (Van Valkenburg, 1955) and OH-Ch and OH-Cl (Yamamoto and Akimoto, 1974, 1977) are represented by +'s in Figure 12: they fall very close to Sahama's calculated values. The effect of substituting Fe + Mn for Mg is to increase <R.I.> by $\sim$ 0.02/mol % Fe + Mn; Ti is expected to have a somewhat greater effect. But of course OH $\rightarrow$ F and $O_{Ti} \rightarrow$ F also affect <R.I.> and 2V, leading to the conclusion that, except for norbergite, which is generally very F-rich and Fe- and Ti-poor, naturally occurring members of the homologous series cannot be distinguished on the basis of refractive indices or 2V (Sahama, 1953; Deer *et al.*, 1962).

Dispersion is $r > v$ for the orthorhombic humites, No and Hu, and according to Deer *et al.* (1962), $r > v$ is "strong" for monoclinic Ch and Cl. The pleochroic formulas, mostly in hues of yellow, are also listed by them, but they are not particularly useful in mineral identification because of the range of $Fe^{2+} \rightarrow$ Mg and concomitant $F \lessgtr OH \lessgtr O_{Ti}$ substitutions which all affect crystal-field splitting parameters.

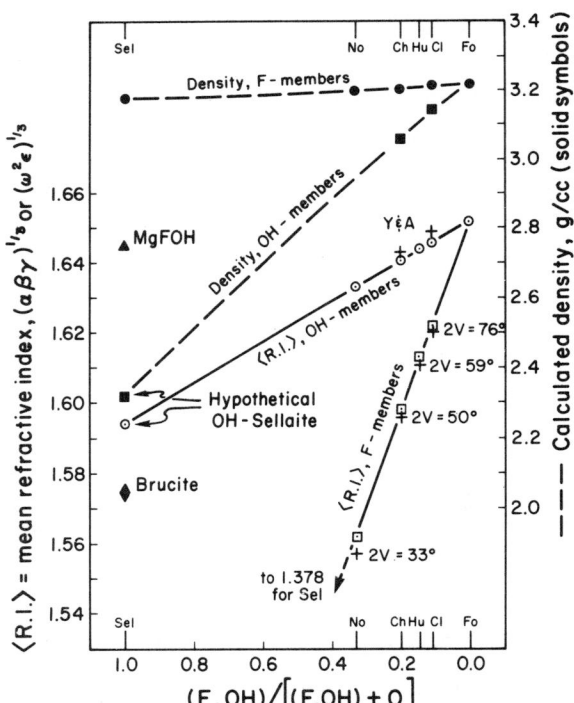

Figure 12

Calculated densities and mean refractive indices of the F- and OH-series of compounds, Sel → No → Ch → Hu → Cl → Fo. Volume data used in density calculations are from Table 6, refractive indices from Sahama (1953), and observed values (+) for F-humites from Van Valkenburg (1955) and for OH-Ch and OH-Cl from Yamamoto and Akimoto (1974, 1977).

Colors range from colorless (with only Mg and Ti present) to yellow, dark orange, red, and brown. These are strikingly different from olivine which, with the same transition metal substituents, is colorless to pale green to brown, the darker colors being attained only with very much higher levels of Fe^{2+} for Mg than is necessary to produce rather dark colors in humites. The colors of the Mn-isotypes are commonly mixtures of red and pink and brown. A zincian sonolite is reported to be dark brown to brown-black (Cook, 1969).

Sahama (1953) showed that the extinction angle $c \wedge \alpha$ on (100) (axial convention as in Table 1) could be used to differentiate between chondrodite (22° - 31°) and clinohumite (9° - 15°) and thus of course distinguish the monoclinic phases from their orthorhombic homologues. These extinction angles reflect the angle between c^* and the serrated octahedral chains parallel to c: $\sim$ 19° in Ch, $\sim$ 11° in Cl.

Specific Gravity

Specific gravities of the F- and OH-series Sel → No → Ch → Hu → Cl → Fo (and brucite for comparison) were calculated from Mg-end-member chemistries and unit cell volumes (see Table 6). In Figure 12 the F-humites are barely distinguishable, but OH has a profound effect on molar volume, as noted earlier (cf. Fig. 11). Counteracting effects of heavier transition metals substituting for Mg and larger OH (and O_{Ti}) substituting for F lead to the conclusion that specific gravity, much like optical properties, is useless as a determinative method. X-ray diffractometry and electron probe microanalysis are obviously the most reliable techniques for identifying and characterizing humite minerals.

Thermal Breakdown Reactions and the Effects of Annealing

In an unpublished Ph.D. dissertation, Butler (1972) found that the principal reaction product of dehydrogenation and/or deflourination attained by heating the humites is forsterite. Single-crystal x-ray studies determined that Fo forms in topotaxic relation to the host phases with $[100]_{Fo}$ and $[010]_{Fo}$ parallel to [100] and [010] of the host, and $[001]_{Fo}$ parallel to [001] of orthohombic No and Hu, or $[014] \equiv c^*$ of monoclinic Ch and Cl. Another orientation of Fo was frequently observed corresponding to other closest-packed planes in

261

Fo, *i.e.* <011> parallel to *c**, and some completely disoriented Fo
was present in most heated specimens. Periclase and the closest-
packed iron oxides, ccp magnetite [maghemite?] and hcp hematite,
also crystallized topotactically within the host phases upon heating
in air. Butler described another breakdown product of Ch as twinned
Cl in which the orientation of the olivine "blocks" was preserved.
Norbergite broke down first to Ch then to Cl and then to Fo. White
(1982a, p. 39) found that high temperature annealing of his synthetic
Hu and Cl resulted in the disappearance of superstructures and a
decrease in the concentration of faults.

TWINNING

The numerous axial labels and choices of space group have led to
confusing descriptions of twinning in humites. [In this discussion I
quote liberally from Jones (1969)]. Deer *et al.* (1962) describe simple
and multiple twinning on (001) for chondrodite and clinohumite; this is
correct if Taylor and West's (1929) axial notation is used, but is not
consistent with the axial notation adopted by Deer *et al.* They also
report twinning on (105) and (305) for chondrodite and on ($\bar{1}$03) for
clinohumite. These are consistent with the notation adopted by the ASTM
and several others for chondrodite, but not with their own choice of
axes. Heinrich (1965) apparently adopts Sahama's choice of axes, yet he
describes twinning on (001), (105), and (305) for chondrodite although
with Sahama's choice of axes it should be (100), (501), and (503). Using
Taylor and West's notations, the convention adopted in this chapter, the
twin planes in chondrodite are (001), (015), and ($0\bar{3}5$); in clinohumite,
(001) and ($0\bar{1}3$). Polysynthetic twinning on (001) is common in chondrodite and provides an example of twinning by reticular pseudomerohedry
(Friedel, 1964) as controlled by the pseudosymmetry of the orthorhombic
superlattice. When parallel twinning of this type occurs, the vector
[$02\bar{1}$]* of one lamella coincides in magnitude and direction with the
vector $2b^*$ of the other lamella (Fig. 13). Since space group extinctions
from each lamella allow diffraction from planes 0*kl* only if $k = 2n$, diffraction spots from one twin are superposed on those from the other. An
0*kl* X-ray photograph thus shows intensities and distribution of spots
consistent with a pseudo-orthorhombic cell. In upper level photographs

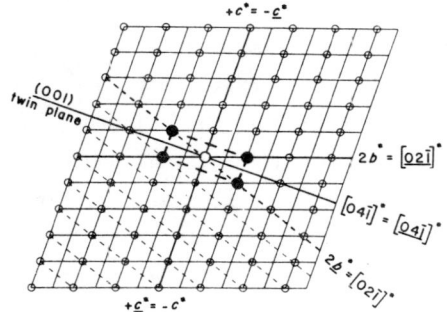

Figure 13. Parallel twinning on (001) in chondrodite as seen on the b^*c^* reciprocal lattice plane. The vectors $[02\bar{1}]$ and $-c^*$ of one twin lamella (solid lines) coincide in magnitude and direction with the vectors $2b^*$ and $+c^*$ of the other twin lamella (dashed). This results in the superposition of diffraction spots from both twin lamellae and a pseudo-orthorhombic cell (outlined) with b_{ortho} parallel to $[04\bar{1}]^*_{mono}$. After Jones (1969, Fig. 1).

of (001)-twinned chondrodite, $2mm$ symmetry is present but systematic extinctions do not correspond to any space group extinction rules. Similar twinning was observed in $Al_5Co[BO_4]_2O_2$, a boron Ch (Capponi and Marezio, 1975) and in leucophoenicite a dimorph of manganhumite (Moore, 1970; also refer to the discussion of leucophoenicite in Chapter 12 in this volume).

When either (015) or ($0\bar{3}5$) are twin planes, diffraction spots from each lamella are distinct. Twinning on (015) and ($0\bar{3}5$), together with polysynthetic twinning on (001), has been observed frequently in chondrodite, and Rogers (1935) reported it in alleghanyite. Tilley (1951) reported such twins in a chondrodite epitaxially intergrown with other hcp orthosilicates in the order Fo|Ch|Hu|Cl|monticellite and found Cl to be twinned on (011) and ($0\bar{1}3$). Yamamoto and Akimoto (1977, p. 309) comment on the origin of (012) and (001) twins in OH-Cl, and Fujino and Takéuchi (1978, p. 537) suggest that the frequency of previously unreported "(100) twins" in Ti-Ch polysynthetic twins increases with increasing Ti content.

White and Hyde (1982a) comment that "microtwinning (twin individuals a few hundred Å or less in width) has not been observed in Mg-humites by electron microscopy/diffraction." But they did see occasional (001) twins in heavily-faulted alleghanyite, recognizing them by "a doubling of the number of reflections in alternate reciprocal lattice rows parallel to c^* in [electron] diffraction patterns with zone axes $[h\bar{1}0]$ when $h = 2n + 1$." Possible structures of the (001) composition plane include a lamellar half unit cell of manganhumite or a "leucophoenicite-type boundary" (see their Figs. 9-12).

263

HUMITE STRUCTURES AS ANION-STUFFED CATION ARRAYS

White (1982a) and White and Hyde (1982a,b) proposed that both the humite and leucophoenicite [see Ch. 12] structural families, which till now have been described as cation-stuffed hcp anion arrays, may be represented more conveniently as anion-stuffed ccp arrays of the majority (M^{2+}) cation. Among the advantages of this approach are (1) although anion arrays are distorted, cation arrays remain quite regular (*cf*. Fig. 14 with Fig. 5, p. 242); and (2) there is a crystallographic shear (CS) or mimetic ("chemical") twin relationship among members of the structural families. "The application of regularly repeated CS to norbergite... generates all members of the humite family, and eventually olivine... Continuing the process yields successive members of the leucophoenicite family..." (White, 1982a, p. 19). The utility of this will be evident in later HRTEM descriptions of these structures.

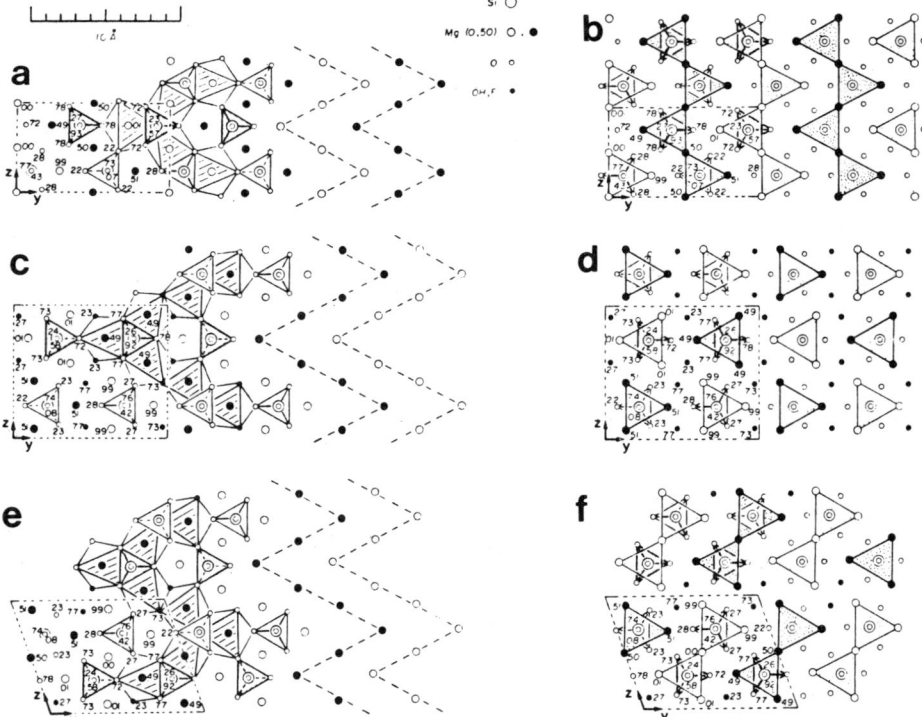

Figure 14. (a) The conventional description of forsterite as serrated chains of edge-sharing MgO$_6$ octahedra. The ...2,(2,2),2... twin bands of Mg atoms are also shown. (b) Forsterite described as the "stuffed" Ni$_2$In endmember of the humite series. On the right the SiMg$_6$ trigonal prisms are emphasized; on the left the SiO$_4$ tetrahedra are also shown (*cf*. Fig. 1, p. 233). (c) The conventional norbergite structure with ...3,(3,3),3... twin bands of Mg atoms. (d) Norbergite described as the "stuffed" Re$_3$B(3) endmember of the humite series. (e) The chondrodite structure showing cation-centered polyhedra, and ...2,(3,2),3... twin bands of Mg atoms. (f) Intergrowth of Ni$_2$In(2) and Re$_3$B(3) structures in chondrodite. Continued on the next page.

264

Briefly, forsterite [$n = \infty$] is viewed as the anion-stuffed Ni_2In end member of the humite series, designated (2,2) or (2^2). Norbergite [$n = 1$] is the Re_3B (3,3) or (3^2) end member (Figs. 14a and b). Chondrodite [$n = 2$] is the (3,2) intergrowth of Ni_2In and Re_3B in the ratio 1:1. Humite [$n = 3$] is the (3,2,2) or (3,2^2) intergrowth, and clinohumite [$n = 4$] is the (3,2,2,2) or (3,2^3) intergrowth. See Figures 14c,d,e. Higher order structures [$n = 5,6,...$], would have (3,2^{n-1}) designations. Parentheses indicate the crystallographic repeat.

In this scheme, the ccp M^{2+} cations form M_6 trigonal prisms containing Si atoms, divalent oxygens occupy SiM_3 tetrahedra, and monovalent (F,OH) anions M_3 triangles. The SiM_6 trigonal prisms form infinite columns parallel to [100], sharing triangular faces in the (100) plane. The columns are isolated in the norbergite (Re_3B) structure and share [100] edges in olivine (Ni_2In). Combinations are formed by crystallographic shearing or "twinning" of these elements, each of which is two or three atoms wide (see Figs. 14-17).

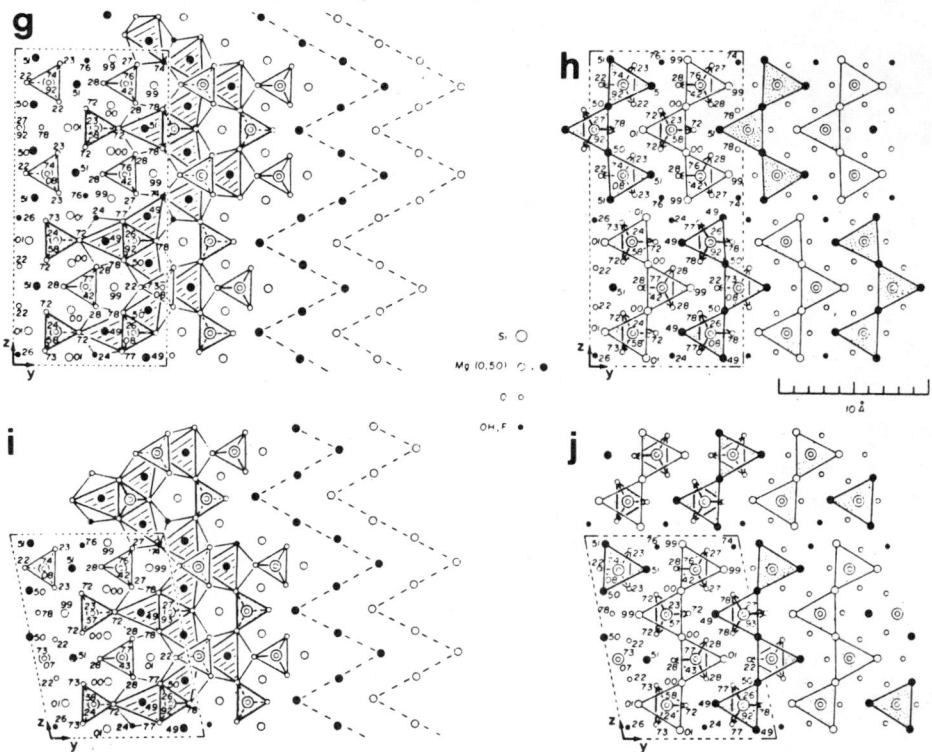

Figure 14, continued. (g) The humite structure showing cation-centred polyhedra and ...2,($3_2$2,2,3,2,2),3... or ...2,(3,2^2,3,2^2),3... twin bands of Mg atoms. (h) Intergrowth of Ni_2In(2^2) and Re_3B(3) structures in humite. (i) The clinohumite structure showing cation-centred polyhedra, and ...2,(3,2,2,2,2),3... twin bands of Mg atoms. (j) Intergrowth of Ni_2In(2^3) and Re_3B(3) structures in clinohumite. From White and Hyde (1982a, Figs. 1-5).

265

FAULTED STRUCTURES, INTERGROWTHS, AND SUPERSTRUCTURES

It was suggested earlier (p. 237F.) that "compostional faults" representing errors in the octahedral chain sequences (and concommitant coupled $\square^{IV} + 4\ F^{III} \rightleftarrows Si^{IV} + 4\ O^{IV}$ substitutions) could account in part for departures in chemistry from the strict stoichiometry demanded by integral values of n in the general formula for humites. Müller and Wenk (1978) found a few such faults in two clinohumites and a chondrodite which they examined by HRTEM; White and Hyde (1982a,b) found many more.

The classification scheme devised by White and Hyde is ideal for describing these compositional faults. Figure 15 is an example, showing four unit cells of $Mg_{13}(SiO_4)_6(F,OH)_2$, the $(3,2^5)$ member [n = 6] of the humite family surrounded by clinohumite, $(3,2^3)$. These higher members only occur in thin lamellae from one to four or five unit cells wide in

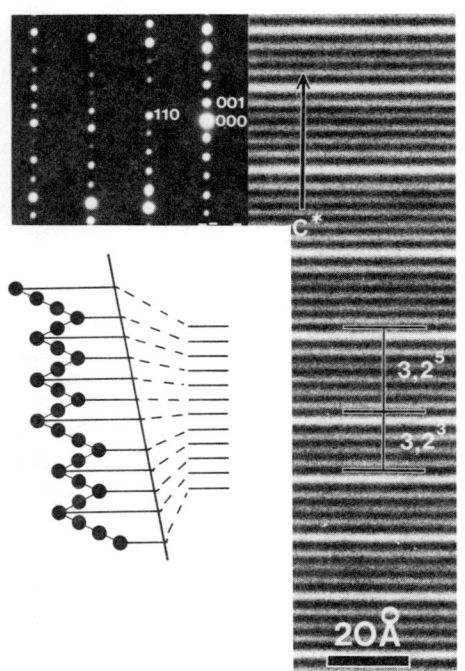

the c^* direction; in general, even values of $n > 4$ predominate over odd values (but see Fig. 16a). "They are apparently unable to exist as stable *macroscopic* entities in nature" or in any synthesis experiments to date (White and Hyde, 1982a, p. 62; Van Valkenburg, 1961; Hinz and Kunth, 1960; Yamamoto and Akimoto, 1974, 1977).

White and Hyde (1982a) found no compositional faulting in any of the chrondite (Ch), norbergite (No) or olivine specimens they investigated. Humites (Hu) were less faulted than clinohumites (Cl), although one Cl from Paragas, Finland was found to be perfect. Coherent intergrowths of massive portions of Hu and Ch were observed "quite frequently" as were Hu specimens containing Ch and/or Cl.

Figure 15. A high resolution TEM image of clinohumite $(3,2^2)$ intergrown with four unit cells of the $(3,2^5)$ member for which n = 6. The sketch to the right shows the Mg atom "twin bands" in relation to the fringes. The specimen is from Brewster, N.Y. From White and Hyde (1982a, Fig. 6).

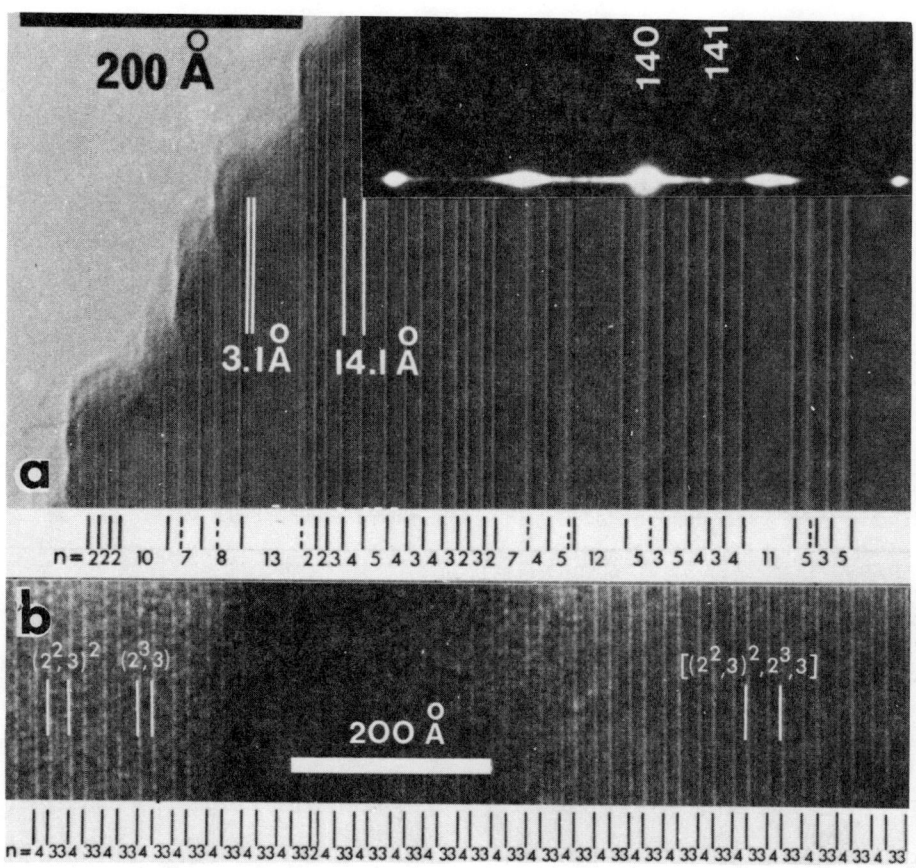

Figure 16. (a) Highly streaked SAD pattern of "alleghanyite + tephroite" from the Benalt Mine, Wales and the HRTEM image, with an interpretation of the n values. The higher n values may be thought of as intergrown tephroite. The average composition of this segment is $n = 4.8$. (b) The superstructure $[3,2^3,(3,2^2)^2] = Mn_{23}(SiO_4)_{10}(OH,F)_6$ consisting of alternate unit cells of manganhumite, $(3,2^2)^2$ $[n = 3]$ and sonolite, $(3,2^3)$ $[n = 4]$ observed in a sonolite from Långban, Sweden. $c \simeq 35.5$ Å. There is one lamellar fault of alleghanyite $[n = 2]$. From White and Hyde (1982b, Fig. 8) and White (1982a, Fig. 4.14).

White and Hyde (1982b) found sonolite (So), like Cl, to be faulted frequently and manganhumite (Mh) to be faulted rarely; some intergrowths of Mh with So were observed, but only one example of intergrown tephroite (Te) was seen. Alleghanyite showed moderate faulting in some specimens, some intergrowths with Te (Fig. 16a), occasional twins on (001), and some regions of completely ordered superstructures. One extraordinary superstructure was found in So from Långban, Sweden (Fig. 16b).

Superstructures

Random faulting in both Mg- and Mn-members of the humite series is the rule, although some crystals, both natural and synthetic, were "observed to have an *almost* regular distribution of faults" (White and Hyde, 1982a, p. 60). Regular superstructures were seen in natural Mn-specimens (Fig. 16b) and in synthetic Mg-humite preparations (Fig. 17). In the latter, the superstructures were found to be $[3,2^5(3,2^3)^x]$ with x = 4,7, and 11 and corresponding d(100) values of 74 Å, 114 Å, and 169 Å. As mentioned previously superstructures disappeared and fault frequencies decreased with annealing at 1300°C for short times, "consistent with the absence of superstructures in natural humites, which have been annealed for long periods at temperatures of less than 1,000°C, and depths of 50 to 150 km (McGetchin *et al.*, 1970)" (White and Hyde, 1982a, p. 62).

Jerrygibbsite and Leucophoenicite

Jerrygibbsite is an orthorhombic dimorph of sonolite with $c_{jg} = 2(c \sin \alpha)_{so}$ (see Dunn *et al.*, 1982, and Chapter 12, this volume). Leucophoenicite is a dimorph of manganhumite (see Chapter 12).

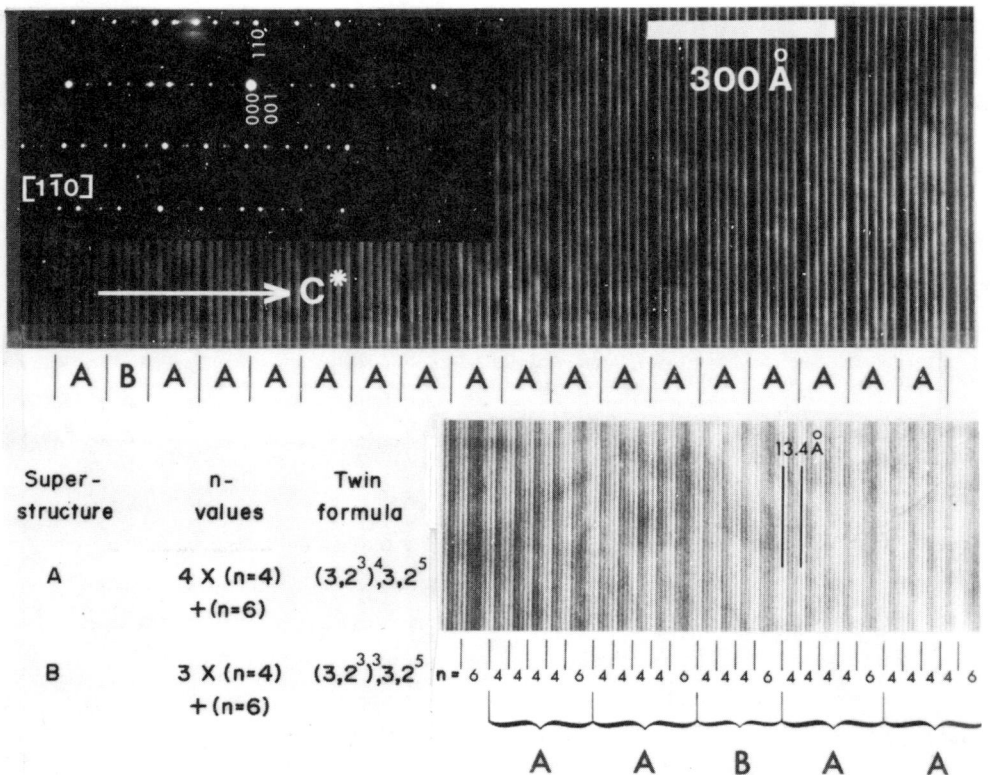

Super-structure	n-values	Twin formula
A	4 X (n=4) +(n=6)	$(3,2^3)^4,3,2^5$
B	3 X (n=4) +(n=6)	$(3,2^3)^3,3,2^5$

Figure 17. Superstructures in a synthetic specimen prepared at 1300°C for 48 hours with a starting composition of $3.5Mg_2SiO_4 \cdot MgF_2$. Courtesy of T. J. White.

APPENDIX: Previously unpublished microprobe analyses
of the humite minerals from Jones (1968)

Table A-1. Original numbers, localities, and donors of chondrodite samples whose analyses appear in Table A-3. From Jones (1968).

Sample numbers New	Original	Locality	Donor★
CHONDRODITE			
1	2	Kafveltorp, Sweden	Th.G. Sahama
2	1	Hangleby, Sibbo, Finland	Th.G. Sahama
3		Limecrest Quarry, Sparta, N.J.	Highton
4		New Jersey	S.G. Fleet
5		Limecrest Quarry, Sparta, N.J.	Highton
6	3	Sparta, N.J.	Th.G. Sahama
7	103384	Sparta, N.J.	U.S. Nat'l Museum
8	95039	Franklin Furnace, N.J.	U.S. Nat'l Museum
9	R14660	Franklin, N.J.	U.S. Nat'l Museum
10	36081	Amity, New York	U.S. Nat'l Museum
11		Limecrest Quarry, Sparta, N.J.	Highton
12		Limecrest Quarry, Sparta, N.J.	Highton
13	13274	Orange Co., N.Y.	U.S. Nat'l Museum
14	4	Edenville, Orange Co., N.Y.	Th.G. Sahama
15		Franklin, N.J.	S.G. Fleet
16	78560	Edenville, Orange Co., N.Y.	S.G. Fleet
17	105488	Amity, N.Y.	U.S. Nat'l Museum
18	103287	Taberg, Wermland, Sweden	U.S. Nat'l Museum
19	R7070	Amity, N.Y.	U.S. Nat'l Museum
20	103384	Sparta, N.J.	U.S. Nat'l Museum
21	TW2528	Tilley Foster Mine, Brewster, N.Y.	S.G. Fleet
22	7	Tilley Foster Mine, Brewster, N.Y.	Th.G. Sahama
23		Unknown	S.G. Fleet
24		Tilley Foster Mine, Brewster, N.Y.	S.G. Fleet
25		Limecrest Quarry, Sparta, N.J.	Highton
26	3199	Tilley Foster Mine, Brewster, N.Y.	U.S. Nat'l Museum
27	R78568	Nya-Kopparberg, Sweden	U.S. Nat'l Museum
28	R3874	Kafveltorp, Sweden	U.S. Nat'l Museum
29	14327	Brewster, N.Y.	U.S. Nat'l Museum
30	8	Kafveltorp, Sweden	Th.G. Sahama

Table A-2. Original numbers, localities, and donors of norbergite, humite and clinohumite samples whose analyses appear in Table A-4. From Jones (1968).

Sample numbers New	Original	Locality	Donor★
NORBERGITE			
1	R12213	Nicol Quarry, Franklin, N.J.	U.S. Nat'l Museum
2	C6298	Nicol Quarry, Franklin, N.J.	U.S. Nat'l Museum
3	115542	Sterling Hill, Ogdensburg, N.J.	U.S. Nat'l Museum
4	1	Norberg, Sweden	Th.G. Sahama
HUMITE			
1	47071	Monte Somma, Vesuvius, Italy	U.S. Nat'l Museum
2	85697	Monte Somma, Vesuvius, Italy	U.S. Nat'l Museum
3	2	Hermala, Lohja, Finland	Th.G. Sahama
4	}	Tor Avaly	S.G. Fleet
5	}	Vesuvius, Italy	S.G. Fleet
6	1	Monte Somma, Vesuvius, Italy	Th.G. Sahama
7	3	Sillböle, Finland	Th.G. Sahama
8	95147	Monte Somma, Vesuvius, Italy	U.S. Nat'l Museum
9		Monte Somma, Vesuvius, Italy	S.G. Fleet
CLINOHUMITE			
1	103786	Aker, Sweden	U.S. Nat'l Museum
2		Santa Lucia Mtns., Monterrey County, California	W.S. Wise
3	2	Ojamo, Lohja, Finland	Th.G. Sahama
4	134	Unknown	S.G. Fleet
5	C5176	Baline, Ala Valley, Piedmont, Italy	U.S. Nat'l Museum
6	C5176	Baline, Ala Valley, Piedmont, Italy	U.S. Nat'l Museum
7		Hämeenkylä, Finland	Th.G. Sahama
8	107384	Buell Park, Arizona	U.S. Nat'l Museum
9	107384	Buell Park, Ariz.	U.S. Nat'l Museum
10	94997	Franscia, Val Malenco, Italy	U.S. Nat'l Museum
11	115658	Dillon, Montana	U.S. Nat'l Museum
12	9891	Dillon, Mont.	Univ. Minnesota

★ Samples from Sahama are discussed in his early definitive monograph (Sahama, 1953); some of those from Fleet bear numbers from the Mineralogical Museum at Cambridge University.

Table A-3. Electron microprobe analyses of chondrodite, $2M_2SiO_4 \cdot M_{1-x}Ti_x(OH,F)_{2-2x}O_{2x}$, in order of increasing Fe. Localities listed in Table A-1. From Jones (1968, Table VI). Boron analyses by Hinthorne and Ribbe (1974).

New No.	SiO$_2$	FeO	MnO	MgO	TiO$_2$	CaO	ZnO	F	OH calc	Total (1)	Si	Fe	Mn	Mg	Ti	Ca	Zn	F	OH	Stoic. (2)	Formula "x"
			Weight Percent Oxide								Atomic Proportions (x10^4)										
1	35.04	0.55	0.05	58.33	0.00	0.00	0.00	5.80	4.71	99.82	5832	77	7	14470	0	0	0	3053	2769	1.002	0.000
2	35.17	0.71	0.17	57.92	0.03	0.01	0.05	6.97	3.61	100.01	5853	98	24	14367	4	2	6	3669	2124	1.009	0.001
3	33.87	1.16	0.10	57.24	0.97	0.01	0.10	7.98	2.32	99.30	5636	161	15	14199	121	2	12	4200	1362	0.971	0.042
4	34.68	1.27	0.12	57.54	0.12	0.03	0.05	8.49	2.21	99.89	5772	177	16	14273	15	5	6	4469	1299	0.996	0.005
5	33.80	1.31	0.06	57.19	0.22	0.03	0.10	8.18	2.40	98.72	5626	183	9	14186	27	5	12	4306	1409	0.975	0.009
6	35.26	1.33	0.17	57.44	0.23	0.03	0.06	6.84	3.64	100.41	5868	184	24	14248	29	5	8	3600	2141	1.012	0.010
7	34.79	1.42	0.18	56.69	0.18	0.04	0.07	7.53	2.93	99.28	5789	197	25	14063	23	7	9	3963	1721	1.010	0.008
8	34.21	2.28	0.13	55.86	0.37	0.03	0.01	8.63	1.81	98.85	5693	317	18	13857	46	5	2	4542	1064	0.999	0.016
9	32.41	2.84	0.10	56.71	0.37	0.03	0.01	10.27	0.53	99.90	5394	396	15	14067	46	5	2	5406	314	0.987	0.016
10	32.88	2.86	0.12	55.98	0.32	0.04	0.00	9.30	1.30	99.34	5472	398	16	13886	40	7	0	4895	764	1.007	0.014
11	34.16	2.88	0.26	55.65	0.23	0.01	0.00	7.80	2.63	99.10	5686	401	36	13804	29	2	0	4106	1545	0.996	0.010
12	33.99	2.97	0.26	55.31	0.80	0.03	0.01	7.75	2.44	99.15	5658	414	36	13722	100	5	2	4079	1432	0.991	0.035
13	34.46	3.15	0.21	55.13	1.10	0.03	0.01	8.50	1.64	99.88	5736	439	29	13676	138	5	2	4474	966	1.004	0.048
14	34.25	3.16	0.14	55.65	1.63	0.01	0.01	6.98	2.90	100.43	5700	440	20	13804	205	2	2	3674	1706	0.985	0.071
15	34.04	3.20	0.21	55.45	1.00	0.01	0.00	7.87	2.30	99.69	5665	446	29	13755	125	2	0	4142	1350	0.986	0.044
16	34.01	3.27	0.12	54.77	1.33	0.03	0.07	8.03	1.93	99.27	5661	455	16	13586	167	5	9	4227	1134	0.994	0.059
17	34.66	3.51	0.10	55.55	0.08	0.04	0.00	7.09	3.35	99.81	5768	489	15	13779	10	7	0	3732	1967	1.008	0.004
18	34.23	3.76	0.18	54.78	0.28	0.07	0.06	8.74	1.71	99.33	5697	523	25	13590	35	12	8	4600	1007	1.003	0.013
19	34.40	3.81	0.27	54.98	0.33	0.01	0.00	7.48	2.86	99.64	5725	530	38	13639	42	2	0	3937	1680	1.004	0.015
20	33.67	4.12	0.23	53.76	1.50	0.04	0.01	6.83	2.86	98.79	5604	573	33	13335	188	7	2	3595	1684	0.991	0.066
21	34.16	4.68	0.26	54.45	0.32	0.00	0.02	6.01	4.17	99.58	5686	652	36	13508	40	0	3	3163	2454	0.998	0.014
22	34.27	4.88	0.30	54.57	0.30	0.00	0.02	5.32	4.76	100.16	5704	679	42	13537	65	0	3	2800	2801	0.995	0.023
23	33.78	5.08	1.61	53.14	0.20	0.07	0.19	7.01	3.28	99.87	5622	707	228	13183	25	12	23	3690	1931	0.991	0.009
24	34.16	5.65	0.30	53.81	0.47	0.00	0.02	5.74	4.35	100.03	5686	786	42	13347	58	0	3	3021	2557	0.998	0.021
25	32.90	6.33	0.27	53.04	0.62	0.06	0.07	7.78	2.42	99.07	5476	881	38	13158	77	10	9	4095	1421	0.966	0.027
26	33.63	6.38	0.25	53.08	0.55	0.00	0.02	5.79	4.21	99.49	5579	888	35	13166	69	0	3	3048	2478	0.988	0.024
27	33.59	6.41	1.55	51.65	0.10	0.06	0.19	7.84	2.44	99.38	5590	892	218	12813	13	10	23	4127	1435	1.001	0.004
28	33.22	6.60	2.67	50.80	0.03	0.11	0.21	8.10	2.22	99.51	5529	919	378	12603	4	20	26	4264	1307	0.991	0.001
29	34.04	6.96	0.14	52.73	0.12	0.00	0.00	7.34	2.96	99.81	5665	969	20	13080	15	0	0	3863	1741	1.006	0.005
30	33.46	7.06	1.77	51.78	0.02	0.08	0.22	7.30	3.07	100.25	5569	983	249	12846	2	15	28	3842	1803	0.986	0.001

(Left margin, corresponding to samples 8–10)

B$_2$O$_3$	B
1.20	344
1.07	306

(1) Total less 0 for F and OH.

(2) Stoic. = $2Si/(2\underline{n}/[2\underline{n}+1])M_{Ti}$

270

Table A-4. Electron microprobe analyses of norbergite, humite and clinohumite, in order of increasing Fe. Localities listed in Table A-2. From Jones (1968, Table VII).

| | Weight Percent Oxide | | | | | | | | | | Atomic Proportions (x10⁴) | | | | | | | | | | |

New No.	SiO₂	FeO	MnO	MgO	TiO₂	CaO	ZnO	F	OH calc	Total (1)	Si	Fe	Mn	Mg	Ti	Ca	Zn	F	OH	Stoic. (2)	Formula "x"

Norbergite $M_2SiO_4 \cdot M_{1-x}Ti_x(OH,F)_{2-2x}O_{2x}$

1	29.74	0.06	0.01	58.73	0.42	0.15	0.05	16.77	1.44	99.63	4949	9	2	14569	52	27	6	8827	847	1.012	0.011
2	29.59	0.06	0.01	59.01	0.30	0.13	0.05	17.25	1.12	99.73	4924	9	2	14639	38	22	6	9080	656	1.004	0.008
3	29.39	0.15	0.04	59.04	0.27	0.11	0.02	18.03	0.45	99.70	4892	21	5	14647	33	20	3	9490	263	0.996	0.007
4	29.31	1.84	0.13	57.85	0.00	0.04	0.00	15.08	3.09	99.54	4878	256	18	14351	0	7	0	7938	1818	1.000	0.000

Humite $3M_2SiO_4 \cdot M_{1-x}Ti_x(OH,F)_{2-2x}O_{2x}$

1	37.46	1.22	0.17	57.27	0.12	0.01	0.01	4.00	3.38	100.37	6234	170	24	14207	15	2	2	2105	1986	1.009	0.007
2	36.30	3.06	0.83	55.16	0.13	0.11	0.09	4.50	2.85	99.80	6042	426	116	13685	17	20	11	2369	1676	0.988	0.008
3	36.45	3.24	1.28	52.89	3.25	0.01	0.00	3.02	2.79	100.35	6067	451	180	13121	407	2	0	1590	1642	1.000	0.201
4	36.60	4.19	1.59	53.67	0.10	0.10	0.01	4.31	2.98	100.34	6092	584	224	13314	13	17	2	2269	1750	1.004	0.006
5	36.75	4.25	1.59	53.62	0.10	0.08	0.00	3.96	3.29	100.42	6117	591	224	13302	13	15	0	2084	1932	1.009	0.006
6	35.66	4.39	0.99	53.41	0.23	0.10	0.09	5.17	2.10	98.97	5935	611	140	13249	29	17	11	2721	1237	0.985	0.014
7	36.43	5.03	0.65	53.84	0.10	0.01	0.00	4.07	3.20	100.11	6064	700	91	13356	13	2	0	2142	1879	0.990	0.006
8	36.35	6.19	0.98	52.26	0.27	0.07	0.06	4.30	2.85	100.18	6049	861	138	12965	33	12	8	2263	1675	1.007	0.017
9	35.66	6.84	1.96	51.30	0.03	0.21	0.01	4.24	2.99	100.04	5935	953	277	12726	4	37	2	2232	1760	0.989	0.002

Clinohumite $4M_2SiO_4 \cdot M_{1-x}Ti_x(OH,F)_{2-2x}O_{2x}$

1	38.36	1.76	0.13	55.80	0.65	0.01	0.00	3.45	2.00	99.77	6384	245	18	13841	81	2	0	1816	1174	1.012	0.052
2	37.95	1.78	0.01	54.92	2.65	0.01	0.00	2.37	2.12	99.81	6316	247	2	13623	332	2	0	1247	1246	1.000	0.210
3	38.08	3.64	0.25	53.89	1.93	0.01	0.00	2.51	2.28	100.46	6338	507	35	13368	242	2	0	1321	1340	1.007	0.154
4	37.50	3.90	1.08	54.04	0.02	0.04	0.00	2.79	2.83	99.70	6242	543	153	13405	2	7	0	1469	1662	0.995	0.001
5	37.18	4.73	0.66	50.99	3.42	0.00	0.00	0.00	3.77	98.98	6188	659	93	12648	428	0	0	0	2217	1.007	0.279
6	37.52	4.79	0.65	50.97	3.20	0.01	0.00	0.00	3.85	99.18	6245	666	91	12644	401	2	0	0	2265	1.018	0.261
7	37.54	5.58	0.62	53.03	0.22	0.01	0.00	3.08	2.46	100.08	6249	777	87	13154	27	2	0	1621	1446	1.001	0.017
8	36.58	9.19	0.14	47.07	4.72	0.00	0.00	0.03	3.09	99.36	6088	1278	20	11677	591	0	0	16	1818	1.010	0.392
9	36.30	10.55	0.21	45.60	5.47	0.00	0.00	0.09	2.69	99.60	6042	1468	29	11311	685	0	0	47	1581	1.008	0.457
10	35.90	11.21	0.50	44.16	5.59	0.01	0.00	0.00	2.64	98.77	5975	1560	71	10953	699	2	0	0	1553	1.012	0.474
11	35.77	12.18	0.14	45.90	2.79	0.00	0.00	2.66	1.52	99.12	5953	1696	20	11385	349	0	0	1400	892	0.996	0.233
12	35.56	12.94	0.14	44.90	2.82	0.01	0.00	1.98	2.06	98.61	5918	1801	20	11139	353	2	0	1042	1211	1.000	0.236

(1) Total less O for F and OH

(2) Stoic. = $2Si/(2\underline{n}/[2\underline{n} + 1])M_{Ti}$

Aoki, K., K. Fujino and M. Akaogi (1976) Titanochondrodite and titanoclinohumite derived from the upper mantle in the Buell Park kimberlite, Arizona, U.S.A. Contrib. Mineral. Petrol. 56, 243-253.

Balk, R. (1954) Petrology section, in Mineral resources of Fort Defiance and Tohatchi quadrangles. Arizona and New Mexico. N. Mex. Bur. Mines Mineral Res. Bull. 36, 192.

Baur, W.H. (1976) Rutile-type compounds. V. Refinement of MnO_2 and MgF_2. Acta Crystallogr. B32, 2200-2204.

Borneman-Starynkevich, I.D. and V.S. Myasnikov (1950) On isomorphous replacements in clinohumite. Dokl. Acad. Sci. USSR 71, 137-144.

Bourne, J.H. (1974) *The Petrogenesis of the Humite Group Minerals in Regionally Metamorphosed Marbles of the Grenville Supergroup.* Ph.D. Dissertation, Queen's University, Ontario, Canada.

Bragg, W.L. and G.B. Brown (1926) Die Struktur des Olivins. Z. Kristallogr. 63, 538.

_____ and G.F. Claringbull (1965) *Crystal Structures of Minerals.* G. Bell and Sons, London.

_____ and J. West (1927) The structure of certain silicates. Proc. Roy. Soc. 114, 450-453.

Buckle, E.R. and H.F.W. Taylor (1958) A calcium analogue of chondrodite. Am. Mineral. 43, 818-823.

Butler, S.A. (1972) *Breakdown Reactions in the Humite Minerals.* Ph.D. Dissertation, Cambridge University, Cambridge, England.

Capponi, J.J. and M. Marezio (1975) The high-pressure synthesis and structural refinement of $Al_4Co(BO_4)_2O_2$, an anhydrous boron chondrodite. Acta Crystallogr. B31, 2440-2443.

Cook, D. (1969) Sonolite, alleghanyite and leucophoenicite from New Jersey. Am. Mineral. 54, 1392-1398.

Deer, W.A., R.A. Howie and J. Zussman (1962) *Rock-forming Minerals, Vol. 1: Ortho and Ring Silicates.* Wiley, New York.

Duffy, C.J. (1977) *Phase Equilibria in the System $MgO-MgF_2-SiO_2-H_2O$.* Ph.D. Dissertation, Univ. of British Columbia, Vancouver, B.C., Canada.

_____ and H.J. Greenwood (1979) Phase equilibria in the system $MgO-MgF_2-SiO_2-H_2O$. Am. Mineral. 64, 1156-1174.

Dunn, P.J., D.R. Peacor and W.B. Simmons, Jr. (1982) Jerrygibbsite, the unit-cell twinned polymorph of sonolite from Franklin, New Jersey. Am. Mineral. (submitted)

Egorov-Tismenko, Yu.K., I.P. Deineko, M.A. Simonov and N.V. Belov (1971) Crystal structure of $Cd_5(SiO_4)_2(OH)_2$ (cadmium chondrodite). Kristallografia 16, 1174-1178.

Engi, M. and D.H. Lindsley (1979) The stability of titanoclinohumite (TICL): an experimental reinvestigation. Trans. Am. Geophys. Union *EOS* 60, 421.

Evans, B.W. and V. Tromsdorff (1978) Breakdown of titanoclinohumite in contact and regional metamorphism, Central Alps (abstr.). Trans. Am. Geophys. Union *EOS* 59, 407.

_____ and _____ (1979) Phase relations of hydroxyl titanoclinohumite in serpentinite. Geol. Soc. Am. Abstr. 11, 422.

Finch, C.B., G.W. Clark and O.C. Kopp (1975) Growth of single-crystal Mn_2SiO_4 (tephroite) by Czochralski and edge-defined film-fed (EFG) techniques. J. Crystal Growth 29, 269-272.

Fleischer, M. and E.N. Cameron (1955) Geochemistry of beryllium. U.S. Atomic Energy Comm. TID-5212, 80-93.

Francis, C.A. (1980) *Magnesium-Manganese Solid Solution in the Olivine and Humite Groups.* Ph.D. Dissertation, Virginia Polytechnic Institute and State University, Blacksburg, Virginia.

_____ and P.H. Ribbe (1978) Crystal structures of humite minerals: V. Magnesian manganhumite. Am. Mineral. 63, 874-877.

Fujino, K. and Y. Takéuchi (1978) Crystal chemistry of titanian chondrodite and titanian clinohumite of high-pressure origin. Am. Mineral. 63, 535-543.

Ganiev, R.M., Yu.A. Kharitonov, V.V. Ilyukhin and N.V. Belov (1969) Crystal Structure of calcium chondrodite, $Ca_5(SiO_4)_2(OH)_2$. Dokl. Akad. Nauk USSR 188, 1281-1283.

Gibbs, G.V. and P.H. Ribbe (1969) The crystal structures of the humite minerals: I. Norbergite. Am. Mineral. 54, 376-390.

_____, _____ and C.W. Anderson (1970) The crystal structures of the humite minerals. II. Chondrodite. Am. Mineral. 55, 1182-1194.

Heinrich, E.W. (1963) Paragenesis of clinohumite and associated minerals from Wolf Creek, Montana. Am. Mineral. 48, 597-613.

Hinthorne, J.R. and C.A. Andersen (1975) Microanalysis for fluorine and hydrogen in silicates with the ion microprobe mass analyzer. Am. Mineral. 60, 143-147.

_____ and P.H. Ribbe (1974) Determination of boron in chondrodite by ion microprobe mass analysis. Am. Mineral. 59, 1123-1126.

Hinz, W. and P.O. Kunth (1960) Phase equilibrium data for the system MgO, MgF_2, SiO_2. Am. Mineral. 45, 1198-1210.

Horiuchi, H., N. Morimoto, K. Yamamoto and S. Akimoto (1978) The crystal structure of $2Mg_2SiO_4 \cdot 3Mg(OH)_2$, a new structure type of high pressure phase $mMg_2SiO_4 \cdot nMn(OH)_2$. Am. Mineral. 63, 000-000.

Jones, N.W. (1968) *Crystal Chemistry of the Humite Minerals*. Ph.D. Dissertation, Virginia Polytechnic Institute, Blacksburg, Virginia.

_____ (1969) Crystallographic nomenclature and twinning in the humite minerals. Am. Mineral. 54, 309-313.

_____, P.H. Ribbe and G.V. Gibbs (1969) Crystal chemistry of the humite minerals. Am. Mineral. 54, 391-411.

Kocman, V. and J. Rucklidge (1973) The crystal structure of a titaniferous clinohumite. Canadian Mineral. 12, 39-45.

Kuznetsova, T.P., N.N. Novskii, V.A. Ilyukhin and N.V. Belov (1980) Refinement of the crystal structure of calcium chondrodite $Ca_5[SiO_4]_2(OH)_2 = Ca(OH)_2 \cdot 2Ca_2SiO_4$. Kristallografia 25, 159-160. *See also Baniev* et al. *(1969)*.

Lyon, S.R. and E.G. Ehlers (1970) The system $MgO-GeO_2-H_2O$. Am. Mineral. 55, 118-125.

McCormack, G.R. (1966) Subsolidus equilibria in the system $MgO-GeO_2-MgF_2$. J. Am. Cer. Soc. 49, 618-621.

McGetchin, T.R., L.T. Silver and A.A. Chodos (1970) Titanoclinohumite: A possible mineralogical site for water in the upper mantle. J. Geophys. Res. 75, 255-259.

Merrill, R.B., J.K. Robertson and P.J. Wyllie (1972) Dehydration reaction of titanoclinohumite reconnaissance to 30 kilobars. Earth Planet. Sci. Lett. 14, 259-262.

Mitchell, R.H. (1978) Manganoan magnesian ilmenite and titanian clinohumite from the Jacupiranga carbonatite, São Paulo, Brazil. Am. Mineral. 63, 544-547.

Möckel, J.R. (1939) Structural petrology of the garnet-peridotite of Alpe Arami (Ticino, Switzerland). Leidse. Mededl. 42, 61-130.

Momoi, H. (1980) Notes on hydrothermal synthesis of alleghanyite group. Mineral. Soc. Japan Bull. 14, 179-187.

Moore, P.B. (1978) Manganhumite, a new species. Mineral. Mag. 42, 133-136.

Müller, W.F. and H.R. Wenk (1978) Mixed-layer characteristics in real humite structures. Acta Crystallogr. A34, 607-609.

Nielsen, T.F.D. and O. Johnsen (1978) Titaniferous clinohumite from Gardiner Plateau Complex, East Greenland. Mineral. Mag. 42, 99-101.

O'Keefe, M. and B.G. Hyde (1981) Role of nonbonded forces in crystals. In *Structure and Bonding in Crystals*, Vol. 1, Academic Press, New York, p. 227-253.

Papike, J.J. and M. Cameron (1976) Crystal chemistry of silicate minerals of geophysical interest. Rev. Geophys. Space Phys. 14, 37-80.

Rankama, K. (1947) Synthesis of norbergite and chondrodite by direct dry fusion. Am. Mineral. 32, 146-157.

Rentzeperis, P.J. (1970) The crystal structure of alleghanyite, $Mn_5[(OH)_2|(SiO_4)_2]$. Z. Kristallogr. 132, 1-18.

_____ (1972) Die kristallstrukturen der humitgruppe. Chimika Chronika, New Series 1, 267-279.

Ribbe, P.H. (1976) Polyhedral chains in the structures of gem orthosilicates: relationships to physical properties. 25th Int'l Geol. Congress Abstr. 2, 593-594.

_____ (1979) Titanium, fluorine and hydroxyl in the humite minerals. Am. Mineral. 64, 1027-1035.

_____, G.V. Gibbs and N.W. Jones (1968) Cation and anion substitution in the humite minerals. Mineral. Mag. 37, 966-975.

_____ and _____ (1971) Crystal structures of the humite minerals: III. Mg/Fe ordering in humite and its relation to other ferromagnesian silicates. Am. Mineral. 56, 1155-1173.

Robinson, K., G.V. Gibbs and P.H. Ribbe (1973a) The crystal structures of the humite minerals. IV. Clinohumite and titanoclinohumite. Am. Mineral. 58, 43-49.

_____, _____ and _____ (1973b) Correction [to 1973a]. *Loc. cit.*, p. 346.

Rogers, A.F. (1935) The chemical formula and crystal system of alleghanyite. Am. Mineral. 20, 25-35.

Ross, M. (1964) Crystal chemistry of beryllium. U.S. Geol. Surv. Prof. Paper 468, 28.

Sahama, Th.G. (1953) Mineralogy of the humite group. Ann. Acad. Sci. Fennicae, III. Geol. Geogr. 31, 1-50.

Schäfer, R.M. (1896) Ueber die metamorphen Gabbro Gesteine des Allalingebietes in Wallis zwischen Zermatt und Saasthal. Tscher. Min. Petr. Mitt. 50, 91.

Shannon, R.D. and C.T. Prewitt (1969) Effective ionic radii in oxides and fluorides. Acta Crystallogr. B25, 925-946.

Smith, D. (1977) Titanochondrodite and titanoclinohumite derived from the upper mantle in the Buell Park kimberlite, Arizona, U.S.A. A discussion. Contrib. Mineral. Petrol. 61, 213-215.

Smyth, J.R. and R.M. Hazen (1973) The crystal structures of forsterite and hortonolite at several temperatures up to 900°C. Am. Mineral. 58, 588-593.

Strunz, H. (1957) *Mineralogische Tabellen, 3rd edition.* Akad. Verlagsges., Leipzig.

Sun, M.-S. (1954) Titanoclinohumite in kimberlitic tuff, Buell Park, Arizona (abstr.). Bull. Geol. Soc. Am. 65. 1311-1312.

Taylor, W.H. and J. West (1928) The crystal structure of the chondrodite series. Proc. Roy. Soc. 117, 517-532.

_____ and _____ (1929) The structure of norbergite. Z. Kristallogr. 70, 461-474.

Tilley, C.E. (1951) The zoned contact-skarns of the Broadford area, Skye: a study of boron-fluorine metasomatism in dolomites. Mineral. Mag. 29, 621-665.

Van Valkenburg, A. (1955) Synthesis of the humites. Am. Mineral. 40, 339.

_____ (1961) Synthesis of the humites $nMg_2SiO_4 \cdot Mg(F,OH)_2$. J. Res. Nat. Bur. Stand., A. Phys. Chem. 65A, 415-428.

Voskresenskaya, V.B., V.V. Koval'skii, K.N. Nikishov and Z.F. Parinova (1965) Discovery of titanolivine in Siberian kimberlites. Zap. Vses. Mineral. Obshch. 94, 600-503. [In Russian].

White, J.G., A. Miller and R.E. Nielson (1965) Fe_3BO_6, a borate isostructural with the mineral norbergite. Acta Crystallogr. 19, 1060-1061.

White, T.J. (1982a) *The Humite and Leucophoenicite Structural Families.* Ph.D. Dissertation, Australian National Univ., Canberra.

_____ (1982b) Crystal data for the fluoro-alleghanyite series, $nMn_2SiO_4 \cdot MnF_2$ (1 < n < 4). J. Appl. Crystallogr. (submitted).

_____ and B.G. Hyde (1982a) Electron microscope study of the humite minerals: I. Mg-rich specimens. Phys. Chem. Minerals 8, 55-63.

_____ and _____ (1982b) Electron microscope study of the humite minerals: II. Mn-rich specimens. Phys. Chem. Minerals 8, (in press).

Winter, G.A., E.J. Essene and D.R. Peacor (1983) Mn-humites from Bald Knob, North Carolina: mineralogy and phase equilibria. Am. Mineral., 68, in press.

Yamamoto, K. (1977) Hydroxyl-chondrodite. Acta Crystallogr. B33, 1481-1485.

_____ and S. Akimoto (1974) High pressure and high temperature investigations in the system $MgO-SiO_2-H_2O$. J. Solid State Chem. 9, 187-195.

_____ and _____ (1977) The system $MgO-H_2O-SiO_2$ at high pressures and temperatures – Stability field for hydroxyl-chondrodite, hydroxyl-clinohumite and 10 A-phase. Am. J. Sci. 277, 288-312.

Yoder, H.S. and Th.G. Sahama (1957) Olivine X-ray determinative curve. Am. Mineral. 42, 475-491.

Chapter 11

OLIVINES and SILICATE SPINELS
G. E. Brown, Jr.

INTRODUCTION

The mineral name *olivine* is often used to designate members of the solid solution series bound by the end-members *forsterite* (Fo: Mg_2SiO_4) and *fayalite* (Fa: Fe_2SiO_4). Naturally occurring members of this series commonly contain minor amounts of other divalent cations, particularly Ca, Mn, and Ni, substituted for Mg and Fe. These minerals crystallize with orthorhombic space group symmetry *Pbnm* and have a relatively dense ($3.2 - 4.4$ gm/cm^3) packing of four formula units per unit cell, which is roughly 5 x 10 x 6 A in dimensions. The olivines, together with the pyroxenes and garnets, are the most important silicate phases in the earth's upper mantle; the transformation of magnesium-rich olivine ($\sim Fo_{90}Fa_{10}$) to spinel and β-spinel phases at depths near 400 km is one of the key phase transformations thought to be responsible for the increase in seismic wave velocities at this depth which marks the top of the mantle's transition zone. In the earth's crust, the magnesium-rich members are important constituents of mafic and ultramafic igneous rocks and are also found in thermally metamorphosed, dolomitic limestones. The iron-rich members are found as minor phases in alkaline, felsic igneous rocks and in metamorphosed, iron-rich sediments.

Crystal chemical characterization of the olivines was begun over fifty years ago when Bragg and Brown (1926) determined the structure of a natural crystal of composition $Fo_{90}Fa_{10}$ and Bowen and Schairer (1935) determined the equilibrium phase relations for the system $MgO-FeO-SiO_2$ at low pressures. The structures of the olivine group minerals are comprised of isolated $[SiO_4]$ tetrahedra cross-linked by chains of edge-sharing, distorted octahedra occupied by divalent cations. Because of the limited number of non-equivalent cation sites in the olivine structure (two octahedral and one tetrahedral site), the compositions of naturally occurring olivines are relatively simple compared with other silicates such as the garnets, amphiboles, and micas. Because of this structural and chemical simplicity, the many studies of physical properties and structures of natural and synthetic olivines and their response to variations in composition, temperature, and pressure have resulted in significant gains in our understanding of the crystal chemistry of silicates in general.

The purpose of this chapter is to organize and critically review current knowledge of the crystal chemistry of the olivines. Silicate spinels have also been included because of their polymorphous relationship with olivine. As with other major mineral groups, a substantial volume of literature exists which treats various aspects of the crystal chemistry and mineralogy of these phases. This review will be selective, focusing on (1) the effects of composition, temperature, and pressure variations on physical properties and crystal structure; (2) the partitioning of cations among olivines and coexisting phases (melts and solids) and between the non-equivalent octahedral sites in olivine; and (3) the role of olivine, spinel, and β-spinel in the earth's mantle. Brief coverage will be given to compositional variations and paragenesis of naturally occurring olivines and the silicate spinels.

THE OLIVINE AND SPINEL STRUCTURE TYPES

The olivine structure type is commonly described as a somewhat distorted hexagonal closest-packed array of anions in which one-eighth of the tetrahedral and one-half of the octahedral interstices are occupied by cations; the spinel structure type is the cubic closest-packed analogue of olivine. These structure types have been designated variously as A_2BX_4, A_2BO_4, or M_2ZX_4, where A (or M) refers to octahedrally coordinated cations, B (or Z) refers to tetrahedrally coordinated cations, and X (or O) refers to anions which are coordinated by three A^{VI} (or M^{VI}) cations and one B^{IV} (or Z^{IV}) cation. The designation AB_2X_4 is also used by some workers (see $e.g.$ Hill et $al.$, 1979), with Si occupying the A site and divalent cations such as Mg, Fe, and Ni occupying the B sites. In order to avoid confusion in this chapter, we will consistently use the $A_2^{VI}B^{IV}X_4$ formula to represent both olivine and normal spinel structure types. The prefixes α and γ are used to distinguish the olivine (α) and spinel (γ) polymorphs of an A_2BX_4 compound.

One of the simplest means of discussing and comparing the various compounds that adopt the olivine and normal spinel structure types is in

276

terms of ideal closest-packed structures and departures therefrom. In
the *olivine structure type* the arrangement of A and B cations in octahedral and tetrahedral interstices reduces the space group symmetry of
the ideal hexagonal closest-packed (hcp) array of anions from $P6_3/mcc$
(hexagonal) to *Pbnm* (orthorhombic). A polyhedral drawing of the ideal
olivine structure type viewed down a is shown in Figure 1A. The designations M1 and M2 refer to the crystallographically nonequivalent octahedra $M(1)O_6$ and $M(2)O_6$. The heavily stippled triangles are isolated
$[SiO_4]$ tetrahedra, one with its apex pointing up and one down.

The major structural feature of olivine is the chain of edge-
sharing octahedra that is parallel to z. The serrated or zig-zag nature
of this chain and its relationship to similar chains of edge-sharing
octahedra in the humite minerals are clearly shown in Figure 1 of the
chapter on the humite minerals (this volume). The importance of these
chains in olivine is consistent with the two observed cleavages {100}
and {010} and the dominance of {hk0} crystal faces. Chains of occupied
octahedra in the yz layer shown are separated from like chains above
and below by the displacement $a/2$ and are related to each other by a
b glide plane. These chains are cross-linked to each other by $[SiO_4]$
tetrahedra.

The M(1) octahedron shares six of its 12 edges with other polyhedra
(two with other M(1) octahedra, two with M(2) octahedra, and two with
tetrahedra), whereas the M(2) octahedron shares only three edges (two
with M(1) octahedra and one with a tetrahedron). Pauling's third rule
predicts that the shared polyhedral edges in olivine will be shorter
than unshared edges such that cation-cation distances across shared
edges are maximized and repulsions are reduced. This rule is indeed
obeyed in the actual olivine structure which is shown in Figure 1B.
The resulting polyhedral distortions are significant; the point group
or site symmetries of the M(1) and M(2) octahedra are reduced from O_h
in the ideal structure to $C_{\bar{1}}$ and C_s, respectively, in the actual structure, and that of the $[SiO_4]$ tetrahedron is reduced from T_d to C_s.
However, for the purpose of spectral analysis, the point symmetries
of the M(1) and M(2) octahedra in olivine are regarded as approximately D_{4h} and C_{3v}, respectively.

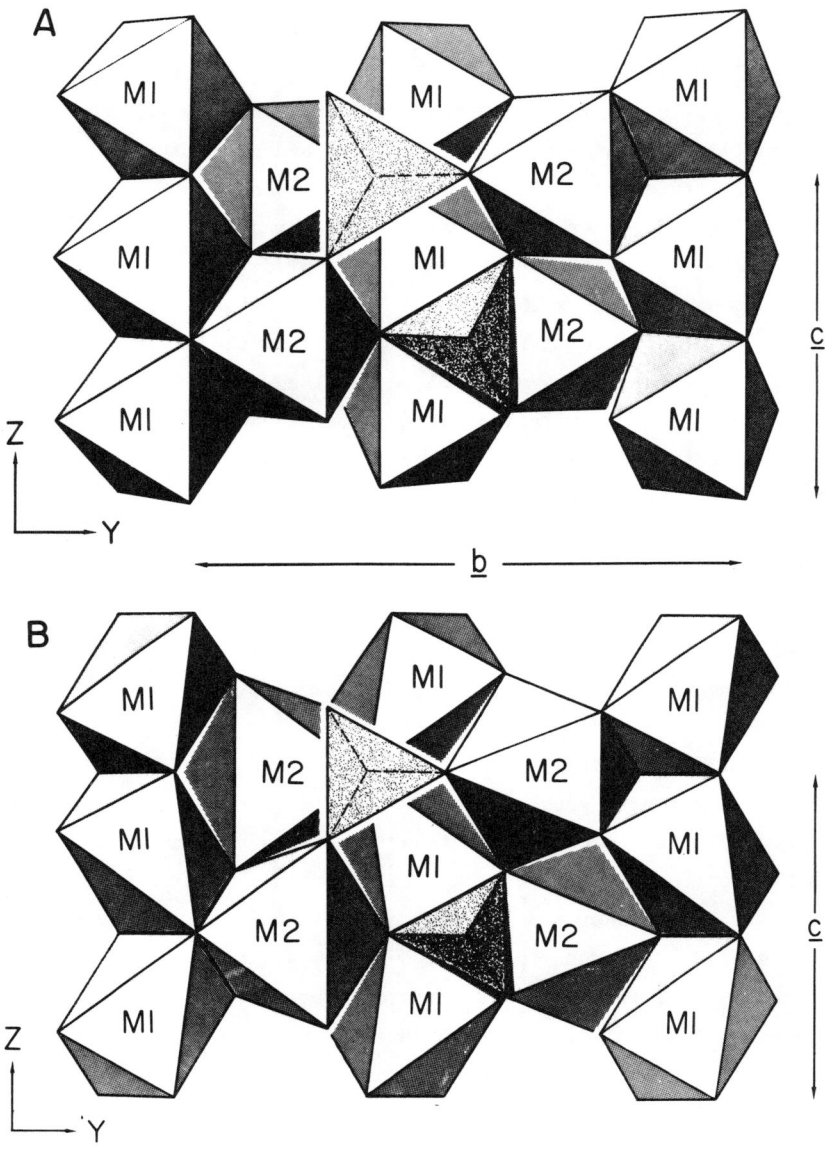

Figure 1. Polyhedral drawings of the idealized (A) and actual (B) structures of forsterite projected on (100), illustrating the distortion from ideal hexagonal closest-packing of anions resulting from cation-cation repulsions across shared edges of the M(1) and M(2) octahedra and tetrahedron. From Brown (1970).

The cause of these distortions has conventionally been attributed to cation-cation repulsions across shared edges following Pauling's reasoning (see *e.g.* Birle *et al.*, 1968; Kamb, 1968; Brown, 1970; Fleet, 1974; Sung and Burns, 1978). However, over the past ten years additional factors have been suggested as causes of the distortions in olivine including cation size effects (Brown, 1970), geometric misfit between octahedra and tetrahedra (Baur, 1972; Vincent *et al.*, 1976), and covalent bonding effects (McLarnan *et al.*, 1979). At the present time, it seems prudent to consider all of the above effects in rationalizing the significant distortions of polyhedra in olivine.

A quantitative comparison of the ideal and actual olivine structures can be made using atomic coordinates and interatomic distances as shown in Tables 1 and 2. Although atomic shifts and bond length differences between ideal and actual forsterite structures are relatively small, large differences are found when comparing angles (see Table A5 in Appendix). These differences, as functions of composition and temperature, will be discussed later.

In the *silicate spinel structure* the arrangement of A and B cations in octahedral and tetrahedral interstices, respectively, of the cubic closest-packed (ccp) array of anions is consistent with cubic space group symmetry $Fd3m$ (Bragg, 1915). The tetrahedral cations occupy equipoint 8a ($x = y = z = 0.0$; point symmetry T_d), the octahedral cations occupy equipoint 16d ($x = y = z = 1/8$; point symmetry D_{3d}), and anions occupy equipoint 32e (x, x, x; point symmetry C_{3v}). Because of the fixed positions that both cations occupy, cation-cation separations are functions only of the unit cell parameter a. Only the oxygen position is variable, with the x coordinate usually designated by the symbol u. When u (or x) = 0.375 (unit cell origin located at equipoint 8a), the anions are arranged in a perfect ccp arrangement.[1] This value of u is rarely observed in spinels, the vast majority of oxide spinels having u values in the range 0.375 to 0.395 (Hill *et al.*, 1979).

As in the olivine structure type, there are no shared polyhedral faces in the silicate spinel structure. However, there are major

---

[1]When the unit cell origin is chosen on the center of symmetry, which is displaced from Wyckoff position a by 0.125, 0.125, 0.125, the value of u for a perfect ccp arrangement of anions becomes 0.25.

Table 1. Comparison of ideal and actual atomic positions
of the six nonequivalent atoms in forsterite (Brown, 1970)

	x	y	z		x	y	z
M(1)				O(1)			
Ideal	0.0	0.0	0.0	Ideal	0.75	0.0833	0.25
Actual	0.0	0.0	0.0	Actual	0.7667	0.0918	0.25
$\|\Delta\|$	0.0	0.0	0.0	$\|\Delta\|$	0.0167	0.0085	0.00
M(2)				O(2)			
Ideal	0.0	0.25	0.25	Ideal	0.25	0.4167	0.25
Actual	0.9896	0.2776	0.25	Actual	0.2202	0.4477	0.25
$\|\Delta\|$	0.0104	0.0276	0.00	$\|\Delta\|$	0.0298	0.0310	0.00
Si				O(3)			
Ideal	0.375	0.0833	0.25	Ideal	0.25	0.1667	0.00
Actual	0.4226	0.0945	0.25	Actual	0.2781	0.1633	0.0337
$\|\Delta\|$	0.0476	0.0112	0.00	$\|\Delta\|$	0.0281	0.0034	0.0337

Table 2. Comparison of selected interatomic distances (in Å) for the ideal
hcp olivine structure type and the idealized and actual forsterite
structures (after Brown, 1970).

	hcp*	Ideal Forsterite**	Actual Forsterite
SiO_4			
Si-O	1.690	1.823	1.637
O-O(s)[1]	2.760	2.977	2.569
O-O(u)[2]	2.760	2.977	2.757
$M(1)O_6$			
M(1)-O	1.952	2.105	2.101
O-O(s,o)[3]	2.760	2.977	2.854
O-O(s,t)[4]	2.760	2.977	2.557
O-O(u)[2]	2.760	2.977	3.169
$M(2)O_6$			
M(2)-O	1.952	2.105	2.135
O-O(s,o)[3]	2.760	2.977	2.854
O-O(s,t)[4]	2.760	2.977	2.593
O-O(u)[2]	2.760	2.977	3.079

* Calculated using the cell parameters $a = 4\sqrt{3}/3(<M-O>) = 4.508$ Å;
$b = 2\sqrt{6}(<M-O>) = 9.563$ Å; $c = 2\sqrt{2}(<M-O>) = 5.521$ Å where $<M-O>$ is the
average M-O distance. The only assumption made was that oxygen anions of
1.38 Å radius are in tangential contact.
** Calculated using the cell parameters $a = 4.861$, $b = 10.312$, $c = 5.954$ Å
which were derived from the above equations.
[1] (s) indicates shared edge. [2] (u) indicates unshared edge. [3] (s,o) indicates
edge shared between octahedra. [4] (s,t) indicates edge shared between octahedron and tetrahedron.

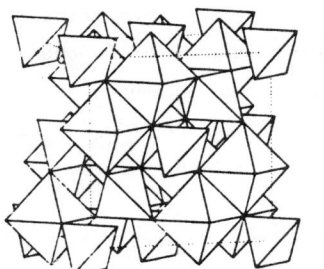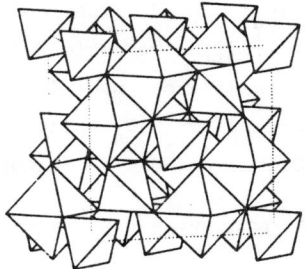

Figure 2. Stereo pair drawing of the spinel structure.
The cubic unit cell is outlined (From Zoltai, 1965).

differences in the type of edge- and corner-sharing between olivine and spinel. Whereas the $[SiO_4]$ tetrahedron in the olivine structure shares three edges with octahedra, the $[SiO_4]$ tetrahedron in spinel shares no edges. As will be discussed later in this chapter, this key difference between olivine and spinel is important in rationalizing the higher pressure stability of silicate spinels relative to their olivine polymorphs. The octahedron in spinel shares six of its edges with other octahedra. The arrangement of edge- and corner-sharing octahedra and corner-sharing octahedra and corner-sharing tetrahedra in the spinel structure is illustrated in the stereo pair drawing in Figure 2.

In spinels with $u > 0.375$, the tetrahedron is larger and the octahedron is smaller than in the undistorted structure with $u = 0.375$. When $u = 0.3875$, octahedral and tetrahedral bond lengths are identical (Hill et $al.$, 1979). Thus spinels with a large u parameter can accomodate relatively large tetrahedral cations. In addition, the length of shared octahedral edges is shorter than unshared edges when u is greater than 0.375. When u falls below 0.375, the tetrahedral site favors relatively small cations and the shared edges between octahedra are $longer$ than unshared edges. Additional details about the spinel structure type, including a procedure for calculating u for any given spinel composition, are contained in the excellent paper by Hill et $al.$ (1979).

The structures of only a few silicate spinels have been determined to date because of the difficulty of preparing single crystals large enough for x-ray structure analysis. Fe_2SiO_4, Co_2SiO_4, and Ni_2SiO_4 spinel structures have been determined by modern methods and were found to have the following u parameters: Fe_2SiO_4: 0.3658 (Yagi et $al.$, 1974), 0.3659 (Finger et $al.$, 1979); Co_2SiO_4: 0.3666 (Morimoto et $al.$,

1974); Ni_2SiO_4: 0.3687 (Yagi *et al*., 1974); 0.3689 (Finger *et al*., 1979).
Hill *et al*. (1979) have estimated a u parameter of 0.3666 for Mg_2SiO_4
spinel. In these experimental studies, small amounts of Si, ranging
from 0.5 percent (Ni_2SiO_4 spinel) to 3.4 percent (Co_2SiO_4 spinel) of
the total Si, were found to occupy the octahedral site.

COMPOSITIONAL RANGE OF OLIVINES AND SILICATE SPINELS

Over the past half century, many compounds with the olivine and
spinel structure types have been synthesized and characterized. Comprehensive listings of these compounds can be found in Muller and Roy
(1974) (olivine and spinel structure types), Ganguli (1977) (olivine
structure type), and Hill *et al*. (1979) (spinel structure type). Table
3 is a listing of compounds known to adopt the olivine structure type,
including the extent of solid solution between end-member compositions
where known. Only a few silicate spinels have been synthesized, including $AlLiSiO_4$, Co_2SiO_4, Fe_2SiO_4, Mg_2SiO_4, and Ni_2SiO_4; the reader
is referred to the extensive listing of oxide and sulfide spinel structures in Hill *et al*. (1979).

Examination of Table 3 shows that the octahedral A (or M) cations
in the olivine structure type can range from monovalent to trivalent
with radii values ranging from 0.53 A (Al^{3+}) to 1.02 A (Na^+) (Shannon
(1976) radii). Tetrahedral B (or T) cations can range from divalent to
pentavalent with sizes ranging from 0.11 A (B^{3+}) to 0.55 A (Sn^{4+}). Although natural olivines contain only oxygen as the anion, so far as is
presently known, olivine isostructures with F, S, and Se as X anion have
been synthesized. There are several notable compounds including Cr_2SiO_4,
Cu_2SiO_4, Zn_2SiO_4, and $CaNiSiO_4$ which have not yet been synthesized in
the olivine structure type, although limited solid solution with several
olivine end-number compositions does occur. Even though some explana-
tions have been offered for these non-existent olivine isostructures
(see *e.g.* Ganguli, 1977), we do not yet understand why these compounds
prefer other structure types.

An efficient way of graphically displaying the range of chemical
substitutions for the olivine and spinel structure types and their
interrelationship is the plot shown in Figure 3. The radius of the
octahedral cation A is plotted *vs* the radius of the tetrahedral cation

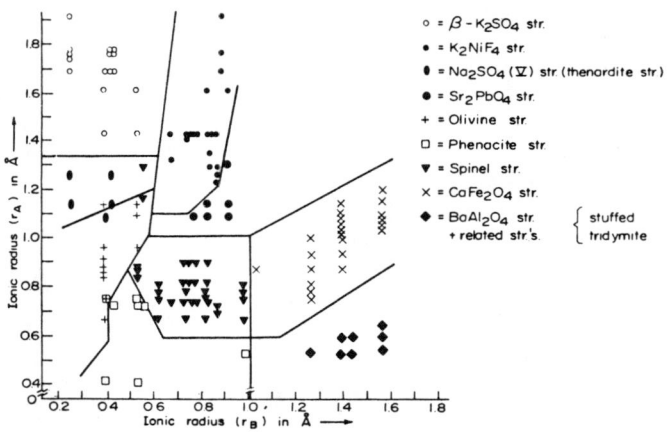

Figure.3. Structure-field map for the most important A$_2$BX$_4$ structures, constructed using Shannon and Prewitt (1969) radii. (From Muller and Roy, 1974).

B for the most important A$_2$BX$_4$ oxide structures. This structure–field map can be thought of as a type of "stability" diagram, illustrating "stability fields" in which a particular A$_2$BX$_4$ structure type predominates. Several general differences are readily observable between the olivine (+ symbols) and spinel (▼ symbols) structure types from this diagram. The olivine structure field falls to the left of the spinel field, indicating that the olivine structure type, with oxygen anions, can accommodate only relatively small cations (less than 0.5 A radius) in the tetrahedral site. On the other hand, the olivine structure can accommodate larger octahedral cations than the spinel structure. Silicate spinels plot on the extreme left side of the spinel structure field shown, whereas olivines of geologic interest plot in a narrow vertical band in the olivine structure field that intersects the abcissa at 0.40 A. In preparing this plot, Muller and Roy (1974) chose a radius value of 0.40 A for $^{IV}Si^{4+}$ rather than the value of 0.26 A recommended by Shannon and Prewitt (1969) and Shannon (1976). These data indicate that the olivine structure type can accommodate a larger size range of octahedral cations than silicate spinels can.

The sharing of edges between tetrahedron and octahedra in the olivine structure type limits the size of ^{VI}A, and particularly, ^{IV}B cations that the structure can accommodate. In contrast the lack of tetrahedral edge sharing in the spinel structure type results in a relatively large size range of cation substitutions in the ^{IV}B site of that structure.

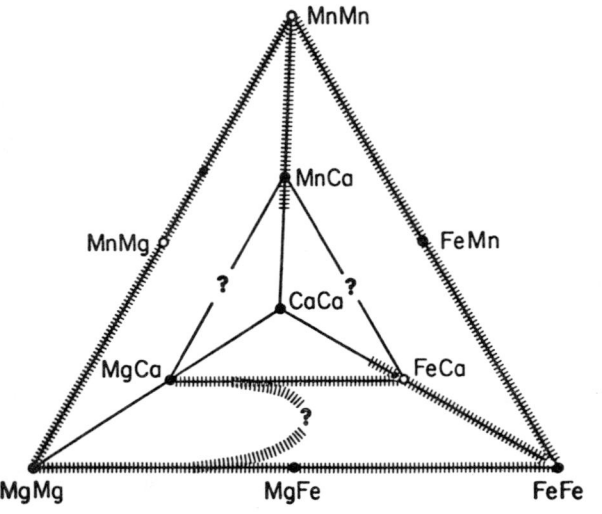

Figure 4. Tetrahedral composition diagram for the Mg_2SiO_4 - Fe_2SiO_4 - Mn_2SiO_4 - Ca_2SiO_4 system illustrating octahedral cation substitutions among the most common members of the olivine group of minerals and the extent of experimentally determined crystalline solutions (indicated by hachured lines) between end members. (From Brown, 1970).

⊪⊪⊪⊪⊪⊪ SOLID SOLUTION

The reader is referred to papers by Kamb (1968), Baur (1972), Tokonami *et al.* (1972), Sung and Burns (1978), and McLarnan *et al.* (1979) for further discussion of the structural differences between olivine and spinel and the structural features and bonding forces that cause the olivine form of Mg_2SiO_4 (α-Mg_2SiO_4) to be stabilized relative to the spinel (γ-Mg_2SiO_4) and β-spinel (β-Mg_2SiO_4) forms at low pressure.

LOW PRESSURE PHASE RELATIONS

The phase relations of various binary, pseudobinary, ternary, and quaternary systems in the MgO-FeO-MnO-CaO-SiO_2 system have been extensively studied since the early 1900's because of their bearing on the origin of mafic and ultramafic rocks and because of their importance in steel-making technology. As indicated by the hachures in Figure 4, continuous solid solution (or crystalline solution) exists along the Mg_2SiO_4-Fe_2SiO_4 (Bowen and Schairer, 1935), Fe_2SiO_4-Mn_2SiO_4 (White, 1943; Riboud and Muan, 1962), Mg_2SiO_4-Mn_2SiO_4 (Kallenberg, 1914; Glasser and Osborn, 1960), and $MgCaSiO_4$-$FeCaSiO_4$ (Schairer and Osborn, 1950) joins at one atmosphere pressure. The question marks along the $MgCaSiO_4$-$MnCaSiO_4$ and $FeCaSiO_4$-$MnCaSiO_4$ joins indicate a lack of experimental data; however, on the basis of crystal chemical reasoning, complete crystalline solution along these joins is also likely at 1 atm. Because the Ca_2SiO_4 end member in this system does not crystallize with the olivine structure under normal conditions (see later discussion of Ca_2SiO_4 polymorphism), only limited crystalline solution occurs between Ca_2SiO_4 and other members of this ternary system. Along the Mn_2SiO_4–Ca_2SiO_4 join, up to 72 mol % Ca_2SiO_4 can replace Mn_2SiO_4 (Tokody, 1928; Glasser, 1961). Up to 63% Ca_2SiO_4 can substitute for Fe_2SiO_4 along the Fe_2SiO_4–Ca_2SiO_4 join (Bowen *et al.*, 1933a). The Mg_2SiO_4–Ca_2SiO_4 join is characterized by a solvus between Mg_2SiO_4 and $MgCaSiO_4$ (Ferguson and Merwin, 1919; Ricker and Osborn, 1954; Biggar and O'Hara, 1969; Yang, 1973; Warner and Luth, 1973), with no appreciable crystalline solution between $MgCaSiO_4$ and Ca_2SiO_4 (Roy, 1956). The most recent solvus determination along the Mg_2SiO_4–$MgCaSiO_4$ join (Warner and Luth, 1973) indicates that up to 22.8 mol % Mg_2SiO_4 substitutes in $MgCaSiO_4$ and up to 21.1 mol % $MgCaSiO_4$ substitutes in Mg_2SiO_4 at 1450°C and 1 atm pressure. The lack of natural olivines much beyond the MgCa–FeCa–MnCa plane shown in Figure 4 is due to the significant size difference between ^{VI}Ca (radius = 1.00 A) and the other divalent cations: ^{VI}Mg (0.72 A), ^{VI}Fe (0.78 A), ^{VI}Mn (0.83 A). Apparently because of the edge sharing between octahedra and tetrahedra in the olivine structure type, the structure cannot stably accommodate more than one Ca cation per formula unit. In later discussion, we will see that Ca can stably occupy the M(2) site in olivine; with more than one Ca per formula unit, Ca is forced into the M(1) site which shares more edges than the M(2) octahedron. The dimensional mismatch between a Ca-filled M(1) site and a Si-filled tetrahedral site must be an important factor in explaining the instability of olivines with compositions beyond the 50 mol % Ca plane in Figure 4.

The extents of crystalline solution between Mg_2SiO_4, Fe_2SiO_4, Mn_2SiO_4, and Ca_2SiO_4 components and other less common olivine components such as Ni_2SiO_4, Co_2SiO_4, Zn_2SiO_4, Cr_2SiO_4, and Cu_2SiO_4 are noted in Table 3 together with references to experimental studies.

The relationship of olivines with other mineral phases has been examined in numerous phase equilibrium studies of multicomponent systems in which Mg_2SiO_4, Fe_2SiO_4, or $(Mg,Fe)_2SiO_4$ are phases, some of which are listed here: forsterite–diopside–silica: Bowen (1914), Atlas (1952),

Table 3. Compounds with the Olivine Structure Type (After Brown, 1970 and Ganguli, 1977)

Silicate Olivine End Members

Mg_2SiO_4	[1][*]
Ca_2SiO_4	[2]
Mn_2SiO_4	[3]
Fe_2SiO_4	[4]
Co_2SiO_4	[5]
Ni_2SiO_4	[6]
$CaMgSiO_4$	[7]
$CaMnSiO_4$	[8]
$CaFeSiO_4$	[9]
$CaCoSiO_4$	[10]
$LiScSiO_4$	[11]
$LiYSiO_4$	[12]
$LiInSiO_4$	[12]
$LiLnSiO_4$ (Ln = Ho-Lu)	[12]
$NaYSiO_4$	[12]
$NaLnSiO_4$ (Ln = Ho-Lu)	[12]

Silicate Olivine Solid Solutions[1]

$(Mg,Mn)_2SiO_4$ (to 100% Mn)	[13]
$(Mg,Fe)_2SiO_4$ (to 100% Fe)	[14]
$(Mg,Ni)_2SiO_4$ (to 100% Ni)	[15]
$(Mg,Co)_2SiO_4$ (< 55% Co)	[16]
$(Mg,Zn)_2SiO_4$ (< 24% Zn)	[17]
$(Mg,Cr)_2SiO_4$ (< 5% Cr)	[18]
$(Mg,Cu)_2SiO_4$ (< 2.5% Cu)	[19]
$(Mg,Ca)_2SiO_4$ (< 17.5% Ca[2])	[20]
$(Mg,Ca)_2SiO_4$ (34-50% Ca)	[20]
$(Fe,Mn)_2SiO_4$ (to 100% Mn)	[21]
$(Fe,Co)_2SiO_4$ (to 100% Co)	[22]
$(Fe,Ca)_2SiO_4$ (< 63% Ca)	[23]
$(Mn,Co)_2SiO_4$ (to 100% Co)	[24]
$(Mn,Ca)_2SiO_4$ (< 72% Ca)	[25]
$Mg_2(Si,Ge)O_4$ (to 100% Ge)	[26]
$Ni_2(Si,Ge)O_4$ (< 42% Ge)	[26]
$(Mg,Ni)_2(Si,Ge)C_4$ (< 46% Ge)	[26]
$Ca_2(Si,Ge)O_4$ (to 100% Ge)	[27]

Germanate Olivine End Members

Mg_2GeO_4	[28]
Ca_2GeO_4	[28]
Mn_2GeO_4	[28]
Cd_2GeO_4	[29]
$MgCaGeO_4$	[30]
$MnMgGeO_4$	[31]
$MnFeGeO_4$	[31]
$MnCoGeO_4$	[31]
$MnZnGeO_4$	[33]
$LiYGeO_4$	[34]
$LiLnGeO_4$ (Ln = Dy-Lu)	[34]
$NaYGeO_4$	[34]
$NaLnGeO_4$ (Ln = Eu-Lu)	[34]

Germanate Olivine Solid Solutions[1]

$(Mg,Co)_2GeO_4$ (< 24% Co)	[35]
$(Mg,Ni)_2GeO_4$ (< 14.5% Ni)	[35]
$(Mg,Cu)_2GeO_4$ (< 7.5% Cu)	[36]
$(Mg,Zn)_2GeO_4$ (< 30% Zn)	[26]

Phosphate Olivines

$LiMgPO_4$	[37]
$LiMnPO_4$	[38]
$LiFePO_4$	[39]
$LiCoPO_4$	[37]
$LiNiPO_4$	[37]
$LiCaPO_4$ (?)[3]	[40]
$LiCuPO_4$ (?)[3]	[40]
$LiCdPO_4$	[41]
$NaMnPO_4$	[42]
$NaCdPO_4$	[43]

Beryllate Olivines

Al_2BeO_4	[44]
Cr_2BeO_4	[45]
$AlCrBeO_4$	[46]
$AlFeBeO_4$	[47]
$CrFeBeO_4$	[48]
$AlGaBeO_4$	[49]
$(Al,Cr)_2BeO_4$ (to 100% Cr)	[45]
$(Al,Fe)_2BeO_4$ (< 50% Fe)	[48]
$(Cr,Fe)_2BeO_4$ (< 50% Fe)	[48]

Arsenate Olivines

$LiMnAsO_4$	[50]
$LiCoAsO_4$	[50]
$LiNiAsO_4$	[50]
$LiCdAsO_4$	[50]
$NaCaAsO_4$	[50]
$NaMnAsO_4$	[50]
$NaCdAsO_4$	[50]

Thiosilicate and Thiogermanate Olivines

Mg_2SiS_4	[51]
Mn_2SiS_4	[52]
Fe_2SiS_4	[53]
Ca_2SiS_4	[51]
Mg_2GeS_4	[54]
Mn_2GeS_4	[52]
Fe_2GeS_4	[55]
Ca_2GeS_4	[54]

Miscellaneous Olivine Isostructures

Mg_2SiSe_4	[54]
Mn_2SiSe_4	[54]
Ca_2SiSe_4	[51]
Mg_2SnSe_4	[54]
Ca_2SnSe_4	[54]
$AlMgBO_4$	[56]
$FeNiBO_4$	[57]
Ag_2SeO_4	[58]
Na_2BeF_4 and $LiNaBeF_4$	[59]

[*]Reference number (see list on right). For each compound, an early and several more recent references (particularly structural studies) are given where possible. Additional references may be found in Muller and Roy (1974).
[1]The extent of solid solution is indicated in parentheses to the right of each compound. These numbers represent mole percent of the cation specified. For example, up to but not more than 5 mole percent Cr^{2+} can replace Mg^{2+} in Mg_2SiO_4. Other solid solutions not included in the table include the following: Mg_2SiO_4 - $CaFeSiO_4$ (100%) (Wyllie, 1960); $CaMgSiO_4$ - $CaFeSiO_4$ (100%) (Schairer and Osborn, 1950); $CaMgSiO_4$ - $CaNiSiO_4$ (< 25 % $CaNiSiO_4$) (Reinen, 1968); $LiYSiO_4$ - $NaYSiO_4$ (< 80% $NaYSiO_4$) (Paques-Ledent, 1976); $CaMgGeO_4$ - $CaCoGeO_4$ (100%) and $CaMgGeO_4$ - $CaNiGeO_4$ (< 50% $CaNiGeO_4$) (Reinen, 1968); $NaCaAsO_4$ - $NaCdAsO_4$ (100%) (Paques-Ledent and Tarte, 1974); $LiFePO_4$ - $LiZnPO_4$ (< 33% $LiZnPO_4$) (Kabalov et al., 1973). Other solid solutions may be possible but are unknown to the author.

Table 3, continued

[1] Bragg and Brown (1926a); Smyth and Hazen (1973); Hazen (1976).
[2] O'Daniel and Tscheischwili (1944); Smith et al. (1965); Czaya (1971); Ghosh et al. (1979).
[3] O'Daniel and Tscheischwili (1944); Francis and Ribbe (1980).
[4] Hanke (1963); Smyth (1975); Hazen (1977).
[5] Gallitelli and Cola (1954); Brown (1970); Morimoto et al. (1974).
[6] Taylor (1930); Brown (1970); Lager and Meagher (1978).
[7] Brown and West (1927); Onken (1964); Onken (1965); Lager and Meagher (1978).
[8] O'Daniel and Tscheischwili (1944); Caron et al. (1965); Brown (1970); Lager and Meagher (1978).
[9] Sahama and Hytonen (1957); Wyderko and Mazanek (1968); Brown (1970).
[10] Cola (1954); Newnham et al. (1966).
[11] Paques-Ledent (1976); Ito (1977).
[12] Paques-Ledent (1976).
[13] Kallenberg (1914); Glasser and Osborn (1960); Brown (1970); Francis and Ribbe (1980).
[14] Bragg and Brown (1926a); Bowen and Schairer (1935); Hanke and Zemann (1963); Hanke (1965);
     Birle et al. (1968); Brown (1970); Finger (1970); Finger and Virgo (1971); Brown and Prewitt
     (1973); Wenk and Raymond (1973); Smyth and Hazen (1973); Basso et al. (1979).
[15] Ringwood (1956); Matsui and Syono (1968); Rajamani et al. (1975).
[16] Matsui and Syono (1968); Ghose and Wan (1974).
[17] Sarver and Hummel (1962).
[18] Scheetz and White (1972).
[19] Schmitz-Dumont et al. (1966).
[20] Ricker and Osborn (1954); Warner and Luth (1973).
[21] White (1943); Brown (1970).
[22] Masse et al. (1966).
[23] Bowen et al. (1933) and references for [9].
[24] Biggars and Muan (1967).
[25] Glasser (1961) and references for [10].
[26] Levin et al. (1964).
[27] Eysel and Hahn (1970).
[28] Durif-Varambon (1959).
[29] Strunz and Jacob (1960).
[30] Hahn (1961).
[31] Ringwood and Reid (1970).
[33] Lyutin et al. (1974).
[34] Paques-Ledent (1976); Blasse and Bril (1967).
[35] Navrotsky (1973).
[36] Hassanein and Azzam (1968).
[37] Thielo (1941); Newnham and Redman (1965).
[38] Geller and Durand (1960); Newnham and Redman (1965); Thielo (1941).
[39] Destenay (1950); Newnham and Redman (1965); Finger and Rapp (1970); Thielo (1941).
[40] Thielo (1941).
[41] Thielo (1941); Paques-Ledent and Tarte (1974).
[42] Moore (1972).
[43] Ivanov et al. (1974).
[44] Bragg and Brown (1926b); Farrell et al. (1963).
[45] Weir and van Valkenburg (1960).
[46] Gjessing et al. (1942).
[47] Gjessing et al. (1942); Newnham et al. (1964).
[48] Newnham et al. (1964).
[49] Gjessing et al. (1942); Hahn (1961).
[50] Paques-Ledent and Tarte (1974).
[51] Weiss and Rocktaschel (1960); Rocktaschel et al. (1964)
[52] Hagenmuller et al. (1964); Hardy et al. (1965).
[53] Vincent et al. (1976).
[54] Rocktaschel et al. (1964).
[55] Vincent and Perrault (1971); Vincent et al. (1976).
[56] Claringbull (1952); Fang and Newnham (1965); Capponi et al. (1973).
[57] Capponi et al. (1973).
[58] Pistorius and Boeyens (1970).
[59] O'Daniel and Tscheischwili (1944); Hanke (1965); Hahn (1961).

[2] Two entries are given for solid solution between Mg_2SiO_4 and $MgCaSiO_4$ because a solvus exists between these two end members. The same is true for the solid solution between Ni_2SiO_4 and Ni_2GeO_4.

[3] Question marks beside these two compounds indicate that they may not be true olivine isostructures. Although Blasse (1963) reported that $LiMgVO_4$ has the olivine structure type, more recent evidence (Paques-Ledent, 1974) indicates that neither this compound nor $LiMnVO_4$, $LiCdVO_4$, $NaCdVO_4$ and $NaCaVO_4$ have the olivine structure type, although they have structural similarities with olivine. Other compounds with similar structures include $[Mg_3 \quad]P_2O_8$ and $[Co_3 \quad]P_2O_8$ (Berthet et al., 1972); $NaSmGeO_4$ (Kharakh et al., 1971). There is also some question about the isostructural relationship of $LiYSiO_4$, $LiYGeO_4$, $NaYSiO_4$ and $NaYGeO_4$ with olivine (see Maksimov et al., 1969, and Koryakina et al., 1972; versus Blasse and Bril, 1967, Chenavas et al., 1969, and Paques-Ledent, 1976). Additional discussion may be found in Paques-Ledent and Tarte (1974). The interesting compound $Fe^{2+}Fe_2^{3+}Si_2O_8$ (Acta Geol. Sinica, 1976, 160-175; see also Min. Abstr. 77-2645, p. 256) shows some structural similarities to olivine.

Schairer and Hytonen (1957), Kushiro and Schairer (1963), Kushiro (1969); forsterite-anorthite-silica: Andersen (1915); forsterite-anorthite-diopside: Osborn and Tait (1952); fayalite-nepheline-albite-silica: Bowen and Schairer (1938); nepheline-diopside-silica: Schairer and Yoder (1960); fayalite-albite: Bowen and Schairer (1936); forsterite-diopside-iron oxide: Presnall (1966); forsterite-diopside-enstatite (to 20 kbar): Kushiro (1964); forsterite-anorthite-albite-silica-H_2O (to 15 kbar): Kushiro (1974); $MgO-FeO-SiO_2$: Bowen and Schairer (1935), Nafziger and Muan (1967); $MgO-FeO-Fe_2O_3-SiO_2$: Muan and Osborn (1956); $MgO-CaO-SiO_2$: Ferguson and Merwin (1919), Ricker and Osborn (1954); $MgO-FeO-CaO-SiO_2$: Schairer and Osborn (1950), Wyllie (1960); $FeO-CaO-SiO_2$: Bowen et $al.$ (1933b), Allen and Snow (1955); $FeO-CaO-Al_2O_3-SiO_2$: Schairer (1942); $MgO-CaO-SiO_2-H_2O$: Franz and Wyllie (1966); $MgO-SiO_2-H_2O$: Bowen and Tuttle (1949), Pistorius (1963); $MgO-CaO-SiO_2-H_2O$ (to 10 kbar): Warner (1973); $MgO-Al_2O_3-SiO_2-H_2O$: Yoder (1952), Roy and Roy (1955); $FeO-Al_2O_3-SiO_2$: Schairer and Yagi (1952); $Mg_2SiO_4-SiO_2$ (to 25 kbar): Chen and Presnall (1975); $MgO-K_2O-SiO_2$: Roedder (1951); $MgO-FeO-Fe_2O_3-CaAl_2Si_2O_8-SiO_2$: Roeder and Osborn (1966); $Mg_2SiO_4-Fe_2SiO_4-CaMgSi_2O_6-CaFeSi_2O_6-KAlSi_3O_8-SiO_2$: Hoover and Irvine (1978): $FeO-Fe_2O_3-SiO_2$: Muan (1955); $Fe-O-SiO_2$: Lindlsey et $al.$ (1968); $Fe-MgO-SiO_2-O_2$: Williams (1971); $MgO-MnO-SiO_2$: Glasser and Osborn (1960); $MnO-CaO-SiO_2$: Glasser (1961); $MgO-SiO_2-H_2O-CO_2$: Johannes (1969); $Mg_2SiO_4-H_2O$: Hodges (1973); and $Mg_2SiO_4-CO_2$: Newton and Sharp (1975). In addition to these studies, the melting of Fe_2SiO_4 to 40 kbar has been determined by Lindsley (1967), Akimoto et $al.$ (1967), and Hsu (1967) and the melting curve of Mg_2SiO_4 to 50 kbar has been determined by Davis and England (1964). The effect of various components on liquidus boundaries involving olivine has been summarized by Kushiro (1975). This listing of references to experimental studies involving olivines is by no means complete. Instead, it is meant to serve as a starting point for readers interested in phase equilibrium studies of systems including olivine.

At high pressures, olivines in the $Mg_2SiO_4-Fe_2SiO_4-Mn_2SiO_4-Ca_2SiO_4$ system break down to other phases ($e.g.$ $MgCaSiO_4$ breaks down to Mg_2SiO_4 + $Ca_3MgSi_2O_8$ (merwinite): Kushiro and Yoder, 1964; Yoder, 1968) or undergo polymorphic transformations (members of the $Mg_2SiO_4-Fe_2SiO_4$ crystalline solution transform to γ-spinel and/or β-spinel phases: Ringwood, 1958;

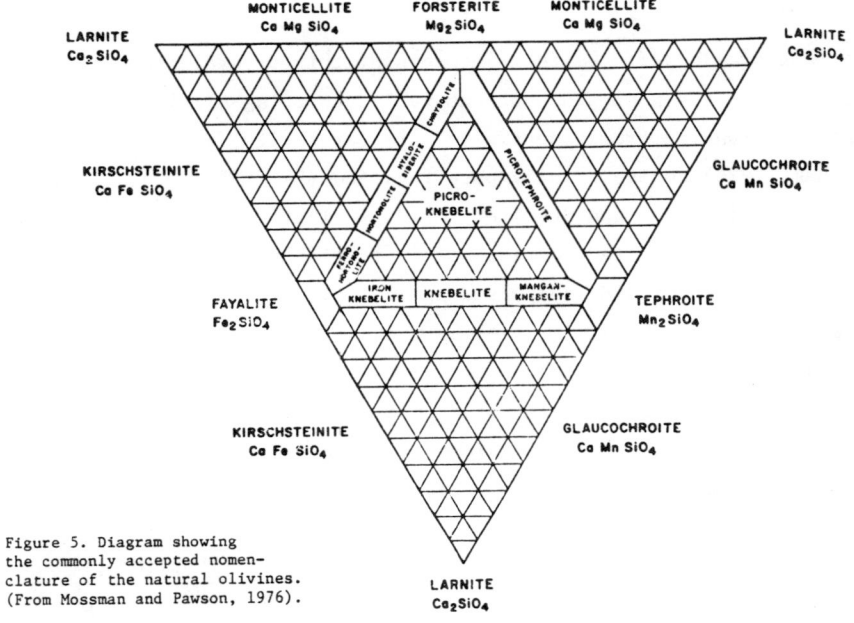

Figure 5. Diagram showing
the commonly accepted nomenclature of the natural olivines.
(From Mossman and Pawson, 1976).

Dachille and Roy, 1960; Ringwood and Major, 1966). An excellent review
of the high pressure phase relations of the olivines is given by Akimoto
et al. (1976).

NOMENCLATURE AND COMPOSITIONAL VARIATIONS OF THE NATURAL OLIVINES

The quaternary system $Mg_2SiO_4-Fe_2SiO_4-Mn_2SiO_4-Ca_2SiO_4$ includes almost
all naturally occurring silicate olivines, when allowance is made for
small quantities (usually <1 wt %) of Ni, Co, Zn, and Cr found in some
olivines. The preferred nomenclature for olivines in this system is shown
in Figure 5. The end member Ca_2SiO_4 (larnite or $\beta-Ca_2SiO_4$) does not normally crystallize with the olivine structure and for practical purposes
should be considered only as a chemical component in olivines. It is
common practice to specify an olivine's composition in terms of the four
end-member components: *Fo* (forsterite), *Fa* (fayalite), *Te* (tephroite),
and *La* (larnite) (see *e.g.* Smith, 1966). It is simpler to use this system of nomenclature for Mg-Fe olivines rather than the terms forsterite,
chrysolite, hyalosiderite, hortonolite, ferrohortonolite, and fayalite
for members of this crystalline solution series. Students should take
care to distinguish between *chrysolite* olivine and *chrysotile*, a fibrous
polymorph of serpentine.

Major Element Chemistry

Forsterite-fayalite series. Members of this crystalline solution series comprise the most abundant naturally occurring olivines, with compositions ranging from essentially pure forsterite ($Fo_{99}Fa_1$: Adams and Graham, 1926) to essentially pure fayalite ($Fa_{99.5}Te_{0.3}La_{0.2}$: Smith, 1966). Prior to Bowen and Schairer's (1935) classic study of this series in which they demonstrated continuous crystalline solution, some workers (see *e.g.* Magnusson, 1918) suspected an immiscibility gap in natural olivines. However, classic studies by Tomkeieff (1939) of olivines from a variety of rock types and by Wager and Deer (1939) of olivines from the Skaergaard intrusive demonstrated that natural olivines are continuously zoned and sometimes show evidence of the reaction relationship (see Bowen and Schairer, 1935). In addition to the analyses of selected members of this series published in Deer *et al.* (1962), the reader is referred to the more modern analyses reported in Smith (1966), Moore and Evans (1967), and Nwe (1976) as examples of major element variations in natural olivines. Although electron microprobe analysis is the preferred method for analyzing olivines, numerous nomographs relating major element composition to optical properties (see *e.g.* Laskowski and Scotford, 1980), cell parameters or d-spacing (see *e.g.* Louisnathan and Smith, 1968; Fisher and Medaris, 1969; Schwab and Kustner, 1977), combinations of optical properties and d-spacing (see *e.g.* Mossman and Pawson, 1976), or vibrational spectra (see *e.g.* Burns and Huggins, 1972) have been published. However, the user of these indirect methods must exercise caution to detect the presence of zoning and to recognize the effect of minor element content on physical properties.

Tephroite-fayalite-forsterite series. As noted earlier, continuous crystalline solution exists between Mn_2SiO_4 and Fe_2SiO_4 and between Mn_2SiO_4 and Mg_2SiO_4. Therefore, olivines of composition intermediate to these three end members are to be expected. Representative analyses of members of this series can be found in *Deer* et al. (1962), Burns and Huggins (1972), and Mossman and Pawson (1976). Figures 6 and 7 are determinative charts relating composition to certain vibrational frequencies and to x-ray and optical parameters. These plots should be used with caution because of the possibilities of compositional zoning within individual olivine crystals and of the presence of significant quantities

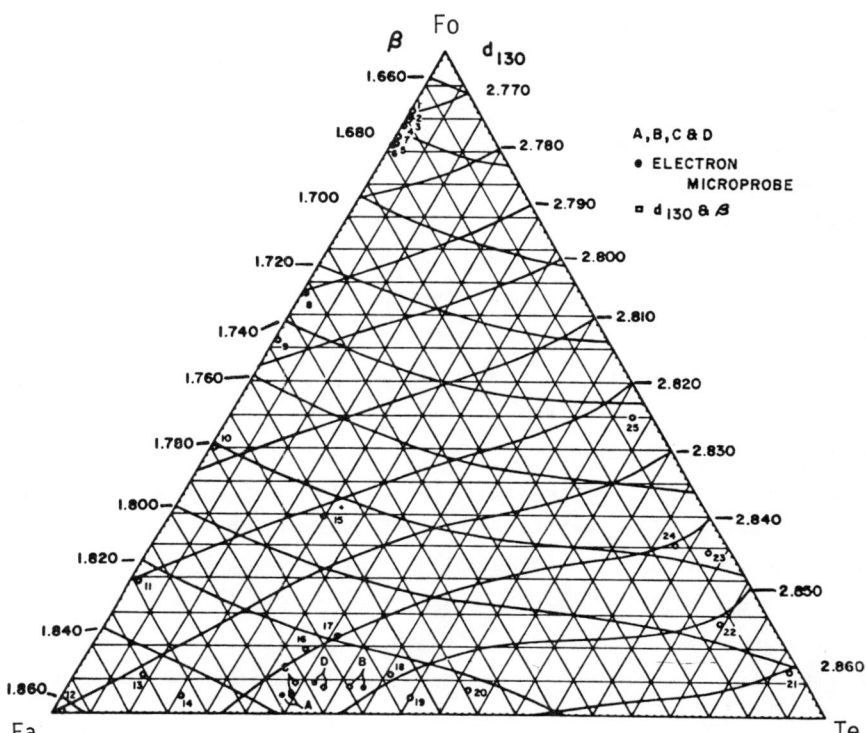

Figure 6. Nomograph for estimating compositions of olivines in the forsterite (Fo) - fayalite (Fa) - tephroite (Te) series using the d-spacing, d_{130},
and the β refractive index. Compositions are in mol %. (From Mossman and
Pawson, 1976).

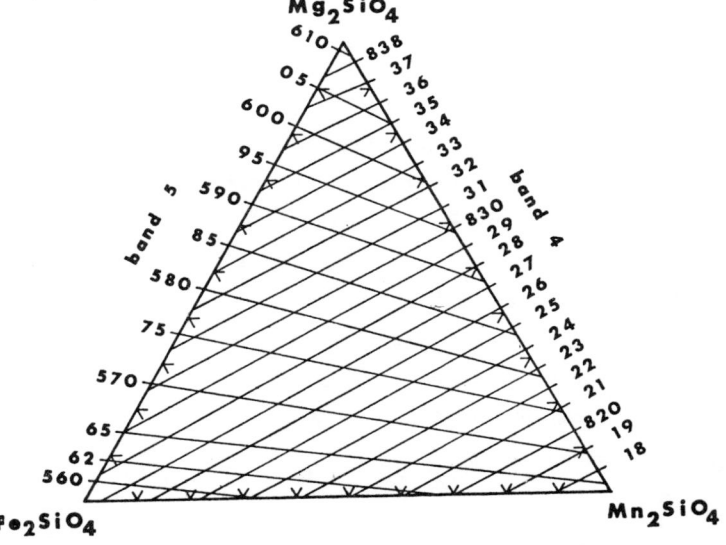

Figure 7. Nomograph relating the positions of the 830 cm^{-1} (band 4) and
the 590 cm^{-1} (band 5) bands to composition in the Fo - Fa - Te series.
Compositions are in mol %. (From Burns and Huggins, 1972).

of elements such as Ca and Zn which are not accounted for in the nomograms. Based on published analyses of Mg-Fe-Mn olivines, picroknebelites (see Fig. 5) are relatively rare.

Early wet chemical analyses of Mg-Fe-Mn olivines sometimes detected Zn, and a variety of these olivines containing high Fe^{2+} and Zn was named *roepperite* (see Palache, 1937). These early analyses should be viewed with some caution because of the possibility that zincite [(Zn,Mn)O] or willemite [Zn_2SiO_4] inclusions were present in these olivines. However, modern microprobe analyses of some picrotephroites show significant quantities of Zn in olivine. For example, Brown (1970) reported 9.66 wt % ZnO in a picrotephroite from Franklin, New Jersey (see Table A1). This specimen contained willemite lamellae, so care had to be taken to avoid these areas with the probe beam. [C. A. Francis, Ph.D. dissertation, Virginia Polytechnic Institute and State University (1980) has discussed Zn in the forsterite-tephroite solid solution series.]

Although limited crystalline solution occurs between Mn_2SiO_4 and Ca_2SiO_4 and between Fe_2SiO_4 and Ca_2SiO_4, natural Fe-Mn olivines rarely contain more than few weight percent CaO. *Glaucochroite* ($CaMnSiO_4$) (see Palache, 1937) and *kirschsteinite* ($CaFeSiO_4$) (see Sahama and Hytonen, 1957) are rare minerals.

Monticellite. Because of the lack of significant crystalline solution between $CaMgSiO_4$ and Ca_2SiO_4 and only limited crystalline solution between $CaMgSiO_4$ and Mg_2SiO_4, natural monticellites tend to have compositions close to the ideal. Even though experimental studies have shown that complete crystalline solution exists between $CaMgSiO_4$ and $CaFeSiO_4$, and is predicted to occur between $CaMgSiO_4$ and $CaMnSiO_4$, minerals of intermediate composition in the $CaMgSiO_4$-$CaFeSiO_4$-$CaMnSiO_4$ series are rare, presumably due to the low level of Mn in most natural systems. Warner and Luth (1973) suggest that the maximum amount of $MgCaSiO_4$ that may be dissolved in forsteritic olivine is less than 5 mol % under conditions likely to exist in natural systems. Representative analyses of natural monticellites are given by Deer *et al.* (1962).

Minor Element Chemistry

Calcium. The calcium content of Mg-Fe olivines rarely exceeds 1.0 wt % and is typically <0.5 wt % based on the survey of minor element

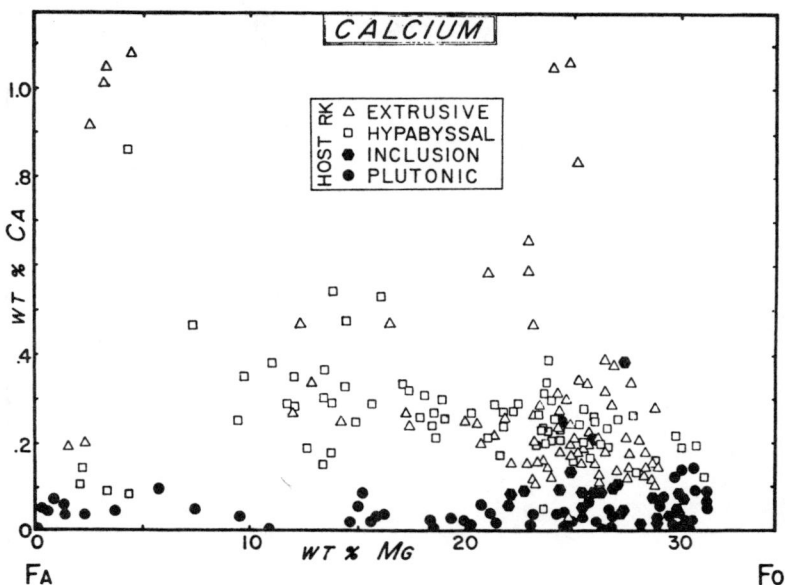

Figure 8. Ca *vs* Mg content in olivines from various rock types. (From Simkin and Smith, 1970).

distribution in olivines from various terrestrial rock types by Simkin and Smith (1970). This study showed no clear correlation between Ca and Mg content but did indicate a significant correlation between Ca content of the olivine and crystallization environment of the host rock (Fig. 8), with plutonic olivines containing less Ca than olivines from hypabyssal and extrusive environments. Careful analysis of available data led Simkin and Smith to rule out crystallization temperature as the factor controlling Ca content, even though the Ca content of forsterite co-existing with monticellite has been shown to be temperature dependent (Ricker and Osborn, 1954; Warner and Luth, 1973). Simkin and Smith also pointed out that although solid state diffusion can result in less calcic olivines in slow-cooling plutonic rocks or in the interiors of lava lakes, a correlation of Ca content with cooling or diffusion rate is unclear because of exceptions to this observation. The clearest correlation is with pressure of crystallization, suggesting that the Ca content of olivines might serve as a useful geobarometer. This correlation is now supported by the experimental studies of Finnerty (1977) and Finnerty and Boyd (1978) who showed that the solubility of Ca in forsterite

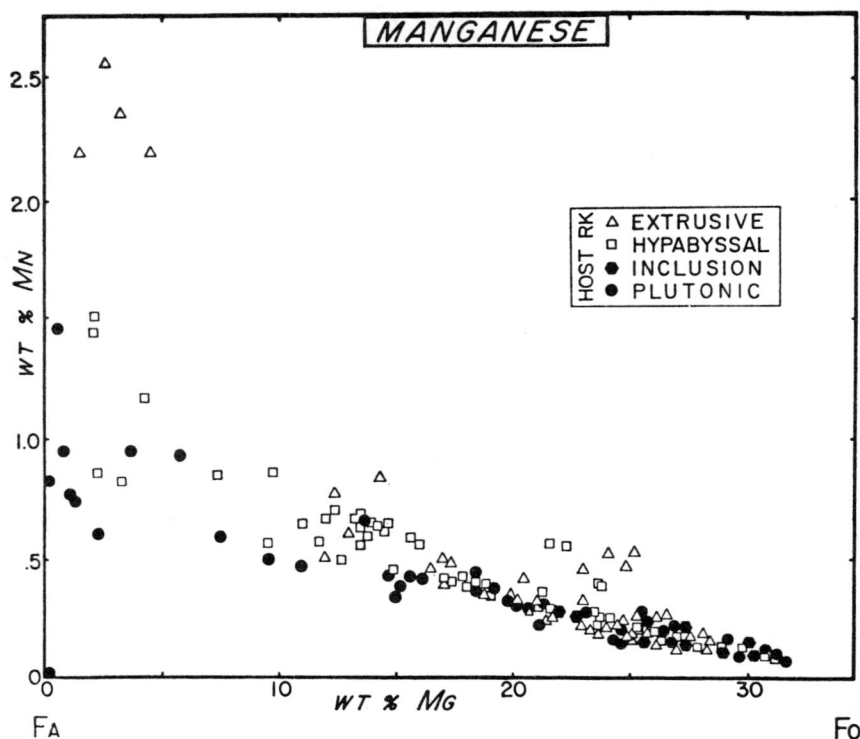

Figure 9. Mn *vs* Mg content in olivines from various rock types. (From Simkin and Smith, 1970).

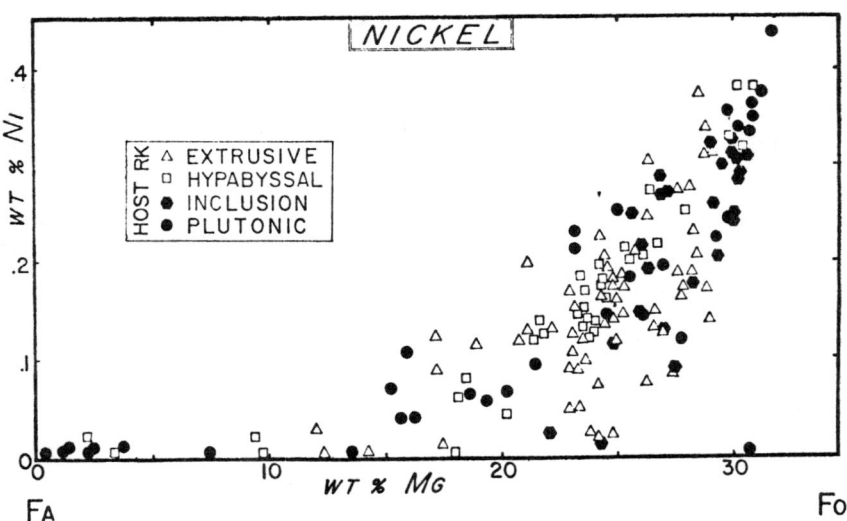

Figure 10. Ni *vs* Mg content in olivines from various rock types. (From Simkin and Smith, 1970).

coexisting with orthopyroxene and clinopyroxene decreases with increasing pressure. Clearly, these observations are related in part to the large size of the Ca cation (r = 1.0 A) relative to Mg, Fe, and Mn.

Other factors besides pressure must still be considered in petro-genetic interpretations of Ca content of olivines. Smith (1974) in a review of lunar olivine chemistry pointed out that metamorphism of olivine, as in a thick ejecta blanket, may cause significant migration of Ca from the olivines, based on available diffusion rate data (see *e.g.* Buening and Buseck, 1973). Thus low Ca content in an olivine may indi-cate either plutonic origin or metamorphism, or both. Furthermore, the effects of metastable crystallization, the interaction of olivine with other Ca-bearing minerals, and the bulk chemistry of the host rock must be considered before any unique petrogenetic conclusions are drawn . The classic studies by Wager and Mitchell (1951) and Wager and Mitchell (1953) on Ca and other trace element distributions between olivine and magma in the Skaergaard intrusion and Hawaiian basalts, respectively, are recommended as sources of data.

Surprisingly, the calcio-olivine Ca_2SiO_4 does exist naturally, although it is quite rare. Bridge (1966) reported the occurrence of this olivine coexisting with larnite and bredigite in the contact zone between a syenite-monzanite intrusion and a dolomite limestone in Texas.

Manganese. Although Mn may be considered a major element in oli-vines from certain parageneses, this element is most often present in Mg-Fe olivines in amounts less than 2.0 wt %. Simkin and Smith (1970) found a positive correlation between Mn and Fe content in olivines from a variety of terrestrial, igneous rock types but no correlation of Mn content with crystallization environment (Fig. 9), suggesting that Mn in olivines is strongly related to major element fractionation. This observation is also borne out by the minor element study by Dodd (1973) of chondritic olivines from the Sharps meteorite. There is some corre-lation between Mn content of olivines and bulk chemistry of host rock, with unusually Mn-rich olivines occurring in undersaturated rocks.

Nickel. The nickel content of most olivines, <0.5 wt %, correlates positively with Mg content but shows no unique correlation with crystal-lization environment (Fig. 10). The former correlation is not surprising in view of the similar sizes of Mg and Ni^{2+}. Olivines from undersaturated rocks are usually poor in Ni, suggesting a correlation between bulk rock chemistry and Ni content of olivines.

The most Ni-rich natural olivine yet reported is the mineral *liebenbergite*, $(Ni_{1.52}Mg_{0.33}Fe_{0.12}Ca_{0.05})Si_{0.99}O_4$, which occurs with other Ni-rich minerals near Barberton, South Africa (deWaal and Calk, 1973).

Chromium. The chromium content of terrestrial olivines is usually below the detectability limit of the electron microprobe (0.0n wt %). However, forsteritic olivines $(Fo_{95}-Fo_{92})$ included in diamonds show concentrations of up to 0.15 wt % Cr_2O_3 (Meyer and Boyd, 1972; Sobolev, 1972[*]; Meyer and Svisero, 1973; Prinz *et al.*, 1973). Because Cr can exist in two oxidation states, the question is whether Cr^{2+} or Cr^{3+}, or both, enter the olivine structure. On the basis of crystal-chemical reasoning and a survey of the available data, Burns (1975) concluded that Cr^{2+} substitutes for Mg in these olivines. An even higher level of Cr is found in lunar olivines, sometimes exceeding 0.4 wt % (Bell, 1970; Haggerty *et al.*, 1970; Smith, 1971, 1974), and tends to correlate positively with Mg content (see Smith, 1974). All present evidence (see especially Haggerty *et al.*, 1970) indicates that Cr in lunar olivines is also divalent, a suggestion in accord with the low oxygen fugacities during magmatic crystallization on the moon. One report of trivalent Cr comes from Dodd's (1973) study of olivine chondrules from the Sharps meteorite, where he found a weak correlation between Cr and Al in olivines and suggested the coupled substitution $Cr^{3+} + Al^{3+} = (Mg,Fe)^{2+} + Si^{4+}$. Cr_2O_3 ranged from 0.02 to 0.2 wt % in these meteoritic olivines $(Fo_{100}-Fo_{70})$. The electron spin resonance study of a synthetic forsterite doped with ∿0.2 wt % Cr_2O_3 (Rager, 1977) showed that Cr^{3+} can occupy the octahedral sites of olivine, although in small amounts. Clearly, the oxygen fugacity in the crystallizing magma, as well as the Cr content of the host melt, are important factors in determining how much Cr enters the olivine structure.

Aluminum. Simkin and Smith (1970) found no Al in terrestrial olivines at the 0.01 wt % level. However, lunar olivines are reported to contain up to about 0.6 wt % Al_2O_3 (Smith, 1974) and chondritic olivines were found to contain up to 0.07 wt % Al_2O_3 (Dodd, 1973). Olivines synthesized by Kushiro *et al.* (1971) (10 kbar, 1290°C) and by Akella and Boyd (1972) (30 kbar, 1350°C) were found to contain 0.07 and 0.43 wt % Al_2O_3, respectively. Although these data on natural and synthetic

....
[*]Sobolev, N.V. (1972) Petrology of xenoliths in kimberlitic pipes and indications of their abyssal origins. Proc. 24th Int'l Geol. Congress, Sect. 2, 297-302.

olivines tend to support the possibility of small amounts of Al^{3+} in lunar olivines, Smith (1974) cautions that ricroprobe samples that were polished with alumina might be contaminated. If these analyses are correct, some type of coupled substitution is necessary to maintain local charge balance in the olivine structure. Because of the lack of significant levels of Cr^{3+} in lunar olivines, the coupled substitution involving Al^{3+} and Cr^{3+} mentioned in the last section is unlikely. Instead, a substitution such as $(x)^{VI}Al^{3+} + (2x)^{IV}Al^{3+} = (2-x)(Mg,Fe)^{2+} + (1-x)(Si^{4+})$ might take place.

Ferric iron. Fe^{3+} is usually undetectable or at very low levels in natural olivines as indicated by a number of ^{57}Fe Mössbauer studies (see *e.g.* Virgo and Hafner, 1972; Duba *et al.*, 1973; Shinno *et al.*, 1974; Rao *et al.*, 1979) and optical absorption studies (see *e.g.* Hazen *et al.*, 1977). However, at detection levels below that of Mössbauer spectroscopy (<0.01 wt % Fe), electron spin resonance spectra of natural (Shcherbakova *et al.*, 1968) and synthetic (Chatelain and Weeks, 1973; Zeira and Hafner, 1974; Weeks *et al.*, 1974) olivines have shown detectable Fe^{3+} occupying octahedral sites. The total iron content of these synthetic forsterites was ≤ 0.01 mol %.

In older wet chemical analyses of olivines reporting Fe_2O_3, the possibility of small, Fe^{3+}-bearing opaque inclusions must be considered. The analyses of olivines reported in Deer *et al.* (1962) show Fe_2O_3 contents as high as 1.69 wt %. It is doubtful that this much Fe^{3+} can substitute into the olivine structure. At present, there is no chemical or structural evidence that Fe^{3+} substitutes for Si^{4+} (see Table 3).

Alkalis. Na_2O and K_2O are sometimes reported in modern microprobe analyses (see *e.g.* Dostal and Capedri, 1975) and older wet chemical analyses (see *e.g.* Deer *et al.*, 1962) of natural olivines, with amounts ranging from below detectability to about 0.05 wt %. Although the possibility of Na substitution in the olivine structure cannot be ruled out, strong crystal-chemical arguments can be made against K^+ substitutions. Reports of K^+ in olivines are probably due to impurities.

Water. H_2O is normally not considered in analyses of nominally anhydrous silicates such as olivine. However, attention has been drawn to the possibility that structural "water" in the form of OH^- ions may be present (Martin and Donnay, 1972; Wilkins and Sabine, 1973; see also Deer *et al.*, 1962). Wilkins and Sabine (1973) report new analyses of 19 nominally anhydrous specimens, including natural olivine, and found 0.008 wt % H_2O in the olivine. There is no present evidence that olivine analogues of the hydrogarnets exist naturally or can be made synthetically.

Uranium and the lanthanides. These elements may be present at ppm or ppb levels in olivines. The reader is referred to the study by Dostal and Capedri (1975) for a survey of uranium concentrations and partition coefficients in olivines. Masuda (1968) made a study of lanthanide concentrations in the olivines present in a pallasitic meteorite and reported values ranging from 37.3 ppb (for La) to 0.67 ppb (for Er).

Titanium. Ti in lunar and chondritic olivines is below the 0.1 wt % level (Smith, 1974; Dodd, 1973).

Defects. Structural defects are considered in this section on minor element chemistry because their concentration is at the trace level and their effect is similar to that of a minor element in that they may participate in coupled, charge-balanced substitutions. The defect chemistry of olivines has been considered theoretically by Pluschkell and Engel (1968), Smyth and Stocker (1975), and Stocker and Smyth (1978) and has been examined experimentally by Sockel (1974) among others. Silicon tetrahedral vacancies and oxygen interstitials are unlikely defects; however, Mg and Fe octahedral vacancies are likely, as are oxygen vacancies. Stocker and Smyth (1978) argue that these vacancies are strongly dependent on enstatite activity and oxygen partial pressure for olivines in the earth's upper mantle and that olivine's defect chemistry in this region may have significant effects on the transport properties of olivine. Similarly, defects may play a significant role in the partitioning of trace elements between olivine and melt (Buseck and Veblen, 1978).

Inclusions in Olivine

Because of the skeletal manner in which some olivine crystals grow from silicate melts, trapping of primary melt inclusions between crystal dendrites is possible. An excellent review of inclusions in lunar and Hawaiian olivines is given by Roedder (1976) who points out that these olivines "contain a full complement of volatile materials and generally

298

are unaffected by later minor deuteric alteration or oxidation of the host olivine." Typical primary inclusions include spinel, globules of CO_2 gas, and immiscible sulfide melt, whereas common daughter phases which nucleate and grow in olivine include "vapor" bubbles, immiscible sulfide melt, pyroxenes, plagioclase, and ilmenite.

Roedder reviews the types of data that studies of such inclusions can provide including (1) melt composition at the time of olivine crystallization; (2) the amount and nature of immiscible sulfide melt; (3) equilibrium temperatures and sequence in melting; (4) relative cooling rates of different lava flows and absolute cooling rates; (5) possible distinctions between phenocrysts and xenocrysts; (6) the presence of a vapor phase at the time of olivine crystal growth; and (7) minimum pressures at the time of olivine growth. An examination of olivine inclusions can be valuable in deciphering the petrogenetic history of the olivine and its host rock. Other inclusions in olivine will be considered in a later discussion of olivine alteration.

GEOLOGIC OCCURRENCE OF OLIVINES

Olivines occur predominantly in igneous rocks but are also found in limited metamorphic parageneses. Mg-Fe olivines occur in significant amounts in at least four classes of igneous rocks including (1) ultramafic intrusions; (2) stratiform mafic intrusions; (3) gabbros, basalts, and dolerites; and (4) felsic rocks. In ultramafic intrusions, forsteritic olivines (Fo_{95}-Fo_{85}) may comprise more than 90 percent of the rock (dunite) or more commonly are present in amounts between 50 and 90 percent (peridotites). In stratiform, mafic intrusions, such as the Stillwater complex in Montana, the Mg-rich olivines may occur either as constituents of the layered peridotites and gabbros, with compositions between Fo_{85} and Fo_{80}, or as irregular masses cross-cutting the layering. In the latter case, the olivines usually fall in the composition range Fo_{90}-Fo_{85} and are probably of secondary hydrothermal origin. In gabbros, basalts, and dolerites, the olivines range in composition between Fo_{80} and Fo_{50} and are usually the first phases to crystallize. A discussion of olivines from stratiform mafic intrusions can be found in Deer and Wager's (1939) classic paper on olivines from the Skaergaard intrusion.

Basaltic olivines have been intensively studied over the past 15 years with modern analytical techniques. The papers by Moore and Evans

(1967) on olivines in Hawaiian basalts and by Hermes and Schilling (1976) on olivines from Icelandic tholeiites are especially recommended because of their emphasis on the role of olivine in these basaltic rocks. The abundant lunar literature contains much discussion of lunar basaltic olivines and an excellent summary of compositional variations and occurrence of olivines in the three major lunar rock types is presented by Smith (1974).

In felsic rocks, including granites, syenites, rhyolites, trachytes, and doleritic pegmatoids, relatively Fe-rich olivine occurs.

Metamorphic olivines are considerably more restricted in their paragenesis relative to olivines of igneous origin. Forsterite is a common product of the thermal and regional metamorphism of slightly siliceous dolomitic limestones, appearing at temperatures slightly higher than diopside forms. Fayalitic olivine, which is less common than forsteritic olivines in metamorphic rocks, forms during the metamorphism of Fe-rich sediments where it can coexist with quartz, hedenbergite, grunerite, and almandite.

Members of the tephroite-fayalite series occur in iron-manganese ore deposits and skarns and in metamorphosed manganese-rich sediments. The rare olivine glaucochroite occurs with tephroite in very restricted parageneses, such as at Franklin, New Jersey.

Monticellite is most commonly found as a constituent of relatively high grade siliceous, magnesian marbles. Its appearance in these rocks marks the lower limit of the sanidinite facies of thermal metamorphism and probably of the granulite facies of regional metamorphism. Much more rarely, it occurs in mafic to ultramafic igneous rocks associated with phlogopite, Mg-Fe olivine, melilite, and augite (C. P. Thornton, written communication, 1965). Kirschsteinite occurs in rare igneous rocks such as the melilite-nephelinites of Nyiragongo, E. Congo, and may have formed in subsolidus reactions involving alkali-bearing melilites and iron ore (Sahama, 1961).

Olivine is a common mineral in certain classes of meteorites. It is the dominant mineral in the bronzite and hypersthene chondrites and is present in most of the carbonaceous chondrites. For further discussion of olivines in meteorites, see Mason (1972). A more detailed discussion of olivine occurrence and paragenesis can be found in Deer *et al.* (1962).

ALTERATION AND OXIDATION OF OLIVINES

Among the common silicate minerals, olivines are especially susceptible to weathering, hydrothermal alteration, and low grade metamorphism
involving hydration, oxidation, silication, and carbonation reactions.
The most common alteration products are serpentine, chlorite, amphibole,
carbonates, iron oxides, talc, "bowlingite," and "iddingsite" (Deer *et al.*,
1962).

Serpentinization of olivines in mafic and ultramafic rocks is a
widespread process that has received considerable attention and has
spawned controversy over whether constant volume or constant composition
is maintained.   In a study of serpentinization of olivine cumulates of
the Stillwater complex, Page (1976) found that olivine + plagioclase
altered to assemblages of lizardite + chrysotile + magnetite + thomp-
sonite $\pm$ carbonate at relatively low temperatures ($\sim$100°C or less) with
large volume expansion (+37 percent) and without much change in bulk
composition, except for the addition of water.   The serpentinization of
Fo_{93} has been studied experimentally by Moody (1976) who found that lizardite replaces olivine before chrysotile which can be explained by the
close topotactic relationship between olivine and lizardite.   Moody's
experimental assemblage closely approximates the observed phase assemblage in serpentinized dunites.

Other common alteration products of olivines are the *coronas* or
kelyphitic rims that mantle olivine in metamorphosed basic igneous rocks.
The coronas usually consist of orthopyroxene + amphibole + spinel +
plagioclase or of orthopyroxene + garnet + plagioclase as the dominant
phases.   England (1974) has suggested that such coronas may form by near
isochemical reactions between olivine and plagioclase with the addition
of only small amounts of H_2O and O.   In the metamorphosed dolerite body
he studied, the coronas developed during upper greenschist to lower amphibolite facies regional metamorphism under almost dry conditions.

The alteration of olivines in basaltic rocks has been studied by
Haggerty and Baker (1967) and Baker and Haggerty (1967).   They found
that "high temperature oxidation of basaltic olivines results in either
exsolution of hematite associated with a more forsteritic olivine, or
formation of a symplectic intergrowth of magnetite and orthopyroxene"

301

and that "processes of oxidation are the prime causes of high temperature deuteric alteration of olivines." These observations have been confirmed and other complexities revealed in later experimental studies of olivine oxidation by Champness and Gay (1968) and Champness (1970). In the latter study, Champness found that oriented hematite- and magnetite-like precipitates plus amorphous silica precipitate in Mg-Fe olivines held at 500 to 800°C. Oxidation at 1000°C produces larger, more equant grains of the Fe-oxides and a more ordered silica phase. These observations have been given a thermodynamic basis by Nitsan (1974) who calculated a $T-f_{O_2}$ phase diagram for olivines of various fayalite content at 1 atm total pressure. Other recent studies of olivine oxidation have been made by Goode (1974), Kohlstedt and Vander Sande (1975), and Putnis (1979). Goode found that olivines in the composition range $Fo_{75}-Fo_{63}$ from Precambrian cumulates are altered to granular orthopyroxene-vermicular magnetite aggregates. Olivines in the range $Fo_{84}-Fo_{76}$ altered to intergrowths of fibrous to granular orthopyroxene and vermicular to granular picotite, a chromian hercynite spinel. Goode also found that olivines in the range $Fo_{89}-Fo_{85}$ were unaltered. These observations show that oxidation of olivines is dependent on composition, a conclusion consistent with Nitsan's calculations. Electron petrographic study of naturally oxidized olivines from lherzolite xenoliths by Kohlstedt and Vander Sande showed that α-tridymite, enstatite, magnetite, and/or hematite nucleated on dislocations in olivines in an oriented fashion. A similar study of olivines from layered peridotites by Putnis showed that clinopyroxenes and magnetite apparently nucleated homogeneously, and not on dislocation as found by Kohlstedt and Vander Sande.

Oxidation of olivines at low temperatures results in the formation of chloritic materials whereas cooling of olivines under non-oxidizing conditions at intermediate temperatures produces chlorite and inter-stratified layer silicates including a smectite (possibly a "bowlingite," which is best described as an interstratified layer silicate) (Baker and Haggerty, 1967). These workers found that alteration under oxidizing conditions at temperatures below about 140°C produces "iddingsite," which is an oriented mixture of goethite and layer silicates. "Iddingsite" is also formed during the weathering of olivine. Further discussion of the three types of iddingsite can be found in Baker and Haggerty (1967). An excellent descriptive petrographic study of olivine alteration in basaltic lavas from the Isle of Mull, which preceded the work described above, was published by Fawcett (1965). Changes in the surface of olivine during weathering and dissolution are discussed by Luce and Parks (1973) and Grandstaff (1978).

PHYSICAL PROPERTIES OF OLIVINES

Color

Natural olivines may range in color from almost colorless to green to black depending on the Fe-Mn content. As usual, color is a poor diagnostic physical property for differentiating among olivines of various compositions. The greenish-yellow to green colors of Mg-Fe olivines are caused by the presence of Fe^{2+}, which has six d-electrons that can undergo crystal field transitions. The optical absorption spectra of Fa_{12} and Fa_{96} olivines shown in Figures 11 and 12 have low absorption in the green and yellowish-orange portions of the visible spectrum, respectively, and therefore, these colors dominate. The crystal-field transitions responsible for the spectrum of Fa_{96} (Fig. 12) are illustrated schematically in Figure 13. This figure from Burns (1970a) should be modified slightly by making the lowest t_{2g} level for the M(1) site doubly degenerate, rather than by the reverse situation as pictured. In deriving this energy diagram, Burns assumed D_{4h} and C_{3v} point symmetry, respectively, for the M(1) and M(2) sites of olivine. Synthetic Ni olivine has an emerald green color as do intermediate members of the Ni-Mg olivine crystalline solution series. Synthetic Co- and Co-Mg olivines have a ruby red color. Pure monticellite is colorless, as is pure forsterite.

Forsteritic olivines show essentially no pleochroism in the visible because of the small differences in absorption along the three mutually perpendicular optical directions in the 4000-7000 Å wavelength range (see Fig. 11). Fayalitic olivines display weak yellow ($\alpha = \gamma$) to orange-yellow (β) pleochorism.

Optical Properties

Refractive indices vary smoothly with Mg-Fe content in the Fo-Fa series as shown in Figure 14 (Deer et $al.$, 1962; Laskowski and Scotford, 1980). Optical properties of the end-member olivines taken from Deer

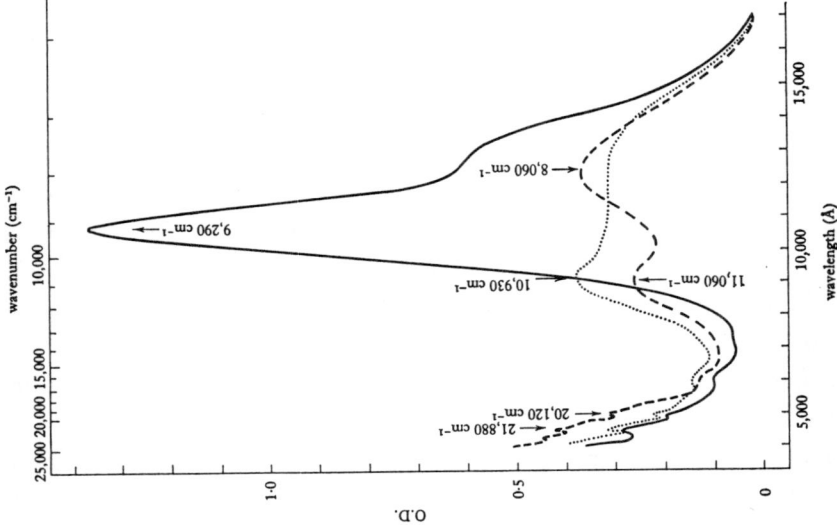

Figure 12. Polarized absorption spectrum of Fa96 from Rockport, Massachusetts. Dotted line is α spectrum, dashed line is β spectrum, and solid line is γ spectrum. Optic orientation is α = b, β = c, γ = a. The various d-d transitions responsible for the spectra are shown in Figure 13. (From Burns, 1970a).

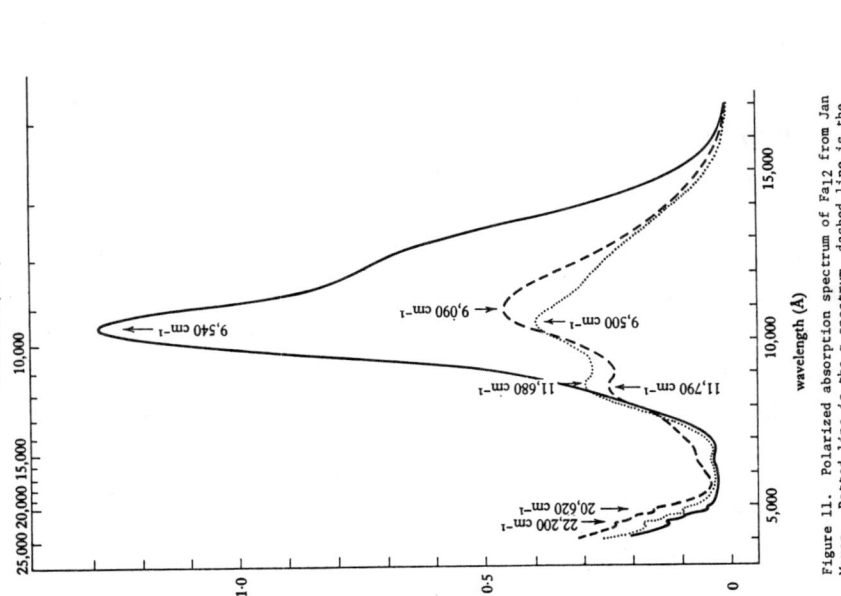

Figure 11. Polarized absorption spectrum of Fa12 from Jan Mayan. Dotted line is the α spectrum, dashed line is the β spectrum, and solid line is the γ spectrum. The visible spectrum is from 4000 to 7000 Å. Optic orientation is α = b, β = c, γ = a. The various d-d transitions responsible for the spectra are shown in Figure 13. (From Burns, 1970a).

et al. (1966) are summarized below. For all compounds $\alpha = y$, $\beta = z$, $\gamma = x$; O.A.P. (001); dispersion $r > v$.

	Forsterite	Fayalite	Tephroite	Knebelite	Monticellite
α	1.635	1.827	1.770	1.815	1.639–1.654
β	1.651	1.869	1.807	1.853	1.646–1.664
γ	1.670	1.879	1.817	1.867	1.653–1.674
$2V_\gamma$	82°	134°	110°	136°	118°–98°

The following regression equations which relate mole percent fayalite to refractive index for olivines belonging to the forsterite-fayalite crystalline solution series were derived by Laskowski and Scotford (1980) who

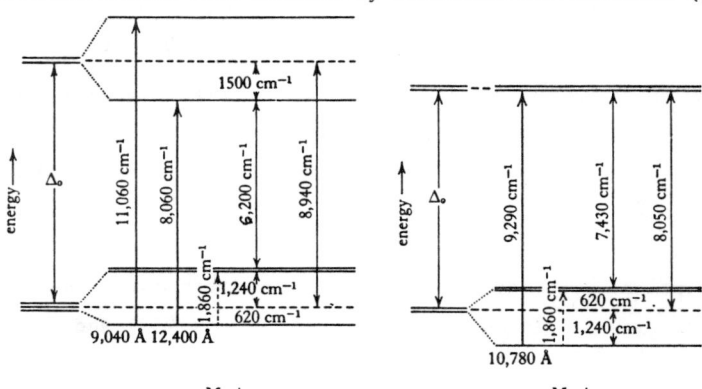

M_1 site M_2 site

Figure 13.   Energy level diagram for Fe^{2+} in the M(1) and M(2) sites of the olivine structure (Fa_{96}). The lowest of the t_{2g} levels of the M(1) site should be doubly degenerate and the next highest level should be singly degenerate rather than as shown. (From Burns, 1970a).

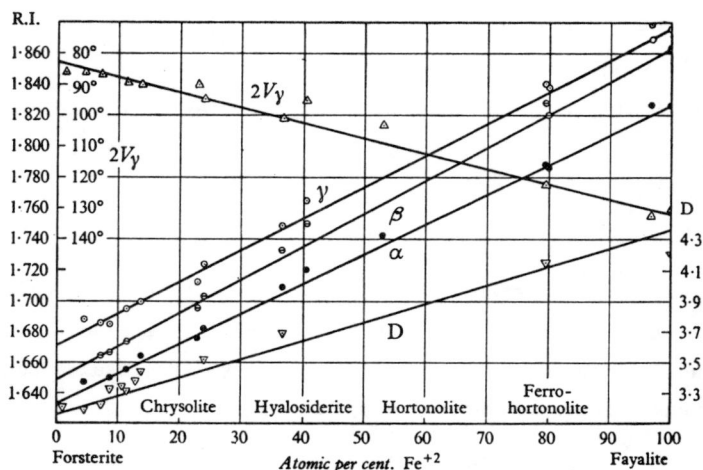

Figure 14.   Variation of refractive indices, 2V, and density with mol % fayalite for the Mg-Fe olivine series. (From Deer *et al.*, 1966).

305

claim that they are accurate to $\pm 2\%$ Fa (correlation coefficients are greater than 0.999):

$$\text{mole \% Fa} = (n_\alpha - 1.6325)/0.0020$$
$$\text{mole \% Fa} = (n_\beta - 1.6490)/0.0022$$
$$\text{mole \% Fa} = (n_\gamma - 1.6651)/0.0022$$

where n_α, n_β, and n_γ are the indices of refraction for sodium D light. Similar equations with refractive indices as dependent variable were also derived by Laskowski and Scotford and are as follows:

$$n_\alpha = 1.6361 + 0.0011\ X_{Fa} + 0.00001\ X_{Fa}^2$$
$$n_\beta = 1.6473 + 0.00159\ X_{Fa} + 0.000006\ X_{Fa}^2$$
$$n_\gamma = 1.6694 + 0.00116\ X_{Fa} + 0.00001\ X_{Fa}^2$$

They claim that these equations, together with the dispersion staining method they employed (see Grabar and Principe, 1963), permit optical determination of olivine compositions on unaltered grains in thin section. Obviously, significant Ca and Mn substitution will affect the utility of this method.

The optical properties of olivines from the forsterite-fayalite-tephroite series have been studied by Henriques (1957) and Mossman and Pawson (1976). Both studies considered the effect of Ca on optical properties. The more recent study combines the compositional dependence of the n_β refractive index and the d_{130} spacing of olivine in a compositional nomograph (see Fig. 6) which they claim is accurate to ± 2 mol % Fo. Hurlbut (1961) also correlated n_β and d_{130} with composition for the forsterite-tephroite series.

Theoretical aspects of the optical properties of the forsterite-fayalite crystalline solution series are discussed by Hauser and Wenk (1976).

Density

The relationship between density and composition for the forsterite-fayalite series has been studied by Bloss (1952) (see also Bloss, 1971[*] p. 349) and by Fisher and Medaris (1969). Bloss' regression equation, based on measurements of natural samples, is:

$$X_{Fa} = -207.754 + 47.6852\ \rho + 5.25529\ \rho^2$$

....
[*]Bloss, F.D. (1971) *Crystallography and Crystal Chemistry, an Introduction.* New York: Holt, Rinehart and Winston, 545 pp.

and the Fisher-Medaris equation for synthetic olivines is:

$$\rho = 4.4048 - 1.1353 \ X_{Fa} - 0.0435 \ X_{Fa}^2.$$

In these equations X_{Fa} is the mole percent fayalite and ρ is density. Both studies found that the relationship between density and composition is consistent with ideal behavior of the forsterite-fayalite series. The variation of density with composition for the Mg-Fe olivines is shown in Figure 14.

Unit Cell Parameters

Numerous studies have been made over the past 30 years of the variation of cell parameters along different compositional joins for the olivine group of minerals. The most notable of these include studies by Henriques (1957) $[Mg_2SiO_4-Fe_2SiO_4-MnSiO_4]$; Bradley *et al.* (1966) $[Mg_2SiO_4-LiMgPO_4]$; Fisher (1967), Louisnathan and Smith (1968), Matsui and Syono (1968), Fisher and Medaris (1969), Schwab and Kustner (1977), and Riekel and Weiss (1978) $[Mg_2SiO_4-Fe_2SiO_4]$; Wyderko and Mazanek (1968) $[Fe_2SiO_4-Ca_2SiO_4]$; Matsui and Syono (1968) $[Mg_2SiO_4-Co_2SiO_4$ and $Mg_2SiO_4-Ni_2SiO_4]$; Nishizawa and Matsui (1972) $[Mg_2SiO_4-Mn_2SiO_4]$; Syono *et al.* (1971) $[Mg_2SiO_4-Zn_2SiO_4$ at pressures ranging from 70 to 90 kbar at 1200°C]; and Warner and Luth (1973) $[Mg_2SiO_4-MgCaSiO_4]$.

A number of studies have also correlated d_{130} with composition for the forsterite-fayalite series (Yoder and Sahama, 1957; Fisher and Medaris, 1969; Schwab and Kustner, 1977). The regression equation of Schwab and Kustner relating mole fraction fayalite (X_{Fa}) to d_{130} (in A) is:

$$X_{Fa} = 7.522 - 14.9071 \ (3.0199 - d_{130})^{1/2}$$

For the forsterite-fayalite series, the variation of b with composition is linear, whereas the variations of a, c, and unit cell volume, V, and d_{130} show slightly positive deviations from linearity (Schwab and Kustner, 1977; Fisher and Medaris, 1969). The report of distinct breaks in slope in plots of a, b, c, and V versus composition for this series reported by Riekel and Weiss (1978) must be viewed with skepticism because of significant differences between their cell parameters and the general concensus values reported by all other workers for this series, particularly the high precision data of Schwab and Kustner (1977).

307

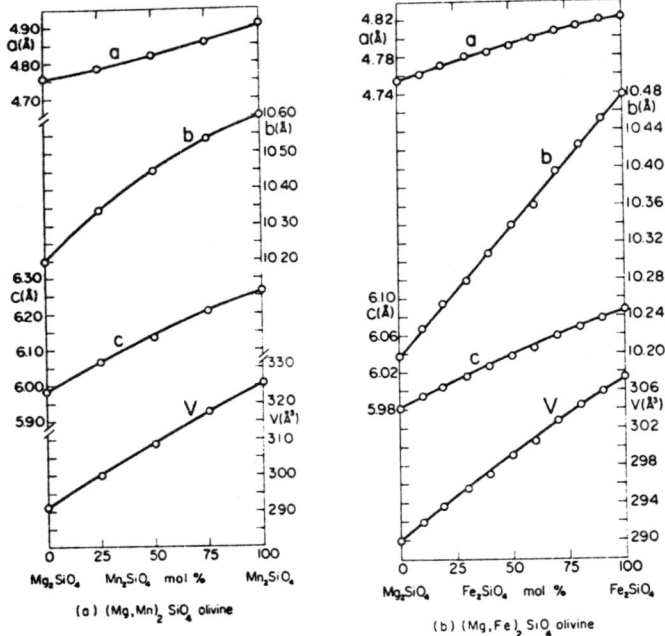

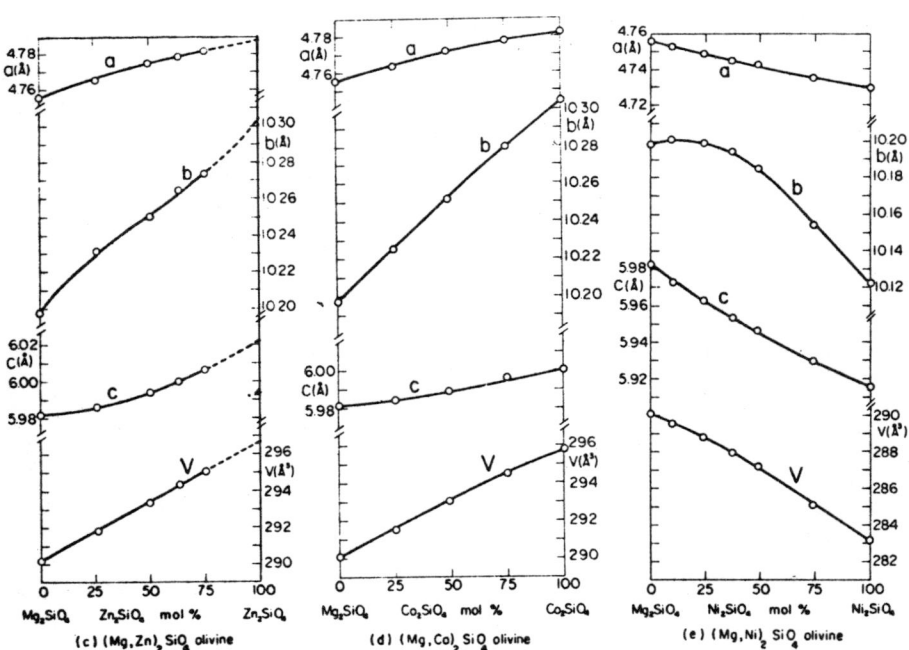

Figure 15. Variation in unit cell parameters for five olivine crystal-line solution series involving Mg. (From Akimoto *et al.*, 1976).

The variation of a, b, c, and V with composition (in mole percent) for the crystalline solution series $(Mg,Mn)_2SiO_4$, $(Mg,Fe)_2SiO_4$, $(Mg,Zn)_2SiO_4$, $(Mg,Co)_2SiO_4$, and $(Mg,Ni)_2SiO_4$ are shown in Figure 15. Significant curvature is seen in all plots but the one for the $(Mg,Fe)_2SiO_4$ series, indicating that Vegard's law is violated in these crystalline solutions. The non-ideal mixing indicated by these data is related to the ordering of octahedral cations in the M(1) and M(2) sites in these olivines. This subject will be discussed in detail in a later section.

As might be expected, an excellent correlation exists between unit cell volume and the size of cations (expressed in A^3) occupying the octahedral sites of olivine. Figure 16 illustrates this correlation for silicate olivines covering the entire composition range shown by natural samples. In preparing this plot, Brown (1970) used the cell parameters of Smith *et al.* (1965) for Ca_2SiO_4. These values have since been revised by Czaya (1971), although the figure has not been redrafted to include this revision. The corrected volume for pure calcio-olivine is 385.3 A^3. Brown's regression equation which relates unit cell volume to cations radius cubed (using Shannon and Prewitt (1969) values) is:

$$V = 188.32\ r_M^3 + 220.17$$

and has a correlation coefficient <0.99. This correlation does not include the unit cell volume of Ca_2SiO_4 and is therefore not affected by the cell parameter error mentioned above for calcio-olivine.

Cell parameters of the end member olivines at 24°C and 1 atm pressure are listed in Table 4.

STRUCTURE AND BONDING IN OLIVINES

Structural Response to Compositional Changes

Since the first crude refinement of the structure of forsterite (Belov *et al.*, 1951), there have been at least 55 modern structural refinements of olivines of different compositions at ambient P,T conditions (see Table 5). These data permit a number of conclusions to be drawn about the effects of composition on the olivine structure. A complete listing of structural data for the 11 olivines studied by Brown (1970) is contained in the Appendix in Tables A4 and A5.

309

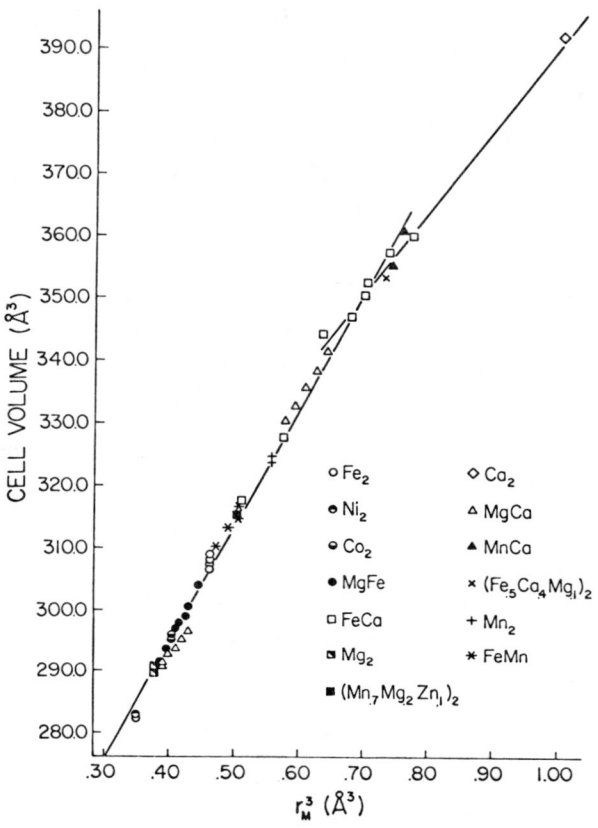

Figure 16. Variation of unit cell volume (A^3) with cation radius cubed. Radii values were taken from Shannon and Prewitt (1969). (From Brown, 1970).

Table 4. Cell Parameters of End-Member Olivines at 24°C and 1 Atm. Pressure*

Olivine	a (Å)	b (Å)	c (Å)	V (Å^3)	Reference
Mg_2SiO_4	4.7540(2)	10.1971(8)	5.9806(6)	289.92(6)	Schwab and Kustner (1977)
Fe_2SiO_4	4.8211(5)	10.4779(7)	6.0889(5)	307.58(8)	Schwab and Kustner (1977)
Mn_2SiO_4	4.904 (1)	10.601 (3)	6.259 (1)	325.4 (2)	Nishizawa and Matsui (1972)
Ca_2SiO_4	5.078 (2)	11.225 (3)	6.760 (2)	385.3	Czaya (1971)
Ni_2SiO_4	4.7274(5)	10.118 (3)	5.9105(8)	282.7 (1)	Brown (1970)
Co_2SiO_4	4.7811(7)	10.2998(9)	6.0004(4)	295.49(5)	Brown (1970)
$(Mg_{.5}Fe_{.5})_2SiO_4$	4.7929(2)	10.3412(3)	6.0380(2)	299.27(3)	Schwab and Kustner (1977)
$(Mg_{.5}Mn_{.5})_2SiO_4$	4.818 (1)	10.447 (2)	6.130 (1)	308.5 (1)	Nishizawa and Matsui (1972)
$(Mg_{.48}Ni_{.52})_2SiO_4$	4.7366(4)	10.1716(13)	5.9374(4)	286.06(4)	Rajamani et al. (1975)
$(Mg_{.5}Zn_{.5})_2SiO_4$	4.775 (1)	10.250 (3)	5.994 (2)	293.3 (3)	Syono et al. (1971)
$(Mg_{.5}Ca_{.5})_2SiO_4$	4.8209(5)	11.0911(9)	6.3726(6)	340.74(4)	Warner and Luth (1973)
$(Fe_{.51}Mn_{.47}Mg_{.02})_2SiO_4$	4.844 (1)	10.577 (4)	6.146 (2)	314.9 (2)	Brown (1970)
$(Fe_{.49}Ca_{.51})_2SiO_4$	4.892 (5)	11.180 (2)	6.469 (4)	353.8 (4)	Brown (1970)
$(Mn_{.5}Ca_{.5})_2SiO_4$	4.944 (4)	11.190 (10)	6.529 (5)	361.2 (5)	Caron et al. (1965)

*Numbers in parentheses are estimated standard errors (1 ô) and refer to the last decimal place.

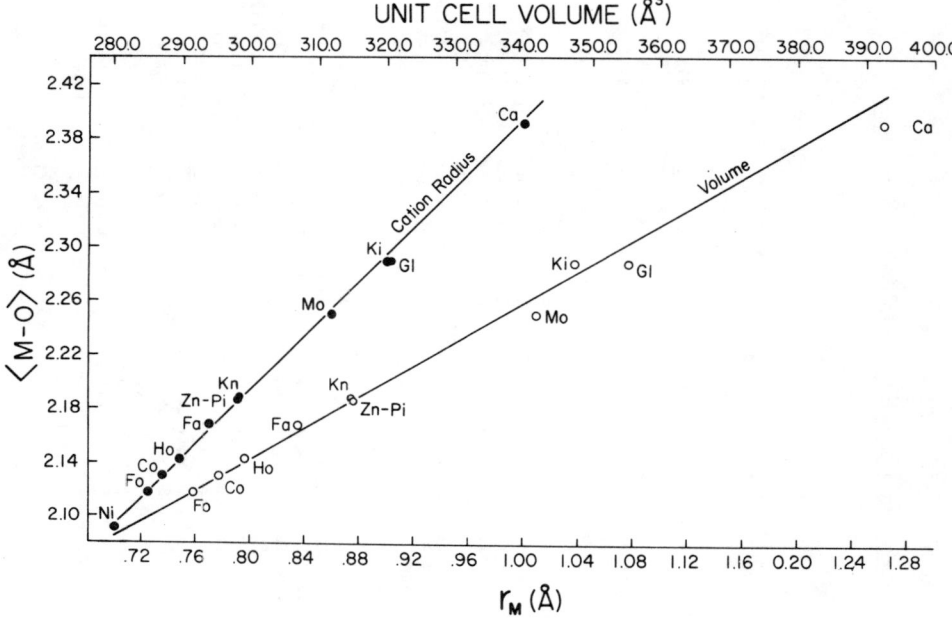

Figure 17. Variation of mean M-O distance with mean octahedral cation radius, r_M, and unit cell volume, V. Abbreviations for mineral names and values of r_M and V are given in Table A1. (From Brown, 1970).

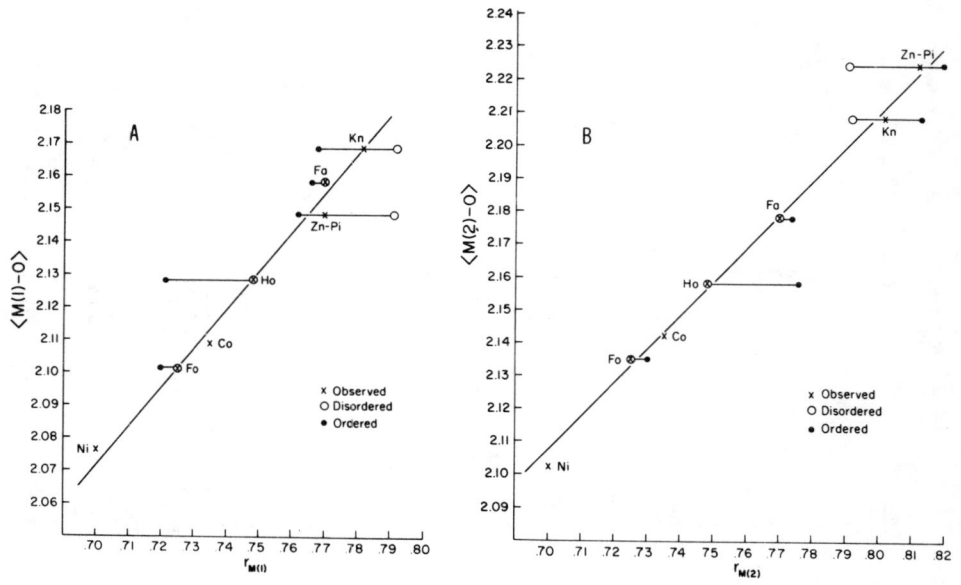

Figure 18. Variation of mean M(1)-O (A) and M(2)-O (B) distances with calculated radius of M(1) and M(2) cations for non-calcium olivines. Data for this plot are contained in Tables A1, A3, and A4 in appendix. (From Brown, 1970).

Table 5. Data for Refined Olivine Structures at 24°C and 1 Atm. Pressure.

FORSTERITE - FAYALITE SERIES

No.[1]	OLIVINE[2]	R-FACTOR[3]	<M - O>[4]	<Si - O>[4]	M(1) OCCUPANCY[5]	M(2) OCCUPANCY[5]	K_D[6]	REFERENCE
1(s)	Fo100	0.048	2.114	1.635	1.0 Mg	1.0 Mg	–	Smyth and Hazen (1973)
2(s)	Fo100	0.048	2.114(4)[9]	1.630	1.0 Mg	1.0 Mg	–	Hazen (1976)
3(t,m)	Fo99	0.023	2.1127(7)	1.6354(8)	.988 Mg + .012 Fe	.982 Mg + .018 Fe	0.66	Wenk and Raymond (1973)
4(t,p)	Fo93.5	0.026	2.120(1)	1.639(2)	.930 Mg + .070 Fe	.937 MG + .063 Fe	1.12	Basso et al. (1979)
5(c,p)	Fo92.5	0.018	2.119(2)	1.637(2)	.925 Mg + .075 Fe	.926 Mg + .074 Fe	1.01	Basso et al. (1979)
6(t,p)	Fo91.5	0.019	2.119(2)	1.636(2)	.915 Mg + .085 Fe	.915 Mg + .085 Fe	1.00	Basso et al. (1979)
7(t,p)	Fo91	0.022	2.119(2)	1.635(2)	.907 Mg + .093 Fe	.907 Mg + .093 Fe	1.00	Basso et al. (1979)
8(t,p)	Fo91	0.025	2.119(2)	1.636(2)	.910 Mg + .090 Fe	.910 Mg + .090 Fe	1.05	Basso et al. (1979)
9(t,p)	Fo91	0.023	2.121(2)	1.637(2)	.910 Mg + .090 Fe	.910 Mg + .090 Fe	1.00	Basso et al. (1979)
10(t,p)	Fo91	0.022	2.116(2)	1.633(2)	.904 Mg + .096 Fe	.914 Mg + .086 Fe	1.13	Basso et al. (1979)
11(t,p)	Fo90.5	0.025	2.119(2)	1.635(2)	.903 Mg + .097 Fe	.905 Mg + .095 Fe	1.02	Basso et al. (1979)
12(t,p)	Fo90.5	0.018	2.120(2)	1.637(2)	.900 Mg + .100 Fe	.912 Mg + .088 Fe	1.15	Basso et al. (1979)
13(t)	Fo90	0.071	2.124	1.631	—[7]	—[7]		Hanke (1965)
14(t,p)	Fo90	0.066	2.118(3)	1.637(2)	.90 Mg + .10 Fe	.90 Mg + .10 Fe	1.00	Birle et al. (1968)[8]
15(t,e)	Fo90	0.022	—[7]	—[7]	.888 Mg + .112 Fe	.896 Mg + .104 Fe	1.09	Will and Nover (1979)
16(t,e)	Fo90(QFM)[10]	0.022	—[7]	—[7]	.883 Mg + .117 Fe	.901 Mg + .099 Fe	1.21	Will and Nover (1979)
17(t,e)	Fo90(QFI)[11]	0.022	—[7]	—[7]	.903 Mg + .097 Fe	.881 Mg + .119 Fe	0.80	Will and Nover (1979)
18(t,p)	Fo89.5	0.021	2.117(2)	1.634(2)	.895 Mg + .105 Fe	.892 Mg + .108 Fe	0.97	Basso et al. (1979)
19(t,m)	Fo89	0.022	2.1189(7)	1.6372(8)	.895 Mg + .105 Fe	.892 Mg + .108 Fe	0.97	Wenk and Raymond (1973)
20(t,e)	Fo88	0.022	—[7]	—[7]	.861 Mg + .139 Fe	.867 Mg + .133 Fe	1.05	Will and Nover (1979)
21(t,e)	Fo88(QFM)[10]	0.022	—[7]	—[7]	.854 Mg + .146 Fe	.874 Mg + .126 Fe	1.19	Will and Nover (1979)
22(t,e)	Fo88(QFI)[11]	0.022	—[7]	—[7]	.862 Mg + .138 Fe	.866 Mg + .134 Fe	1.03	Will and Nover (1979)
23(t,i)	Fo82.5Fa17.2La1.3	0.038	—[7]	—[7]	.807 Mg + .192 Fe	.846 Mg + .147 Fe + .004 Ca	1.38	
24(1,e)	Fo82	0.023	2.126(1)	1.639(2)	.814 Mg + .186 Fe	.826 Mg + .164 Fe	1.15	Finger and Virgo (1971)
25(1,e)	Fo73	0.048	2.128(2)	1.634(2)	.729 Mg + .271 Fe	.734 Mg + .258 Fe	1.06	Brown and Prewitt (1973)
26(t,m)	Fo71	0.029	2.130(2)	1.637(2)	.708 Mg + .292 Fe	.712 Mg + .288 Fe	1.02	Finger (1970)
27(1,e)	Fo69.5	0.025	2.133(2)	1.638(2)	.675 Mg + .325 Fe	.705 Mg + .295 Fe	1.19	Brown and Prewitt (1973)
28(1,e)	Fo69.5	0.022	2.1322(6)	1.6362(6)	.626 Mg + .374 Fe	.648 Mg + .352 Fe	1.10	Brown and Prewitt (1973)
29(1,e)	Fo67.2	0.022	2.1293(7)	1.6353(7)	.666 Mg + .334 Fe	.678 Mg + .322 Fe	1.06	Wenk and Raymond (1973)
30(t,p)	Fo54Fa45Te1	0.08	2.140(3)	1.634(4)	—[7]	—[7]	—[7]	Birle et al. (1968)
31(1,i)	Fo49.7	0.026	2.144(1)	1.636(1)	.475 Mg + .525 Fe	.520 Mg + .480 Fe	1.20	Ghose et al. (1976)
32(t,i)	Fo49Fa49Te1La1	0.068	2.143(3)	1.636(3)	.50 Mg + .50 Fe	.48 Mg + .48 Fe + .02(Mn,Ca)	1.00	Birle et al. (1968)[8]
33(t,e)	Fo49	0.026	2.146(3)	1.634(3)	.477 Mg + .523 Fe	.503 Mg + .490 Fe	1.13	Finger (1970)
34(t,m)	Fo38	0.031	2.151	1.629	.361 Mg + .639 Fe	.389 Mg + .611 Fe	1.13	Smyth and Hazen (1973)
35(t,e)	Fo4Fa92Te4	0.045	2.168(2)	1.638(1)	.04 Mg + .92 Fe + .04 Mn	.04 Mg + .92 Fe + .04 Mn	1.00	Birle et al. (1968)[8]
36(t)	Fa100	0.075	2.172	1.633	1.0 Fe	1.0 Fe	0.0	Hanke (1965)
37(s)	Fa100	0.047	2.168(4)	1.628(4)	1.0 Fe	1.0 Fe	0.0	Smyth (1975)

FORSTERITE - FAYALITE - TEPHROITE SERIES

No.[1]	Composition[2]	R[3]	$\langle M(2)\text{-}O\rangle$[4]	$\langle M(1)\text{-}O\rangle$[4]	M(2) occupancy[5]	M(1) occupancy[5]	K_D[6]	Reference
38(s,1000°C)	$Fo_{53}Te_{47}$	0.059	—[7]	—[7]	.724 Mg + .276 Mn	.339 Mg + .661 Mn	0.20	Ghose and Weidner (1974)
39(t)	$Fo_{51}Te_{49}$	0.029	2.162(2)	—[7]	.921 Mg + .079 Mn	.110 Mg + .890 Mn	0.011	Francis and Ribbe (1981)
40(t)	Fo_9Te_{91}	0.038	2.206(2)	—[7]	.172 Mg + .828 Mn	1.0 Mn	0.00	Francis and Ribbe (1981)
41(t,m)	$Fo_{2.5}Fa_{51.5}Te_{46}$	0.032	2.188(2)	1.636(2)	.05 Mg + .66 Fe + .29 Mn	.37 Fe + .63 Mn +	0.26	Brown (1970)
42(t,m)	$Fo_{17.2}Fa_{6.4}Te_{65.0}Zn_{11.4}$[12]	0.039	2.186(2)	1.632(2)	.35 Mg + 0.4 Fe + .43Mn + .18 Zn[13]	.08 Fe + .87 Mn + .05 Zn	–	Brown (1970)

NICKEL - COBALT OLIVINES

No.[1]	Composition[2]	R[3]	$\langle M(2)\text{-}O\rangle$[4]	$\langle M(1)\text{-}O\rangle$[4]	M(2) occupancy[5]	M(1) occupancy[5]	K_D[6]	Reference
43(s)	Ni_2SiO_4	0.052	2.089(2)	1.640(2)	1.0 Ni	1.0 Ni	0.0	Brown (1970)
44(s)	Ni_2SiO_4	0.037	2.089(4)	1.639(5)	1.0 Ni	1.0Ni	0.0	Lager and Meagher (1978)
45(s)	$Ni_{1.03}Mg_{.97}SiO_4$	0.029	2.101(2)	1.638(2)	.767 Ni + .233 Mg	.263 Ni + .737 Mg	9.22	Rajamani et al. (1975)
46(s)	Co_2SiO_4	0.052	2.130(2)	1.636(2)	1.0 Co	1.0 Co	0.0	Brown (1970)
47(s)	Co_2SiO_4	0.046	2.134(4)	1.627(4)	1.0 Co	1.0 Co	0.0	Morimoto et al. (1974)
48(s)	$Co_{1.1}Mg_{.9}SiO_4$	0.044	2.124(3)	1.636(3)	.730Co + .270 Mg	.370 Co + .630 Mg	4.60	Ghose and Wan (1974)

Ca - RICH OLIVINES

No.[1]	Composition[2]	R[3]	$\langle M(2)\text{-}O\rangle$[4]	$\langle M(1)\text{-}O\rangle$[4]	M(2) occupancy[5]	M(1) occupancy[5]	K_D[6]	Reference
49(t,m)	$MgCaSiO_4$	0.037	2.250(2)	1.641(2)	1.0 Mg	1.0 Ca	0.0	Onken (1965)[8]
50(t,m)	$Mg_{.93}Fe_{.07}Ca_{1.0}SiO_4$	0.030	2.248(4)	1.637(4)	.93 Mg + .07 Fe	1.0 Ca	0.0	Lager and Meagher (1978)
51(t,m)	$Mg_{.09}Mn_{.93}Ca_{.98}SiO_4$	0.050	2.288(2)	1.635(2)	.09 Mg + .91 Mn	.02 Mn + .98 Ca	0.0	Brown (1970)
52(t,m)	$Mg_{.10}Mn_{.87}Zn_{.05}Ca_{.98}SiO_4$	0.024	2.288(4)	1.640(6)	.10 Mg + .85 Mn + .052 Zn	.02 Mn + .98 Ca	0.0	Lager and Meagher (1978)
53(s)	$Fe_{.87}Ca_{1.13}SiO_4$	0.106	2.289(3)	1.629(3)	.85 Fe + .15 Ca	.02 Fe + .98 Ca	0.004	Brown (1970)
54(s)	Ca_2SiO_4	0.054	2.391(5)	1.646(3)	1.0 Ca	1.0 Ca	0.0	Smith et al. (1965)[8]
55(s)	Ca_2SiO_4	0.07	2.369	1.645	1.0 Ca	1.0 Ca	0.0	Czaya (1971)

---

[1] The symbols in parentheses refer to the olivine's paragenesis. (s) indicates synthetic, (t,) indicates terrestrial, (l,) indicates lunar, (,m) indicates metamorphic, (,p) indicates plutonic, (,i) indicates hypabyssal, (,e) indicates extrusive volcanic.

[2] Composition is given in terms of Fo content for Fo-Fa olivines. When other components are present in significant amounts, they are listed.

[3] R-factor is defined as $R = (\Sigma|Fo|-|Fa|)/\Sigma|Fo|$ and can be thought of as a reliability index. Generally, the lower the R-factor, the more reliable the structure refinement.

[4] $\langle M\text{-}O\rangle$ and $\langle Si\text{-}O\rangle$ indicate mean M-O and Si-O bond lengths in Angstroms.

[5] Occupancy values were determined in most cases by direct refinement of occupancy parameters using explicit chemical constraints (see Finger, 1969).

[6] K_D, a distribution coefficient, is defined as $K_D = [Mg(M2)Fe(M1)]/[Mg(M1)Fe(M2)]$

[7] Not reported.

[8] These values come from Brown's (1970) re-refinement of data from the source cited.

[9] Number in parentheses indicates estimated standard error in the least significant digit quoted.

[10] Indicates that the olivine was heat treated at 750° at the $p(O_2)$ fixed by the quartz (Q)-fayalite(F)-magnetic (M) buffer $P(O_2) = 10^{-16}$ bar .

[11] Indicates that the olivine was heat treated at 750° at the $p(O_2)$ fixed by the quartz (Q)-fayalite(F)-iron(I) buffer $P(O_2) = 10^{-21}$ bar .

[12] $Zn_{11.4}$ indicates that 11.4 mole percent Zn_2SiO_4 is present as a component in this olivine.

[13] The site occupancies in this case with four species distributed over four sites are not uniquely determinable. Mg was assumed fixed in M(1). (Mn+Fe) was treated as one species and refined against Zn.

(1)   The mean octahedral cation-oxygen distance, <M-O>, increases regularly as the size of the octahedral cation increases. Figure 17 demonstrates the near linear dependence of <M-O> on octahedral cation radius for the olivines studied by Brown (1970). The increase in <M-O> with increasing r_M also manifests itself in an increase in unit cell volume. Similarly, <M(1)-O> and <M(2)-O> increase almost linearly as the size of the M(1) and M(2) cations increase, as is shown in Figure 18 for the non-calcium olivines studied by Brown (1970). Also shown in this figure are the effects of octahedral cation order and disorder on the <M(1)-O> and <M(2)-O> distances. In certain cases where the difference in size of the two major octahedral cations is relatively large (*e.g.* Mg and Fe) and more than 10 percent of each is present, mean M-O distance plotted against octahedral cation radius is sensitive to site occupancy.

(2)   No correlation can be made between individual or mean Si-O distance and octahedral cation size. The value of <Si-O> for Ni-olivine, which has the smallest octahedral cation, is 1.640 A, whereas <Si-O> reported for Ca_2SiO_4, which has the largest cation, is only 1.645 A. This difference is probably well within the $2\hat{\sigma}$ limit.

(3)   <Si-O> is amazingly constant. For the 45 olivines listed in Table 5 with reported Si-O distances, the grand mean Si-O distance is 1.636 A, with a standard deviation of $\pm$0.004 A. This deviation is roughly twice the $1\hat{\sigma}$ least-squares precision level reported for most of these distances. This finding is somewhat surprising in view of the range of cation properties exhibited by octahedral cations in silicate olivines. For example, the relatively electronegative Ni^{2+} cation ($\chi = 1.9$) has about the same effect on individual Si-O distances as does the least electronegative cation considered (Ca: $\chi = 1.0$).

(4)   Shared polyhedral edges, O$\cdots$O, are shorter than unshared edges in all olivines refined to date. Table A4 in the Appendix lists individual O$\cdots$O lengths for the three unique polyhedra in the 11 olivines studied by Brown (1970). Individual distances in this table and the angles listed in Table A5 are referenced to Figures 19 and 20, which show the two possible types of shared polyhedral edges in the olivine structure.

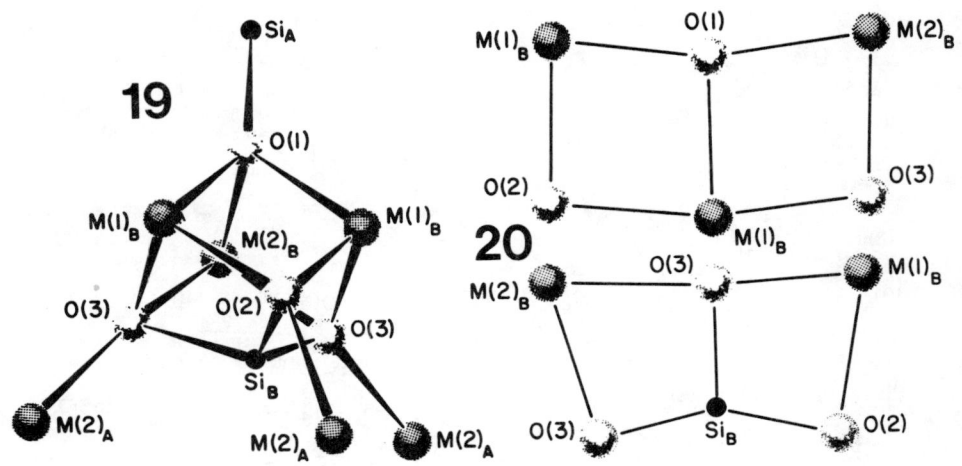

Figure 19 (left). Perspective view of a portion of the olivine structure illustrating oxygen ligation to M(1), M(2) and Si with atoms labelled to conform to the designations in Table A4 and A5 in the Appendix.

Figure 20 (right). Portions of the distorted cube shown in the center of Figure 19. These are "flattened" to illustrate angular distortions from ideal hcp geometry due to cation - cation repulsion across shared edges [*cf*. Fig. 6b in the chapter on humites (Ribbe, this volume) and see Fig. 21].

(5) Mean O-O distances for each structure studied by Brown (1970) and listed in Tables A1 through A5 correlate linearly with $\langle M\text{-}O\rangle$, $\langle M(1)\text{-}O\rangle$, and $\langle M(2)\text{-}O\rangle$.

(6) No clear correlation exists between individual angles and cation size, although in some cases the Ca and non-Ca olivines comprise two distinct populations in plots of individual angle versus cation radius.

(7) Comparison of bond angles from the refined structures detailed in Table A5 with the angles expected in the ideal hexagonal closest-packed structure (also listed in Table A5) shows that deviations from the ideal values (referred to here as "bond angle strain") are greatest for angles such as $O(3)\text{-}M(2)_B\text{-}O(3)$ (see Fig. 20), which are opposite shared octahedral-tetrahedral edges, are less for angles such as $O(2)\text{-}M(1)_B\text{-}O(1)$, which are opposite shared octahedral-octahedral edges, and are least for angles not involved in shared edges. For example, the "bond angle strain" at $M(1)_B$ and $M(2)_B$ is two to four times larger when an O-O edge is shared with a Si-tetrahedron than when shared with another octahedron. In all cases, strain at oxygens is positive, whereas strain at cations is negative, in agreement with the commonly held notion that cation-cation

315

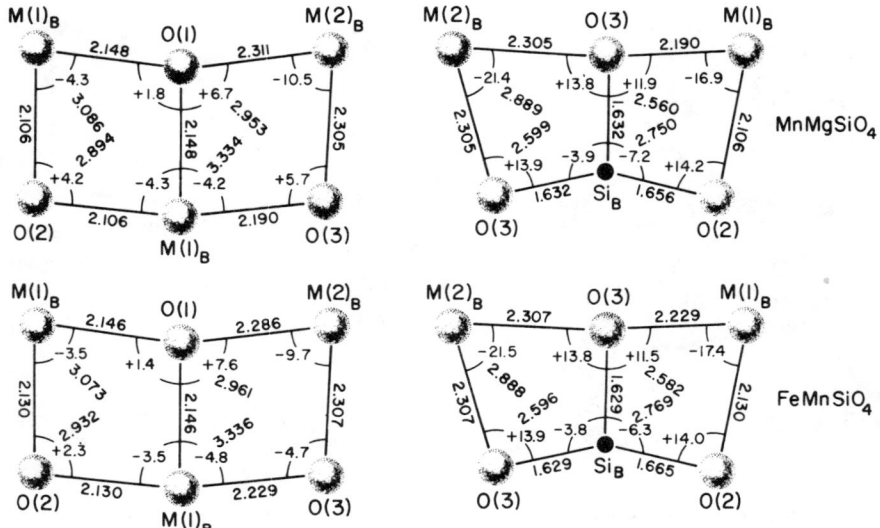

Figure 21. Planar projections of the unique M-O-M-O and M-O-Si-O faces of the distorted cube of nearest neighbor cations and oxygens in a zinc picrotephroite and knebelite, showing "bond angle strain" values (in degrees) and bond distances (in Å). (From Brown, 1970).

interactions are generally greater than anion-anion interactions. Figure 21 shows indiviudal "bond angle strain values" for two of the olivine structures studied by Brown (1970).

(8) Bond angle strains (as defined above) vary with cation size as shown in Figure 22. This figure plots "strain" at the M(1)(A) and M(2)(B) octahedral cations versus octahedral cation radius. The smallest strains are found in structures with the smallest octahedral cations. Three populations are apparent -in this figure, corresponding to the disordered, non-Ca olivines; the ordered, non-Ca olivines; and the Ca olivines studied by Brown (1970).

The structural data for Ca_2SiO_4 in Tables A4 and A5 should be replaced by the data of Czaya (1971) in future correlations. The Czaya structure exhibits a considerably less distorted tetrahedron than does the Smith *et al.* (1965) structure or Brown's (1970) re-refinement of Smith's data. However, the use of the older Ca_2SiO_4 data does not affect any of the conclusions reached above.

Octahedral distortions. Over the past ten years, various attempts have been made to quantify the distortion of polyhedra, such as the

316

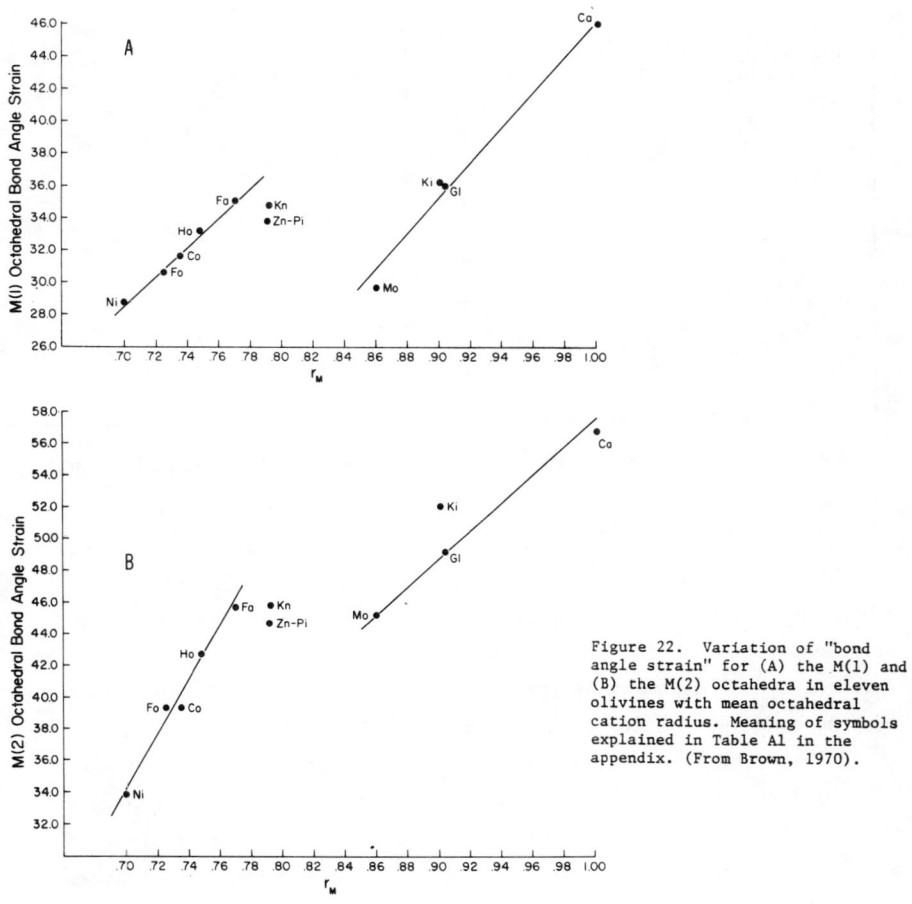

Figure 22. Variation of "bond angle strain" for (A) the M(1) and (B) the M(2) octahedra in eleven olivines with mean octahedral cation radius. Meaning of symbols explained in Table A1 in the appendix. (From Brown, 1970).

octahedra in olivines, because a knowledge of such distortions is impor- tant in rationalizing intracrystalline cation partitioning as well as the partitioning behavior of cations such as Cu^{2+} and Cr^{2+} between melts and crystals. In olivines, the reduction from O_h to approximate D_{4h} site symmetry in the M(1) octahedron can be thought of as arising from a lengthening of the M(1)-O(3) bonds relative to the approximately equal M(1)-O(1) and M(2)-O(2) bonds. The M(1)-O(3) bond length is, therefore, a crude measure of M(1) octahedral distortion and increases from Ni oli- vine to Fe olivine smoothly. A comparison of relative distortions of the M(2) octahedron is more difficult if we consider only M(2)-O dis- tances and the C_{3v} local symmetry approximation. A true measure of distortion should involve both distance and angle factors.

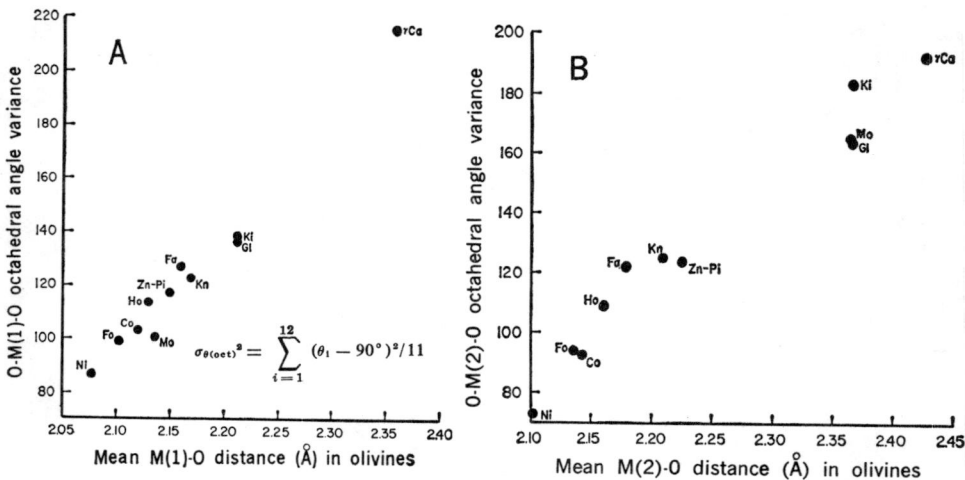

Figure 23. Plot of octahedral angle variance $[\sigma_\theta(\text{oct})^2 = \Sigma\ (\theta_i - 90°)^2/11$; where the sum is over the 12 unique O-M-O angles] *versus* mean $\bar{M}(1)$-O and $M(2)$-O distances in the olivines studied by Brown (1970). The symbols are explained in Table A1. (From Robinson *et al.*, 1971).

One such measure of distortion is the angle variance introduced by Robinson *et al.* (1971). They analyzed octahedral distortions in the olivine structures studied by Brown (1970) using this parameter and found that octahedral distortions correlate in an approximately linear fashion with mean M(1)-O and M(2)-O distances, i.e., with increasing cation size (Fig. 23). They also found that the M(1) octahedron in these olivines is more distorted (has a larger angle variance value) than the M(2) octahedron (see Fig. 23), a finding opposite to that reached by Brown (1970) who used a much cruder distortion index. Other measures of octahedral distortion for the olivines have been proposed by Dollase (1974), Fleet (1974), Vincent *et al.* (1976), and Gaite (1980) and give essentially the same result. The effects of differences in distortion between the M(1) and M(2) sites in olivines on crystal field stabilization energy of Fe^{2+} have been considered by Walsh *et al.* (1974).

Computer simulation of the olivine structure. Useful insights about the olivine structure have been provided by computer simulation of the structure using the distance least-squares (DLS) method. This method permits the simulation of different structures under different P,T conditions not easily attained in the laboratory and has, therefore, been used to predict high pressure phases of olivine for the most part (Baur, 1972; Dempsey and Strens, 1976). Baur (1972) concluded from an ambient

318

condition DLS simulation of olivine that the shared edge shortening is caused primarily by a dimensional misfit between octahedra and tetrahedra, and that metal-metal repulsions of the type proposed by Pauling's third rule are only partly responsible. McLarnan *et al.* (1979) also carried out DLS calculations on the forsterite structure together with semi-empirical molecular orbital (CNDO/2) calculations on a tetrahedral-octahedral cluster modeled after the arrangement in forsterite. Their conclusions, which contradict Baur's, suggest that distortions in the olivine structure are due to short range forces of both the covalent and electrostatic type.

It is fair to state that we still do not really understand why cations such as Mg, Fe, Mn, Ca, and Si and the oxygen anion adopt the olivine structure under certain P,T conditions. Although clear correlations can be made between polyhedral distortions in olivine-type compounds and the relative size differences of octahedra and tetrahedra (Vincent *et al.*, 1976), it is misleading to state that steric details are "caused" by dimensional mismatch between polyhedra. One might also be tempted to conclude from the correlations of structural details with cation size that the latter is a causative factor in determining steric details. However, cation size is a poorly understood concept which embodies many of the physical factors we are presently ignorant about. Clearly, a more fundamental reason for the distortions in olivine-type structures must be sought.

The structural data base for olivines at ambient P,T conditions appears to be adequate in terms of quality and compositional range for the types of structural correlations made above. In the future, more complete data sets and more accurate data on end-member, synthetic olivines are needed to advance our understanding of the bonding forces that are responsible for the atomic arrangement adopted by compounds of olivine stoichiometry and for olivine's stability (or metastability) over a relatively wide range of P,T conditions. Such studies aimed at determining the valence electron density distribution in olivines and other compounds are just beginning in several mineralogical laboratories in the United States.

A discussion of intracrystalline cation ordering in olivines will follow in a later section.

Table 6. Linear thermal expansion coefficients (°C^{-1} x 10^5) of mean M-O distances and unit cell parameters in six olivines (After Lager and Meagher, 1978)

Olivine	$\langle M(1)-O\rangle$	$\langle M(2)-O\rangle$	a	b	c
Ni Olivine	1.54(2)	1.31(10)	1.$\overline{18}$(5)	1.$\overline{09}$(4)	1.$\overline{11}$(3)
Fo(100)	1.93(24)	1.72(8)	0.87(11)	1.54(2)	1.33(2)
Fo(69)	1.17(2)	1.22(7)	0.61(1)	0.96(7)	0.97(10)
Fa(100)	1.26(13)	1.50(9)	0.99(8)	0.95(9)	1.19(9)
Monticellite	1.74(5)	1.33(9)	1.01(8)	0.99(2)	1.13(4)
Glaucochroite	1.54(20)	1.52(4)	0.87(11)	1.02(3)	1.45(8)

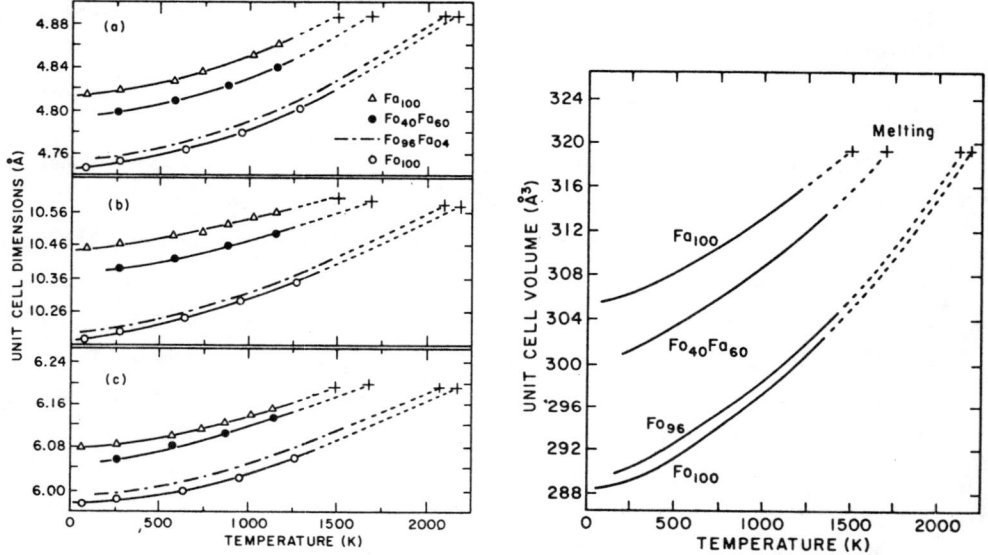

Figure 24. Unit cell dimensions and unit cell volume *versus* temperature for several Mg - Fe olivines. (From Hazen, 1977a).

Structural Response to Temperature Change

High temperature structural studies of olivine were begun over 40 years ago with the measurement of thermal expansion data (Kozu *et al.*, 1934; Rigby *et al.*, 1945, 1946; Skinner, 1962; Singh and Simmons, 1976). Within the last ten years, development of high temperature furnaces for use with x-ray single-crystal diffractometers has permitted the study of structural details of olivine at temperatures up to 1000°C. Olivines studied by this method include the following: Fa$_{31}$ (Brown and Prewitt, 1973); Fo$_{100}$ (Smyth and Hazen, 1972); Fo$_{38}$Fa$_{55}$Te$_7$ (Smyth and Hazen, 1973; also see revision by Hazen, 1977a); Fa$_{100}$ (Smyth, 1975); Fo$_{100}$ (Hazen, 1976); Ni$_2$SiO$_4$, Ca$_{1.0}$Mg$_{0.93}$Fe$_{0.07}$SiO$_4$ (monticellite), and Ca$_{0.98}$Mn$_{0.87}$Mg$_{0.10}$Zn$_{0.05}$SiO$_4$ (glaucochroite) (Lager and Meagher, 1978).

Unit cell expansion. The a, b, and c cell parameters and unit cell volume, V, of Fe-Mg olivines expand in smooth though non-linear fashion with increasing temperature (Fig. 24). Mg/Fe ratio appears to have little effect on the linear or volume thermal expansion of ferromagnesian olivines (Hazen, 1977a). Thermal expansion curves for olivines of other compositions are shown in Figure 25, and linear thermal expansion coefficients of mean M-O distances and unit cell parameters for six olivines of various compositions are given in Table 6.

Structural expansion. All of the high-temperature structural studies of olivines yielded consistent results which may be summarized as follows:

(1) Mean and individual Si-O distances remain essentially constant with increasing temperature. Thus compositional and temperature changes have no effect on Si-O bond length in the olivines. This finding is consistent with the results of other high-temperature studies of silicates which show no significant increases in Si-O distance with increasing T (see Hazen and Prewitt, 1977).

(2) Mean M(1)-O and M(2)-O distances increase with increasing temperature, with Mg-O distances exhibiting the greatest thermal expansion (see Fig. 26a).

(3) The M(2) cation position in olivines is temperature dependent, with greatest displacement found for non-transition metal cations (Lager and Meagher, 1978).

(4) Octahedral distortions increase with increasing temperature, with the M(1) octahedron remaining more distorted than the M(2) octahedron (Brown and Prewitt, 1973; Smyth, 1975; Smyth and Hazen, 1973; Lager and Meagher, 1978).

With the exception of (3) above, these conclusions are essentially the same as those reached in the last section on the effect of composition on structure. This observation led Hazen (1977b)[*] to conclude that temperature and composition are structurally analogous variables.

Structural view of olivine melting. Forsterite melts congruently at 1890°C whereas fayalite melts at 1205°C, both at 1 atm pressure. One common way of structurally rationalizing the lower melting temperature of fayalite is by noting that M-O bonds are longer in fayalite (2.168 A)
....
[*]Hazen, R.M. (1977b) Temperature, pressure and composition: structurally analogous variables.
     Phys. Chem. Minerals, 1, 83-94.

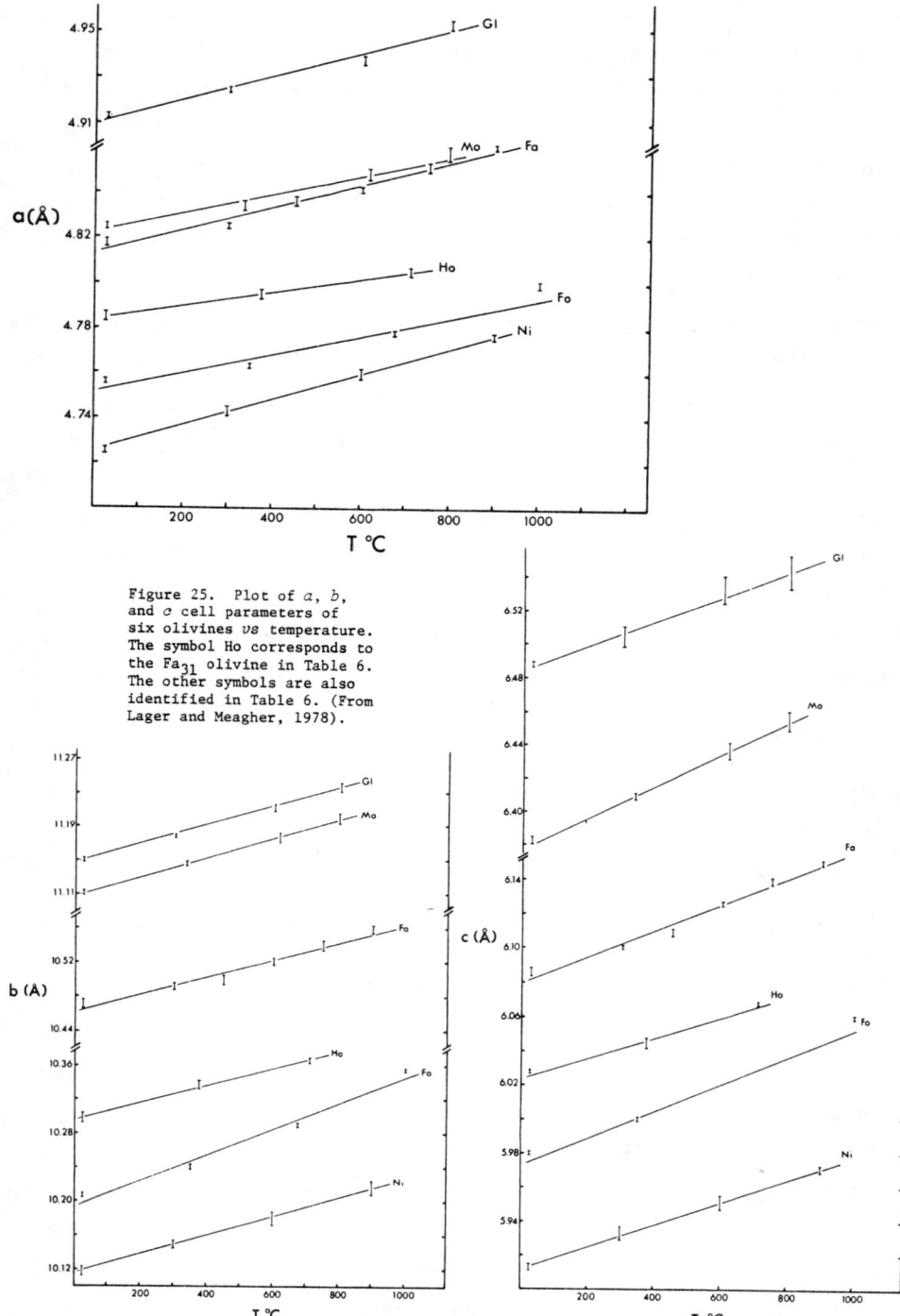

Figure 25. Plot of a, b, and c cell parameters of six olivines *vs* temperature. The symbol Ho corresponds to the Fa$_{31}$ olivine in Table 6. The other symbols are also identified in Table 6. (From Lager and Meagher, 1978).

322

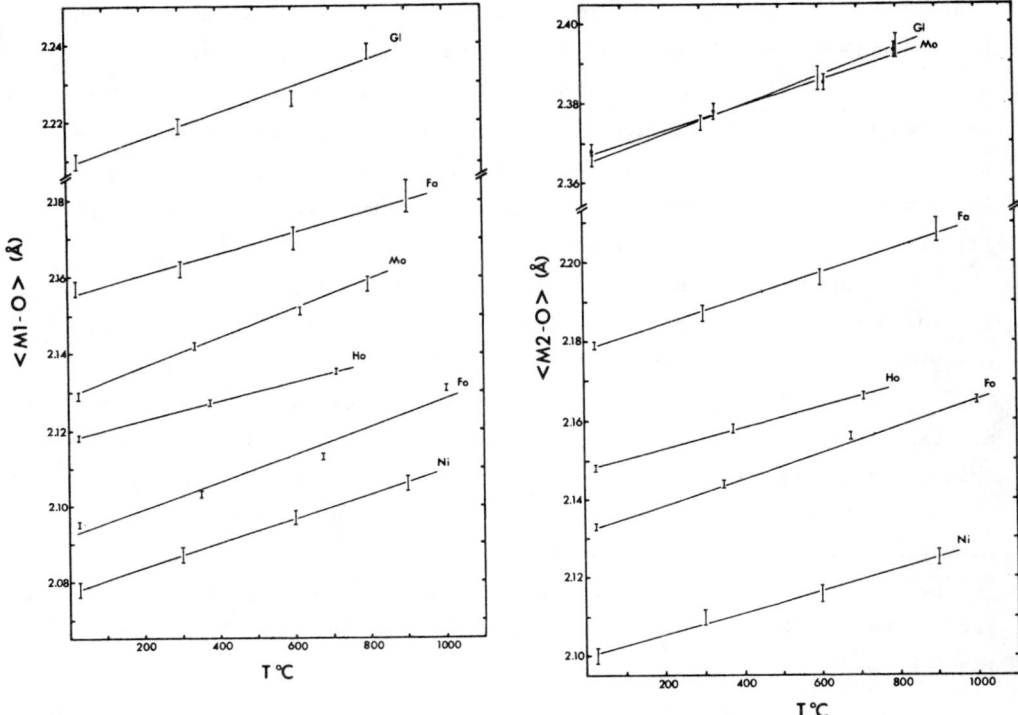

Figure 26a. Plots of mean M(1)-O and M(2)-O distances *versus* temperature for the six olivines listed in Table 6. (From Lager and Meagher, 1978).

than in forsterite (2.114 A), which manifests itself in larger unit cell parameters for fayalite. A rule of thumb concerning dependence of physical properties on structure and composition is that the weakest bonds tend to dictate physical properties such as hardness, melting point, etc. In the olivine structure, the M-O bonds are the weakest cation-oxygen bonds, and they probably break before the stronger Si-O bonds at the melting point. If we make the gross assumption that the longer Fe-O bonds in fayalite are weaker than the shorter Mg-O bonds in forsterite, and further assume that all other factors are essentially the same (which is most likely not the case), then the lower melting point of fayalite can be rationalized.

A more satisfying structural rationalization of olivine melting was recently presented by Hazen (1977) who extrapolated the cell parameters and M-O distances of ferromagnesian olivines to their melting points and found that they all have similar cell parameters (a = 4.89, b = 10.6, c = 6.19 A), cell volume ($V \sim 319$ A^3), <M(1)-O> (2.19 A) and <M(2)-O> (2.22 A)

distances at their respective melting temperatures (see Fig. 24). Hazen
concluded that the solidus of ferromagnesian olivines represents an isostructural line above which olivine of a given composition is not stable
with respect to a more Fe-rich olivine melt (except at end-member compositions). Hazen speculated that "perhaps due to misfit of expanding octahedra with rigid tetrahedra, the solidus structure represents a critical
structural limit for ferromagnesian olivines." It is interesting to note
that these upper distance limits for ferromagnesian olivines at their
melting point are exceeded in olivines at room temperature only when Ca
occupies the M(2) site. Both the <M(1)-O> and <M(2)-O> distances in
Ca_2SiO_4 exceed these limits yet the silicon tetrahedron is of the same
dimensions in this structure as in all other silicate olivines. Following Hazen's reasoning, one might speculate that a contributing factor
to the rarity of calcio-olivine in nature is the apparent dimensional
mismatch between the rigid, small tetrahedron, and the more pliable,
"overstuffed" Ca octahedra. However, we cannot consider these speculations to be anything but manifestations of more fundamental physical
reasons.

Ca_2SiO_4 polymorphism. The common olivines undergo no polymorphic
changes at low pressures before they melt. However, the calcio-olivine
Ca_2SiO_4 undergoes a complex series of structural changes before melting
as is shown below:

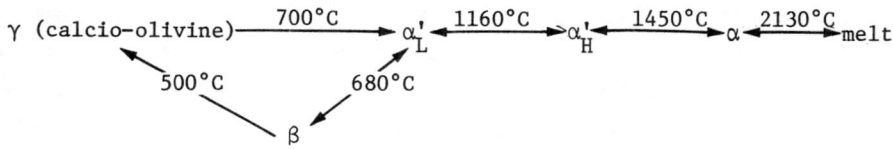

The calcio-olivine structure is given the Greek prefix γ in most
literature. This practice does not follow the standard U. S. convention
of labeling the lowest temperature polymorph α, the next highest temperature polymorph β, etc. Furthermore, this system of nomenclature con-
flicts with the usage of the prefixes α and γ in this chapter and in
the geophysical literature where γ indicates the spinel phase and α indicates the olivine phase. Larnite, the β-form of Ca_2SiO_4, is metastably
formed as the α' phase is cooled below about 700°C because of the closer
structural similarity between the α' and β forms than between the α' and

324

γ forms (Smith et $al.$, 1965). This metastable persistence of the hydratable β form is fortunate for all of us in a very practical sense because the β phase must be preserved in the manufacturing of cement clinker; when the β phase reverts to the olivine form in this clinker, it is quite inert and constitutes a failure.

Eysel and Hahn (1970) present a lucid review of Ca_2SiO_4 and Ca_2GeO_4 polymorphism. Other pertinent literature on the polymorphism of Ca_2SiO_4 includes papers by Bredig (1950), Foster (1968), Forest (1971), Kazak et $al.$ (1975), and Ghosh et $al.$ (1979). The last paper places emphasis on the uses of Ca_2SiO_4 polymorphs in the cement industry.

As pointed out in an earlier section on minor element chemistry, calcio-olivine, larnite (β-phase), and bredigite (α' phase) can coexist in nature.

Structural Response to Pressure Changes

Considerably less work has been done on the effects of pressure on the olivine structure because high pressure single-crystal x-ray or spectroscopic experiments are typically an order of magnitude more difficult than room or high-temperature experiments. Nonetheless, high pressure cells developed during the past decade at various labs in the U. S. have permitted olivine structural data to be taken for synthetic forsterite (Hazen, 1976: 50 kbar) and synthetic fayalite (Hazen, 1977a: 31 and 42 kbar). In addition, the effects of pressure on the cell parameters of powdered peridot (Olinger and Duba, 1971; Schock et $al.$, 1972; Olinger and Halleck, 1974) and synthetic fayalite (Yagi et $al.$, 1975; also see Adams, 1931) were determined. The high-pressure cell parameters for forsterite and fayalite are listed in Tables 7 and 8, respectively.

Table 7. Forsterite (Fo_{100}) unit cell parameters at different pressures (T = 23°C). Numbers in parentheses are standard errors referring to the last decimal place. (From Hazen, 1976).

P	a(Å)	b(Å)	c(Å)	V(Å³)
1 atm.	4.7535(4)	10.1943(5)	5.9807(4)	289.80(5)
20 kbar	4.743 (5)	10.09 (1)	5.954 (6)	285.0 (5)
40 kbar	4.734 (5)	10.02 (1)	5.940 (6)	281.8 (5)
50 kbar	4.712 (5)	9.97 (1)	5.955 (6)	279.7 (5)
1 atm.*	4.749 (5)	10.19 (1)	5.980 (6)	289.5 (5)

*
After high pressure experiments.

Table 8. Fayalite (Fa$_{100}$) unit cell parameters at different
pressures (T = 23°C). Numbers in parentheses are standard
errors referring to last decimal place. (From Yagi *et al.*, 1975).

P (kbar)	Fe$_2$SiO$_4$ olivine							
	V (A^3)	V/V$_0$	a (A)	a/a$_0$	b (A)	b/b$_0$	c (A)	c/c$_0$
0	307.9(7)	1.0000	4.817(6)	1.0000	10.49(1)	1.0000	6.091(8)	1.0000
19	303.2(15)	0.9848	4.806(8)	0.9977	10.43(2)	0.9934	6.052(11)	0.9936
33	300.5(3)	0.9761	4.771(1)	0.9905	10.37(1)	0.9881	6.073(2)	0.9970
44	297.9(11)	0.9678	4.772(6)	0.9907	10.33(1)	0.9847	6.042(8)	0.9920
53	296.7(10)	0.9638	4.776(5)	0.9915	10.31(1)	0.9824	6.026(7)	0.9893
23	301.5(23)	0.9793	4.803(13)	0.9971	10.36(2)	0.9875	6.057(17)	0.9944
42	298.5(19)	0.9696	4.788(10)	0.9940	10.32(2)	0.9838	6.038(14)	0.9913
60	295.7(7)	0.9604	4.769(4)	0.9900	10.28(1)	0.9798	6.030(5)	0.9900
71	293.0(15)	0.9518	4.762(8)	0.9886	10.25(2)	0.9764	6.005(11)	0.9859
29	301.6(15)	0.9796	4.786(8)	0.9936	10.38(2)	0.9892	6.071(11)	0.9967
47	296.7(25)	0.9637	4.796(14)	0.9956	10.29(3)	0.9807	6.011(18)	0.9869
61	294.7(16)	0.9572	4.770(9)	0.9902	10.26(2)	0.9772	6.024(12)	0.9890
73	292.5(17)	0.9500	4.750(9)	0.9861	10.23(2)	0.9746	6.020(13)	0.9883
62	293.8(16)	0.9543	4.767(8)	0.9896	10.25(2)	0.9767	6.012(10)	0.9870

In view of the difficulty in obtaining these data, it is not surprising
that major discrepancies exist between fayalite cell parameters of Yagi
et al. (1975) (taken at 44 kbar) and those reported by Hazen (1977a)
(taken at 42 kbar). Although data are limited, the following conclusions
have been made by Hazen (1977a) on the effects of pressure on ferromagnesian olivines:

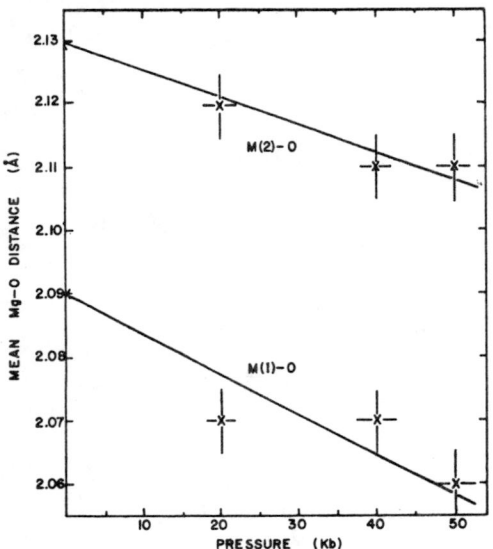

Figure 26b. Variation of mean M(1)-O and
M(2)-O distances in forsterite with increasing pressure. (From Hazen, 1976).

(1) The [SiO$_4$] tetrahedra in forsterite and fayalite are essentially constant in dimensions
with increasing pressure, i.e.,
tetrahedra are very incompres-
sible.

(2) The M(1) and M(2) octahedra of
forsterite and fayalite undergo
significant compressions with
increasing pressure, as shown
in Figure 26b for forsterite.

(3) Pressure appears to have little
effect on octahedral distortions
in forsterite. However, more
accurate data are needed to
confirm this conclusion and to

make more detailed conclusions on the effect of pressure on the olivine structure.

The unit cell volume compression of forsterite and fayalite can be rationalized structurally by considering the reduction in volume of filled and empty octahedra. In the case of forsterite a total volume reduction of about 10 A^3 occurs at 50 kbar pressure. Hazen's (1976) study showed that the M(1) and M(2) octahedra undergo volume reductions of about 0.7 and 0.5 A^3, respectively, at 50 kbar. There are four M(1) and four M(2) octahedra per unit cell. Thus, about 4.8 A^3 of the unit cell compression of forsterite can be accounted for by contraction of filled octahedra. The remaining 5.2 A^3 of contraction can be accounted for by compression of octahedral and tetrahedral voids in the olivine structure. A similar explanation of the unit cell expansion of forsterite with heating has been made by Hazen (1976) who found that about 30 percent of the 13 A^3 expansion over the range -196 to 1000°C is accounted for by M(1) and M(2) octahedral expansion, with the remaining 70 percent due to volume expansion of unoccupied octahedral and tetrahedral sites in the approximately hexagonal closest-packed array of oxygens.

Using polyhedral compression data from high-pressure structural studies such as those by Hazen on olivine, the bulk modulus of individual polyhedra in a variety of structure types has been related to polyhedral volumes (Hazen and Prewitt, 1977; Hazen and Finger, 1979). The bulk modulus (K_p) of [SiO_4] tetrahedra in olivines is on the order of 2.5 Mbar or greater whereas K_p for [MgO_6] octahedra is 1.5 Mbar (Hazen and Finger, 1979). A K_p value of 1.8 Mbar for [FeO_6] octahedra in fayalite was calculated using the inverse relationship between bulk modulus of a polyhedron and polyhedral volume compressibility (β_v) ($K_p = 1/\beta_v$: Hazen, 1976) and the β_v value for [FeO_6] octahedra (0.56 $Mbar^{-1}$) from Hazen and Finger (1977). As pointed out by these authors, the compressibility of polyhedra in silicate structures is greater than the study by Hazen and Prewitt (1977) indicates. The polyhedral bulk moduli for olivines are larger than the macroscopic bulk moduli of forsterite (1.35 Mbar: Hazen, 1976) and fayalite (1.24 Mbar: Yagi *et al.*, 1975), indicating that compression of vacant polyhedra makes an important contribution to the bulk compression of olivines.

Hazen (1977a) combined his unit cell data for forsterite and fayalite
as a function of temperature and pressure to construct the following equation of state for ferromagnesian olivines:

$$V = (290 + 0.17X_{Fa} + 0.006T + 0.000006T^2)[1 - P/(1350. - 0.16T]$$

where V is unit cell volume (in A^3), X_{Fa} is mole fraction fayalite, T is
temperature in °C, and P is pressure in kbar. Using this equation, Hazen
predicted the unit cell volume of $Fo_{90}Fa_{10}$ under P,T conditions predicted
at a depth of 100 km (approximately 1000°C and 30 kbar). The value predicted (295 A^3) is almost identical to that of a pure forsterite at 600°C
and 1 atm pressure. This equation will probably undergo revision when
more accurate high-pressure cell parameters become available.

Summary of P, T, X Effects on the Olivine Structure

The structural studies of olivines of different compositions and at
different temperatures have shown that increases in octahedral cation size
over the range 0.69 A (radius of Ni^{2+}) to 1.00 A (Ca^{2+}) and in temperature
over the range 20 to 1000°C have qualitatively the same effects on the
olivine structure. There is some evidence that $[MgO_6]$ octahedra in olivines expand more than Mn-, Fe-, Ni-, or Ca-containing octahedra over a
given temperature interval (see Table 6); however, this seems to be a
relatively minor effect relative to octahedral thermal expansion in
general. The parallelism between the effects of temperature and composition is in part due to the fact that the dimensions of $[SiO_4]$ tetrahedra in olivines are essentially unaffected by changes in temperature
or composition. In contrast, the M(1) and M(2) octahedra expand significantly, up to critical limits, with increasing temperature or increasing
cation size. Increases in temperature and octahedral cation size also
cause an increase in octahedral distortions as defined using the bond
angle variance parameter.

The effects of pressure on the olivine structure cannot be fully
assessed until additional, higher quality, high-pressure structural data
become available. The high-pressure studies of forsterite and fayalite
structures by Hazen showed that $[SiO_4]$ tetrahedra are essentially incom-
pressible. This finding is consistent with the results of such studies
of other silicates. $[SiO_4]$ tetrahedra have a mean linear compressibility
$(\bar{\beta}_{Si-O})$ of only 0.13 $Mbar^{-1}$ in silicate structures, including olivine

(Hazen and Finger, 1978). Olivine octahedra are quite compressible, with $[MgO_6]$ octahedra ($\bar{\beta}_{Mg-O}$ = 167 Mbar^{-1}) predicted to be less compressible than $[FeO_6]$ ($\bar{\beta}_{Fe-O}$ = 181 Mbar^{-1}), $[MnO_6]$ ($\bar{\beta}_{Mn-O}$ = 194 Mbar^{-1}), or $[CaO_6]$ ($\bar{\beta}_{Ca-O}$ = 243 Mbar^{-1}) octahedra ($\bar{\beta}$ values from Hazen and Prewitt, 1977). Present data are insufficient to make further general conclusions concerning the pressure effect.

The picture that emerges from the available olivine structural data as a function of P, T, and X is one in which relatively rigid $[SiO_4]$ tetrahedra are interconnected by pliable octahedra which share edges with the tetrahedra. Because of this polyhedral edge-sharing, there are limits to (1) the size of octahedral cations that can substitute in the M(1) and M(2) sites; (2) the thermal stability of olivines before they melt or, in the case of Ca_2SiO_4, undergo polymorphic transformations to less dense phases; and (3) the pressure stability of olivines before they undergo polymorphic transformations to denser phases or break down to other phases. In other words, the olivine structure can tolerate limited dimensional mismatch between $[SiO_4]$ tetrahedra and $[MO_6]$ octahedra; however, beyond certain critical values of octahedral cation size, temperature, and pressure, the octahedron-tetrahedron dimensional mismatch becomes too great for the structure to exist stably. Because of these studies, we now have some notions, though not any fundamental physical reasons, why the olivine structure field on the A_2BO_4 structure field map (Fig. 3) is limited and why olivines of different compositions have different melting points and different polymorphic transformation pressures.

Bonding in Olivines

Our current understanding of the bonding forces in olivines, or in any silicate mineral, is still rather primitive. Several generations of mineralogists since Pauling's classic 1929 paper on the principles determining the structure of complex ionic crystals have thought of silicate and other oxide minerals as assemblages of cations and anions held together by relatively short-range electrostatic forces. While this is a very useful conceptual model, Pauling clearly did not intend for us to consider minerals only in an ionic context as is apparent from his often overlooked 1948 paper on the electroneutrality principle (Pauling, 1948) and his recent paper (Pauling, 1980) reiterating his opinion that the Si-O bond has roughly 50 percent covalent character.

One way of avoiding the ionic *versus* covalent controversy, which still arises (see *e.g.* Stewart *et al.*, 1980 *versus* Pauling, 1980), is to think of bonds in minerals in terms of relative "bond strengths" as defined by Pauling (1929). The Si-O bond is assigned a strength near 1.0 and other "weaker" cation-oxygen bonds, such as Mg-O, Fe-O, and Ca-O bonds in olivine, are assigned strengths near 0.33 valence units. This bond strength concept has been extended during the last decade such that account can be taken of the effect of bond length variation on bond strength, the idea being that long Si-O bonds have strengths less than 1.0 whereas short Si-O bonds have strengths greater than 1.0 (see *e.g.* Brown and Shannon, 1973; Ferguson, 1974). Although such empirical bond strength-bond length relationships are quite valuable for making structural predictions, they offer little insight about the nature of bonding forces that hold atoms together in a structure like olivine.

Among the first studies of bonding in olivines from a covalent viewpoint were the papers by Louisnathan and Gibbs (1972a,b) which presented Extended Hückel Molecular Orbital (EHMO) calculations of valence MO energies and bond overlap populations for isolated $[SiO_4]$ tetrahedra with the geometry of those present in olivines of various compositions. The conclusion from these studies, which are rather crude by today's standards, was that part of the bond length variations of $[SiO_4]$ tetrahedra in olivines can be rationalized using a purely covalent model. One criticism of these studies is that the calculations considered only isolated $[SiO_4]$ tetrahedra, without surrounding octahedral cations, as a model of the olivine structure.

More recently, Gibbs, Tossell, and co-workers have used more sophisticated molecular orbital theories (including SCF-NEMO, CNDO/2, and SCF-Xα) to study details of the olivine structure, particularly the shared tetrahedron-octahedron edge conformation in olivine (Tossell and Gibbs, 1976, 1977; Tossell, 1977; McLarnan *et al.*, 1979). The study by Tossell and Gibbs (1976) was successful in predicting the correct O-Si-O angle (∿103°) opposite the edge shared between tetrahedron and octahedron using an $SiMgO_8H_{10}$ molecule as a model. They concluded that an important contribution to the minimization of total energy of this molecule at the above angle is covalent overlap repulsion between Si and Mg across the shared edge. The CNDO/2 MO study of shared-edge distortions in olivine by McLarnan et al. (1979) made use of a cluster of $SiO_3(OH)Mg_3(OH)_{10}^{7-}$ composition, which consisted of an $[SiO_4]$ tetrahedron sharing edges with three $[MgO_6]$ octahedra. This "better" model of the olivine structure resulted in a minimum energy geometry similar to the results obtained in the Tossell-Gibbs study. However, this calculation indicates that repulsive forces between Si and Mg, and Mg and Mg across shared edges are not the determinants in causing the minimum energy geometry of the molecule. Instead, these workers found that the equilibrium geometry with shortened shared edges is the result of a complex misture of short-range covalent and electrostatic forces and that repulsive Si-Mg and Mg-Mg interactions favor long shared edges. These workers also concluded, based on a combination of MO and distance-least-squares (DLS) calculations, that dimensional misfit between tetrahedra and octahedra is not responsible for shared-edge distortions in olivine-type structures. As pointed out earlier, this conclusion is in contradiction to the suggestions of Baur (1972) and Vincent et al. (1976) who based their opinion solely on the results of DLS calculations. The conclusions of McLarnan et al. (1979) are based on a more rigorous bonding model than those of Baur and Vincent and co-workers and, therefore, should carry more weight for the present. The important point to remember is that shared-edge distortions in olivine cannot be simply interpreted using Pauling's third rule, according to the best bonding calculation to date.

The interesting study by Tossell (1977) compared and theoretically analyzed the x-ray emission (Kuroda and Iguchi, 1971) and x-ray photoelectron (Nefedov et al., 1972) spectra of olivine and quartz. The comparison between observed and predicted valence MO energies for an isolated $[SiO_4]$ tetrahedron with T_d point symmetry is shown in Table 9. The quantitative agreement between calculated and observed MO energies is not perfect but is remarkably good in light of the fact that an isolated $[SiO_4]$ tetrahedral molecule was used as a model for olivine. This result suggests that silicon XES and XPS spectra are not affected greatly by second nearest-neighbor metal cations. However, as shown by the calculations of McLarnan et al. (1979), second nearest-neighbor interactions are very significant in modeling equilibrium geometries using MO methods. Using observed orbital energies and the method of Kowalczyk et al. (1974),

Table 9. Comparison of experimental and calculated MO energies of olivine (after Tossell, 1977).

	molecular orbital energies (in eV)					
	$4a_1$	$3t_2$	$5a_1$	$4t_2$	$1e, 5t_2$	$1t_1$
XES and XPS values for olivine	-20.2	-	-6.4	-3.0	-1.1	0
Calculated values*	-17.2	-14.4	-7.3	-3.1	-1.2	0

* Calculated using SCF-Xα-Scattered Wave method. d(Si-O) = 1.634 A.

Tossell concluded that the Si-O bond in olivine is weaker and more ionic (63% ionic character) than in quartz (58% ionic character). Tossell also concluded that Mg-O bonds in olivine play a significant role in stabilizing olivine relative to SiO_2 + MgO, with Mg-O bonds in olivine stronger than those in MgO (periclase).

One final result of these MO calculations worth mentioning is the predicted charge distribution in olivine. The CNDO/2 charges for Si (+0.86), M(1) (+1.38), M(2) (+1.31), O(1) (-0.74), O(2) (-0.82), and O(3) (-0.81) calculated by McLarnan *et al.* (1979) are considerably smaller than the nominal formal charges (+4 for Si, +2 for Mg and -2 for oxygen) in agreement with Pauling's (1948) electroneutrality principle which states that electrons distribute themselves among atomic centers because of the partial ionic-covalent character of cation-oxygen bonds such that the charge on an anion or cation is usually less than +1.0. The partial covalent character of bonds in olivine and the distribution of charge indicated by the CNDO/2 calculations above are reflected in experimentally-determined electron density difference maps (total electron density minus densities of spherical atoms) of forsterite (G. Lager, pers. comm., 1978).

Although MO calculations of "olivine-like" molecules including $[FeO_6]$ octahedra have not yet been carried out, at least one rigorous calculation has been made for an isolated $[FeO_6]$ octahedron using the SCF-Xα-scattered wave MO method by Tossell (1976). The results of this study can serve as an approximate model for Fe-O bonding in Fe-bearing olivines for the present. Tossell found that as Fe-O distances are reduced, as they are with increasing pressure, the separation between the t_{2g} and the e_g crystal field levels of Fe^{2+} increase as predicted by the

R^{-5} law of crystal-field theory. In addition, the covalency of Fe-O bonds and the width (in energy) of the valence electron region of the [FeO$_6$] cluster increase with decreasing Fe-O distance from 2.17 to 1.95 A. This result is consistent with the predictions of Huggins (1975, 1976) concerning the effect of pressure on the covalency of Fe^{2+}-O bonds based on high pressure Mössbauer studies. It is interesting to note that reducing Mg-O distances in an isolated [MgO$_6$] cluster from 2.12 to 1.92 A has minimal effect on the electronic structure of this cluster, in contrast with the finding for [FeO$_6$] clusters.

In addition to the bond strength and molecular orbital models of bonding in olivines, a great deal has been written about the crystal-field model for first-row transition metal cations in the distorted octahedra of olivines (Farrell and Newnham, 1965; White and Keester, 1966; Reinen, 1968; Burns, 1970a, 1974, 1976; Runciman *et al.*, 1973, 1974; Wood, 1974; Walsh *et al.*, 1976, among others). Though this is a purely electrostatic model of the interaction of d-electrons with point charge ligands, its predictions are useful for interpreting optical absorption spectra of transition metal-bearing olivines and for rationalizing the intra- and inter-crystalline cation partitioning, as has been demonstrated in the literature cited above. In spite of this success, details of energy level splitting predicted by the crystal-field model, especially at high pressures, are not borne out by the SCF-Xα results of Tossell (1976).

A second type of purely electrostatic model which has been applied to olivines is the "lattice" (or structure or Madelung) energy model. In this model, all atoms are treated as positive or negative point charges, using the full nominal valence of each, and the energy is calculated using the relationship $U = (-\alpha_M e^2/R)(1 - \rho/R)$, where α_M is the Madelung constant, e is the electronic charge, R is the minimum cation-anion separation, and ρ is a repulsive parameter derived from the compressibility, K, of the phase ($\rho = R/[9VR/\alpha_M e^2 K) + 2]$; where V is the molecular volume). Successful applications involving olivines include (1) prediction of the equilibrium M(2) position in forsterite (Born, 1964); (2) prediction of the relative stabilities of Mg-, Fe-, and Ni-olivines and spinels (Gaffney and Ahrens, 1970); (3) comparison of the theoretical and experimental (Born-Haber) structure energies for forsterite (Raymond, 1971); (4) prediction of the relative stabilities of the olivine, spinel, and β-spinel

333

polymorphs of Co_2SiO_4 (Tokonami *et al.*, 1972); (5) calculation of the thermal vibration ellipsoids at the M(1) and M(2) sites in Mg-olivine (Ohashi and Finger, 1973); (6) rationalization of the sequence of crystallization of silicates from silicate melts (Ohashi, 1976); and (7) rationalization of the cation distributions in olivine-type structures (Alberti and Vezzalini, 1978). The apparent success of these calculations should not be interpreted as indicating that a point charge or ionic model for olivines is valid. As pointed out by Phillips and Williams (1965), there are several self-compensating features of the structure energy model, such as the use of observed interatomic distances, which are the result of all bonding forces in a crystal.

In summary, the bonding in olivines should not be thought of in purely ionic or covalent terms even though mostly ionic descriptions of olivines have been popular, and useful, over the past few decades. The most rigorous bonding calculation to date on a sizeable atomic cluster modeled after the equilibrium geometry of forsterite (McLarnan *et al.*, 1979) indicates that this geometry is caused by a complex mixture of short-range covalent *and* electrostatic forces. Though ionic models like crystal field theory and the structure energy model are useful for crystal-chemical predictions or rationalizations of stability differences and cation partitioning, they should not be taken literally. Similarly, the predictive success of the modified Pauling bond strength model should *not* be interpreted as implying that it embodies an accurate description of bonding in olivines. An accurate description of bonding in olivines and in other minerals must await further detailed calculations using the most rigorous bonding models available and clusters of atoms large enough to represent the basic structure. The study by McLarnan *et al.* (1979) approaches this description. However, the method they employed (CNDO/2) is not the most rigorous because a full *ab initio* calculation with extended basis set on a cluster of the size required is presently not economically feasible.

INTRACRYSTALLINE CATION PARTITIONING IN OLIVINES

In the early 1900's metallurgists first recognized that atoms such as Cu and Au in alloys can be arranged in ordered, partially ordered, and fully disordered arrays and that such ordering is a temperature-dependent phenomenon. Some years later, mineralogists suggested that cations, such

as Si and Al, may occupy two or more geometrically similar, but crystallographically nonequivalent sites in a mineral in an ordered or disordered
fashion (*e.g.*, Barth (1934)[*]suggested that the Al-Si distribution in feldspars could be ordered or disordered).

Ghose (1962) first recognized the possibility of Mg-Fe ordering in
olivine and predicted that the larger Fe^{2+} cation would preferentially
occupy the larger M(2) octahedral site. Ghose's (1965) later work on
hypersthene confirmed Mg-Fe ordering in orthopyroxenes and appeared to
strengthen his prediction concerning olivine. However, the first modern
x-ray refinements of Mg-Fe olivines (Hanke, 1965; Birle *et al.*, 1968)
showed these cations to be disordered over the M(1) and M(2) sites.

Observations from X-ray and Spectroscopic Studies. The ordering
of various octahedral cations in the silicate olivines has been investigated by numerous workers using a variety of techniques including x-ray
diffraction (see *e.g.* Finger, 1970; other references listed in Table 5);
neutron diffraction (Caron *et al.*, 1965; Newnham *et al.*, 1966); Mössbauer
spectroscopy (Bancroft *et al.*, 1967; Malysheva *et al.*, 1969; Bush *et al.*,
1970; Virgo and Hafner, 1972; Duncan and Johnston, 1973; Shinno, 1974;
Shinno *et al.*, 1974); electron paramagnetic resonance spectroscopy
(Shcherbakova *et al.*, 1968; Michoulier *et al.*, 1969; Weeks *et al.*, 1974;
Ziera and Hafner, 1974; Niebuhr, 1975; Rager, 1977); electronic absorption spectroscopy (Reinen, 1968; Grum-Grzhimailo *et al.*, 1969; Burns,
1970a,b; Runciman *et al.*, 1973; Wood, 1974); and vibrational spectroscopy
(Duke and Stephens, 1964; Huggins, 1973). The results of these studies
are summarized and are compared with ordering predicted on the basis of
cation size in Table 10. Figure 27 shows the variation of distribution
coefficient, K_D, with effective ionic radius for transition metal-Mg
olivines and orthopyroxenes.

Factors Affecting Octahedral Cation Distributions in Olivines.
Examination of Table 10 (see also Table 5) and Figure 27 suggests that
the ordering of cations in the M(1) and M(2) sites of olivine is affected
by factors in addition to cation and site size differences. The most
obvious additional factor for olivines containing transition metal
cations is the crystal field stabilization energy (CFSE) these cations
gain by occupying the smaller and more distorted of the two octahedral
sites. For example, in Mg-Co olivine, the larger Co^{2+} cation is ordered
....

[*]Barth, T.F.W. (1934) Polymorphic phenomena and crystal structure. Am. J. Sci., 227, 273-286.

Table 10. Comparison of predicted and observed octahedral cation ordering in the olivines and pyroxenes.

M-Cations	$\Delta r(A)$*	Predicted** M(1)	M(2)	Olivines*** M(1)	M(2)	Pyroxenes*** M(1)	M(2)
Mg – Ni	0.03	Ni	Mg	Ni	Mg	Ni	Mg
Mg – Co	0.025	Mg	Co	Co	Mg	Mg	Co
Mg – Fe	0.06	Mg	Fe	Fe	Mg	Mg	Fe
Mg – Mn	0.11	Mg	Mn	Mg	Mn	Mg	Mn
Mg – Zn	0.02	Mg	Zn	Zn	Mg	Mg	Zn
Mg – Ca	0.28	Mg	Ca	Mg	Ca	Mg	Ca
Fe – Mn	0.05	Fe	Mn	Fe	Mn	–	–
Fe – Ca	0.22	Fe	Ca	Fe	Ca	Fe	Ca
Mn – Ca	0.17	Mn	Ca	Mn	Ca	–	–

*Calculated using Shannon's (1976) effective ionic radii.
**Predicted solely on the basis of difference in cation and site size.
***Observations from x-ray site refinements. Pyroxene distributions from Ghose *et al.* (1974). Distribution for Mg-Zn olivine from Ghose (pers. comm., 1975).

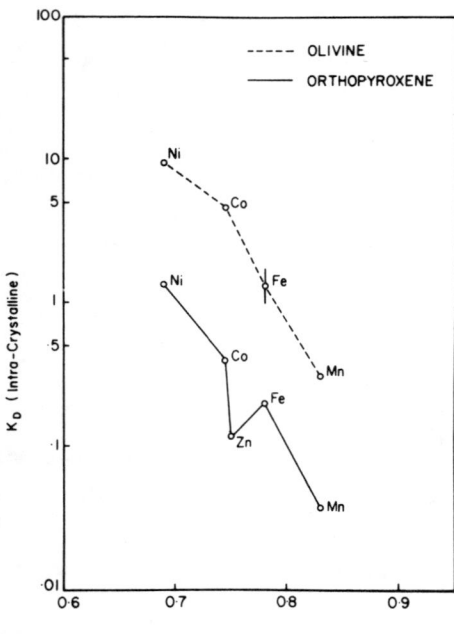

Figure 27. Log K_D *vs* cation radius for olivines and orthopyroxenes containing Mg and the transition metals Mn, Fe, Co, Ni, and Zn. Data are from the following sources: Ni-Mg olivine (Rajamani *et al.*, 1975); Co-Mg olivine (Ghose *et al.*, 1974); Fe-Mg olivine (Brown and Prewitt, 1973); Fe-Mn olivine (Brown, 1970); Ni-Mg, Co-Mg, Zn-Mg, Fe-Mg, and Mn-Mg orthopyroxenes (Ghose *et al.*, 1974). The vertical line through the point for Fe-Mg olivines expresses the range in K_D values. (From Rajamani *et al.*, 1975).

336

preferentially into the smaller M(1) site. Similarly, in Mg-Fe olivines, the larger Fe^{2+} shows a slight preference for the M(1) site (see Table 5). Measurement of the CFSE of Fe^{2+} in the M(1) and M(2) sites of a Fo_{88} using optical absorption spectroscopy (Burns, 1970a) showed that the CFSE gained by Fe at M(1) is 12.9 kcal/mole, whereas that gained in M(2) is 13.1 kcal/. mole. Other estimates of the CFSE gained by Fe in olivine suggest that the difference in CFSE between M(1) and M(2) is on the order of 140 kcal with Fe favoring M(1) (Huggins, as quoted in Ghose *et al.*, 1976). Additional optical absorption measurements on well-characterized Fe-Mg olivines are needed to resolve this discrepancy. Although no CFSE measurements have been made for Co in olivines, measurements have been made for Ni^{2+} by Wood (1974) who found that Ni gains 27.3 kcal/gm atom in M(1) and 25.7 kcal/gm atom in M(2). Walsh *et al.* (1976) calculated a difference in CFSE for the M(1) and M(2) sites in Ni-bearing olivines of 1.9 kcal, which compares reasonably with the 1.6 kcal measured by Wood. These workers also estimated differences in ‑CFSE at M(1) and M(2) for Co^{2+} and Fe^{2+} to be about 1.0 and 0.8 kcal, respectively, indicating that crystal field effects cause strong ordering of Ni^{2+} in M(1) and less ordering of Co^{2+} and Fe^{2+} in the M(1) site of olivine.

Ghose *et al.* (1976) suggested that the greater degree of covalent bonding predicted for the M(1) site of olivine on the basis of Mössbauer isomer shift values is also a factor which favors cations of relatively high electronegativity (Fe, Co, Ni, and Zn) in the M(1) site of olivines. This seems to be a reasonable explanation for Zn^{2+} which gains no CFSE in either M(1) or M(2) because of its d^{10} electronic configuration and may be a factor in the observed preferential ordering of Fe, Co, and Ni in M(1).

In light of the above data, the effect of certain crystal-chemical factors on the distribution of octahedral cations in the M(1) and M(2) sites of olivines may be summarized as follows:

(1) In the absence of crystal field (Mn^{2+}) or covalency (Zn^{2+}) effects, the smaller cation is preferred in the smaller M(1) site. This rule is consistent with the observed distribution of cations in Mg-Mn, Mg-Ca, and Mn-Ca olivines.

(2) The transition metal cations Fe^{2+}, Co^{2+}, and Ni^{2+} prefer the smaller and more distorted M(1) site of olivine because of the CFSE they

gain relative to M(2). The measured, or estimated, CFSE values
discussed above are consistent with the observed ordering of these
cations in olivine.

(3) The more electronegative transition metal cations, such as Zn^{2+},
prefer the M(1) site relative to Mg because this site allows a
greater degree of covalent bonding.

In addition to these crystal-chemical factors, variables such as
equilibration temperature and oxygen fugacity have been shown to have a
measurable effect on K_D in Fe-Mg olivines by Shinno (1974) and Will and
Nover (1979). Their interesting findings are summarized below.

Shinno (1974) carried out ^{57}Fe Mössbauer measurements on synthetic
olivines equilibrated at different temperatures in order to study the
temperature dependence of Mg-Fe distributions. He found that in a sample
equilibrated at 1150°C, Fe^{2+} is strongly partitioned into M(1) [K_D = 3.16];
further equilibrations at 950 and 800°C resulted in distribution coefficients of 1.85 and 1.32, respectively, indicating an increase in disorder.
This result is surprising in light of the normally expected increase in
cation disorder with increasing equilibration temperature, as is found
in the orthopyroxenes. Shinno found that redistribution of Fe and Mg
continued to temperatures as low as 600°C in his experiments.

Will and Nover (1979) buffered slightly ordered volcanic olivines
(K_d = 1.06 - 1.09) of compositions Fo_{90} and Fo_{88} at 10^{-16} and 10^{-21} bars
pO_2 and measured the Fe-Mg site distributions using x-ray site-refinement
techniques. Buffering at a pO_2 of 10^{-16} bar caused increased ordering
of Fe^{2+} in M(1) in both olivines (K_D = 1.2), whereas buffering at a pO_2
of 10^{-21} bar reduced K_D to 0.80 in both samples, indicating a reversal
in ordering with Fe^{2+} now showing slight preference for the M(2) site.

These experimental studies suggest that natural Mg-Fe olivines
equilibrated at high temperatures (i.e. those of volcanic origin) and
at relatively high oxygen fugacities should have K_D values significantly
greater than 1.0. Olivines equilibrated at low temperatures (i.e., those
of plutonic or metamorphic origin which have cooled slowly) and at relatively low oxygen fugacities should have K_D values slightly below or
near 1.0, indicating slight ordering of Fe^{2+} on M(2) or disorder.

Although there are some exceptions to these generalizations found
in Table 5 for Fe-Mg olivines from various parageneses, metamorphic

338

olivines tend to have low K_D values whereas volcanic olivines tend to have higher K_D values. Lunar olivines which formed under conditions of low oxygen fugacity have relatively high K_D values centered at 1.13. Thus, it appears that the effect of equilibration temperature (or cooling rate) is more important than oxygen fugacity in determining Fe-Mg distributions. The effect of Fe-Mg content on ordering is less clear than the effects of equilibration temperature or oxygen fugacity. However, Shinno *et al.* (1974) and Ghose *et al.* (1976) have found that in olivines with more than 20 mol % Fa, Fe^{2+} tends to prefer the M(1) site with this site preference decreasing with decreasing equilibration temperature. They also generalized that in Mg-rich olivines, Fe^{2+} prefers the M(1) site at high temperature but may prefer the M(2) site at low temperature. These suggestions are consistent with the Fe-Mg distributions from two lunar olivines of different compositions (Fa_{33} and Fa_{18}) which are believed to be from the same lunar rock (#12018). Finger and Virgo (1971) found that the Fa_{33} olivine had a K_D of 1.75, whereas Brown and Prewitt (1973) found that the Fa_{18} sample had a K_D of 1.15. These lunar olivines, which presumably experienced similar cooling histories, show different distribution coefficients perhaps because of their difference in composition. This must be viewed as a tentative conclusion, as are those by Ghose *et al.* (1976). There is a clear need for further study of Fe-Mg distributions in olivines.

Among other possible factors affecting the exchange of Fe and Mg between the M(1) and M(2) sites of olivines is the difference in interdiffusion of these cations as a function of equilibration temperature, oxygen fugacity, pressure, and composition. The studies by Buening and Buseck (1973) and Misener (1974) of Fe-Mg interdiffusion in olivines as a function of these variables show that diffusion increases as a function of temperature, Fe content, and pO_2, but decreases as a function of pressure (Misener, 1974). The minimum temperature at which Fe-Mg exchange between sites takes place is at least 600°C and is probably lower. Although the results of the above diffusion studies are not directly applicable to exchange between sites, they do show a direct dependence of Mg and Fe diffusion on the variables listed.

In spite of the lack of Fe-Mg ordering found in early studies of olivines, it now appears that Fe-Mg distributions are variable as a function of equilibration temperature, oxygen fugacity, composition, and (possibly) pressure. However, it is unlikely that the intracrystalline distribution of these cations will become widely used as an indicator of thermal histories or oxygen fugacities of the rocks containing them.

Before we compare intracrystalline cation ordering in olivines and orthopyroxenes, the results of one additional study of cation ordering in silicate olivine of composition $LiScSiO_4$ merits some mention. The x-ray structure study of this phase by Steele *et al.* (1976) showed that Li^+ occupies the M(1) site and Sc^{3+} occupies the M(2) site This finding is consistent with the ordering of monovalent cations (Li and Na) in the M(1) sites and of cations of higher charge in the M(2) sites of lithiophilite [$LiMnPO_4$] (Geller and Durand, 1960); triphylite [Li(Fe,Mn)PO_4] (Finger and Rapp, 1969); and natrophilite [$NaMnPO_4$] (Moore, 1972), which are olivine isostructures. Although Alberti and Vezzalini (1978) attempted to rationalize this ordering using Madelung energy calculations, their results cannot be considered very meaningful because they utilized observed structures, with ordered cation distributions, as the basis for their calculations. Thus the calculations had a built-in bias for the ordering predicted. At the present time, no adequate explanation exists for the preference of monovalent cations for M(1) in certain olivines.

On the basis of EPR experiments on Fe^{3+}-doped synthetic Mg_2SiO_4, Weeks *et al.* (1974) suggested that what little Fe^{3+} exists in natural forsterites should occur in the M(2) site. On the other hand, the EPR study by Ziera and Hafner (1974) found Fe^{3+} disordered over the M(1) and M(2) sites. Because Fe^{3+} has a d^5 electronic configuration, it gains no CFSE. Therefore, crystal-field effects have nothing to with the site distribution of Fe^{3+} in this case. Consideration of size differences between Mg^{2+} (0.72 A) and high-spin Fe^{3+} (0.645 A) leads to the prediction that Fe^{3+} should occur in the smaller M(1) site. However, this prediction is not consistent with observation. Perhaps the same as yet unknown factors that result in Sc^{3+} ordering into the M(2) site of $LiScSiO_4$ (Steele *et al.*, 1976) are operative here, too. Cr^{3+}, which can gain CFSE in distorted octahedral sites, was found to order more strongly into M(1) than M(2) (Rager, 1977).

Comparison of Intracrystalline Cation Partitioning in Olivines and Orthopyroxenes. As shown in Table 10 and in Figure 27, there are some

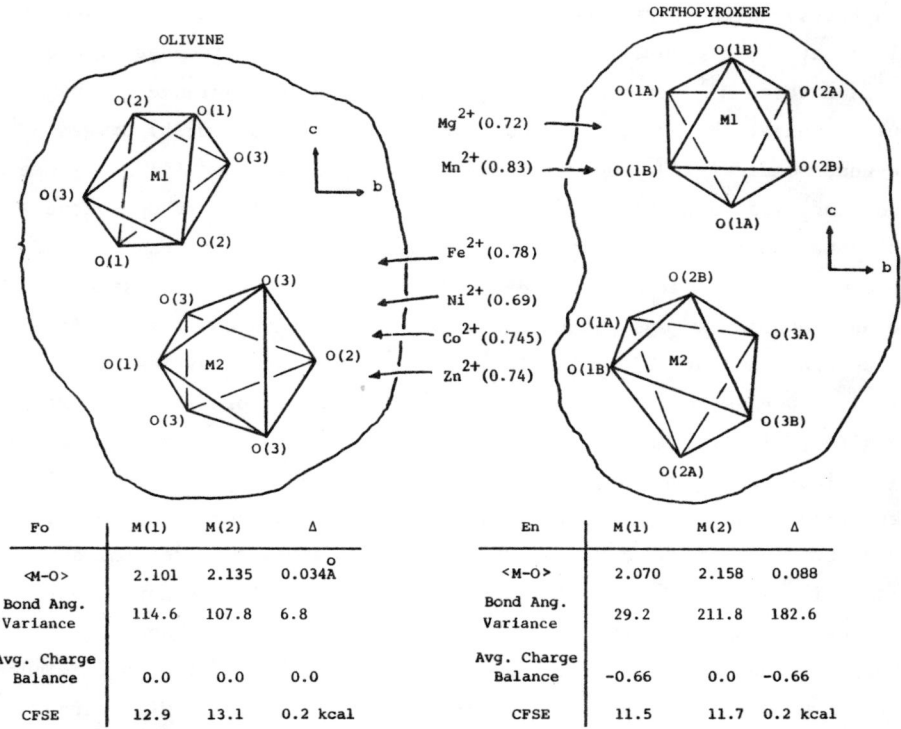

Fo	M(1)	M(2)	Δ		En	M(1)	M(2)	Δ
⟨M-O⟩	2.101	2.135	0.034Å		⟨M-O⟩	2.070	2.158	0.088
Bond Ang. Variance	114.6	107.8	6.8		Bond Ang. Variance	29.2	211.8	182.6
Avg. Charge Balance	0.0	0.0	0.0		Avg. Charge Balance	-0.66	0.0	-0.66
CFSE	12.9	13.1	0.2 kcal		CFSE	11.5	11.7	0.2 kcal

Figure 28. Comparison of some of the factors that affect inter- and intra-crystalline cation partitioning in forsterite and enstatite. CFSE values from Burns (1970).

striking differences in the observed ordering of octahedral cations in olivines and orthopyroxenes. Although Ni-Mg and Mg-Mn olivines and orthopyroxenes show qualitatively similar ordering, Co-Mg, Fe-Mg, and Zn-Mg olivines and orthopyroxenes exhibit opposite ordering schemes. In order to aid in understanding these differences, some of the characteristics of the M(1) and M(2) sites of olivine and orthopyroxene are compared in Figure 28. The discussion below follows that of Brown (1970) and Rajamani *et al*. (1975).

In spite of the large differences in the M(1) and M(2) sites in orthopyroxene relative to olivine, Ni is only slightly enriched in the M(1) site of orthopyroxene. In addition, Fe, Co, and Zn exhibit strong ordering into the M(2) site of orthopyroxenes in contrast to the preference of these cations for the M(1) site in olivines. In the case of Ni^{2+}, if cation size, site size, or crystal-field stabilization alone, or in combination, were mainly responsible for cation site preference, then Ni should exhibit stronger ordering in the orthopyroxene than in the olivine structure. Another factor, or factors, must be preventing the ordering. A key difference between the olivine and orthopyroxene structures, which is related to the above observations, is the valence balance at the oxygens. In olivine, each oxygen is surrounded by one Si^{4+} and three M^{2+} cations, leading to formal valence balance in the Pauling sense. However, in orthopyroxene the O(2A) and O(2B) oxygens are underbonded (-1/3) and O(3A) and O(3B) are overbonded (+1/3), whereas O(1A) and O(1B) are charge balanced (see Fig. 28). The orthopyroxene M(2) site has as many overbonded as it has underbonded oxygens and is, therefore, charge balanced on the average. However, the M(1) site has two underbonded oxygens and four charge balanced oxygens, leading to a net underbonding of -2/3. Considering the electronegativities of Mg (χ = 1.3) *versus* Fe, Co, Ni, and Zn ($\chi \sim 1.8$), Mg is predicted to prefer the site [M(1)] with the higher proportion of underbonded oxygens because it can lose electrons more easily to them than the more electronegative transition metal cations can. Aside then from differences in the sizes and distortions of the M(1) and M(2) sites in orthopyroxenes, the difference in "ionicity" of these sites leads to the relative enrichments observed. Using different reasoning than that presented above, Ghose (1962), Burnham *et al.* (1971) and O'Nions and Smith (1973) also concluded that the M(2) site of pyroxene is more "covalent" than the M(1) site.

INTERCRYSTALLINE CATION PARTITIONING BETWEEN OLIVINE AND OTHER SOLID PHASES

Olivine - Orthopyroxene

Considerable interest has been shown in the partitioning of cations between coexisting solid phases (intercrystalline partitioning) since the classic study by Ramberg and DeVore (1951) which examined the partitioning of Mg and Fe between coexisting olivines and orthopyroxenes in rocks of different origins. They found, as shown in Figure 29, that Mg is strongly partitioned into orthopyroxene for Fe-rich bulk compositions and shows no preference to a slight preference for olivine in the more Mg-rich bulk compositions $[(X_{Mg}^{ol} + X_{Mg}^{px})/2 > 0.65]$. These observations have been confirmed by numerous studies of Mg-Fe exchange between olivines and orthopyroxenes in natural assemblages (see Medaris, 1969, for references to this work). More recently, these and other data have been thermodynamically analyzed (see *e.g.*, Saxena, 1969; Grover and Orville, 1969; Matsui and Nishizawa, 1974; Sack, 1980). In addition, the temperature dependence of the exchange reaction

342

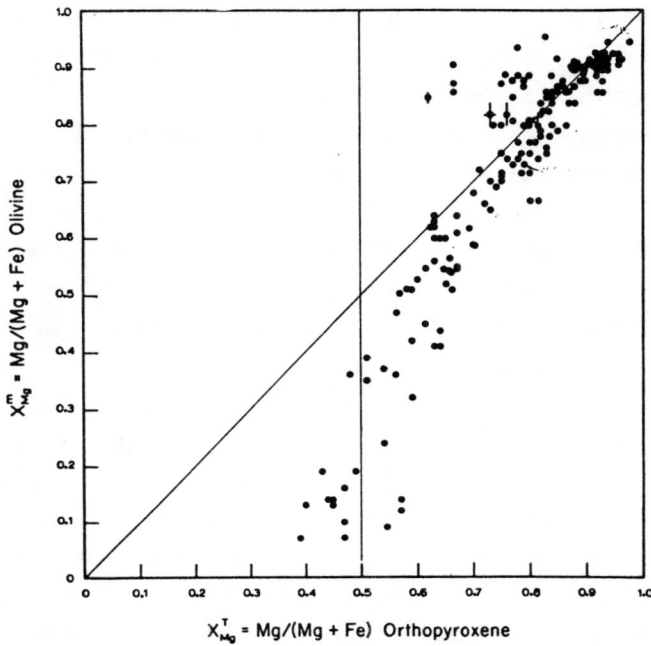

Figure 29. Distribution of Mg between coexisting olivine and orthopyroxene in natural and synthetic samples. (From Grover and Orville, 1969).

$$Mg_2SiO_4 + 2FeSiO_3 = Fe_2SiO_4 + 2MgSiO_3$$

has been investigated in an experimental study by Medaris (1969). Unfortunately, Medaris found that this reaction is not appreciably temperature-sensitive, at least in his analysis, over the range 700 to 1300°C. Because it is not pressure-sensitive either (Ramberg and Devore, 1951), this exchange has not yet proved to be useful in geothermometry or geobarometry. However, Sack (1980) has recently re-evaluated the thermodynamic formulations relating exchange and ordering for this pair. He found that in assemblages containing olivines in the composition range Fo_{100} to Fo_{80} or Fo_{40} to Fa_{100}, a newly calibrated orthopyroxene-olivine geothermometer is potentially useful.

Matsui and Banno (1969) studied the partitioning of other element pairs besides Mg-Fe in coexisting olivines and orthopyroxenes from lherzolites and garnet peridotites and found that Ni and Co are strongly partitioned into the olivine, that Zn shows a slight preference for olivine, and that Mn is strongly partitioned into orthopyroxene. This last observation was confirmed by the experimental study of Mg-Mn partitioning

343

between olivine and orthopyroxene by Nishizawa and Matsui (1972), who applied their data to the estimation of pressure and temperature conditions during the formation of peridotites.

These observations can be rationalized using the same crystal chemical reasoning that was employed earlier in discussions of the intra-crystalline distribution of these cations in olivines and in orthopyroxenes (see Fig. 28). In the absence of crystal-field effects, the larger cation is preferred in orthopyroxene, which has the larger M(2) site of the pair. The strong partitioning of Mn^{2+} into orthopyroxene bears this prediction out. Mg^{2+} partitions preferentially into the M(1) site of orthopyroxene, relative to the M(1) and M(2) sites of coexisting olivine and the M(2) site of orthopyroxene, because of the higher "ionicity" of this site as discussed earlier. Ni^{2+} and Co^{2+} are strongly partitioned into the M(1) site and, to a lesser extent, the M(2) site of olivine because of the gain in CFSE they receive relative to the orthopyroxene sites. Because Zn^{2+} gains no CFSE in either phase and usually prefers to form covalent bonds, it partitions into the phase where it can covalently bond most strongly.

Olivine - Clinopyroxene

In coexisting olivine-orthopyroxene pairs from natural assemblages, Fe^{2+} partitions preferentially into the olivine in Mg-rich bulk compositions. No data exist for Fe-rich bulk compositions (Obata et al., 1974). Not surprisingly, this exchange is similar to that found in the olivine-orthopyroxene pair (the major difference between orthopyroxene and clinopyroxene is that the M(2) site in the latter is considerably more distorted than in the former). Although this exchange has not been studied experimentally, it has been used empirically as a geothermometer by Obata et al. (1974). Powell and Powell (1974) thermodynamically analyzed this geothermometer and attempted a calibration using groundmass olivine and clinopyroxene in lavas for which temperatures had been established using the iron-titanium oxide geothermometer. However, Wood (1976) concluded that this geothermometer is likely to be unreliable as formulated because of the limited temperature range over which it was calibrated and the form of the expression for temperature.

Perhaps a more reliable geothermometer, though one that is more restricted in use, is the one involving partitioning of Ni between coexisting olivine and clinopyroxene. Hakli and Wright (1967) formulated and calibrated this geothermometer by measuring the Ni partitioning among olivine, augite, and groundmass glass using samples collected from the modern Makaopuhi lava lake in Hawaii and for which equilibration temperatures had been measured. Using this active, natural laboratory, they found that Ni^{2+} is strongly partitioned into the olivine relative to glass and augite and less strongly partitioned into augite relative to glass. They also found that this partitioning is temperature sensitive. This geothermometer was later applied to a mafic intrusive (Hakli, 1968) and to the prehistoric Makaopuhi lava lake in Hawaii (Evans, 1969).

One geobarometer involving olivine coexisting with clinopyroxene and orthopyroxene merits brief discussion. The pressure dependence of Ca solubility in forsterite coexisting with diopside and enstatite has been studied experimentally by Finnerty (1977) and Finnerty and Boyd (1978). They found that Ca solubility is reduced with increasing pressure and decreasing temperature. They tested this olivine thermobarometer against pressure estimates for garnet lherzolite xenoliths from kimberlites made with the Al-orthopyroxene geobarometer and obtained encouraging results. This finding that Ca solubility decreases with increasing pressure confirms the suggestion by Simkin and Smith (1970) discussed earlier in the section on minor element chemistry. However, this thermobarometer needs further work before it can be considered reliable.

In addition to the above studies, there have been recent measurements of partitioning of various elements between coexisting olivine and garnet, between olivine and spinels, and between olivines and sulfides. The results of these studies are briefly summarized below.

Olivine - Garnet

The partitioning of Mg and Fe^{2+} between olivines and garnets has been investigated at high temperatures and pressures by Kawasaki and Matsui (1977) and O'Neill and Wood (1979). The earlier study found that Fe^{2+} is partitioned into garnets $[(Mg,Fe^{2+})_3Al_2Si_3O_{12}]$ relative to olivine at all Fe/Mg ratios from 0.0 to 1.0. The more detailed study by O'Neill and Wood confirmed the findings of Kawasaki and Matsui but also found that the partitioning of Fe into garnet increases with increasing Mg and Ca contents. These exchange reactions are between octahedral sites in olivine and the distorted dodecahedral (8-coordinated) sites in garnets.

Optical absorption spectral measurements by Burns (1970a) on almandine-pyrope garnets indicate that the CFSE of Fe^{2+} in garnet's dodecahedral site varies from 12.4 (Alm_1) to 11.7 (Alm_{100}) kcal/mole. Because these values are lower than the CFSE of Fe^{2+} in either site in forsteritic olivines, the crystal-chemical reasons for the observed Fe-Mg partitioning between olivine and almandine-pyrope garnets are not clear.

Finnerty (1977) calculated Ca-Mg and Mn-Mg exchange between garnet and olivine based on measured exchange between other coexisting pairs involving garnet and, separately, olivine. The estimated $\Delta H°$ and $\Delta S°$ values for Ca-Mg and Mn-Mg exchange between garnet and olivine are 19.00 and 1.22 (Ca-Mg) and -3.76 and 0.49 (Mn-Mg) (all in kcal/gfw).

Olivine - Spinel

The exchange of Mg and Fe^{2+} between olivine and spinel of composition $(Mg,Fe^{2+})(Al_x Cr_y Fe_z^{3+})O_4$ has been investigated in a number of studies before 1976 which are summarized by Fujii (1977). Fe^{2+} partitions preferentially into olivine at mole fractions of Cr^{3+} less than about 0.5 and into spinel at mole fractions of Cr^{3+} above 0.5 at temperatures above 1200°C. At lower temperatures (550 to 700°C), Fe^{2+} is partitioned preferentially into spinel at all but the lowest mole fractions of Cr^{3+}. Thus, as suggested by Irvine (1965), $Mg-Fe^{2+}$ partitioning between olivine and spinel is a function of mole fraction Cr^{3+} and temperature.

The olivine-spinel geothermometer developed by Jackson (1969) assumes ideal mixing of Fe^{2+} and Mg and gives reasonable temperatures in several studies including those of the chromitite layers in the Stillwater complex and of some ultramafic intrusions (Loney *et al.*, 1971). However, Evans and Wright (1972) found that it gave a large overestimate of the liquidus temperatures of Hawaiian tholeiitic lavas. One of the major problems with Jackson's geothermometer is the difficulty of obtaining accurate Fe^{2+}/Fe^{3+} ratios for spinels. A more recent experimental study of Fe-Mg exchange between coexisting olivine and spinel (Engi, 1978) showed that exchange ceases below about 800°C.

The exchange of Mg and Fe^{2+} between coexisting $(Mg,Fe)_2SiO_4$ spinel and olivine has been experimentally investigated at pressures up to 90 kbar by Nishizawa and Akimoto (1973). They found that Fe^{2+} partitions preferentially into the spinel polymorph.

346

Olivine - Sulfide

Studies of Fe-Ni partitioning between olivines and Fe-Ni sulfides include those by Clark and Naldrett (1972) and Binns and Groves (1976). Fe^{2+} is strongly partitioned into olivine and Ni^{2+} strongly prefers the sulfide phase. This partitioning has a pronounced temperature dependence and is, therefore, of potential use as a geothermometer.

Summary

The intercrystalline cation partitioning between olivine and other crystalline phases is temperature-dependent in certain cases, as discussed above and, therefore, is of potential use in geothermometry. Although several geothermometers have been proposed over the past decade, most of them have marked flaws, as discussed by Wood (1976)*and Wood and Fraser (1976). These problems are due in part to a general lack of knowledge about the effect of other minor components in the exchange reactions on their temperature dependence, and in part to the assumption, in some models, of ideal cation mixing. Detailed experiments would be required to sort out these effects. It should be emphasized that temperatures predicted by these various geothermometers represent the minimum temperatures before which cations exchange ceases and that this exchange can proceed to temperatures of 800°C and below.

Crystal-chemical rationalization of the cation exchange behavior is possible for the olivine-orthopyroxene and olivine-clinopyroxene pairs using the same arguments as we did for intracrystalline cation partitioning in the last section. However, simple crystal-field arguments based on Burn's (1970a) measured CFSE for Fe^{2+} in olivines and garnets are not consistent with the observed enrichment of Fe^{2+} in garnets. The observed enrichment of Ni in olivine-(Fe,Ni) sulfide pairs is consistent with the greater CFSE gained by Ni^{2+} in sulfides due to π-bond formation (Burns, 1970a).

MELT GROWTH OF OLIVINES AND OLIVINE - MELT CATION PARTITIONING

The primary purpose of this section is to complete the discussion of cation partitioning between olivines and other phases. We will briefly review available data on the partitioning of cations between olivine and melt and discuss how melt structure plays an important part in
....

*Wood, B.J. (1976) An olivine-clinopyroxene geothermometer. Contrib. Mineral. Petrol., 56, 297-303.

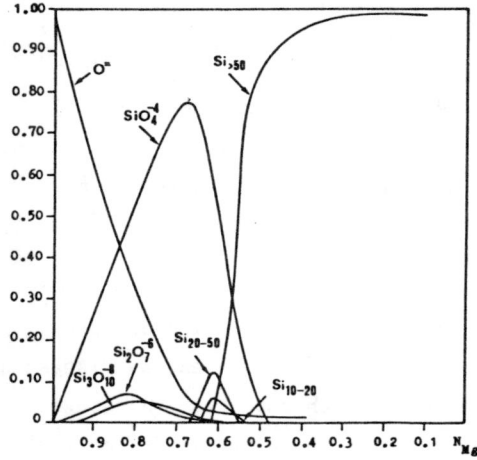

Figure 30. Polymer species distribution for melts in the system MgO-SiO$_2$ calculated using the Monte Carlo method. The ordinate represents the fraction of oxygen atoms associated with the different polymer species. The abcissa is mole fraction MgO. The composition of forsterite falls at N_{MgO} of 0.67. Si$_{20-50}$ and Si$_{>50}$ represent large polymeric units with between 20 and 50 Si and greater than 50 Si, respectively. (From Borgiani and Granati, 1979).

this partitioning. However, before we tackle this task, it seems appropriate to briefly discuss the structure of olivine and more complex composition melts in order to provide a bssis for understanding melt-crystal partitioning of cations. We will also briefly consider the growth of olivine crystals from melts and the morphological variations these crystals display.

The Nature of Olivine Composition Melts

The most direct structural information on olivine composition melts that is currently available comes from x-ray radial distribution studies of silicate melts in the systems MgO-SiO$_2$ and FeO-SiO$_2$ (Waseda and Toguri, 1977, 1978). In melts at or near olivine compositions in these systems, x-ray studies have shown that Si occurs in isolated [SiO$_4$] tetrahedra, on the average, and that Mg and Fe are irregularly coordinated by approximately four oxygen ligands, on the average. In addition to these experimental studies, recent Monte Carlo calculations (Borgiana and Granati, 1979) were used to predict the distributions of silicate species in MgO-SiO$_2$ and FeO-SiO$_2$ melts. The calculation for the MgO-SiO$_2$ system predicted isolated [SiO$_4$] tetrahedra as the predominant species at the olivine composition (see Fig. 30). However, the calculation for FeO-SiO$_2$ melts is not consistent with the observed melt structure for fayalite composition melts.

The remarkable x-ray studies by Waseda and Toguri (1978) of fayalite composition melts at various temperatures (1250 to 1400°C) and oxygen fugacities (p0$_2$ = 10^{-7} to 10^{-11} bars) showed that little structural change occurs in fayalite melt as a function of these variables. Thus we may conclude that a fayalite melt, just above the freezing temperature (1205°C) where fayalite crystals first precipitate, consists primarily of isolated [SiO_4] tetrahedra with Fe^{2+} cations in irregular tetrahedral sites, on the average. The description of MgO-SiO_2 melts containing 44 mol % SiO_2 at 1200°C is similar. The lack of significant polymerization of [SiO_4] tetrahedra is consistent with the observation that olivine melts are not easily quenched to glass (Jeanloz et al., 1977).

Melt Growth of Olivines

Although significant gains have been made in our understanding of melt growth of crystals from a macroscopic-thermodynamic viewpoint over the last 20 years (see review by Kirkpatrick, 1975), we still do not have an adequate model of melt growth from an atomistic-structural viewpoint. However, in the case of olivine growth from olivine composition melts, it seems safe to conclude that portions of the melt may have an approximate olivine structural arrangement. This conclusion is based on the melt-structure model for fayalite which was derived from x-ray data taken only 45°C above the greezing temperature. This work suggests that Mg and Fe must increase their coordination from irregular tetrahedral to octahedral just above or at the freezing temperature.

In basaltic melts, it is well known that forsteritic olivines are among the first minerals to crystallize and that as crystallization proceeds and temperature drops, tetrahedral polymerization in the precipitating minerals and remaining melt increases. Osborn (1954) rationalized these changes, including the early precipitation of olivines, using simple crystal-chemical principles. More recently, Ohashi (1976) went through a similar exercise using the electrostatic lattice energy per tetrahedral site as a basis for rationalizing the sequence of appearance of various phases from basaltic melts. Although these studies do not directly address melt growth, they give us some insights about the crystal-chemical factors responsible for differences in thermal stability of different structural arrangements in the crystallizing phases and melt.

Much useful information has come from experimental studies of the melt growth of olivine, although they have added little to our structural understanding of the melt growth process. The most notable experimental studies of olivine growth are those of Donaldson (1975, 1976, 1979). In his 1976 study, Donaldson experimentally reproduced many of the olivine crystal morphologies recognized in mafic and ultramafic rocks (see *e.g.*, the classic paper by Drever and Johnston, 1957). He found that with increasing cooling rate and increasing degree of supercooling below the liquidus, there are systematic changes in olivine morphology from granular to skeletal. An important conclusion reached by Donaldson is that the "skeletal olivines in picrites, olivine-rich basalts, and Archaean 'spinifex' rocks are not due to rapid cooling but to rapid olivine growth caused by the high normative olivine content of the magma." Some of the growth habits produced by Donaldson in his experiments and observed by other workers in natural samples have been structurally rationalized by Fleet (1975) and 'T Hart (1978a,b).

In a later study, Donaldson considered the mechanism of olivine nucleation in a melt-structure context. He speculated that olivines probably homogeneously nucleate as small volumes of melt with an approximate olivine structure. In complex composition melts such as the ones studied by Donaldson, a certain fraction of Si atoms is predicted to occupy isolated tetrahedra by Flory-Huggins polymer theory, especially at high temperatures. Conceivably, small melt volumes will attain compositions and structures similar to the phase or phases on the liquidus. These melt volumes could then serve as nuclei for crystal growth.

Element Partitioning Between Olivine and Melt

The exchange of cations between melts and coexisting crystals is intimately related to the availability of suitable sites in the melt and crystal which can stably accommodate the cations. There are two somewhat distorted octahedral sites in olivine into which cations may partition from a variety of sites in the melt. However, because we do not have much knowledge of cation sites in melts as a function of temperature, pressure, composition, and oxygen fugacity, we cannot fully rationalize observations from studies of cation partitioning between olivine and melt pairs.

As an example of how melt structure might affect the partitioning of a cation between olivine and melt, consider the partitioning of Ni and Mg between olivine and a simplified basaltic melt. Burns and Fyfe (1966) and Whittaker (1967) analyzed the anomalous enrichment of Ni^{2+} in early-formed olivines from basaltic rocks and concluded that solidus-liquidus

relations in the binary Ni_2SiO_4-Mg_2SiO_4 crystalline solution series are
inverted. According to the Bernal liquid structure model (see Whittaker,
1967) there are certain numbers of tetrahedral, octahedral, and larger-
than-octahedral-sites predicted in a given volume of silicate melt. Ni^{2+}
was predicted by Burns and Fyfe to preferentially occupy octahedral sites
because of its high octahedral site preference energy. Mg, on the other
hand, gains no CFSE in either octahedral or tetrahedral sites but is predicted to preferentially enter octahedral sites because of its favorable
size. With high concentrations of large cations such as Ca, Na, and K in
a typical basaltic melt, Whittaker speculated that larger-than-octahedral
sites as well as some of the octahedral sites would be filled by these
larger cations, forcing Ni^{2+} and Mg into tetrahedral sites. Thus, when
olivine with two available octahedral sites nucleates and begins to grow,
Ni^{2+} would gain octahedral CFSE by partitioning from the tetrahedral sites
in the melts to the available octahedral sites in olivine. This series
of speculations should not be taken too literally in light of recent
criticisms of the Bernal model for liquid structure (see Whittaker, 1978).
However, this example does illustrate how melt structure may affect melt-
mineral cation partitioning.

Measurements of partitioning of various cations between olivine and
melt have been made on natural samples (*e.g.*, Hakli and Wright, 1967;
Henderson and Dale, 1969; Gunn, 1971; Mysen, 1975; Cawthorne and McCarthy,
1977) and on synthetic olivine-glass pairs (see Irving, 1978, for a review). Typical variations of distribution coefficient with temperature
for Ni^{2+} and Mn^{2+} partitioning determined experimentally are shown in
Figures 31 and 32. Experimental data such as these have clearly established that exchange of cations between olivine and melt is dependent on
temperature, composition, and oxygen fugacity, in those cases where variable oxidation state cations are involved. Not enough data exist from
high pressure experiments to draw conclusions concerning the pressure
dependence of the partitioning of cations between olivine and melt except
in a few cases. For example, Ni partitioning into olivine is reduced
with increasing pressure (Mysen and Kushiro, 1979). In general, the
magnitude of depletion of metal cations in the melt caused by the crystallization of olivine is in the order Ni > Mg > Co > Fe > Mn (Roeder,
1974; Takahashi, 1978). Not surprisingly, this sequence is in the order
of their octahedral site preference energy, with the exception of Mg.

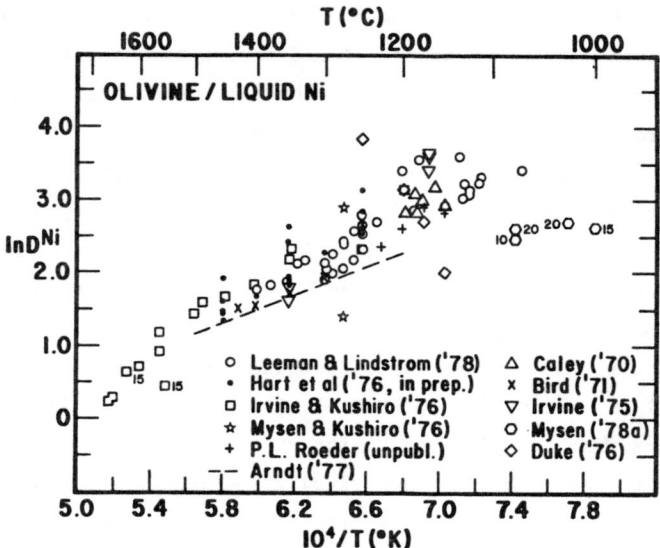

Figure 31. Variation of lnD for olivine vs reciprocal temperature from a number of experimental studies of Ni partitioning. The distribution coefficient D is defined as follows: $D = C(M)_{ol}/C(M)_{liq}$ where the C(M) terms are the concentrations of the metal cation, M, in olivine and liquid, respectively. (From Irving, 1978).

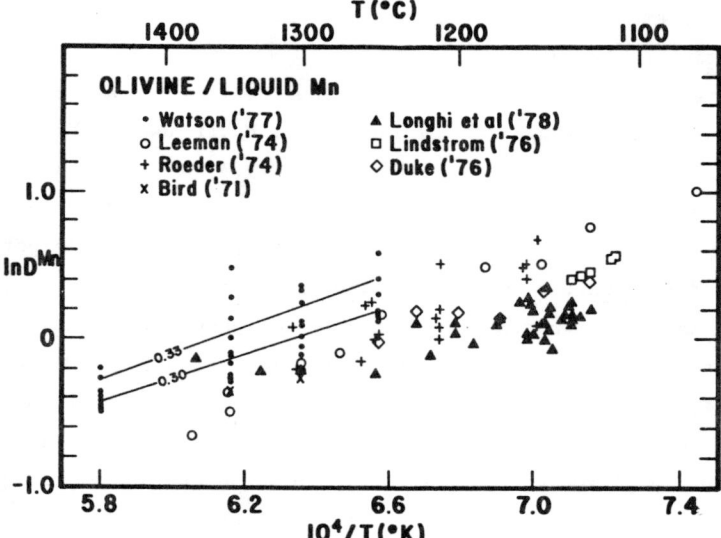

Figure 32. Variation of lnD for olivines vs reciprocal temperature from a number of experimental studies of Mn-partitioning. (From Irving, 1978).

More recent studies not discussed in the Irving review include experimental investigations of Ni partitioning (Takahashi, 1978; Mysen and Kushiro, 1979; Mysen, 1979; Nabelek, 1980), Ca partitioning (Watson, 1979), and Mg, Mn, Fe, and Co partitioning (Takahashi, 1978).

The partitioning of Ni^{2+} between olivine and melt has been the most extensively studied exchange reaction in part because of the controversy concerning whether Ni behaves according to Henry's Law when partitioned between olivine and liquid. Mysen (1979) and Nabelek (1980) review this literature and apparently have resolved the controversy. The concensus opinion, which has been confirmed by Nabelek's study, is that Ni partitioning obeys dilute solution behavior in olivines and liquids over the Ni concentration range found in most natural basalts (100-300 ppm).

As pointed out by Takahashi (1978) and Leeman and Scheidegger (1977), the distribution coefficients for transition metal cations between olivine and melt can be used to test for crystal-liquid equilibrium.

THE OLIVINE → SPINEL TRANSITION

Olivine of approximate composition Fo_{90} is considered to be the most abundant phase in the Earth's upper mantle. The known polymorphic transformations of Fo_{90} to the β-spinel and γ-spinel phases at pressures near 118 and 115 kbar, respectively, and at 1000°C, result in density increases of approximately 7.7 and 10.2%, respectively. Therefore, these transformations are key elements in explaining the discontinuous increase in seismic wave velocities at a depth near 400 km.

In this section, we will examine the olivine to spinel transformation from a structural viewpoint and offer a structural rationalization for the low pressure stability of the olivine structure type and the high pressure stabilities of β-spinel and γ-spinel. The excellent review of the high pressure crystal chemistry of orthosilicates by Akimoto *et al.* (1976) is recommended, as is the chapter on high pressure transformations of A_2BO_4 compounds in Ringwood (1975). It is appropriate that a natural silicate spinel of composition $(Mg_{.7}, Fe_{.3})_2 SiO_4$ from the Tenham meteorite has been named ringwoodite in honor of Prof. Ringwood who has contributed so much in this area of research (Binns *et al.*, 1969).

High Pressure Phase Relations in the System Mg_2SiO_4-Fe_2SiO_4

Ringwood and Major (1966) were the first to publish the high pressure phase diagram for the Mg_2SiO_4-$FeSiO_4$ system shown in Figure 33. They discovered a new, non-isotropic phase near the Mg-rich end of this diagram and termed this phase β-spinel after they suggested that it had a "modified"

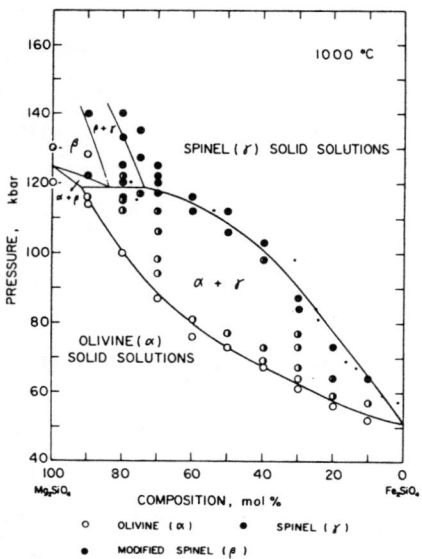

Figure 33. Phase relationships in the system Mg₂SiO₄ – Fe₂SiO₄ at pressures up to 140 kbar and at 1000°C. (From Akimoto *et al.*, 1976).

spinel structure. This discovery has since been confirmed in several other laboratories around the world.

Structures of the β-Mg₂SiO₄ and γ-Mg₂SiO₄ Polymorphs

The structure of β-$(Mg_{0.9}Ni_{0.1})$ SiO₄, which is isostructural with β-$(Mg_{0.9}Fe_{0.1})_2$SiO₄, was first determined by Moore and Smith (1969, 1970) using powder diffraction data. They found that the essential features of the structures are the approximate cubic closest packing of oxygens and corner-shared [SiO₄] tetrahedra forming [Si₂O₇] sorosilicate groups. Mg and Ni occupy octahedral sites as in olivine and each octahedron shares six

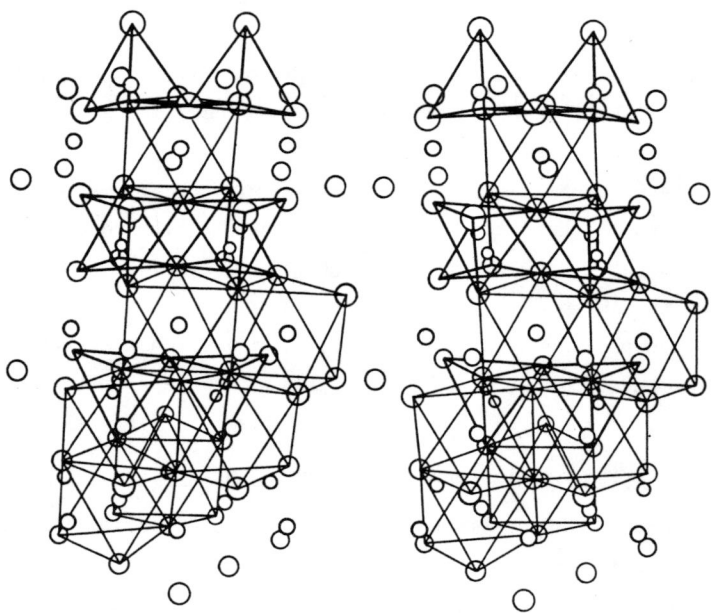

Figure 34. Stereo-pair view of the structure of β-Co₂SiO₄ projected on (100). Largest circles, oxygen; intermediate-sized circles, cobalt; smallest circles, silicon. (From Akimoto *et al.*, 1976).

edges with other octahedra but none with tetrahedra. The Moore–Smith structure has since been confirmed and refined by Morimoto et $al.$ (1974) for β-Co_2SiO_4 which is pictured in Figure 34.

Although no crystals of γ-Mg_2SiO_4 suitable for single-crystal x-ray work have yet been synthesized, the u parameter of this spinel has been estimated by interpolation using the well refined structures for γ-Ni_2SiO_4 (Yagi et $al.$, 1974; Ma, 1975), γ-Co_2SiO_4 (Morimoto et $al.$, 1974), and γ-Fe_2SiO_4 (Yagi et $al.$, 1974; Finger et $al.$, 1979). These values were reported earlier in the section on "Olivine and Spinel Structure Type." In addition, the u value of γ-Mg_2SiO_4 can be predicted using the regression model of Hill et $al.$ (1979) and is found to be 0.3666. The structure of γ-Ni_2SiO_4 has been studied at pressures up to 38 kbar by Finger et $al.$ (1979) who find the u parameter to be essentially constant with increasing pressure, as is the Si-O distance.

Structural Rationalization of the Stabilities of the α-, β-, and γ-Polymorphs

Kamb (1968) rationalized the difference in stability between the olivine and γ-spinel polymorphs of $(Mg,Fe)_2SiO_4$ by noting that the shared edges in olivine are shortened relative to the unshared edges, whereas those in γ-$(Mg,Fe)_2SiO_4$ are predicted to be lengthened in violation of Pauling's third rule. He also noted that the types of shared edges in olivines (both octahedron-octahedron and octahedron-tetrahedron) relative to γ-spinel (only octahedron-octahedron edges) favors the γ-spinel form at low pressures because of reduced cation-cation repulsions. However, the shared edge shortening in olivines counterbalances this destabilizing effect. Another manifestation of these observations recognized by Kamb (1968) is found in the ratio of mean octahedral bond length, d_A, and mean tetrahedral bond length, d_B, in A_2BO_4 compounds of the olivine and γ-spinel types. Kamb found that those A_2BO_4 compounds with $d_A/d_B < 1.9$ preferred the γ-spinel structure type whereas those with $d_A/d_B > 1.9$ preferred the olivine structure type. This prediction is borne out by Mg_2SiO_4 and Fe_2SiO_4 which have ratios of 1.30 and 1.33, respectively. As pressure is increased on olivine, we now know that the $[(Mg,Fe)O_6]$ octahedra compress significantly whereas the $[SiO_4]$ tetrahedron remains constant. Thus the distance ratio is reduced and eventually falls below 1.9 at which point the γ-spinel polymorph is formed. We can restate this in another way which follows our earlier discussion of the effect of pressure

on the olivine structure. As pressure increases on the olivine polymorph, the octahedra are compressed but the tetrahedra are not. Because
of shared edges between octahedra and tetrahedra, the dimensional mismatch
will eventually reach a point with increasing pressure such that the olivine arrangement is unstable relative to the γ-spinel or β-spinel arrangements, neither of which have shared octahedral-tetrahedral edges.

A more elaborate structural rationalization of the stability differ-
ences among the α-, β-, and γ-polymorphs of M_2SiO_4 compounds has been offered by Sung and Burns (1978). However, its essential details are similar
to those presented above. Using quite a different approach based on differences in structure energy among the α-, β-, and γ-polymorphs, Tokonami
et al. (1972) found that the calculated internal energies suggest decreasing stability in the order $\alpha > \beta > \gamma$. Their estimates of entropies
of the polymorphs increase in the order $\gamma < \beta < \alpha$ as do molar volumes.
Therefore, the β-phase is predicted to have a field of stability at high
temperatures.

The effects of crystal-field stabilization on the olivine $\rightarrow$ spinel
transformation has been examined by Syono *et al*. (1971), Mao and Bell
(1972), and most recently by Burns and Sung (1978). Syono *et al*. first
suggested that the excess CFSE that certain transition metal cations gain
in spinel relative to olivine leads to remarkably lower transformation
pressures to the γ-spinel phase. Mao and Bell considered this effect in
Fe_2SiO_4 and calculated a lowering of about 98 kbar caused by crystal-
field effects. In a more detailed study, Burns and Sung found that due
to crystal-field effects, the olivine to γ-spinel transformation in
$(Mg_{1-x}Fe_x)_2SiO_4$ composition is lowered in pressure by about $50x$ kbar
They concluded that this effect is equivalent to having the olivine-
spinel boundary raised in the upper mantle by about 15 km. The β-spinel
polymorph would presumably show this effect as well, but possibly to a
lesser extent than γ-spinels.

Navrotsky *et al*. (1979) carried out high-temperature calorimetric
measurements on α- and γ-Fe_2SiO_4 and α-, β-, and γ-Co_2SiO_4 and from these
data calculated the phase relations at high P and T. Their results agree
qualitatively with the stability relations proposed by Tokonami *et al*.
(1972). Their study shows that entropy effects are quite significant
in determining the relative stabilities of the three polymorphs and are

356

not constant with changing composition. This indicates that structural arguments such as those made earlier may not be adequate to rationalize stability differences among these polymorphs. They also found that small amounts of M^{2+}-Si^{4+} disorder in the γ-spinel or β-spinel phase may have a significant effect on phase boundaries.

Geophysical Consequences of the Olivine $\rightarrow$ (γ,β)-Spinel Transformation

Although detailed discussion of the geophysical consequences of these transformations is beyond the scope of this review, they may be summarized briefly as follows:

(1) The olivine to γ- and β-spinel transformation for the $(Mg_{0.9}Fe_{0.1})_2SiO_4$ composition is probably responsible for the 400 km discontinuity in seismic wave velocities (Ringwood, 1969).

(2) The depth at which this transformation takes place is raised in the superadiabatic, descending lithospheric slabs relative to the surrounding mantle, causing an increase in relative density and, therefore, acts to increase the downward pull on the slab (Ringwood, 1972; Schubert *et al.*, 1975; Ringwood, 1976).

(3) This transformation, especially for Fe-rich compositions and in the presence of mineralizers, may trigger deep focus earthquakes (Ringwood, 1972; Schubert *et al.*, 1975; Sung and Burns, 1978).

TRANSPORT PROPERTIES OF OLIVINES

The transport of cations (diffusion), charged species (conductivity), and heat (radiative transfer or lattice thermal conductivity) has been measured in numerous studies of olivines. A brief summary of this literature is given below.

Cation Diffusion

Several measurements of lattice diffusion of cations in olivines have been made as functions of composition, temperature, pO_2, pressure, and crystallographic orientation (Fe-Mg: Buening and Buseck, 1973; Misener, 1974), (Co-Mg: Morioka, 1980), (Ni-Mg: Clark and Long, 1971). The studies by Buening and Buseck and by Misener showed that interdiffusion of Fe and Mg in olivine (1) decreases with increasing Mg content, (2) decreases with increasing pressure, and (3) increases with increasing pO_2. In addition, they found that interdiffusion is sensitive to crystallographic orientation with the greatest interdiffusion parallel to [001], the direction of the octahedral chain in olivine. The diffusion anisotropy in olivine was theoretically analyzed by Ohashi and Finger (1974) who explained the highest diffusion rate along c by zig-zag jumps alternating along the M(1)-M(2)-M(1)-M(2)$\cdots$ zig-zag chain rather than straight through the M(1) chain. Misener's calculation of electrical conductivity in olivines, based on his diffusion measurements, agrees with observed conductivity measurements.

Morioka measured interdiffusion coefficients for Co-Mg as a function of temperature and crystallographic orientation and found that Co-Mg mobility in olivines is low compared to the Fe-Mg interdiffusion observed by Buening and Buseck and by Misener.

Clark and Long (1971) found Ni-Mg interdiffusion to be temperature dependent and to be greatest parallel to the c axis. The interdiffusion coefficient is least for Ni-Mg in forsterite and is greatest for Fe-Mg interdiffusion, with Co-Mg interdiffusion intermediate. The order of these relative interdiffusion coefficients can be rationalized by noting that the order of octahedral site preference energy of Ni^{2+}, Co^{2+}, and Fe^{2+} in olivines decreases in that order.

The diffusion of Mg and Fe in olivines can result in zoning. Onorato et $al.$ (1978) have proposed a kinetic model for zoning in olivines which takes into account diffusion in both solid and liquid during crystallization, as well as diffusion in the solid after crystallization is complete. This model has been used to predict zoning profiles and cooling rates for olivines and has been referred to by these authors as an "olivine-cooling speedometer."

Conductivity

There has been a great deal of interest in the electrical conductivity of olivines of mantle composition and its temperature-, pressure- and pO_2-dependence because of the need for these data in constructing geothermal profiles of the Earth. The papers by Duba and co-workers covering the period 1972 to 1976 (Duba, 1972; Duba and Nicholls, 1973; Duba et $al.$, 1973, 1974, 1976) are recommended as an introduction to the conductivity literature on olivines and the effect of the above variables on olivine conductivity. The papers by Will et $al.$ (1979) and Cemic et $al.$ (1980) contain a more up-to-date summary of the past literature on olivines and present new measurements on olivine, including Ni_2SiO_4, as functions of

P, T, X, and pO_2. The trends in conductivity as functions of these variables are generally too detailed to present here. However, it is important to point out that ferric iron and defects in olivines that are otherwise similar can cause differences in conductivity of several orders of magnitude.

One of the most interesting studies of conductivity in olivines was the one carried out by Mao and Bell (1972) at pressures above 100 kbar. They found that the absorption edges of the olivine and γ-spinel forms of Fe_2SiO_4 shifted rapidly with pressure from the near UV to the lower energy IR region with a simultaneous exponential increase in electrical conductivity. They attributed this increase to a new, efficient charge transfer process.

Radiative Heat Transfer

The red shift of absorption observed by Mao and Bell (1972) has important implications about radiative heat transfer in the earth's mantle. If this phenomenon is operative in the mantle, heat transfer by radiation would be severely blocked at depths greater than 500 km. Radiative heat transfer in olivines at low pressure has been studied experimentally by Shankland *et al.* (1979) and discussed by Mao (1976). These two papers serve as good introductions to this type of transport process in olivines.

Lattice Thermal Conductivity

Thermal conductivity (or thermal diffusivity) in olivines has been studied by Holt (1975) and by Kieffer (1976). Changes in lattice thermal conductivity accompany the olivine→spinel transition and the spinel→post-spinel phase changes in the mantle.

ELECTRICAL AND MECHANICAL PROPERTIES OF OLIVINES

A very extensive literature exists which discusses the mechanical and elastic properties of olivines. An understanding of these properties is important for the derivation and utilization of olivine equations-of-state and for an understanding of the flow properties and deformation of olivines under mantle conditions. The recent paper by Jeanloz (1980) is recommended as a source of information and further references on shock effects in olivines.

Several interesting and unexpected results were found in a shocked synthetic olivine (Jeanloz *et al.*, 1977) and in a naturally shocked olivine from the Coorara chondrite (Mason *et al.*, 1968). Jeanloz *et al.* reported the discovery of glass of olivine composition in an experimentally shocked-deformed single crystal of natural peridot (Fo_{88}) recovered from peak pressures of about 56×10^9 pascals. This is the first report of olivine glass; however, closer examination of naturally shocked olivine may show that olivine glass exists in nature. The study by Mason *et al.* of naturally shocked olivine from a chondrite showed garnet of approximate composition $Mg_3Fe_2Si_3O_{12}$ replacing olivine. These workers suggested that an olivine of composition Fo_{75} underwent transformation to garnet because of some shock event. The transformation of Mn_2SiO_4 (olivine) to $MnSiO_3$ (garnet) plus MnO has been performed experimentally (Ringwood, 1975). This transition may be significant in the earth's mantle.

The hardness of olivine and its temperature dependence have been studied extensively by Evans and Goetze (1979), and their paper is recommended as a source of further references. Finally, a discussion of the elastic properties of the olivine, γ-spinel, and β-spinel polymorphs of silicate and germanate olivines can be found in the paper by Lieberman (1975).

TABLE A1. OLIVINE FORMULA UNITS, SPECIFIC GRAVITY, UNIT CELL DIMENSIONS AND CATION RADIUS (From Brown, 1970)

ABBREV.	FORMULA UNIT	SPECIFIC GRAVITY	CALCULATED DENSITY	a	b	c	V	r_m	REF.
(Ni)	Ni_2SiO_4	4.88	4.914	4.7274(5)	10.118(3)	5.9105(8)	282.7(1)	0.700	1
(Fo)	$(Mg_{.90}Fe_{.10})_2SiO_4$		3.342	4.762(5)	10.225(10)	5.994(6)	291.9(5)	0.726	2
(Co)	Co_2SiO_4		4.716	4.7811(7)	10.2998(9)	6.0004(4)	295.49(5)	0.735	1
(Ho)	$(Mg_{.49}Fe_{.49}Mn_{.01}Ca_{.01})_2SiO_4$		3.827	4.787(5)	10.341(10)	6.044(6)	299.2(5)	0.748	2
(Fa)	$(Fe_{.92}Mg_{.04}Mn_{.04}Ca_{.002})_2SiO_4$		4.343	4.816(5)	10.469(10)	6.099(6)	307.5(5)	0.770	2
(Zn-Pi)	$(Mn_{.65}Mg_{.17}Zn_{.11}Fe_{.06})_2SiO_4$	3.96	4.082	4.8334(5)	10.567(3)	6.1732(2)	315.29(9)	0.791	1
(Kn)	$(Fe_{.51}Mn_{.46}Mg_{.02})_2SiO_4$	4.16	4.248	4.844(1)	10.577(4)	6.146(2)	314.9(2)	0.792	1
(Mo)	$(Mg_{.50}Ca_{.50})_2SiO_4$		3.038	4.822(1)	11.108(3)	6.382(2)	341.8(2)	0.860	3
(Ki)	$(Ca_{.57}Fe_{.43})_2SiO_4$	3.50	3.487	4.892(5)	11.180(2)	6.469(4)	353.8(4)	0.901	1
(Gl)	$(Ca_{.49}Mn_{.46}Mg_{.05}Fe_{.004})_2SiO_4$	3.46	3.448	4.9131(6)	11.1466(4)	6.4885(2)	355.34(5)	0.903	1
(α-Ca)	Ca_2SiO_4		2.912	5.091(10)	11.371(20)	6.782(10)	392.6(12)	1.000	4

(1) Brown (1970)   (2) Birle et al. (1968)   (3) Onken (1965)   (4) Smith et al. (1965)

TABLE A2. OLIVINE POSITIONAL PARAMETERS, ISOTROPIC TEMPERATURE FACTORS, AND R.M.S. EQUIVALENTS[a] (From Brown, 1970)

		HCP	Ni	Fo	Co	Ho	Fa	Zn-Pi	Kn	Mo	Ki	Gl	α-Ca
M(1)	x	0.0	0.0	0.0	0.0	0.0	0.0	0.0	0.0	0.0	0.0	0.0	0.0
	y	0.0	0.0	0.0	0.0	0.0	0.0	0.0	0.0	0.0	0.0	0.0	0.0
	z	0.0	0.0	0.0	0.0	0.0	0.0	0.0	0.0	0.0	0.0	0.0	0.0
	B		0.13(2)	0.34(4)	0.21(3)	0.34(3)	0.43(1)	0.66(2)	0.39(1)	0.35(4)	0.55(3)	0.36(3)	0.85(4)
	<μ>		0.041	0.066	0.052	0.066	0.074	0.091	0.070	0.067	0.083	0.068	0.104
M(2)	x	0.0	0.9925(1)	0.9896(3)	0.9915(1)	0.9870(2)	0.9864(1)	0.9873(1)	0.9888(1)	0.9770(2)	0.9802(4)	0.9799(2)	0.9893(5)
	y	0.25	0.2735(1)	0.2776(1)	0.2764(1)	0.2790(1)	0.2801(1)	0.2796(1)	0.2799(1)	0.2767(1)	0.2783(1)	0.2780(1)	0.2809(2)
	z	0.25	0.25	0.25	0.25	0.25	0.25	0.25	0.25	0.25	0.25	0.25	0.25
	B		0.14(2)	0.36(4)	0.21(3)	0.45(3)	0.40(1)	0.64(2)	0.41(1)	0.45(3)	0.47(3)	0.34(3)	0.80(4)
	<μ>		0.042	0.068	0.052	0.075	0.071	0.090	0.072	0.075	0.077	0.066	0.101
Si	x	0.375	0.4277(3)	0.4226(2)	0.4282(3)	0.4282(3)	0.4305(1)	0.4250(3)	0.4277(2)	0.4101(2)	0.4181(4)	0.4156(3)	0.4287(6)
	y	0.0833	0.0946(1)	0.0945(1)	0.0949(1)	0.0960(1)	0.0972(1)	0.0932(2)	0.0952(1)	0.0811(1)	0.0846(2)	0.0861(2)	0.0973(3)
	z	0.25	0.25	0.25	0.25	0.25	0.25	0.25	0.25	0.25	0.25	0.25	0.25
	B		0.02(3)	0.20(3)	0.13(3)	0.10(3)	0.30(1)	0.42(2)	0.28(2)	0.40(4)	0.25(4)	0.22(3)	0.71(5)
	<μ>		0.016	0.050	0.041	0.036	0.062	0.073	0.060	0.071	0.056	0.053	0.095
O(1)	x	0.75	0.7689(7)	0.7667(6)	0.7656(6)	0.7676(7)	0.7674(2)	0.7580(7)	0.7618(4)	0.7447(6)	0.7508(14)	0.7453(8)	0.7495(15)
	y	0.0833	0.0931(4)	0.0918(3)	0.0918(4)	0.0915(4)	0.0920(2)	0.0877(4)	0.0904(2)	0.0777(3)	0.0822(5)	0.0833(4)	0.0872(9)
	z	0.25	0.25	0.25	0.25	0.25	0.25	0.25	0.25	0.25	0.25	0.25	0.25
	B		0.24(5)	0.33(5)	0.32(5)	0.29(6)	0.50(2)	0.78(5)	0.54(5)	0.68(7)	0.95(9)	0.49(7)	0.91(12)
	<μ>		0.055	0.065	0.064	0.061	0.080	0.099	0.083	0.093	0.110	0.079	0.107
O(2)	x	0.25	0.2192(7)	0.2202(6)	0.2162(6)	0.2144(6)	0.2110(3)	0.2227(7)	0.2141(4)	0.2465(6)	0.2248(13)	0.2270(9)	0.2035(16)
	y	0.4167	0.4452(3)	0.4477(3)	0.4484(3)	0.4505(4)	0.4532(1)	0.4511(4)	0.4518(2)	0.4484(3)	0.4516(5)	0.4541(4)	0.4635(7)
	z	0.25	0.25	0.25	0.25	0.25	0.25	0.25	0.25	0.25	0.25	0.25	0.25
	B		0.19(5)	0.35(5)	0.27(5)	0.12(6)	0.51(2)	0.71(5)	0.41(3)	0.70(7)	0.36(9)	0.56(7)	0.94(12)
	<μ>		0.049	0.067	0.058	0.039	0.080	0.095	0.072	0.094	0.068	0.084	0.109
O(3)	x	0.25	0.2748(5)	0.2781(4)	0.2811(4)	0.2845(5)	0.2883(2)	0.2836(5)	0.2881(3)	0.2729(4)	0.2877(10)	0.2806(6)	0.2981(9)
	y	0.1667	0.1625(2)	0.1633(2)	0.1638(2)	0.1635(2)	0.1654(1)	0.1605(3)	0.1627(1)	0.1472(2)	0.1524(3)	0.1531(3)	0.1619(4)
	z	0.0	0.0295(3)	0.0337(4)	0.0340(4)	0.0339(5)	0.0362(2)	0.0391(4)	0.0388(2)	0.0452(4)	0.0513(6)	0.0486(4)	0.0570(8)
	B		0.10(4)	0.40(4)	0.29(4)	0.29(5)	0.53(1)	0.70(4)	0.54(3)	0.59(5)	0.60(7)	0.52(5)	0.78(8)
	<μ>		0.036	0.071	0.061	0.061	0.082	0.094	0.083	0.086	0.087	0.081	0.099

[a]Estimated standard errors are in brackets and refer to the last decimal place.

TABLE A3. OCTAHEDRAL SITE OCCUPANCIES[*]

OLIVINE	M(1) OCCUPANCY	M(2) OCCUPANCY	σ[**]
Ni	Ni	Ni	
Fo	0.90Mg + 0.10Fe	0.90Mg + 0.10Fe	0.01
Co	Co	Co	
Ho	0.50Mg + 0.50Fe	0.50Mg + 0.50Fe	0.01
Fa	0.92Fe + 0.04Mg + 0.04Mn	0.92Fe + 0.04Fe + 0.04Mn	
Zn-Pi	0.43Mn+0.34Mg+0.18Zn+0.04Fe	0.87Mn+0.08Fe+0.05Zn	0.07
Kn	0.66Fe + 0.29Mn + 0.04Mg	0.37Fe + 0.63Mn	0.04
Mo	Mg	Ca	
Ki	0.85Fe + 0.15Ca	0.02Fe + 0.98Ca	0.01
Gl	0.91Mn + 0.09Mg	0.02Mn + 0.98Ca	0.01
α-Ca	Ca	Ca	

[*]Occupancy values from the study of Brown (1970)
[**]Estimated standard error of the occupancy value

Table A4. Olivine Bond Lengths

	Ni	Fo	Co	Ho	Fa	Zn-Pi	Kn	Mo	Ki	Gl	Ca
TETRAHEDRON											
[1]* Si-O(1)	1.613(3)**	1.620(4)	1.613(3)	1.625(4)	1.623(2)	1.610(3)	1.619(2)	1.615(4)	1.628(8)	1.620(4)	1.637(8)
[1] Si-O(2)	1.663(3)	1.656(3)	1.659(3)	1.652(4)	1.655(2)	1.656(4)	1.665(2)	1.656(3)	1.643(5)	1.630(5)	1.665(8)
[2] Si-O(3)	1.641(2)	1.636(3)	1.637(2)	1.633(3)	1.636(1)	1.632(2)	1.629(2)	1.640(2)	1.622(4)	1.645(3)	1.642(5)
<Si-O>	1.640(1)	1.637(2)	1.636(1)	1.636(2)	1.638(1)	1.632(1)	1.636(1)	1.641(1)	1.629(3)	1.635(2)	1.646(3)
[1] O(1)-O(2)	2.750(4)	2.747(5)	2.736(4)	2.730(5)	2.724(3)	2.732(4)	2.732(3)	2.771(5)	2.747(9)	2.734(6)	2.702(11)
[2] O(1)-O(3)	2.765(4)	2.762(4)	2.756(3)	2.758(4)	2.759(2)	2.746(3)	2.745(2)	2.736(4)	2.720(8)	2.742(5)	2.778(9)
[2] O(2)-O(3)[a]	2.556(4)	2.557(4)	2.570(3)	2.561(4)	2.576(2)	2.560(4)	2.582(2)	2.569(4)	2.587(5)	2.575(5)	2.610(8)
[1] O(3)-O(3)[a]	2.607(4)	2.593(6)	2.593(5)	2.612(6)	2.608(3)	2.599(4)	2.596(3)	2.615(5)	2.571(8)	2.614(6)	2.618(10)
<O-O>	2.666(2)	2.663(2)	2.664(2)	2.663(2)	2.667(1)	2.657(2)	2.664(1)	2.666(2)	2.655(3)	2.664(2)	2.683(4)
M(1) OCTAHEDRON											
[2] M(1)-O(1)	2.065(2)	2.088(2)	2.098(2)	2.101(3)	2.123(1)	2.148(2)	2.146(2)	2.194(2)	2.224(5)	2.252(3)	2.343(6)
[2] M(1)-O(2)	2.062(2)	2.075(2)	2.092(2)	2.101(3)	2.122(2)	2.106(2)	2.130(2)	2.091(2)	2.173(4)	2.166(3)	2.308(6)
[2] M(1)-O(3)	2.102(2)	2.141(2)	2.166(2)	2.181(3)	2.230(2)	2.190(2)	2.229(2)	2.120(2)	2.235(4)	2.216(3)	2.418(5)
<M(1)-O>	2.076(1)	2.101(1)	2.119(1)	2.128(1)	2.158(1)	2.148(1)	2.168(1)	2.135(1)	2.211(2)	2.211(1)	2.356(2)
[2] O(1)-O(3)[b]	2.813(4)	2.853(5)	2.882(3)	2.895(4)	2.930(3)	2.953(3)	2.961(2)	2.966(4)	3.028(8)	3.037(5)	3.200(9)
[2] O(1)-O(3')	3.075(3)	3.122(4)	3.143(3)	3.156(4)	3.221(2)	3.178(4)	3.221(2)	3.132(4)	3.273(6)	3.277(5)	3.526(9)
[2] O(1)-O(2)[b]	2.845(5)	2.854(4)	2.884(4)	2.877(5)	2.902(3)	2.894(4)	2.932(3)	2.844(5)	2.952(9)	2.968(6)	3.121(12
[2] O(1)-O(2')	2.990(1)	3.032(3)	3.038(1)	3.064(3)	3.098(2)	3.119(1)	3.114(1)	3.206(2)	3.259(2)	3.273(1)	3.449(3)
[2] O(2)-O(3')	3.289(4)	3.353(4)	3.396(4)	3.432(5)	3.510(3)	3.452(4)	3.514(2)	3.336(4)	3.569(7)	3.546(5)	3.942(8)
[2] O(2)-O(3)[a]	2.556(4)	2.557(4)	2.570(3)	2.561(4)	2.576(2)	2.560(4)	2.582(2)	2.569(4)	2.587(5)	2.575(5)	2.610(8)
<O-O>	2.928(2)	2.962(2)	2.986(1)	2.998(2)	3.040(1)	3.026(1)	3.054(1)	3.009(2)	3.111(3)	3.113(2)	3.308(4)
M(2) OCTAHEDRON											
[1] M(2)-O(1)	2.110(4)	2.176(3)	2.187(4)	2.205(4)	2.234(2)	2.311(4)	2.286(2)	2.478(3)	2.463(6)	2.452(4)	2.520(9)
[1] M(2)-O(2)	2.041(3)	2.057(3)	2.071(3)	2.081(4)	2.110(2)	2.144(3)	2.120(2)	2.309(4)	2.278(6)	2.308(4)	2.350(8)
[2] M(2)-O(3)	2.178(2)	2.221(3)	2.224(2)	2.272(3)	2.293(2)	2.305(2)	2.307(2)	2.411(3)	2.428(4)	2.414(3)	2.453(6)
[2] M(2)-O(3")	2.051(2)	2.067(3)	2.072(2)	2.058(3)	2.069(2)	2.139(2)	2.113(2)	2.289(2)	2.299(4)	2.303(3)	2.390(6)
<M(2)-O>	2.102(1)	2.135(1)	2.142(1)	2.158(1)	2.178(1)	2.224(1)	2.208(1)	2.364(1)	2.366(2)	2.366(1)	2.426(3)
[2] O(1)-O(3")	2.974(4)	3.027(4)	3.041(3)	3.061(4)	3.084(2)	3.209(4)	3.160(2)	3.593(4)	3.555(6)	3.519(5)	3.542(9)
[2] O(1)-O(3)[b]	2.813(4)	2.854(4)	2.882(3)	2.895(4)	2.930(3)	2.953(3)	2.961(2)	2.966(4)	3.028(8)	3.037(5)	3.200(9)
[2] O(2)-O(3)	3.155(4)	3.196(4)	3.220(3)	3.260(4)	3.304(2)	3.354(4)	3.341(3)	3.595(4)	3.597(6)	3.610(5)	3.703(9)
[2] O(2)-O(3"')	2.886(3)	2.935(4)	2.927(3)	2.927(4)	2.954(2)	3.019(2)	2.979(2)	3.147(4)	3.118(6)	3.161(4)	3.261(8)
[1] O(3)-O(3)[a]	2.607(4)	2.593(6)	2.593(5)	2.612(6)	2.608(3)	2.599(4)	2.596(3)	2.615(5)	2.571(8)	2.614(6)	2.618(10)
[2] O(3)-O(3")	2.974(4)	2.996(3)	3.006(3)	3.016(4)	3.022(2)	3.110(3)	3.082(2)	3.371(4)	3.345(5)	3.332(4)	3.331(7)
[1] O(3")-O(3"')	3.304(4)	3.401(6)	3.408(5)	3.432(6)	3.491(4)	3.574(4)	3.550(4)	3.771(5)	3.898(8)	3.875(6)	4.166(10)
<O-O>	2.960(1)	3.001(1)	3.013(1)	3.030(1)	3.057(1)	3.122(1)	3.099(1)	3.311(1)	3.313(2)	3.317(1)	3.405(3)
O(1) TETRAHEDRON											
[2] O(1)-M(1)_B	2.065(2)	2.088(2)	2.098(2)	2.101(3)	2.123(1)	2.148(2)	2.146(2)	2.194(2)	2.224(5)	2.252(3)	2.343(6)
[1] O(1)-M(2)_B	2.110(4)	2.176(3)	2.187(4)	2.205(4)	2.234(2)	2.311(4)	2.286(2)	2.478(3)	2.463(6)	2.452(4)	2.520(9)
[1] O(1)-Si_A	1.613(3)	1.620(4)	1.613(3)	1.625(4)	1.623(2)	1.610(3)	1.619(2)	1.615(4)	1.628(8)	1.620(4)	1.637(8)
<O(1)-M>	1.963(1)	1.993(1)	1.999(1)	2.008(2)	2.026(1)	2.054(1)	2.049(1)	2.120(1)	2.135(3)	2.144(2)	2.211(4)
[1] M(1)_B-M(1)_B	2.955(1)	2.997(2)	3.000(1)	3.022(2)	3.050(3)	3.086(1)	3.073(1)	3.193(2)	3.234(2)	3.244(1)	3.392(2)
[2] M(1)_B-M(2)_B	3.138(1)	3.210(2)	3.218(1)	3.257(2)	3.306(1)	3.334(1)	3.336(1)	3.466(1)	3.506(2)	3.499(1)	3.618(2)
[2] M(1)_B-Si_A	3.228(1)	3.261(3)	3.268(1)	3.280(3)	3.299(2)	3.329(1)	3.325(1)	3.385(3)	3.412(4)	3.435(1)	3.545(4)
[1] M(2)_B-Si_A	3.227(1)	3.270(3)	3.279(1)	3.276(3)	3.291(2)	3.359(1)	3.347(1)	3.493(3)	3.508(4)	3.501(2)	3.537(4)

	Ni	Fo	Co	Ho	Fa	Zn-Pi	Kn	Mo	Ki	Gl	Ca
O(2) TETRAHEDRON											
[2] O(2)-M(1)$_B$	2.062(2)	2.075(2)	2.092(2)	2.101(3)	2.122(2)	2.106(2)	2.130(2)	2.091(2)	2.173(4)	2.166(3)	2.308(6)
[1] O(2)-M(2)$_A$	2.041(3)	2.057(3)	2.071(3)	2.081(4)	2.110(2)	2.144(3)	2.120(2)	2.309(4)	2.278(6)	2.308(4)	2.346(8)
[1] O(2)-Si$_B$	1.663(3)	1.656(3)	1.659(3)	1.652(4)	1.655(2)	1.656(4)	1.665(2)	1.656(3)	1.643(5)	1.630(5)	1.665(8)
<O(2)-M>	1.957(1)	1.966(1)	1.978(1)	1.984(1)	2.002(1)	2.003(1)	2.011(1)	2.037(1)	2.067(2)	2.068(2)	2.157(4)
[1] M(1)$_B$-M(1)$_B$	2.955(1)	2.997(2)	3.000(1)	3.022(2)	3.050(3)	3.086(1)	3.073(1)	3.193(2)	3.234(2)	3.244(1)	3.392(2)
[2] M(1)$_B$-M(2)$_A$	3.632(1)	3.650(2)	3.669(1)	3.679(2)	3.707(2)	3.734(1)	3.730(1)	3.882(2)	3.902(2)	3.910(1)	3.982(3)
[2] M(1)$_B$-Si$_B$	2.681(1)	2.703(2)	2.720(1)	2.733(2)	2.768(2)	2.750(1)	2.769(1)	2.697(2)	2.774(2)	2.779(1)	2.978(3)
[1] M(2)$_A$-Si$_B$	3.270(2)	3.265(2)	3.302(2)	3.304(2)	3.344(2)	3.340(2)	3.359(2)	3.426(2)	3.461(3)	3.473(2)	3.624(4)
O(3) TETRAHEDRON											
[1] O(3)-M(1)$_B$	2.102(2)	2.141(2)	2.166(2)	2.181(3)	2.230(2)	2.190(2)	2.229(2)	2.120(2)	2.235(4)	2.216(3)	2.418(5)
[1] O(3)-M(2)$_B$	2.178(2)	2.221(3)	2.224(2)	2.272(3)	2.293(2)	2.305(2)	2.307(2)	2.411(3)	2.428(4)	2.414(3)	2.453(6)
[1] O(3)-M(2)$_A$	2.051(2)	2.067(2)	2.072(2)	2.058(3)	2.069(2)	2.139(2)	2.113(2)	2.289(3)	2.299(4)	2.303(3)	2.390(5)
[1] O(3)-Si$_B$	1.641(2)	1.636(3)	1.637(2)	1.633(3)	1.636(1)	1.632(2)	1.629(2)	1.640(2)	1.622(4)	1.645(3)	1.642(5)
<O(3)-M>	1.993(1)	2.016(1)	2.025(1)	2.036(1)	2.057(1)	2.066(1)	2.070(1)	2.115(1)	2.146(3)	2.144(1)	2.226(3)
[1] M(1)$_B$-M(2)$_B$	3.138(1)	3.210(2)	3.218(1)	3.257(2)	3.306(1)	3.334(1)	3.336(1)	3.466(1)	3.508(1)	3.499(1)	3.618(2)
[1] M(1)$_B$-M(2)$_A$	3.585(1)	3.585(2)	3.616(2)	3.598(2)	3.621(2)	3.655(1)	3.659(1)	3.742(2)	3.788(2)	3.783(1)	3.911(3)
[1] M(1)$_B$-M(2)$_A$	2.681(1)	2.703(2)	2.720(1)	2.733(2)	2.768(2)	2.750(1)	2.769(1)	2.697(2)	2.774(2)	2.779(1)	2.978(3)
[1] M(2)$_B$-M(2)$_A$	3.814(1)	3.869(2)	3.875(1)	3.901(2)	3.936(2)	3.969(1)	3.964(1)	4.045(2)	4.104(2)	4.117(1)	4.299(2)
[1] M(2)$_B$-Si$_B$	2.741(2)	2.799(2)	2.803(1)	2.836(2)	2.870(2)	2.889(1)	2.888(1)	3.015(2)	3.046(3)	3.026(2)	3.061(4)
[1] M(2)$_A$-Si$_B$	3.257(1)	3.284(2)	3.294(1)	3.299(2)	3.320(2)	3.381(1)	3.358(1)	3.578(2)	3.592(2)	3.594(1)	3.677(3)
METAL-METAL DISTANCES											
M(1)$_B$-M(1)$_B$	2.955(1)	2.997(2)	3.000(1)	3.022(2)	3.050(3)	3.086(1)	3.073(1)	3.193(2)	3.234(2)	3.244(1)	3.392(2)
M(1)$_B$-M(2)$_B$	3.138(1)	3.210(2)	3.218(1)	3.257(2)	3.306(1)	3.334(1)	3.336(1)	3.466(1)	3.508(1)	3.499(1)	3.618(2)
M(1)$_B$-Si$_B$	2.681(1)	2.703(2)	2.720(1)	2.733(2)	2.768(2)	2.750(1)	2.769(1)	2.697(2)	2.774(2)	2.779(1)	2.978(3)
M(2)$_B$-Si$_B$	2.741(2)	2.799(2)	2.803(2)	2.836(2)	2.870(2)	2.889(2)	2.888(1)	3.015(2)	3.046(3)	3.026(2)	3.061(4)
M(2)$_A$-M(1)$_B$	3.632(1)	3.650(2)	3.669(1)	3.679(2)	3.707(2)	3.734(1)	3.730(1)	3.882(2)	3.902(2)	3.910(1)	3.982(3)
M(2)$_A$-M(2)	3.814(1)	3.869(2)	3.875(1)	3.901(2)	3.936(2)	3.969(1)	3.964(1)	4.045(2)	4.104(2)	4.117(1)	4.299(2)
M(1)$_B$-M(2)	3.585(1)	3.585(2)	3.616(2)	3.598(2)	3.621(2)	3.655(1)	3.659(1)	3.742(2)	3.778(2)	3.783(1)	3.911(3)
Si-Si	3.586(2)	3.634(2)	3.646(2)	3.681(3)	3.727(2)	3.732(2)	3.740(1)	3.767(2)	3.832(3)	3.860(2)	4.116(4)

*Multiplicity
**Estimated standard errors referring to the last decimal place

[a]Edge shared between an octahedron and tetrahedron
[b]Edge shared between two octahedra

Table A5. Olivine Bond Angles

	HCP	Ni	Fo	Co	Ho	Fa	Zn-Pi	Kn	Mo	Ki	Gl	Ca
TETRAHEDRON												
[1] O(1)-Si-O(2)	109.5	114.1(2)	114.0(2)	113.5(2)	112.8(2)	112.4(1)	113.5(2)	112.6(1)	115.8(2)	114.2(3)	114.5(2)	109.8(5)
[1] O(1)-Si-O(3)	109.5	116.4(1)	116.1(1)	116.0(1)	115.7(1)	115.7(1)	115.8(1)	115.4(1)	114.5(1)	113.6(2)	114.2(1)	115.8(3)
[2] O(2)-Si-O(3)[a]	109.5	101.3(1)	102.0(1)	102.5(1)	102.4(1)	103.0(1)	102.3(1)	103.2(1)	102.4(1)	104.8(3)	103.7(2)	104.2(3)
[1] O(3)-Si-O(3)[a]	109.5	105.2(2)	104.8(2)	104.7(2)	106.2(2)	105.7(1)	105.6(2)	105.7(1)	105.8(1)	104.8(3)	105.2(2)	105.7(4)
<O-Si-O>	109.5	109.1(1)	109.2(1)	109.2(1)	109.2(1)	109.2(1)	109.2(1)	109.2(1)	109.2(1)	109.3(1)	109.2(1)	109.2(2)

	HCP	Ni	Fo	Co	Ho	Fa	Zn-Pi	Kn	Mo	Ki	Gl	Ca
M(1) OCTAHEDRON												
[2] O(1)-M(1)-O(3)[b]	90.0	84.9(1)	84.9(1)	85.0(1)	85.1(1)	84.6(1)	85.8(1)	85.2(1)	86.9(1)	85.5(2)	85.6(1)	84.4(2)
[2] O(1)-M(1)-O(3')	90.0	95.1(1)	95.1(1)	95.0(1)	94.9(1)	95.4(1)	94.2(1)	94.8(1)	93.1(1)	94.5(2)	94.4(1)	95.6(2)
[2] O(1)-M(1)-O(2)[b]	90.0	87.2(1)	86.5(1)	87.0(1)	86.4(1)	86.3(1)	85.7(1)	86.5(1)	83.1(1)	84.3(2)	84.4(1)	84.3(2)
[2] O(1)-M(1)-O(2')	90.0	92.8(1)	93.5(1)	93.0(1)	93.6(1)	93.7(1)	94.3(1)	93.5(1)	96.9(1)	95.7(2)	95.6(1)	95.7(2)
[2] O(2)-M(1)-O(3')	90.0	104.3(1)	105.3(1)	105.8(1)	106.6(1)	107.5(1)	106.9(1)	107.4(1)	104.8(1)	108.1(2)	108.0(1)	113.0(2)
[2] O(2)-M(1)-O(3)[a]	90.0	75.7(1)	74.7(1)	74.2(1)	73.4(1)	72.5(1)	73.1(1)	72.6(1)	75.2(1)	71.9(2)	72.0(1)	67.0(2)
<O-M(1)-O>	90.0	90.0(1)	90.0(1)	90.0(1)	90.0(1)	90.0(1)	90.0(1)	90.0(1)	90.0(1)	90.0(1)	90.0(1)	90.0(1)

	HCP	Ni	Fo	Co	Ho	Fa	Zn-Pi	Kn	Mo	Ki	Gl	Ca
M(2) OCTAHEDRON												
[2] O(1)-M(2)-O(3")	90.0	91.2(1)	91.0(1)	91.1(1)	91.7(1)	91.5(1)	92.2(1)	91.7(1)	97.8(1)	96.5(1)	95.4(1)	92.3(2)
[2] O(1)-M(2)-O(3)[b]	90.0	82.0(1)	80.9(1)	81.6(1)	80.6(1)	80.7(1)	79.5(1)	80.3(1)	74.7(1)	76.5(2)	77.2(1)	80.1(2)
[2] O(2)-M(2)-O(3)	90.0	96.8(1)	96.6(1)	97.1(1)	96.9(1)	97.2(1)	97.8(1)	97.9(1)	99.2(1)	99.6(2)	99.7(1)	101.0(2)
[2] O(2)-M(2)-O(3''')	90.0	89.7(1)	90.7(1)	89.9(1)	90.0(1)	90.0(1)	89.6(1)	89.5(1)	86.4(1)	85.9(1)	86.6(1)	87.0(2)
[1] M(2)-O(3)[a]	90.0	73.5(1)	71.4(1)	71.3(1)	70.2(1)	69.3(1)	68.6(1)	68.5(1)	65.7(1)	63.9(2)	65.5(1)	64.5(2)
[2] O(3)-M(2)-O(3")	90.0	89.3(1)	88.6(1)	88.7(1)	88.2(1)	87.5(1)	88.7(1)	58.3(1)	91.6(1)	90.0(1)	89.8(1)	86.9(1)
[1] O(3)-M(2)-O(3"')	90.0	107.3(1)	110.7(2)	110.6(1)	112.9(2)	115.0(1)	113.3(1)	114.3(1)	110.9(1)	115.9(2)	114.6(2)	121.3(2)
<O-M(2)-O>	90.0	89.9(1)	89.8(1)	89.9(1)	89.8(1)	89.8(1)	89.8(1)	89.8(1)	89.7(1)	89.7(1)	89.8(1)	90.0(1)

	HCP	Ni	Fo	Co	Ho	Fa	Zn-Pi	Kn	Mo	Ki	Gl	Ca
O(1) TETRAHEDRON												
[1] M(1)$_B$-O(1)-M(1)$_B$	90.0	91.4(1)	91.7(1)	91.3(1)	92.0(1)	91.8(1)	91.8(1)	91.4(1)	93.4(1)	93.3(3)	92.2(1)	92.8(3)
[2] M(1)$_B$-O(1)-M(2)$_B$	90.0	97.5(1)	97.6(1)	97.4(1)	98.3(1)	98.7(1)	96.7(1)	97.6(1)	95.6(1)	96.8(2)	96.0(1)	96.1(2)
[2] M(1)$_B$-O(1)-Si$_A$	125.3	122.2(1)	122.7(1)	122.9(1)	122.8(1)	122.9(1)	124.1(1)	123.4(1)	124.8(1)	123.2(2)	124.2(1)	125.0(3)
[1] M(2)$_A$-O(1)-Si$_A$	125.3	119.5(2)	118.2(2)	118.5(2)	116.8(2)	116.2(1)	116.8(2)	117.0(1)	115.6(2)	116.1(3)	117.1(2)	115.0(5)
<M-O(1)-M>	107.6	108.4(1)	108.4(1)	108.4(1)	108.5(1)	108.5(1)	108.4(1)	108.4(1)	108.3(1)	108.2(1)	108.3(1)	108.3(1)

	HCP	Ni	Fo	Co	Ho	Fa	Zn-Pi	Kn	Mo	Ki	Gl	Ca
O(2) TETRAHEDRON												
[2] M(1)$_B$-O(2)-M(2)$_A$	131.8	124.5(1)	124.1(1)	123.6(1)	123.2(1)	122.3(1)	122.9(1)	122.7(1)	123.8(1)	122.5(2)	121.8(1)	117.6(2)
[1] M(2)$_A$-O(2)-Si$_B$	125.3	123.7(2)	122.8(2)	124.2(2)	124.1(2)	124.9(1)	122.5(2)	124.6(1)	118.6(2)	123.1(3)	122.8(1)	128.4(4)
[2] M(1)$_B$-O(2)-Si$_B$	78.9	91.4(1)	92.2(1)	92.2(1)	92.7(1)	93.4(1)	93.1(1)	92.9(1)	91.3(1)	92.2(2)	93.0(2)	95.7(3)
[1] M(1)$_B$-O(2)-M(1)$_B^-$	90.0	91.5(1)	92.5(1)	91.7(1)	92.0(1)	91.9(1)	94.2(1)	92.3(1)	99.5(1)	96.2(2)	97.0(2)	94.6(3)
<M-O(2)-M>	106.2	107.8(1)	108.0(1)	107.9(1)	108.0(1)	108.1(1)	108.1(1)	108.0(1)	108.0(1)	108.1(1)	108.2(1)	108.3(1)

	HCP	Ni	Fo	Co	Ho	Fa	Zn-Pi	Kn	Mo	Ki	Gl	Ca
O(3) TETRAHEDRON												
[1] M(1)$_B$-O(3)-M(2)$_B$	90.0	94.3(1)	94.8(1)	94.3(1)	94.0(1)	93.9(1)	95.7(1)	94.7(1)	99.6(1)	97.5(2)	98.1(1)	95.9(2)
[1] M(1)$_B$-O(3)-M(2)$_A$	131.8	119.4(1)	116.9(1)	117.1(1)	116.1(1)	114.7(1)	115.2(1)	114.8(1)	116.1(1)	112.9(2)	113.7(1)	108.9(2)
[1] M(1)$_B$-O(3)-Si$_B$	78.9	90.6(1)	90.4(1)	90.2(1)	90.4(1)	90.1(1)	90.8(1)	90.4(1)	90.8(1)	90.5(2)	90.8(1)	92.3(2)
[2] M(2)$_A$-O(3)-M(2)$_A$	131.8	128.8(1)	128.9(1)	128.8(1)	128.5(1)	128.9(1)	126.5(1)	127.4(1)	118.8(1)	120.5(2)	121.5(1)	125.2(2)
[1] M(1)$_B$-O(3)-Si$_B$	78.9	90.6(1)	91.8(1)	91.8(1)	91.7(1)	92.3(1)	92.8(1)	92.8(1)	94.2(1)	95.4(2)	94.5(1)	94.7(2)
[1] M(2)$_A$-O(3)-Si$_B$	125.3	123.4(1)	124.5(1)	124.9(1)	126.3(2)	126.8(1)	126.9(1)	127.2(1)	130.4(1)	131.9(3)	130.4(2)	130.7(3)
<M-O(3)-M>	106.2	107.8(1)	107.9(1)	107.8(1)	107.8(1)	107.8(1)	108.0(1)	107.9(1)	108.3(1)	108.1(1)	108.2(1)	108.0(1)

364

Adams, F. D., and R. F. D. Graham (1926) On some minerals from the Ruby mining district of Mogok, Upper Burma. Trans. Royal Soc. Canada, 3rd Ser., 20, 113.

Adams, L. H. (1931) The compressibility of fayalite and the velocity of elastic waves in peridotite with different iron-magnesium ratios. Beitr. Geophys., 31, 315-321.

Akella, J., and F. R. Boyd (1972) Partitioning of Ti and Al between pyroxenes, garnets, oxides, and liquid. *The Apollo 15 Lunar Samples*, Lunar Sci. Inst., 14-18.

Akimoto, S., E. Komada, and I. Kushiro (1967) Effect of pressure on the melting of olivine and spinel polymorph of Fe_2SiO_4. J. Geophys. Res., 72, 679-689.

Akimoto, S., Y. Matsui, and Y. Syono (1976) High-pressure crystal chemistry of ortho silicates and the formation of the mantle transition zone. In, *The Physics and Chemistry of Minerals and Rocks*, (Ed., R. J. G. Strens), New York: John Wiley & Sons, 327-363.

Alberti, A., and G. Vezzalini (1978) Madelung energies and cation distribution in olivine type structures. Z. Kristallogr., 147, 167-175.

Allen, W. C., and R. B. Snow (1955) The orthosilicate-rich oxide portion of the system CaO-"FeO"-SiO_2. J. Amer. Ceram. Soc., 38, 264-280.

Anderson, O. (1915) The system anorthite-forsterite-silica. Amer. J. Sci., 39, 407-454.

Atlas, L. (1952) Polymorphism of $MgSiO_3$ and solid state equilibria in the system $MgSiO_3-CaMgSi_2O_6$. J. Geol., 60, 125-147.

Bancroft, G. M., R. G. Borne, and A. G. Maddock (1967) Application of the Mössbauer effect to silicate mineralogy. Part I Iron silicates of known crystal structures. Geochim. Cosmochim. Acta, 31, 2219-2246.

Baker, I., and S. E. Haggerty (1967) The alteration of olivine in basaltic and associated lavas. Part II. Intermediate and low-temperature alteration. Contr. Mineral. Petrol., 16, 258-273.

Basso, R., A. Dal Negro, A. Della Guista, and G. Rossi (1979) Fe/Mg distribution in the olivine of ultrafemic nodules from Assab (Ethiopia). N. Jahrb. Mineral. Monat., 197-202.

Baur, W. H. (1972) Computer-simulated crystal structures of observed and hypothetical Mg_2SiO_4 polymorphs of low and high density. Amer. Mineral., 57, 704-731.

Bell, P. M. (1970) Analysis of olivine crystals in Apollo 12 rocks. Carnegie Inst. Wash. Year Book, 69, 228-229.

Berthet, G., J. C. Joubert, and E. F. Bertant (1972) Vacancies ordering in new metastable orthophosphates $[Co_3 \square]P_2O_8$ and $Mg_3 \square]P_2O_8$ with olivine related structure. Z. Kristallogr., 136, 98-105.

Biggar, G. M., and M. J. O'Hara (1969) Monticellite and forsterite crystalline solution. J. Amer. Cer. Soc., 52, 249-252.

Biggars, J. V. and A. Muan (1969) Activity-composition relations in orthosilicate and metasilicate solid solutions in the system $MnO-CaO-SiO_2$. J. Amer. Cer. Soc., 50, 230-235.

Binns, R. A., and D. I. Graves (1976) Iron-nickel partition in metamorphosed olivine-sulfide assemblages from Perserverance, Western Australia. Amer. Mineral., 61, 781-787.

Binns, R. A., R. J. Davis and S. J. B. Reed (1969) Ringwoodite, natural $(Mg,Fe)_2SiO_4$ spinel in the Tenham meteorite. Nature, 221, 943-944.

Birle, J. D., G. V. Gibbs, P. B. Moore, and J. V. Smith (1968) Crystal structures of natural olivines. Amer. Mineral., 53, 807-824.

Blasse, G. (1963) Die Kristallstruktur einiger Verbindungen vom Typ $LiMe^{3+}Me^{4+}O_4$ und $LiMe^{2+}Me^{5+}O_4$. J. Inorg. Nucl. Chem., 25, 230-231.

Blasse, G. and A. Bril (1967) Structure and Eu^{3+}-fluorescence of lithium and sodium lanthanide silicates and germanates. J. Inorg. Nucl. Chem., 29, 2231-2241.

Bloss, F. D. (1952) Relationship between density and composition in mole per cent for some solid solution series. Amer. Mineral., 37, 966-981.

Birgiani, C., and P. Granati (1979) Monte Carlo calculations of ionic structure in silicate and alumino-silicate melts. Metal. Trans. B., 108, 21-25.

Born, L. (1964) Eine "gitterenergetische Verfeinerung" der freien Mg-Position in Olivin. N. Jahrb. Mineral. Monat., 1964, 81-95.

Bowen, N. L. (1914) The ternary system: diopside-forsterite-silica. Amer. J. Sci., 38, 207-264.

Bowen, N. L., and J. F. Schairer (1935) The system, MgO-FeO-SiO$_2$. Amer. J. Sci., 29, 151-217.

Bowen, N. L., and J. F. Schairer (1936) The system albite-fayalite. Proc. Nat. Acad. Sci. U.S.A., 22, 345.

Bowen, N. L., and J. F. Schairer (1938) Crystallization equilibrium in nepheline-albite-silica mixtures with fayalite. J. Geol., 46, 397.

Bowen, N. L., J. F. Schairer, and E. Posnjak (1933a) The system Ca$_2$SiO$_4$-Fe$_2$SiO$_4$. Amer. J. Sci., 25, 273-297.

Bowen, N, L., J. F. Schairer, and E. Posnjak (1933b) The system CaO-FeO-SiO$_2$. Amer. J. Sci., 193-284.

Bowen, N. L., and O. F. Tuttle (1949) The system MgO-SiO$_2$-H$_2$O. Bull. Geol. Soc. Amer., 60, 439.

Bradley, R. S., P. Engel, and D. C. Munro (1966) Subsolidus solubility between R$_2$SiO$_4$ and LiR"PO$_4$: a hydrothermal investigation. Mineral. Mag., 35, 742-755.

Bragg, W. H. (1915) The structure of the spinel group of crystals. Philos. Mag., 30, 305-315.

Bragg, W. L., and G. B. Brown (1926a) Die Struktur des Olivins. Z. Kristallogr., 63, 538-556.

Bragg W. L., and G. B. Brown (1926b) Die Kristallstruktur von Chyrsoberyl (BeAl$_2$O$_4$). Z. Kristallogr., 63, 122-143.

Bredig, M. A. (1950) Polymorphism of calcium orthosilicate. J. Amer. Cer. Soc., 33, 188-192.

Bridge, T. E. (1966) Bredigite, larnite, and γ dicalcium silicates from Marble Canyon. Amer. Mineral., 51, 1766-1774.

Brown, G. E. (1970) *The Crystal Chemistry of the Olivines*. Ph.D. Thesis, Virginia Polytechnic Institute and State University, Blacksburg, Virginia, 121 p.

Brown, G. E., and C. T. Prewitt (1973) High-temperature crystal chemistry of hortonolite. Amer. Mineral., 58, 577-587.

Brown, G. B., and J. West (1927) The structure of monticellite MgCaSiO$_4$. Z. Kristallogr., 66, 154-161.

Brown, I. D., and R. D. Shannon (1973) Empirical bond-strength-length curves for oxides. Acta Crystallogr., 429, 266-282.

Buening, D. K., and P. R. Buseck (1973) Fe-Mg lattice diffusion in olivine. J. Geophys. Res., 78, 6852-6862.

Burnham, C. W., Y. Ohashi, S. S. Hafner, and D. Virgo (1971) Cation distribution and atomic thermal vibrations in an iron-rich orthopyroxene. Amer. Mineral., 56, 850-876.

Burns, R. G. (1970a) *Mineralogical Applications of Crystal Field Theory*. Cambridge, England: Cambridge University Press, 224 p.

Burns, R. G. (1970b) Crystal field spectra and evidence of cation ordering in olivine minerals. Amer. Mineral., 55, 1608-1632.

Burns, R. G. (1974) The polarized spectra of iron in silicates: olivine. A discussion of neglected contributions from Fe^{3+} ions in M(1) sites. Amer. Mineral., 59, 625-629.

Burns, R. G. (1975) On the occurrence and stability of divalent chromium in olivines included in diamonds. Contr. Mineral. Petrol., 51, 213-221.

Burns, R. G., and W. S. Fyfe (1966) The behavior of nickel during magmatic crystallization. Nature, 210, 1147-1148.

Burns, R. G., and F. E. Higgins (1972) Cation determinative curves for Mg-Fe-Mn olivines from vibrational spectra. Amer. Mineral., 57, 967-985.

Burns, R. G., and C. M. Sung (1978) The effect of crystal field stabilization in the olivine → spinel transition in the system $Mg_2SiO_4-Fe_2SiO_4$. Phys. Chem. Minerals, 2, 349-364.

Buseck, P. R., and D. Veblen (1978) Trace elements, crystal defects, and high resolution electron microscopy. Geochim. Cosmochim. Acta, 42, 669-678.

Bush, W. R., S. S. Hafner, and D. Virgo (1970) Some ordering of iron and magnesium at the octahedrally coordinated sites in a magnesium-rich olivine. Nature, 227, 1339-1341.

Capponi, J. J., J. Chenavas, and J. C. Joubert (1973) Synthese hydrothermale a tres haute pression de deux borates de type olivine, $AlMgBO_4$ et $FeNiBO_4$. Mater. Res. Bull., 8, 275-282.

Caron, L. G., R. P. Santoro, and R. E. Newnham (1965) Magnetic structure of $CaMnSiO_4$. J. Phys. Chem. Solids, 26, 927-930.

Cawthorne, R. G., and T. S. McCarthy (1977) Partitioning of nickel between immiscible picritic liquids. Earth Planet. Sci. Lett., 37, 339-346.

Cemic, L., G. Will, and E. Hinze (1980) Electrical conductivity measurements on olivines $Mg_2SiO_4-Fe_2SiO_4$ under defined thermodynamic conditions. Phys. Chem. Mineral., 6, 95-107.

Champness, P. E. (1970) Nucleation and growth of iron oxides in olivines, $(Mg,Fe)_2SiO_4$. Mineral. Mag., 37, 790-800.

Champness, P. E., and P. Gay (1968) Oxidation of olivines. Nature, 218, 157-158.

Chatelain, A., and R. A. Weeks (1973) Electron paramagnetic resonance of Fe^{3+} in forsterite (Mg_2SiO_4). J. Chem. Phys., 58, 3722-3726.

Chen, C. H., and D. C. Presnall (1975) The system $Mg_2SiO_4-SiO_2$ at pressures up to 25 kilobars. Amer. Mineral., 60, 398-406.

Chenavas, J., A. Waintal, J. J. Capponi, and M. Gondrand (1969) Etude sons haute pression et haute temperature des Composes $NaTGeO_4$ et $NaTSiO_4$ (t = terres rares + yttrium). Mat. Res. Bull., 4, 425-432.

Claringbull, G. F. (1952) Sinhalite ($MgAlBO_4$), a new mineral. Amer. Mineral., 37, 700.

Clark, A. M., and J. V. P. Long (1970) The anisotropic diffusion of nickel in olivine. In, *Thomas Graham Memorial Symposium on Diffusion Processes*, London: Gordon and Breach, 511-521.

Clark, T., and A. J. Naldrett (1972) The distribution of Fe and Ni between synthetic olivine and sulfide at 900°C. Econ. Geol., 67, 939-952.

Cola, M. (1954) Sintesi e porprietá cristallografiche ottiche e strutturali del composto $CaCoSiO_4$ (tipo monticellite $MgCaSiO_4$). Att. Accad. Nazl. Lincei. Rend. Classe Sci. Fis. Mat. Nat., 17, 258-264.

Czaya, R. (1971) Refinement of the structure of $\gamma-Ca_2SiO_4$. Acta Crystallogr., B27, 848-849.

Dachille, F., and R. Roy (1960) High pressure studies of the system $Mg_2GeO_4-Mg_2SiO_4$ with special reference to the olivine-spinel transition. Amer. J. Sci., 258, 225-246.

Davis, B. T. C., and J. L. England (1964) The melting of forsterite up to 50 kilobars. J. Geophys. Res., 69, 1113-1116.

Deer, W. A., R. A. Howie, and J. Zussman (1962) *Rock Forming Minerals: Vol. 1 Ortho- and Ring Silicates.* London: Longmans, Green and Co., Ltd.

Deer, W. A., R. A. Howie, and J. Zussman (1966) *An Introduction to the Rock-Forming Minerals.* London: Longmans, Green and Co., Ltd.

Deer, W. A., and L. R. Wager (1939) Olivines from the Skaergaard intrusion; Kangerd-lugssuaq, east Greenland. Amer. Mineral., 24, 18.

Dempsey, M. J., and R. J. Strens (1976) Modelling crystal structures. In, *The Physics and Chemistry of Minerals and Rocks* (Ed. R. G. J. Strens), New York: John Wiley & Sons, 443-458.

Destenay, D. (1950) Structure cristalline de la triphiline. Mém. soc. roy. sci. Liége, 10, 28.

Dodd, R. T. (1973) Minor element abundance in olivines of the Sharps (H-3) chondrite. Contr. Mineral. Petrol., 42, 159-167.

Dollase, W. A. (1974) A method for determining the distortion of coordination poly-hedra. Acta Crystallogr., A30, 513-517.

Donaldson, C. H. (1975) Calculated diffusion coefficients and the growth rate of olivine in a basalt magma. Lithos, 8, 163-174.

Donaldson, C. H. (1976) An experimental investigation of olivine morphology. Contr. Mineral. Petrol., 57, 187-213.

Donaldson, C. H. (1979) An experimental investigation in nucleating olivine in mafic magmas. Contr. Mineral. Petrol., 69, 21-32.

Dostal, J. and S. Capedri (1975) Partition coefficients of uranium for some rock-forming minerals. Chem. Geol., 15, 285-294.

Drever, H. I., and R. Johnston (1957) Crystal growth of forsteritic olivine in magmas and melts. Trans. Roy. Soc. (Edinburgh), 63, 289-315.

Duba, A. (1972) Electrical conductivity of olivine. J. Geophys. Res., 77, 2483-2495.

Duba, A., H. C. Heard, and R. N. Schock (1974) Electrical conductivity of olivine at high pressure and under controlled oxygen fugacity. J. Geophys. Res., 79, 1667-1673.

Duba, A., J. Ito, and J. C. Jamieson (1973) The effect of ferric iron on the electrical conductivity of olivine. Earth Planet. Sci. Lett., 18, 279-284.

Duba, A., and I. A. Nicholls (1973) The influence of oxidation state on the electrical conductivity of olivine. Earth Planet. Sci. Lett., 18, 59-64.

Duba, A., A. J. Piwinskii, H. C. Heard, and R. N. Schode (1976) The electrical conduc-tivity of forsterite, enstatite. and albite. In, *The Physics and Chemistry of Minerals and Rocks* (Ed. R.G.J. Strens), New York: John Wiley & Sons, 249-260.

Duke, D. A., and J. D. Stephens (1964) Infrared investigation of the olivine group minerals. Amer. Mineral., 49, 1388-1406.

Durif-Varambon, A. (1959) Étude de la substitution du silicium dans quelques types d'orthosilicates. Bull. Soc. franc. Mineral. Crist.   82, 285-314.

Engi, M. (1978) Olivine-spinel geothermometry: an experimental study of the magnesium-iron exchange (abstr.) EØS. Trans. Amer. Geophys. Union, 59, 401.

England, R. N. (1974) Corona structures formed by near-isochemical reaction between and plagiocase in a metamorphosed dolerite. Mineral. Mag., 39, 816-818.

Evans, B., and C. Goetze (1979) The temperature variation of hardness of olivine and its implication for polycrystalline yield stress. J. Geophys. Res., 54, 5505-5524.

Evans, B. W. (1969) The nickel partition geothermometer applied to the prehistoric Makaopuhi Lava Lake, Hawaii. Geochim. Cosmochim. Acta, 33, 409-411.

Evans, B. W., and T. L. Wright (1972) Composition of liquidus chromite from 1959 (Kilanea Iki) and 1965 (Makaopuhi) eruptions of Kilauea Volcano, Hawaii. Amer. Mineral. 57, 217-230.

Eysel, W., and T. Hahn (1970) Polymorphism and solid solution of Ca_2GeO_4 and Ca_2SiO_4. Z. Kristallogr., 131, 322-341.

Fang, J. H., and R. E. Newnham (1965) The crystal structure of sinhalite. Mineral. Mag., 35, 196-199.

Farrell, E. F., J. H. Fang, and R. E. Newnham (1963) Refinement of the chrysoberyl structure. Amer. Mineral., 48, 804-810.

Fawcett, J. J. (1965) Alteration products of olivine and pyroxene in basalt lavas from the Isle of Mull. Mineral. Mag., 35, 55-68.

Ferguson, R. B. (1974) A cation-anion distance-dependent method for evaluating valence-bond distributions in ionic structures and results for some olivines and pyroxenes. Acta Crystallogr., B30, 2527-2539.

Ferguson, J. B., and H. E. Merwin (1919) Ternary system CaO-MgO-SiO$_2$. Amer. J. Sci., 48, 81-]23.

Finger, L. W. (1969) Determination of cation distribution by least-squares refinement of single-crystal x-ray data. Carnegie Inst. Wash. Year Book, 67, 216-217.

Finger, L. W. (1970) Fe/Mg ordering in olivines. Carnegie Inst. Wash. Year Book, 69, 302-305.

Finger, L. W., R. M. Hazen, and T. Yagi (1979) Crystal structures and electron densities of nickel and iron silicate spinels at elevated temperatures and pressures. Amer. Mineral., 64, 1002-1009.

Finger, L. W., and G. R. Rapp, Jr. (1970) Refinement of the crystal structure of triphylite. Carnegie Inst. Wash. Year Book, 68, 290-292.

Finger, L. W., and D. Virgo (1971) Confirmation of Fe/Mg ordering in olivines. Carnegie Inst. Wash. Year Book, 70, 221-225.

Finnerty, T. A. (1977) Exchange of Mn, Ca, Mg, and Al between synthetic garnet, ortho-pyroxene, clinopyroxene, and olivine. Carnegie Inst. Wash. Year Book, 76, 572-579.

Finnerty, A. A., and F. R. Boyd (1978) Pressure-dependent solubility of calcium in forsterite coexisting with diopside and enstatite. Carnegie Inst. Wash. Year Book, 77, 713-717.

Fisher, G. W. (1967) Fe-Mg solid solutions. Carnegie Inst. Wash. Year Book, 65, 209-217.

Fisher, G. W., and L. G. Medaris, Jr., (1969) Cell dimensions and x-ray determinative curve for synthetic Mg-Fe olivines. Amer. Mineral., 54, 741-753.

Fleet, M. E. (1974) Distortions in the coordination polyhedra of M site atoms in olivines, clinopyroxenes, and amphiboles. Amer. Mineral., 59, 1083-1093.

Fleet, M. E. (1975) The growth habits of olivine - a structural interpretation. Canadian Mineral., 13, 293-297.

Forest, J. (1971) Connaissance de l'orthosilicate de calcium. Bull. Soc. fr. Mineral. Cristallogr., 94, 118-137.

Foster, W. R. (1968) Comment on the Ca$_2$SiO$_4$ phase diagram. J. Amer. Cer. Soc., 51, 353.

Francis, C. A., and P. H. Ribbe (1980) The forsterite-tephroite series: II crystal structure refinements. Amer. Mineral. (in press).

Franz, G. W., and P. J. Wyllie (1966) Melting relations in the system CaO-MgO-SiO$_2$-H$_2$O at 1 kilobar pressure. Geochim. Cosmochim. Acta, 30, 9-22.

Fujii, T. (1977) Fe-Mg partitioning between olivine and spinel. Carnegie Inst. Wash. Year Book, 76, 563-569.

Gaite, J. M. (1980) Pseudo-symmetries of crystallographic coordination polyhedra. Applications to forsterite and comparison with some EPR results. Phys. Chem. Mineral., 6, 9-17.

Gallitelli, P., and M. Cola (1954) Sintesi, proprietà, cristallegrafiche e strutturali del composto Co$_2$SiO$_4$ (tipo dell'olivina). Atti. Accad. Nazl. Lincei. Rend. Classe Sci. Fis. Mat. Nat., 17, 172-177.

Ganguli, D. (1977) Crystal chemical aspects of olivine structures. N. Jahrb. Miner. Abh., 130, 303-318.

Geller, S., and J. L. Durand (1960) Refinement of the structure of LiMnPO$_4$. Acta Crystallogr., 13, 325-331.

Ghose, S. (1962) The nature of $Mg^{2+}-Fe^{2+}$ distribution in some ferromagnesian silicate minerals. Amer. Mineral., 47, 388-394.

Ghose, S. (1965) $Mg^{2+}-Fe^{2+}$ distribution in metamorphic and volcanic orthopyroxenes. Z. Kristallogr., 125, 1-6.

Ghose, S., F. P. Okamura, C. Wan, and H. Ohashi (1974) Site preference of transition metal ions in pyroxenes and olivine (abstr.). EOS, Trans. Amer. Geophys. Union, 55, 467.

Ghose, S., and C. Wan (1974) Strong site preference of Co^{2+} in olivine $Co_{1.10}Mg_{0.90}SiO_4$. Contr. Mineral. Petrol., 47, 131-140.

Ghose, S., C. Wan, and I. S. McCallum (1976) $Fe^{2+}-Mg^{2+}$ order in an olivine from the lunar anorthosite 67075 and the significance of cation order in lunar and terrestrial olivines. Indian J. Earth Sci., 3, 1-8.

Ghose, S., and J. R. Weidner (1974) Site preference of transition metal ions in olivine (abstr.). Geol. Soc. Amer. Abstr. with Prog., 6, 751.

Ghosh, S. N., P. B. Rao, A.K. Paul, and K. Raira (1979) Review: The crystal chemistry of dicalcium silicate minerals. J. Mat. Sci., 14, 1554-1566.

Gjessing, L., T. Larsson, and H. Major (1942) Isomorphous substitution for $Al^{...}$ in the compound Al_2BeO_4. Norsk Geol. Tideskr., 22, 92-99.

Glasser, F. P. (1961) The system $Ca_2SiO_4-Mn_2SiO_4$. Amer. J. Sci., 259, 46-59.

Glasser, F. P., and E. F. Osborn (1960) The ternary system $MgO-MnO-SiO_2$. J. Amer. Cer. Soc., 43, 132-140.

Goode, A. D. T. (1974) Oxidation of natural olivines. Nature, 248, 500-501.

Grabar, D. G., and A. H. Principe (1963) Identification of glass fragments by measurement of refractive indices and dispersion. J. Forensic Sci., 8, 54-67.

Grandstaff, D. E. (1978) Changes in surface area and morphology and the mechanism of forsterite dissolution. Geochim. Cosmochim. Acta, 42, 1899-1901.

Grover, J. E., and P. M. Orville (1969) The partitioning of cations between coexisting single- and multi-site phases with application to the assemblages: orthopyroxene-clinopyroxene and orthopyroxene-olivine. Geochim. Cosmochim. Acta, 33, 205-226.

Grum-Grzhimailo, S. V., O. N. Boksha, and T. M. Varino (1969) The absorption spectrum of olivine. Sov. Phys. Crystallogr., 14, 272-274.

Gunn, B. M. (1971) Trace element partition during olivine fractionation of Hawaiian basalts. Chem. Geol., 8, 1-13.

Hagenmuller, P., G. Perez, J. Serment, and A. Hardy (1964) Manganese orthothiosilicate and orthothiogermanate. Compt. Rend., 259, 4689.

Haggerty, S. E., and I. Baker (1967) The alteration of olivine in basaltic and associated lavas. Part I. High-temperature alteration. Contr. Mineral. Petrol., 16, 233-257.

Haggerty, S. E., F. R. Boyd, P. M. Bell, L. W. Finger, and W. B. Bryan (1970) Opaque minerals and olivine in lavas and breccias from Mare Tranquillitatis. *Proc. Apollo 11th Lunar Sci Conf.*, Geochim. Cosmochim. Acta, Suppl. 1, 1, 513-558.

Hakli, T. A. (1968) An attempt to apply the Makaspuhi nickel fractionation data to the temperature determination of a basic intrusive. Geochim. Cosmochim. Acta, 32, 449-460.

Hakli, T. A., and T. L. Wright (1967) The fractionation of nickel between olivine and augite as a geothermometer. Geochim. Cosmochim. Acta, 31, 877-884.

Hanke, K. (1963) Verfeinerung der Kristallstruktur des Fayalits von Bad Harzburg. N. Jahrb. Mineral. Monat., 8, 192-194.

Hanke, K. (1965) Beitrage zu Kristallstrukturen vom Olivin-typ. Beitr. Mineral. Petrog., 11, 535-558.

Hanke, K. and J. Zemann (1963) Verfeinerung der Kristallstruktur von Olivin. Naturwissenschaften, 3, 91-92.

Hardy, A., G. Perez, and J. Serment (1965) Crystal structure of manganese ortho-thiosilicate and orthothiogermanate. Bull. Soc. Chim. France, 1965, 2638.

Hassanein, M., and A. M. Azzam (1968) The light absorption of copper (II) and host lattices with trirutiles and olivine structures. Z. anorgan. allegem. Chem., 362, 331-336.

Hauser, J., and H. R. Wenk (1976) Optical properties of composite crystals (submicroscopic domains, exsolution lamellae, solid solutions). Z. Kristallogr., 143, 188-219.

Hazen, R. M. (1976) Effects of temperature and pressure on the crystal structure of forsterite. Amer. Mineral., 61, 1280-1293.

Hazen, R. M. (1977) Effects of temperature and pressure on the crystal structure of ferromagnesian olivine. Amer. Mineral., 62, 286-295.

Hazen, R. M., and L. W. Finger (1977) Compression models for oxides and silicates (abstr.). Geol. Soc. Amer. Abstr. Prog., 9, 1008-1009.

Hazen, R. M., and L. W. Finger (1978) Crystal chemistry or silicon-oxygen bonds at high pressure: implications for the earth's mantle mineralogy. Science, 201, 1122-1123.

Hazen, R. M., and L. W. Finger (1979) Bulk modulus - volume relationship for cation - anion polyhedra. J. Geophys. Res., 84, 6723-6728.

Hazen, R. M., H. K. Mao, and P. M. Bell (1977) Comparison of absorption spectra of lunar and terrestrial olivines. Carnegie Inst. Wash. Year Book, 76, 508-512.

Hazen, R. M., and C. T. Prewitt (1977) Effects of temperature and pressure on interatomic distances in oxygen-based minerals. Amer. Mineral., 62, 309-315.

Henderson, P., and I. M. Dale (1969) The partitioning of selected transition element ions between olivine and groundmass of oceanic basalts. Chem. Geol., 5, 267-274.

Henriques, A. (1957) The effect of cations on the optical properties and the cell dimensions of knebelite and olivine. Arkiv Mineral. Geol., 2, 304-313.

Hermes, O. D., and J. G. Schilling (1976) Olivine from Reykjanes Ridge and Iceland tholeiites, and its significance to the two-mantle source model. Earth Planet. Sci. Lett., 28, 345-355.

Hill, R. J., J. R. Craig, and G. V. Gibbs (1979) Systematics of the spinel structure type. Phys. Chem. Minerals, 4, 317-339.

Hodges, F. N. (1973) Solubility of H_2O in forsterite melt at 20 kbar. Carnegie Inst. Wash. Year Book, 72, 495-497.

Holt, J. B. (1975) Thermal diffusivity of olivine. Earth Planet. Sci. Lett., 27, 404-408.

Hoover, J. D., and T. N. Irvine (1978) Liquidus relations and Mg-Fe partitioning on part of the system Mg_2SiO_4-Fe_2SiO_4-$CaMgSi_2O_6$-$CaFeSi_2O_6$-$KAlSi_3O_8$-SiO_2. Carnegie Inst. Wash. Year Book, 77, 774-784.

Hsu, L. C. (1967) Melting of fayalite up to 40 kilobars. J. Geophys. Res., 72, 4235-4244.

Huggins, F. E. (1973) Cation order in olivines: evidence from vibrational spectra. Chem. Geol., 11, 99-109.

Huggins, F. E. (1975) The effect of pressure on the covalency of Fe^{2+}-oxygen bonds. Carnegie Inst. Wash. Year Book, 74, 551-555.

Huggins, F. E. (1976) Mössbauer studies of iron minerals under pressure of up to 200 kilobars. In, *The Physics and Chemistry of Minerals and Rocks*, (Ed. R. G. J. Strens), New York: John Wiley & Sons, 613-640.

Hurlburt, C. S. (1961) Tephroite from Franklin, New Jersey. Amer. Mineral., 46, 549-559.

Irving, A. J. (1978) A review of experimental studies of crystal/liquid trace-element partitioning. Geochim. Cosmochim. Acta, 42, 743-770.

371

Ito, J. (1977) Crystal synthesis of a new olivine. Amer. Mineral., 62, 356-361.

Ivanov, Y. A., M. A. Simonov, and N. V. Belov (1974) Crystal structure of the Na,Cd orthophosphate NaCd[PO$_4$]. Sov. Phys. Crystallogr., 19, 96-97.

Jackson, E. D. (1969) Chemical variation in coexisting chromite and olivine in chromitite zones of the Stillwater complex. Econ. Geol. Monogr., 4, 41-71.

Jeanloz, R. (1980) Shock effects in olivine and implications for Hugoniot data. J. Geophys. Res., 85, 3163-3176.

Jeanloz, R., T. J. Ahrens, J. S. Lally, G. L. Nord, Jr., J. M. Christie, and A. H. Heuer (1977) Shock-produced olivine glass: First observation. Science, 197, 457-458.

Johannes, W. (1969) An experimental investigation of the system MgO-SiO$_2$-H$_2$O-CO$_2$. Amer. J. Sci., 267, 1083.

Kabolov, Y. K., M. A. Simonov, B. I. Ivanov, O. K. Melnikov, and N. V. Belov (1973) Crystal structure of Li(Fe,Zn)PO$_4$. Sov. Phys. Dohl., 18, 106-107.

Kallenberg, S. (1914) Untersuchungen uber die Binären Systeme: Mn$_2$SiO$_4$-Ca$_2$SiO$_4$, Mn$_2$SiO$_4$-Mg$_2$SiO$_4$ und MnSiO$_3$-FeSiO$_3$. Z. anorg. u. allgem. Chem., 88, 355-363.

Kamb, B. (1968) Structural basis of the olivine-spinel stability relation. Amer. Mineral., 53, 1439-1455.

Kawasaki, T., and Y. Matsui (1977) Partitioning of Fe^{2+} and Mg^{2+} between olivine and garnet. Earth Planet. Sci. Lett., 37, 159-166.

Kazak, V. F., A. I. Domanskii, A. I. Boikova, V. V. Ilyukhin, and N. V. Belov (1975) Crystal chemical aspects of the polymorphic transformations of dicalcium silicate. Sov. Phys. Crystallogr., 19, 733-736.

Kharakh, E. A., A. V. Chichagov, and N. V. Belov (1971) Crystal structure of sodium samarium orthogermanate. Sov. Phys. Crystallogr., 15, 924-925.

Kieffer, S. W. (1976) Lattice thermal conductivity within the earth and considerations of a relationship between the pressure dependence of the thermal diffusivity and volume dependence of the Gruneisen parameter. J. Geophys. Res., 81, 3025-3030.

Kirkpatrick, R. J. (1975) Crystal growth from the melt: a review. Amer. Mineral., 60, 798-814.

Kohlstedt, D. L., and J. B. Vander Sande (1975) An electron microscopy study of naturally occurring oxidation produced precipitates in iron-bearing olivines. Contr. Mineral. Petrol., 53, 13-24.

Koryakina, N. S., V. A. Kuznetsov, and N. V. Belov (1971) Hydrothermal crystallization in the system Li$_2$O-Sc$_2$O$_3$-SiO$_2$-H$_2$O. Sov. Phys. Crystallogr., 17, 191-192.

Kowalezyk, S. P., L. Ley, F. K. McFeely, and D. A. Shirley (1974) An ionicity scale based on x-ray photoemission valence band spectra of A^{n}B^{8-n} and A^{n}B^{8-n} type crystals. J. Chem. Phys., 61, 2850-2856.

Kozu, S., J. Veda, and S. Tsurumi (1934) Thermal expansion of olivine. Proc. Imp. Acad. Japan, 10, 83-86.

Karoda, Y., and Y. Iguchi (1971) A soft x-ray study of olivines. Mineral. Soc. Japan Spec. Paper, 1, 247-249.

Kushiro, I. (1964) The system diopside-forsterite-enstatite at 20 kb. Carnegie Inst. Wash. Year Book, 63, 101-108.

Kushiro, I. (1969) The system forsterite-diopside-silica with and without water at high pressures. Amer. J. Sci., 267A, 269-294.

Kushiro, I. (1974) The system forsterite-anorthite-albite-silica-H$_2$O at 15 kbar and the genesis of andesitic magmas in the upper mantle. Carnegie Inst. Wash. Year Book, 73, 244-248.

Kushiro, I. (1975) On the nature of silicate melt and its significance in magma genesis: regularities in the shift of the liquidus boundaries involving olivine, pyroxene, and silica minerals. Amer. J. Sci., 275, 411-431.

Kushiro, I., Y. Nakamura, K. Kitayama, and S. Akimoto (1971) Petrology of some Apollo 12 crystalline rocks. *Proc. 2nd Lunar Sci. Conf.*, Geochim. Cosmochim. Acta, Suppl. 2, 1, 481-495.

Kushiro, I., and J. F. Schairer (1963) New data on the system diopside-forsterite-silica. Carnegie Inst. Wash. Year Book, 62, 95-103.

Kushiro, I., and H. S. Yoder, Jr. (1964) Breakdown of monticellite and akermanite at high pressures. Carnegie Inst. Wash. Year Book, 63, 81-83.

Lager, G. A., and E. P. Meagher (1978) High temperature structural study of six olivines. Amer. Mineral., 63, 365-377.

Laskowski, T. E., and D. M. Scotford (1980) Rapid determination of olivine compositions in thin section using dispersion staining methodology. Amer. Mineral. 65, 401-403.

Leeman, W. P., and K. F. Scheidegger (1977) Olivine/liquid distribution coefficients and a test for crystal-liquid equilibrium. Earth Planet. Sci. Lett., 35, 247-257.

Levin, E. M., C. R. Robbins, and H. F. McMurdie (1964) *Phase Diagrams for Ceramists.* (Ed. M. K. Reser). Columbus, Ohio: Amer. Cer. Soc., 601 p.

Liebermann, R. C. (1975) Elasticity of olivine (α), beta (β) and spinel (γ) polymorphs of germanates and silicates. Geophys. J. Royal Astron. Soc., 42, 899-929.

Lindsley, D. H. (1967) Pressure-temperature relations in the system FeO-SiO$_2$. Carnegie Inst. Wash. Year Book, 66, 226-230.

Lindsley, D. H., D. H. Speidel, and R. H. Nafziger (1968) P-T-f$_{O_2}$ relations in the system Fe-O-SiO$_2$. Amer. J. Sci., 266-342-361.

Loney, R. A., G. R. Himmelberg, and R. G. Coleman (1971) Structure and petrology of the alpine-type peridotite at Burro Mountain, California, U.S.A. J. Petrol., 12, 245-309.

Louisnathan, S. J., and G. V. Gibbs (1972a) The effect of tetrahedral angles on Si-O bond overlap populations for isolated tetrahedra. Amer. Mineral., 57, 1614-1642.

Louisnathan, S. J., and G. V. Gibbs (1972b) Variation of Si-O distances in olivines sodamelilite, and sodium metasilicate as predicted by semi-empirical molecular orbital cacluations. Amer. Mineral., 57, 1643-1663.

Louisnathan, S. J., and J. V. Smith (1968) Cell dimensions of olivine. Mineral. Mag., 36, 1123-1134.

Luce, R. W., and G. A. Parks (1973) Point of zero charge of weathered forsterite. Chem. Geol., 12, 147-153.

Lyutin, V. I., E. A. Kuzman, V. V. Ilyukhin, and N. V. Belov (1974) Crystal structure of mixed zinc-manganese orthogermanate, ZnMnGeO4. Sov. Phys. Dokl, 19, 10-11

McLarnan, T. J., R. J. Hill, and G. V. Gibbs (1979) A CNDO/2 molecular orbital study of shared tetrahedral edge conformations in olivine-type compounds. Australian J. Chem., 32, 949-959.

Ma, C. B. (1975) Structure refinement of high-pressure Ni$_2$SiO$_4$ spinel. Z. Kristallogr., 141, 126-137.

Magnusson, N. H. (1918) Beitrag zur Kenntnis der optischen Ergenschaften der Olivingruppe. Geol. För. Förh. Stockholm, 40, 601-626.

Maksimov, B. A., B. N. Litvin, V. V. Ilyukhin, and N. V. Belov (1969) Hydrothermal crystallization in the system A$_2$O-TR$_2$O$_3$-SiO$_2$-H$_2$O. I. Synthesis of crystals of alkali metal yttrium silicates. Sov. Phys. Crystallogr., 14, 407-410.

Malysheva, T. V., V. V. Kurash, and A. N. Ermakov (1969) Study of isomorphic replacement of magnesium and iron (II) in olivines by Mössbauer γ-resonance spectroscopy. Geokhimiya, 11, 1405-1408.

Mao, H. K. (1976) Charge-transfer processes at high-pressure. In, *Physics and Chemistry of Minerals and Rocks,* (Ed. R. G. J. Strens). New York: John Wiley & Sons, 573-581.

Mao, H. K., and P. M. Bell (1972) Electrical conductivity and the red shift of absorption in olivine and spinel at high pressures. Science, 176, 403-406.

Mao, H. K., and P. M. Bell (1972) Crystal-field stabilization of the olivine-spinel transition. Carnegie Inst. Wash. Year Book, 71, 527-528.

Martin, R. F., and G. Donnay (1972) Hydroxyl in the mantle. Amer. Mineral., 57, 554-570.

Mason, B. (1972) The mineralogy of meteorites. Meteoritics, 7, 309-326.

Mason, B., J. Nelen, and J. S. White, Jr. (1968) Olivine-garnet transformation in a meteorite. Science, 160, 66-67.

Masse, D. P., E. Rosen, and A. Muan (1966) Activity-composition relations in Co_2SiO_4-Fe_2SiO_4 solid solutions at 1180°C. J. Amer. Cer. Soc., 49, 328-329.

Masuda, A. (1968) Lanthanide concentrations in the olivine phase of the Brenham pallasite. Earth Planet. Sci. Lett., 5, 59-62.

Matsui, Y., and S. Banno (1969) Partition of divalent transition metals between coexisting ferromagnesian minerals. Chem. Geol., 5, 259-265.

Matsui, Y., and O. Nishizawa (1974) Iron (II) - magnesium exchange equilibrium between olivine and calcium-free pyroxene over a temperature range 800°C to 1300°C. Bull. Soc. fr. Mineral. Cristallogr., 97, 122-130.

Matsui, Y., and Y. Syono (1968) Unit cell dimensions of some synthetic olivine group solid solutions. Geochem. J., 2, 51-59.

Medaris, L. G. (1969) Partitioning of Fe^{++} and Mg^{++} between coexisting synthetic olivine and orthopyroxene. Amer. J. Sci., 267, 945-968.

Meyer, H. O. A., and F. R. Boyd (1972) Composition and origin of crystalline inclusions in natural diamonds. Geochim. Cosmochim. Acta, 36, 1255-1273.

Meyer, H. O. A., and D. P. Srisero (1973) Mineral inclusions in Brazilian diamonds (abstr.). Intern. Conf. Kimberlites, 225-228.

Michonlier, J., J. Gaite, and B. Maffeo (1969) Resonance paramagnetique electronique de l' ion Mn^{2+} dans un monocristal de forsterite. C. R. Acad. Sc. Paris, 269, 535-538.

Misener, D. J. (1974) Cationic diffusion in olivine to 1400°C and 35 kbar. In, *Geochemical Transport and Kinetics*, (Ed. A. W. Hoffman, B. J. Gilletti, H. S. Yoder, Jr., and R. A. Yund). Carnegie Inst. Wash., Pub. 634, 117-129.

Moody, J. B. (1976) An experiemental study on the serpentinization of iron-bearing olivines. Canadian Mineral., 14, 462-478.

Moore, J. G., and B. W. Evans (1967) The role of olivine in the crystallization of the prehistoric Makaopuhi tholeiite lava lake, Hawaii. Contr. Mineral. Petrol., 15, 202-223.

Moore, P. B. (1972) Natrophilite, $NaMn(PO_4)$, has ordered cations. Amer. Mineral., 57, 1333-1344.

Moore, P. B., and J. V. Smith (1969) High pressure modification of Mg_2SiO_4: crystal structure and crystallochemical and geophysical implications. Nature, 221, 653-655.

Moore, P. B., and J. V. Smith (1970) Crystal structure of β-Mg_2SiO_4: crystal-chemical and geophysical implications. Phys. Earth Planet. Inter., 3, 166-177.

Morimoto, N., M. Tokonami, M. Watanabe, and K. Koto (1974) Crystal structures of three polymorphs of Co_2SiO_4. Amer. Mineral., 59, 475-485.

Morioka, M. (1980) Cation diffusion in olivines-I. cobalt and magnesium. Geochim. Cosmochim. Acta, 44, 759-762.

Mossman, D. J., and D. J. Pawson (1976) X-ray and optical characterization of the forsterite-fayalite, tephroite series with comments on knebelite from Bluebell Mine, British Columbia. Canadian Mineral., 14, 479-486.

Muan, A. (1955) Phase equilibria in the system FeO-Fe_2O_3-SiO_2. A. I. M. E. Trans., 203, 965-976.

Muan, A., and E. F. Osborn (1956) Phase equilibria at liquidus temperatures in the system MgO-FeO-Fe_2O_3-SiO_2. J. Amer. Cer. Soc., 39, 121-140.

Muller, O., and R. Roy (1974) *The Major Ternary Structure Families.* New York: Springer-Verlag.

Mysen, B. O. (1975) Partitioning of iron and magnesium between crystals and partial melts in peridotite upper mantle. Contr. Mineral. Petrol., 52, 69-76.

Mysen, B. O. (1979) Nickel partitioning between olivine and silicate melt: Henry's Law revisited. Amer. Mineral., 64, 1107-1114.

Mysen, B. O., and I. Kushiro (1979) Pressure dependence of nickel partitioning between forsterite and aluminous silicate melts. Earth Planet, Sci. Lett., 42, 383-388.

Nabelek, P. I. (1980) Nickel partitioning between olivine and liquid in natural basalts: Henry's Law behavior. Earth Planet Sci. Lett., 48, 293-302.

Nafziger, R. H., and A. Muan (1967) Equilibrium phase compositions and thermodynamic properties of olivines and pyroxenes in the system MgO-"FeO"-SiO_2. Amer. Mineral., 52, 1364-1385.

Navrotsky, A. (1973) The thermodynamic relations among olivine, spinel and phencite structures in silicates and germanates: I. Volume relations and the systems NiO-MgO-GeO_2 and CoO-MgO-GeO_2. J. Solid State Chem., 6, 21-41.

Navrotsky, A., F. Pintchovski, and S. Akimoto (1979) Calorimetric study of the stability of high pressure phases in the systems CoO-SiO_2 and "FeO"-SiO_2 and calculation of phase diagrams in MO-SiO_2 systems. Phys. Earth Planet. Int., 19, 275-292.

Nefedov, V. I., V. S. Urusov, and M. M. Kokhana (1972) X-ray photoelectron spectroscopy of bonds in Na, Mg, Al, and Si minerals. Geochem. Int., 9, 9-13.

Newnham, R. E., L. G. Caron, and R. P. Santoro (1966) Magnetic properties of $CaCoSiO_4$ and $CaFeSiO_4$. J. Amer. Cer. Soc., 49, 284-285.

Newnham, R. E., and M. J. Redman (1965) Crystallographic data for $LiMgPO_4$, $LiCoPO_4$, and $LiNiPO_4$. J. Amer. Cer. Soc., 49, 547.

Newnham, R. E., R. Santoro, J. Pearson, and C. Jansen (1964) Ordering of Fe and Cr in chysoberyl. Amer. Mineral., 49, 427-430.

Newton, R. C., and W. E. Sharp (1975) Stability of forsterite + CO_2 and its bearing on the role of CO_2 in the mantle. Earth Planet. Sci. Lett., 26, 239-244.

Niebuhr, H. H. (1975) Electron spin resonance of ferric iron in forsterite (Mg_2SiO_4). Acta Crystallogr., A31, 274-275.

Nishizawa, O., and S. Akimoto (1973) Partition of magnesium and iron between olivine and spinel, and between pyroxene and spinel. Contr. Mineral. Petrol., 41, 217-230.

Nishizawa, O., and Y. Matsui (1972) An experimental study on partition of magnesium and manganese between olivine and orthopyroxene. Phys. Earth Planet. Int., 6, 377-384.

Nitsan, U. (1974) Stability field of olivine with respect to oxidation and reduction. J. Geophys. Res., 79, 706-711.

Nwe, Y. Y. (1976) Electron-probe studies of the earlier pyroxenes and olivines from the Skaergaard intrusion, east Greenland. Contr. Mineral. Petrol., 55, 105-126.

Obata, M., S. Banno, and T. Mori (1974) The iron-magnesium partitioning between naturally occurring coexisting olivine and Ca-rich clinopyroxene: an application of the simple mixture model to olivine solid solution. Bull. Soc. fr. Mineral. Cristallogr., 97, 101-107.

O'Daniel, H., and L. Tscheischwili (1944) Strukturuntersuchungen am Tephroit (Mn_2SiO_4) Glaukochroit $(Mn,Ca)_2SiO_4$ und Willemit (Zn_2SiO_4) von Franklin Furnace. Z. Kristallogr., 105, 273-278.

Ohashi, Y. (1976) Lattice energy of some silicate minerals and the effect of oxygen bridging in relation to crystallization sequence. Carnegie Inst. Wash. Year Book, 75, 644-648.

Ohashi, Y., and L. W. Finger (1973) Thermal vibration ellipsoids and equipotential surfaces at the cation sites in olivine and clinopyroxenes. Carnegie Inst. Wash. Year Book, 72, 547-551.

Ohashi, Y., and L. W. Finger (1974) Diffusion anisotropy in olivine--model calculations. Carnegie Inst. Wash. Year Book, 73, 403-405.

Olinger, B., and A. Duba (1971) Compression of olivine to 100 kilobars. J. Geophys. Res., 76, 2110-2616.

Olinger, B., and P. M. Halleck (1975) Redetermination of the relative compression of the cell edges of olivine. J. Geophys. Res., 74, 5535-5536.

O'Nions, R. K., and D. G. W. Smith (1973) Bonding in silicates: an assessment of bonding in orthopyroxene. Geochim. Cosmochim. Acta, 37, 249-257.

Onorato, P. I. K., D. R. Uhlmann, L. A. Taylor, R. A. Coish, and R. P. Gamble (1978) Olivine cooling speedometers. Proc. 9th Lunar Planet. Sci. Conf., 613-628.

O'Neill, H. S., and B. J. Wood (1979) An experimental study of Fe-Mg partitioning between garnet and olivine and its calibration as a geothermometer. Contr. Mineral. Petrol., 70, 59-70.

Onken, H. (1964) Verfeinerung der Kristallostruktur von Monticellite. Naturwissenschaften, 51, 334.

Onken, H. (1965) Verfeinerung der Kristallstruktur von Monticellite. Tscherm. Mineral. Petrog. Mitt., 10, 34-44.

Osborn, E. F. (1954) Segregation of elements during the crystallization of magma. J. Amer. Cer. Soc., 33, 219-224.

Osborn, E. F., and D. B. Tait (1952) The system diopside-forsterite-anorthite. Amer. J. Sci., Bowen Vol., 413-436.

Page, N. J. (1976) Serpentinization and alteration in a olivine cummulate from the Stillwater Complex, Southwestern Montana. Contr. Minteral. Petrol., 54, 127-137.

Palache, C. (1937) The minerals of Franklin and Sterling Hill, Sussex County, New Jersey. U. S. Geol. Survey Prof. Paper 180.

Paques-Ledent, M. T. (1974) Non-olivine structure of $LiMgVO_4$: evidence from x-ray diffractometry and vibrational spectroscopy. Chem. Phys. Lett., 24, 231-233.

Paques-Ledent, M. T. (1976) Vibrational studies of olivine-type compounds – III. Orthosilicates and germanates $A^IB^{III}X^{IV}O_4$. Spectrochim. Acta, 32A, 383-395.

Paques-Ledent, M. T., and P. Tarte (1974) Vibrational studies of olivine-type compounds-II: Orthophosphates-arsenates and –vanadates $A^IB^{II}X^{IV}O_4$. Spectrochim. Acta, 30A, 673-689.

Pauling, L. (1929) The principles determining the structure of complex ionic crystals. J. Amer. Chem. Soc., 51, 1010-1026.

Pauling, L. (1948) The modern theory of valency. J. Chem. Soc., 1948, 1461-1467.

Pauling, L. (1980) The nature of silicon-oxygen bonds. Amer. Mineral., 65, 321-323.

Pistorius, C. W. F. T. (1963) Some phase relations in the systems $CoO-SiO_2-H_2O$, $NiO-SiO_2-H_2O$ and $ZnO-SiO_2-H_2O$ to high pressures and temperatures. N. Jahrb. Mineral. Monat., 31-57.

Pistorius, C. W. F. T., and J. C. A. Boyens (1970) Crystallographic aspects of the polymorphism of silver chromate and selenate at high pressures and temperatures. Z. anorgan. allgem. Chem., 372, 263-267.

Pluschkell, W., and H. J. Engell (1968) Ionen- und Elektronen-leitung in Magnesium orthosilikat. Ber. Dtsch. Keram. Ges., 4S, 388.

Powell, M., and R. Powell (1974) An olivine-clinopyroxene geothermometer. Contr. Mineral. Petrol., 48, 249-263.

Presnall, D. C. (1966) The join forsterite-diopside-iron oxide and its bearing on the crystallization of basaltic and ultramafic magmas. Amer. J. Sci., 264, 753-809.

Prinz, M., D. V. Manson, P. F. Hlava, and K. Keil (1975) Inclusions in diamonds: garnet lherzolite and eclogite assemblages (abstr.). Intern. Conf. Kimberlites, 267-269.

Putnis, A. (1979) Electron petrography of high-temperature oxidation in olivine from Rhum layered intrustion. Mineral. Mag., 43, 293-296.

Rager, H. (1977) Electron spin resonance of trivalent chromium in forsterite, Mg_2SiO_4. Phys. Chem. Mineral., 2, 371-378.

Rajamani, V., G. E. Brown, and C. T. Prewitt (1975) Cation ordering in Ni-Mg olivine. Amer. Mineral., 60, 292-299.

Ramberg, H., and G. W. DeVore (1951) The distribution of Fe^{2+} and Mg^{2+} in coexisting olivines and pyroxenes. J. Geol., 59, 193-210.

Rao, V. U. S., F. E. Huggins, and G. P. Huffman (1979) Study of Fe-rich superparamagnetic clusters in olivine $(Mg,Fe)_2SiO_4$ by Mössbauer spectroscopy. J. Appl. Phys., 50, 2408-2410.

Raymond, M. (1971) Madelung constants for several silicates. Carnegie Inst. Wash. Year Book, 70, 225-227.

Reiner, D. (1968) Ge^{4+} - und Si^{4+}-haltige Olivinphasen. Zeit. anorgan. allgem. Chem., 356, 182-187.

Ribouel, P. V., and A. Muan (1962) Phase equilibria in a part of the system "FeO"-$MnO-SiO_2$. Trans. A.I.M.E., 224, 27-33.

Ricker, R. W., and E. F. Osborn (1974) Additional phase equilibrium data for the system $CaO-MgO-SiO_2$. J. Amer. Cer. Soc., 37-133-139.

Rickel, C., and A. Weiss (1978) Cation-ordering in synthetic $Mg_{2-x}Fe_xSiO_4$-Olivines. Z. Naturforsch., 33b, 731-736.

Rigby, G. R., G. H. B. Lovell, and A. T. Green (1945) The reversible thermal expansion and other properties of some calcium ferrous silicates. Trans. Brit. Ceram. Soc., 44, 37-52.

Rigby, G. R., G. H. B. Lowell, and A. T. Green (1946) The reversible thermal expansion and other properties of some magnesian ferrous silicates. Trans. Brit. Ceram. Soc., 45, 237-250.

Ringwood, A. E. (1956) Melting relationships of Ni-Mg olivines and some geochemical implications. Geochim. Cosmochim. Acta, 10, 297-303.

Ringwood, A. E. (1958) Constitution of the mantle - II. Further data on the olivine-spinel transition. Geochim. Cosmochim. Acta, 15, 18-29.

Ringwood, A. E. (1969) Phase transformations in the mantle. Earth Planet. Sci. Lett., 5, 401-412.

Ringwood, A. E. (1972) Phase transformation of mantle dynamics. Earth Planet. Sci. Lett., 14, 233-241.

Ringwood, A. E. (1975) *Composition and Petrology of the Earth's Mantle*. New York: McGraw-Hill.

Ringwood, A. E., and A. Major (1966) Synthesis of Mg_2SiO_4-Fe_2SiO_4 spinel solid solutions. Earth Planet. Sci. Lett., 1, 241-245.

Ringwood, A. E., and A. F. Reid (1970) Olivine-spinel transformation in $MgMnGeO_4$, $FeMnGeO_4$, $CoMnGeO_4$. J. Phys. Chem. Solids, 31, 2791-2793.

Robinson, K., G. V. Gibbs, and P. H. Ribbe (1971) Quadratic elongation: a quantititative measure of distortion in coordination polyhedra. Science, 172, 567-570.

Rocktäschel, G., W. Ritter and A. Weiss (1964) Ternary chalcogenides with group IV-A elements and their olivine structure. Z. Naturforsch. 19b, 958.

Roedder, E. W. (1951) The system $K_2O-MgO-SiO_2$. Amer. J. Sci., 249, 81-130, 224-248.

Roedder, E.W. (1976) Petrologic data from experimental studies on crystallized silicate melt and other inclusions in lunar and Hawaiian olivine. Amer. Mineral., 61, 684-690.

Roeder, P. L. (1974) Activity of iron and olivine solubility in basaltic liquids. Earth Planet. Sci. Lett., 23, 397-410.

Roeder, P. L., and E. F. Osborn (1966) Experimental data for the system $MgO-FeO-Fe_2O_3-CaAl_2Si_2O_4-SiO_2$ and their petrologic implications. Amer. J. Sci., 264, 443.

Roy, D. M. (1956) Subsolidus data for the join $Ca_2SiO_4-CaMgSiO_4$ and the stability of merwinite. Min. Mag., 31, 187-

Roy, D. M., and R. Roy (1955) Synthesis and stability of minerals in the system $MgO-Al_2O_3-SiO_2-H_2O$. Amer. Mineral., 40, 147.

Runciman, W. A., D. Sengupta, and J. T. Gormley (1973) The polarized spectra of iron in silicates. II. Olivine. Amer. Mineral., 58, 451-456.

Runciman, W. A., D. Sengupta, and J. T. Gormley (1974) The polarized spectra of iron in silicates: II Olivine: a reply. Amer. Mineral., 59, 630-631.

Sack, R. O. (1980) Some constraints on the thermodynamic mixing properties of Fe-Mg orthopyroxenes and olivines. Contr. Mineral. Petrol., 71, 257-269.

Sahama, Th. G. (1961) Thermal metamorphism of the volcanic rocks of Mt. Nyiragongo (Eastern Congo). Finland Commission Geologique Bull., 196, 151-175.

Sahama, Th. G., and K. Hytonen (1957) Kirschsteinite, a natural analogue to synthetic iron monticellite, from the Belgium Congo. Mineral. Mag., 31, 698.

Sarver, J. V., and F. A. Hummel (1962) Solid solubility and eutectic temperature in the system $Zn_2SiO_4-Mg_2SiO_4$. J. Amer. Cer. Soc., 45, 304-

Saxena, S. K. (1969) Silicate solid solutions and geothermometry. 2. Distribution of Fe^{2+} and Mg^{2+} between coexisting olivine and pyroxene. Contr. Mineral. Petrol., 22, 147-156.

Schairer, J. F. (1942) The system $CaO-FeO-Al_2O_3-SiO_2$: I. Results of quenching experiments on five joins. J. Amer. Cer. Soc., 25, 248-252.

Schairer, J. F., and E. F. Osborn (1950) The system $CaO-MgO-FeO-SiO_2$: I, preliminary data on the join $CaSiO_3-MgO-FeO$. J. Amer. Cer. Soc., 33, 160-167.

Schairer, J. F., and K. Yagi (1952) The system $FeO-Al_2O_3-SiO_2$. Amer. J. Sci., Bowen Vol., 471.

Schairer, J. F., and H. S. Yoder, Jr. (1960) The nature of residual liquids from crystallization with data on the system nepheline-diopside-silica. Amer. J. Sci., 258A, 273-283.

Scheetz, B. E., and W. B. White (1972) Synthesis and optical absorption spectra of Cr^{2+}-containing orthosilicates. Contr. Mineral. Petrol., 37, 221-227.

Schmitz-Dumont, O., H. Fendel, M. Hassanein, and H. Weissenfeld (1966) Farbe und konstitution bei anorganischen Feststoffen, 14. Mitt. Die lichtabsorption des zweiwertigen kupfers in oxidischen wirtsgittern. Mh. Chem., 97, 1660-1695.

Schock, R. N., B. Olinger, and A. Duba (1972) Additional data on the compression of olivine to 140 kilobars. J. Geophys. Res., 77, 383-384.

Schubert, G., D. A. Yuen, and D. L. Turcotte (1975) Role of phase transitions in a dynamic mantle. Geophys. J. Royal Astron. Soc., 42, 705-735.

Schwab, R. B., and D. Kustner (1977) Präzisionsgitterkonstantenbestimmung zur festlegung röntgenographischer Bestimmungskurven für synthetische Olivine der Mischkristallreihe Forsterit-Fayalit. N. Jahrb. Mineral. Mn., 1977, 205-215.

Shankland, T. J., U. Nitsan, and A. Duba (1979) Optical absorption and radioactive heat transport in olivine at high temperature. J. Geophys. Res., 84, 1603-1610.

Shannon, R. D. (1976) Revised effective ionic radii and systematic studies of interatomic distances in halides and chalcogenides. Acta Crystallogr., A32, 751-767.

Shannon, R. D., and C. T. Prewitt (1969) Effective ionic radii in oxides and fluorides. Acta Crystallogr., B25, 925-946.

Sheherbakova, M. Y., L. D. Shipilov, V. I. Sinyakov, and V. E. Istomin (1968) Study of Mn^{2+} and Fe^{2+} in the structure of monticellite by the electron paramagnetic resonance method. J. Struct. Chem., 9, 877-882.

Shinno, I. (1974) Mössbauer study of olivine -- the relation between Fe^{2+} site occupancy number T_{Mi} and interplanar distrance d_{130}. Mem. Geol. Soc. Japan, 1, 11-17.

Shinno, I., M. Hayashi, and Y. Kuroda (1974) Mössbauer studies of natural olivines. Mineral. J. (Japan), 7, 344-358.

Simkin, T., and J. V. Smith (1970) Minor element distribution in olivine. J. Geol., 78, 304-325.

Singh, H. P., and G. Simmons (1976) X-ray determination of thermal expansion of olivines. Acta Crystallogr., A32, 771-773.

Skinner, B. J. (1962) Thermal expansion of ten minerals. U. S. Geol. Surv. Prof. Paper 450D, 109-112.

Smith, D. K., A. Majumdar, and F. Ordway (1965) The crystal structure of γ-dicalcium silicate. Acta Crystallogr., 18, 787-795.

Smith, J. V. (1966) X-ray emission microanalysis of rock-forming minerals II. olivines. J. Geol., 74, 1-16.

Smith, J. V. (1971) Minor elements in Apollo 11 and Apollo 12 olivine and plagioclase. Proc. 2nd Lunar Sci. Conf., Geochim. Cosmochim. Acta, Suppl. 2, 1, 143-150.

Smith, J. V. (1974) Lunar Mineralogy: a heavenly detective story. Presidential address, part I. Amer. Mineral., 59, 231-243.

Smyth, D. M., and R. L. Stocker (1975) Point defects and nonstoichiometry in forsterite. Phys. Earth Planet. Inst., 10, 183.

Smyth, J. R. (1975) High-temperature crystal chemistry of fayalite. Amer. Mineral., 60, 1092-1097.

Smyth, J. R., and R. M. Hazen (1973) The crystal structures of forsterite and hortonolite at several temperatures up to 900°C. Amer. Mineral., 58, 588-593.

Sockel, H. G. (1974) Defect structure and electrical conductivity of crystalline ferrous silicate. In, Defects and Transport in Oxides (Ed. N. S. Seltzer and R. I. Jaffee), New York: Plennum Press, 341 p.

Steele, I. M., J. J. Pluth, and J. Ito (1976) Crystal structure of synthetic LiScSiO4. Olivine-comparison with Mg_2SiO_4 and $LiFePO_4$ (abstr.). Amer. Crystallogr. Assoc. Summer Mtg., 68.

Stewart, R. F., M. A. Whitehead, and G. Donnay (1980) The ionicity of the Si-O bond in low quartz. Amer. Mineral., 65, 324-326.

Stocker, R. L., and D. M. Smyth (1978) Effect of enstatite activity and oxygen partial pressure on the point defect chemistry of olivine. Phys. Earth Planet. Inst., 16, 145-156.

Strunz, H., and P. Jacob (1960) Germanate mit Phenacit-und Olivinstruktur. N. Jahrb. Mineral. Abhandl., 78-79.

Sung, C. M., and R. G. Burns (1976) Kinetics of the olivine → spinel transition: implications to deep-focus earthquake genesis. Earth Planet. Sci. Lett., 32, 165-170.

Sung, C. M., and R. G. Burns (1978) Crystal structural features of the olivine → spinel transition. Phys. Chem. Minerals, 2, 177-197.

Syono, Y., S. Akimoto, and Y. Matsui (1971) High pressure transformations in zinc silicates. J. Sol. State Chem., 3, 369-380.

Syono, Y., M. Tokonami, and Y. Matsui (1971) Crystal field effect on the olivine-spinel transformation. Phys. Earth Planet. Inst., 4, 347-352.

Takahashi, E. (1978) Partitioning of Ni^{2+}, Co^{3+}, Fe^{2+}, Mn^{2+}, and Mg^{2+} between olivine and silicate melts: compositional dependence of partition coefficients. Geochim. Cosmochim. Acta, 42, 1829-1844.

Taylor, N. W. (1930) Die Kristallstrukturen der Verbindungen Zn_2TiO_4, Zn_2SnO_4, Ni_2SiO_4 u. $NiTiO_3$. Z. physik. Chem. (Leipzig), B9, 241-264.

'T Hart, J. (1978a) The structural morphology of olivine. I. A qualitative derivation. Canadian Mineral., 16, 175-186.

'T Hart, J. (1978b) The structural morphology of olivine. II. A quantitative derivation. Canadian Mineral., 16, 547-560.

Thielo, E. (1941) Uber die Isotypie zwischen Phosphaten der Allgemienen zusammensetzung MeLi[PO$_4$] und Silikaten der Olivin=Monticellit-Reihe. Naturwissenschaften,
29, 239.

Tokody, L. (1928) The binary system; Mn$_2$SiO$_4$-Ca$_2$SiO$_4$. Z. anorg. allgem. Chem., 169,
51-56.

Tokonami, M., N. Morimoto, S. Akimoto, Y. Syono, and A. Takeda (1972) Stability
relations between olivine, spinel, and modified spinel. Earth Planet. Sci. Lett.,
14, 65-69.

Tomkeieff, S. I. (1918) Zoned olivines and their petrogenetic significance. Mineral.
Mag., 25, 229.

Tossell, J. A. (1976) Electronic structures of iron-bearing oxide minerals at high-
pressures. Amer. Mineral., 61, 130-144.

Tossell, J. A. (1977) A comparison of silicon-oxygen bonding in quartz and magnesium
olivine from x-ray spectra and molecular orbital calculations. Amer. Mineral.,
62, 136-141.

Tossell, J. A., and G. V. Gibbs (1976) A molecular orbital study of shared-edge distortions in linked polyhedra. Amer. Mineral., 61, 287-294.

Tossell, J. A., and G. V. Gibbs (1977) Molecular orbital studies of geometries and
spectra of minerals and inorganic compounds. Phys. Chem. Minerals, 2, 21-57.

Vincent, H., E. F. Bertaut, W. H. Baur, and R. D. Shannon (1976) Polyhedral deformations
in olivine-type compounds and the crystal structure of Fe$_2$SiS$_4$ und Fe$_2$GeS$_4$. Acta
Crystallogr., B32, 1749-1755.

Vincent, H., and G. Perrault (1971) Structure cristalline des orthothiogermanates de
magnesium et de fer. Bull. Soc. fr. Mineral. Crist., 94, 551-555.

Virgo, D., and S. S. Hafner (1972) Temperature-dependent Mg,Fe distribution in a lunar
olivine. Earth Planet. Sci. Lett., 14, 305-312.

deWaal, S. A., and L. C. Calk (1973) Nickel minerals from Barberton, South Africa:
VI. liebenbergite, a nickel olivine. Amer. Mineral., 58, 733-735.

Walsh, D., G. Donnay, and J. D. H. Donnay (1974) Jahn-Teller effects in ferromagnesian
minerals. Bull. Soc. fr. Mineral. Crist., 97, 170-183.

Walsh, D., G. Donnay, and J. D. H. Donnay (1976) Ordering of transition metal ions in
olivine. Canadian Mineral., 14, 149-150.

Wager, L. R., and W. A. Deer (1939) Geological investigation in east Greenland Part
III. The petrology of the Skaergaard intrusion, Kangerdlugssuaq, east Greenland.
Meddel. on Grønland, 105, no. 4.

Wager, L. R., and R. L. Mitchell (1951) The distribution of trace elements during strong
fractionation of basic magma. Geochim. Cosmochim. Acta, 1, 129-208.

Wager, L. R., and R. L. Mitchell (1953) Trace elements in Hawaiian lavas. Geochim.
Cosmochim. Acta, 3, 217.

Warner, R. D. (1973) Liquidus relations in the system CaO-MgO-SiO$_2$-H$_2$O at 10 kb P$_{H_2O}$
and their petrologic significance. Amer. J. Sci., 273, 925-946.

Warner, R. D., and W. C. Luth (1973) Two-phase data for the join monticellite (CaMgSiO$_4$)-
forsterite (Mg$_2$SiO$_4$): experimental results and numerical analysis. Amer. Mineral.,
58, 998-1008.

Waseda, Y., and J. M. Toguri (1977) The structure of molten binary silicate systems
CaO-SiO$_2$ and MgO-SiO$_2$. Metal Trans. B., 8B, 563-568.

Waseda, Y., and J. M. Toguri (1978) The structure of the molten FeO-SiO$_2$ system. Metal.
Trans. B., 9B, 595-601.

Watson, E. B. (1979) Calcium content of forsterite coexisting with silicate liquid
in the system Na$_2$O-CaO-MgO-Al$_2$O$_3$-SiO$_2$. Amer. Mineral., 64, 824-829.

Weeks, R. A., J. C. Pigg, and C. B. Finch (1974) Charge-transfer spectra of Fe^{3+} and Mn^{2+}
in synthetic forsterite (Mg$_2$SiO$_4$). Amer. Mineral., 59, 1259-1266.

Weir, C. E., and A. van Valkenburg (1960) Studies of beryllium chromite and other
    beryllia compounds with R_2O_3 oxides. J. Res. Nat. Bur. Stand., A64, 103-106.

Weiss, A., and G. Rocktäschel (1960) Zur Kenntnis von Thiosilicaten. Z. anorg. Chem.,
    307, 1-6.

Wenk, H. R., and K. N. Raymond (1973) Four new structure refinements of olivine.
    Z. Kristallogr., 137, 86-105.

White, J. (1943) The physical chemistry of open-hearth slags. Iron Steel Inst. J.,
    148, 579-694.

White, W. B., and K. L. Keester (1966) Optical absorption spectra of iron in the
    rock-forming silicates. Amer. Mineral., 51, 774-791.

Whittaker, E. J. W. (1967) Factors affecting element ratios in the crystallization
    of minerals. Geochim. Cosmochim. Acta, 31, 2275-2288.

Whittaker, E. J. W. (1978) The cavities in a random close-packed structure. J. Non-
    crystal. Sol., 28. 293-304.

Wilkins, R. W. T., and W. Sabine (1973) Water contents of some nominally anhydrous
    silicates. Amer. Mineral., 58, 508-516.

Will, G., L. Cernic, E. Hinze, K. F. Seifert, and R. Voight (1979) Electrical conduc-
    tivity measurements of olivines and pyroxenes under defined thermodynamic activities
    as a function of temperature and pressure. Phys. Chem. Minerals, 14, 189-197.

Will, G., and G. Nover (1979) Influence of oxygen partial pressure on the Mg/Fe distri-
    bution in olivine. Phys. Chem. Minerals, 4, 199-208.

Williams, R. J. (1971) Reaction constants in the system $Fe-MgO-SiO_2-O_2$ at 1 atm between
    900° and 1300°: experimental results. Amer. J. Sci., 270, 334-360.

Wood, B. J. (1974) Crystal field spectrum of Ni^{2+} in olivine. Amer. Mineral., 59,
    244-248.

Wood, B. J. (1976) An olivine-clinopyroxene geothermometer: a discussion. Contr.
    Mineral. Petrol., 56, 297-303.

Wood, B. J., and D. G. Fraser (1976). *Elementary Thermodynamics for Geologists*. Oxford,
    England: Oxford University Press.

Wyderko, M., and E. Mazanek (1958) The mineralogical characteristics of calcium-iron
    olivines. Mineral. Mag., 36, 955-961.

Wyllie, P. J. (1960) The system $CaO-MgO-FeO-SiO_2$ and its bearing on the origin of ultra-
    basic and basic rocks. Mineral. Mag., 28, 459-470.

Yagi, T., Y. Ida, Y. Sato, and S. Akimoto (1975) Effect of hydrostatic pressure on the
    lattice parameters of Fe_2SiO_4 olivine up to 70 kbar. Phys. Earth Planet. Int.,
    10, 348-354.

Yagi, T., F. Marumo, and S. Akimoto (1974) Crystal structures of spinel polymorphs of
    Fe_2SiO_4 and Ni_2SiO_4. Amer. Mineral., 59, 486-490.

Yang, H. (1973) New data or forsterite and monticellite solid solutions. Amer. Mineral.,
    58, 343-345.

Yoder, H. S., Jr. (1952) The $MgO-Al_2O_3-SiO_2-H_2O$ system and the realted metamorphic
    facies. Amer. J. Sci., Bowen vol., 569.

Yoder, H. S., Jr. (1968) Akermanite and related melilite-bearing assemblages. Carnegie
    Inst. Wash. Year Book, 66, 471-477.

Yoder, H. S., Jr., and T. G. Sahama (1957) Olivine x-ray determinative curve. Amer.
    Mineral., 42, 475-491.

Ziera, S., and S. S. Hafner (1974) The location of Fe^{3+} ions in forsterite (Mg_2SiO_4)
    Earth Planet. Sci. Lett., 21, 201-208.

Zoltai, T. (1965) *Stereoscopic Drawings of Polyhedral Mineral-Structure Models*. Minne-
    apolis, Minnesota: University of Minnesota, 84 p.

Chapter 12

MISCELLANEOUS ORTHOSILICATES

J.A. Speer & P.H. Ribbe

In this chapter we present brief descriptions of minerals which are
not discussed elsewhere in this volume and whose crystal structures are
known to fit into the arbitrary confines that were set in the preface of
the first edition of this volume (Ribbe, 1980). Specifically, those
minerals are included in which SiO_4 groups are *not* polymerized to other
SiO_4 groups by corner-sharing *nor* are they polymerized to other tetrahedral groups containing cations such as Be, B, Al or Zn.

In order that this chapter be usable as an index to this the
second edition, we have, in addition, listed alphabetically all the
species names of those minerals which are classified in Chapters 2
through 11 as isostructural, polymorphous or homologous with the major
orthosilicate groups.

We make no claim to have found *all* the known silicates which fit
this description, but we have searched the lists of Povarennykh (1972),
Strunz (1977), Ramdohr and Strunz (1980) and Fleischer (1980, 1981).

REFERENCES

Fleischer, M. (1980) *1980 Glossary of Mineral Species*. Mineralogical
    Record, P. O. Box 35565, Tucson, Arizona, 192 p.

Fleischer, M. (1981) *The Ford-Fleischer File of Mineralogical References*.
    U. S. Geological Survey Open-File Report 81-1169 (microfiche).

Povarennykh, A. S. (1972) *Crystal Chemical Classification of Minerals,
    Volume 1*. Plenum Press, New York, 458 p.

Ramdohr, P. and H. Strunz (1980) *Klockmanns Lehrbuch der Mineralogie*,
    16th edition, updated to 1980. Ferdinand Enke Verlag, Stuttgart,
    931 p.

Ribbe, P. H., editor (1980) *Orthosilicates*. Reviews in Mineralogy 5,
    381 p.

Strunz, H. (1977) *Mineralogische Tabellen*, 6th edition. Akad.
    Verlagsges. Geest & Portig K. G., Leipzig.

● AFWILLITE, $Ca_3(SiO_3OH)_2 \cdot 2H_2O$.

Monoclinic, Cc; a = 16.278, b = 5.6321, c = 13.236 Å; β = 134.9°; Z = 4.

Afwillite occurs in cavities with calcite and bultfonteinite in dolerite and shale enclaves in the Dutoitspan Mine kimberite, South Africa (Parry and Wright, 1925) and in the contact metamorphic marbles of the Scawt Hill dolerite, Ireland. However, it is of most interest because of its being one of the cement minerals.

The crystal structure of afwillite was determined by Megaw (1952) and the hydrogen atoms were located by Malik and Jeffery (1976). The structure is comprised of isolated (SiO_3OH) tetrahedra sharing corners and edges with irregular CaO_7 polyhedra; together they form sheets of composition $[Ca_3Si_2O_4]^{6+}$ parallel to $(\overline{1}01)$. The sheets are joined by a few Ca-O-Si and hydrogen bonds, leading to a well developed $(\overline{1}01)$ cleavage.

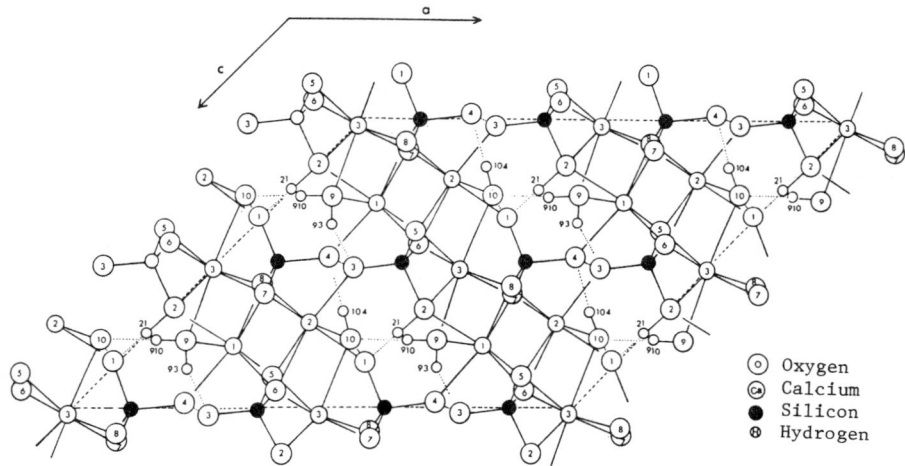

The crystal structure of afwillite projected on (010). From Malik and Jeffery (1976).

- - -

Malik, K.M.A. and J.W. Jeffery (1976) A re-investigation of the structure of afwillite. Acta Crystallogr. B32, 475-480.

Megaw, H.D. (1952) The structure of afwillite, $Ca_3(SiO_3OH)_2 \cdot 2H_2O$. Acta Crystallogr. 5, 477-491.

Parry, J. and Wright, F.F. (1925) Afwillite, a new hydrous calcium silicate from the Dutoitspan Mine, Kimberley, South Africa. Mineral. Mag. 20, 277-286.

● ALLEGHANYITE, $2Mn_2SiO_4 \cdot Mn(F,OH)_2$.   The Mn-analog of chondrodite; see Chapter 10.

● ALMANDINE, $Fe_3Al_2Si_3O_{12}$.   A garnet; see Chapter 2.

● ANDRADITE, $Ca_3Fe_2Si_3O_{12}$.   A garnet; see Chapter 2.

● BAKERITE.   See *DATOLITE*, this chapter.

● Beta-URANOPHANE.   See section on *URANYL SILICATES*, this chapter.

● BOLTWOODITE.   See section on *URANYL SILICATES*, this chapter.

● BRAUNITE, $Mn^{2+}Mn_6^{3+}SiO_{12}$.

   Tetragonal, $I4_1/acd$; a = 9.41, c = 18.67 Å; Z = 8.

● BRAUNITE-II, $(Ca,Mn)^{2+}Mn_{14}^{3+}SiO_{24}$.

   Tetragonal, $I4_1/acd$; a = 9.44, c = 37.76; Z = 8.

   Braunite can be found in manganese ores formed at low temperatures as well
as in Mn-rich rocks over a range of metamorphic grades in both regional and con-
tact metamorphism.   It has the general formula $(Mn^{2+},Ca,Mg)_{1\pm x}(Mn^{3+},Al,Fe^{3+})_{6\pm 2x}$
$Si_{1\pm x}O_{12}$ (Abs-Wurmbach, 1980).   The largest chemical variation is in Fe which
depends on the coexisting mineral assemblage, bulk-rock chemistry and temperature of equilibration (Dasgupta and Manickavasagam, 1981); 95% of the iron is
ferric (Seifert and Dasgupta, 1982).   Muan (1959a,b), on the basis of experimental studies, concluded that the SiO_2 content of braunite can vary between
0 and 40 wt %, but chemical analyses of braunites and subsequent experimental
work (Abs-Wurmbach, 1980) failed to confirm this.   A braunite with half the
silica content of other braunites has been designated braunite-II by DeVilliers
and Herbstein (1967).   Rather than a complete solid solution, DeVilliers (1975)
suggests the $Mn^{3+}Mn^{3+} \rightleftarrows Mn^{2+}Si^{4+}$ coupled substitution produces discrete, ordered
compounds between Mn_2O_3 (hausmannite) and $3Mn_2O_3 \cdot MnSiO_3$ (braunite).

   The structure of braunite was determined by DeVilliers (1975) and Moore
and Araki (1976) and is related to that of fluorite.   It is comprised of a
packing of edge- and corner-shared distorted Mn(1) cubes, Mn(2), Mn(3), Mn(4)
octahedra and SiO_4 tetrahedra.   Moore and Araki (1976) describe the structure as
a stacking of two types sheets along [001]: the A sheet is made up of Mn(2) and
Mn(3) octahedra arranged in a checkerboard by corner- and edge-sharing (figure a),
the B sheet is checkerboard of corner- and edge-sharing Mn(1) cubes, Mn(4) octahedra and SiO_4 tetrahedra (figure b).   The stacking is ... $[AB]_4$...

   The structure of braunite-II is unknown, but DeVilliers (1975) suggests it is
comprised of two Mn_2O_3 cells plus an braunite cell along [001].   Moore and Araki
(1976) suggested a similar arrangement with the notation ... $[AA'AA'AB]_4$ ...

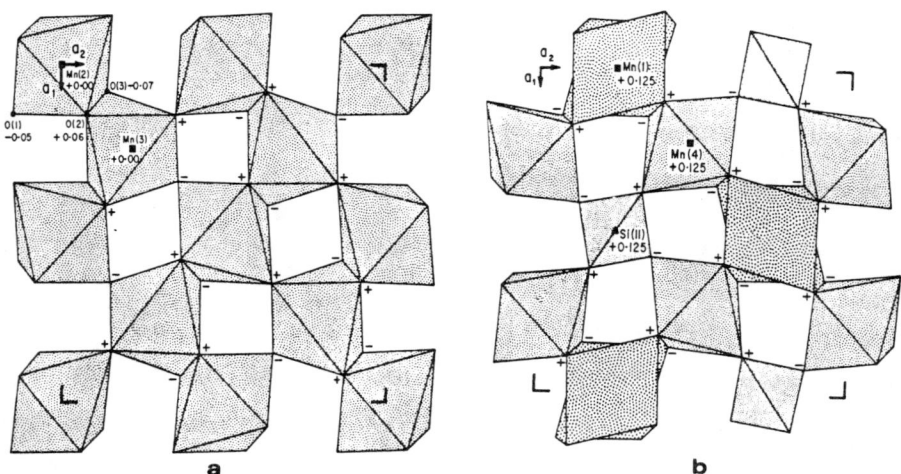

(a) The A sheet in the braunite crystal structure. (b) The B sheet. Note tetra-hedral, octahedral, and cubic coordination for Si, Mn(4) and Mn(1), respectively. From Moore and Araki (1976), Figs. 1A and 2A.

- - -

Abs-Wurmbach, I. (1980) Miscibility and compatibility of braunite, $Mn^{2+}Mn_6^{3+}O_8$-SiO_4, in the system Mn-Si-O at 1 atm in air. Contr. Miner. Petrol. 71, 393-399.

Dasgupta, H.C. and R.M. Manickavasagam (1981) Chemical and X-ray investigation of braunite from the metamorphosed manganiferous sediments of India. N. Jahrb. Mineral. Abh. 142, 149-160.

DeVilliers, J.P.R. (1975) The crystal structure of braunite with reference to its solid solution behavior. Am. Mineral. 60, 1098-1104.

_____ and F.H. Herbstein (1967) Distinction between two members of the braunite group. Am. Mineral. 52, 20-30.

Moore, P.B. and T. Araki (1976) Braunite: Its structure and relationship to bixbyite and some insights on the genealogy of fluorite derivative struc-tures. Am. Mineral. 61, 1226-1240.

Muan, A. (1959a) Phase equilibria in the system manganese oxide - SiO_2 in air. Am. J. Sci. 257, 297-315.

_____ (1959b) Stability relations among some manganese minerals. Am. Mineral. 44, 946-960.

Seifert, F. and H.C. Dasgupta (1982) A note on the Mössbauer spectrum of ^{57}Fe in braunite. N. Jahrb. Mineral. Mh., 11-15.

● BREDIGITE, $Ca_{1.75}Mg_{0.25}[SiO_4]$.

Orthorhombic, $P2nn$; a = 10.909, b = 18.34, c = 6.739 Å; Z = 2.

Bredigite occurs in calcsilicate rocks which have been contact-metamorphosed in the spurrite-merwinite facies and, more commonly, in slags. Much of the information regarding bredigite is based on material from slags and synthetic bredigite.

Based on an occurrence in slag, bredigite was originally considered to be a calcium orthosilicate in which Ca can be replaced by Mg, Mn and Ba (Tilley and Vincent, 1948). Subsequent studies of synthetic bredigite (Gutt, 1961; Biggar, 1971; Lin and Foster, 1975) and microprobe analyses of bredigite from Scawt Hill, Ireland (Midgley and Bennett, 1971; Sarkar and Jeffery, 1978) and Christmas Mtns., Texas (Joesten, 1974) show that the magnesium is essential with compositions ranging between $Ca_{1.67}Mg_{0.33}SiO_4$ and $Ca_{1.8}Mg_{0.2}SiO_4$.

The figure below illustrates an ideal configuration on which the structure of a slag bredigite of composition $Ca_{24.6}Ba_{1.2}Mg_{4.8}Mn_{1.4}[SiO_4]_{16}$ is based. It consists of edge-sharing isolated polyhedra of ideal formula $X^{12}X_2^9Y_4^{10}M^6[T^4O_4]_4$ and has symmetry of $Pmcb$. But ordering of Ba and Mn and tilting of the tetrahedra lowers the symmetry to $P2nn$ and the actual structural formula is (X^{10}, X^6) $X_2^8Y_{10}^9Y^8Y^8Y^7M^6[T^4O_4]_4$ with Ba present in X^{10} and Mg+Mn in M^6.

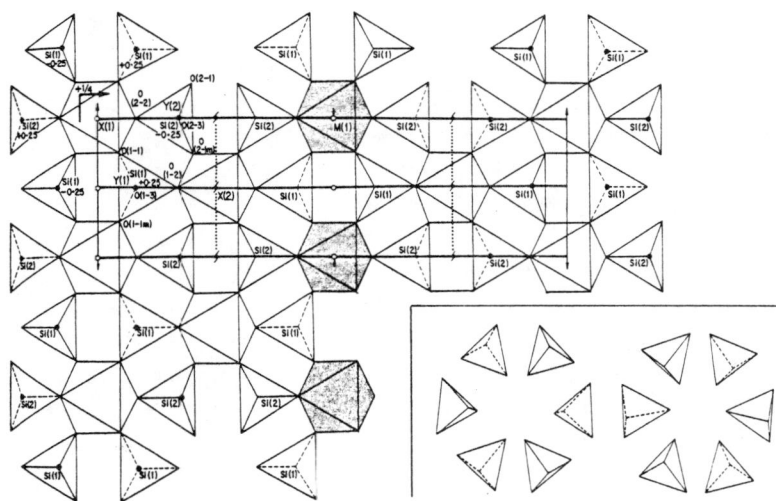

Structure ideal for the (001) projection of the $Pmcb$ arrangement to which bredigite is related. The symmetry elements and the unit cell are outlined. The non-equivalent atom positions are labelled, and the octahedral sites are stippled. In the real structure the tetrahedra are tilted substantially (see inset, lower right). After Moore and Araki (1976, Figs. 2 and 3).

● BREDIGITE, continued

Biggar, G.M. (1971) Phase relationship of bredigite ($Ca_5MgSi_3O_{12}$) and of the quaternary compound ($Ca_6MgAl_8SiO_{21}$) in the system $CaO-MgO-Al_2O_3-SiO_2$. Cement Concr. Res. 1, 493-513.

Gutt, W. (1961) A new calcium magnesiosilicate. Nature 190, 339-340.

Joesten, R. (1974) *Metasomatism and magmatic assimilation at a gabbro-limestone contact, Christmas Mtns., Big Bend region, Texas.* Ph.D. thesis, Calif. Inst. Tech., Pasadena, Calif. 397pp.

Lin, H.C. and W.R. Foster (1975) Stability relations of bredigite ($5CaO \cdot MgO \cdot 3SiO_2$). J. Am. Ceram. Soc. 58, 73.

Midgley, H.G. and M. Bennett (1971) A microprobe analysis of larnite and bredigite from Scawt Hill, Larne, N. Ireland. Cem. Concr. Res. 1, 413-418.

Moore, P.B. and T. Araki (1976) The crystal structure of bredigite and the genealogy of some alkaline earth orthosilicate. Am. Mineral. 61, 74-87.

Sarkar, S.L. and J.W. Jeffery (1978) Electron microprobe analysis of Scawt Hill bredigite-larnite rock. J. Am. Ceram. Soc. 61, 177-178.

Tilley, C.E. and H.C.G. Vincent (1948) Occurrence of an orthorhombic high-temperature form of Ca_2SiO_4 (bredigite) in the Scawt Hill contact zone and as a constituent of slags. Mineral. Mag. 28, 255-271.

● BRITHOLITE.   See section on *SILICATE APATITES*, this chapter.

● BRITHOLITE-Y.   See section on *SILICATE APATITES*, this chapter.

● BULTFONTEINITE, $Ca_4[SiO_2(OH_{\frac{1}{2}})_2]_2 \cdot F_2 \cdot 2H_2O$.

Triclinic, $P\bar{1}$, a = 10.992, b = 8.185, c = 5.671 Å; α = 93°57', β = 91°19', γ = 89°51'; Z = 2.

Bultfonteinite occurs with afwillite and calcite in enclaves in the Dutoitspan Mine kimberlite, South Africa (Parry *et al.*, 1932) and in the contact metamorphic marbles of Crestmore, California.

The structure of bultfonteinite was suggested to be related to afwillite by Megaw and Kelsey (1955) and does contain similar elements as shown by McIver (1963). The structure is comprised of isolated $[SiO_2(OH_{\frac{1}{2}})_2]$ tetrahedra and 7-coordinated Ca atoms which share edges to form double columns of composition $[Ca_4Si_2O_4]^{8+}$ parallel to (100). Similar double columns are present in afwillite, but they are joined by sharing edges to give the sheets (See the figure). These columns are linked by Ca-O-Ca, Ca-O-Si and hydrogen bonds as well as Ca-F-Ca bonds.

Figure on next page.

McIver, E.J. (1963) The structure of bultfonteinite, $Ca_4Si_2O_{10}F_2H_6$. Acta Crystallogr. 6, 551-558.

Megaw, H.D. and C.H. Kelsey (1955) An accurate determination of the cell dimensions of bultfonteinite, $Ca_4Si_2O_{10}H_6F_2$. Mineral. Mag. 30, 569-573.

Parry, J., A.F. Williams and F.H. Wright (1932) On bultfonteinite, a new fluorine-bearing hydrous calcium silicate from South Africa. Mineral. Mag. 23, 145-162.

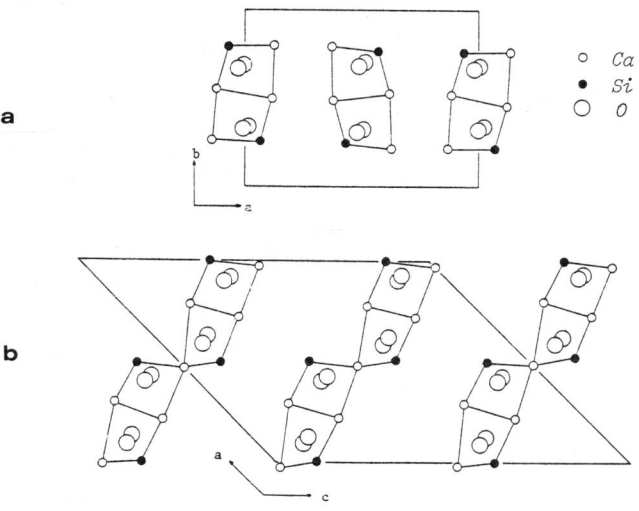

(a) Arrangement of the $(Ca_4Si_2O_4)^{8+}$ strips in bultfonteinite. (b) Arrangement of the $(Ca_4Si_2O_4)^{8+}$ strips in afwillite. In this case they are joined together to form infinite sheets with composition $(Ca_3Si_2O_4)^{6+}$. From McIver (1963). Fig. 7.

●CALDERITE, $Mn_3Fe_2Si_3O_{12}$. A garnet; see Chapter 2.

●CERITE, $(Ce,Ca)_9(Mg,Fe^{+2})Si_7(O,OH,F)_{28}$.

Trigonal, $R3c$; hexagonal cell: a = 10.78, c = 38.03 Å; Z = 6.

Cerite occurs in pegmatites and hydrothermal, carbonate-bearing veins associated with alkali syenites and granites. It is enriched in light REE's and is notably Ce-rich. Keppler (1968) has shown that cerite is isotypic with whitlockite, and Ito (1968) has shown that synthetic cerite can be considered as $M_3^{2+}Ln_7Si_7O_{27}OH \cdot H_2O$ or $M_2^{2+}Ln_8Si_7O_{28} \cdot 3H_2O$. Based on the crystal structure of whitlockite, Calvo and Gopal (1975) suggest the former, namely $(Ca,Mg,Fe^{+2})_3RE_7Si_7O_{27}OH \cdot H_2O$. The compositional range for cerite and its relationship to the RE-silicate apatites (britholites) in terms of ionic radii of the rare earth and divalent elements is shown in the figure. See also the section below on *SILICATE APATITES*.

Compositional stability range of cerites. Abscissa is the ionic radii for divalent cations given by Wells (1962); ordinate is rare earth radii of Templeton and Dauben (1954). From Ito (1968, his Fig. 1).

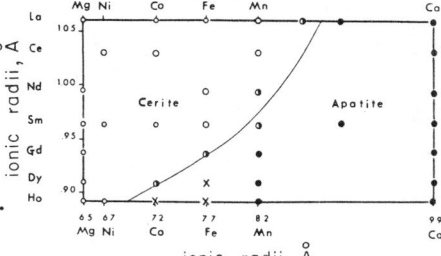

- - -

Calvo, C. and R. Gopal (1975) The crystal structure of whitlockite from the Palermo Quarry. Am. Mineral. 60, 120-133.

Ito, J. (1968) Synthesis of cerite. Jour. Res. National Bureau Standards - A, Phys. & Chem. 72A, 355-358.

Keppler, U. (1968) Structural investigations of "calcium phosphate" and isotopic structures. Bull. Soc. Chimique France 1968, 1774-1777.

● CHANTALITE, $CaAl_2(OH)_4SiO_4$.

Tetragonal, $I4_1/a$, $a = 4.952$, $c = 23.275$ Å, $Z = 4$.

Chantalite was discovered in rodingitic dykes in an ophiolitic zone in the Taurus Mountains, Turkey, occurring with vuagnatite, prehnite, hydrogrossular, chlorite and calcite (Sarp *et al.*, 1977). Leistner and Chatterjee (1978) succeeded in synthesizing chantalite. Sarp *et al.* (1977) suggested "une analogie cristallogchimiqie" with flinkite, $Mn_2^{2+}Mn^{3+}[AsO_4](OH)_4$, and retzian, $Mn_2^{2+}Y^{3+}[AsO_4]$ $(OH)_4$ (Moore, 1967), and for that reason it was considered an orthosilicate, a conclusion subsequently borne out by Liebich *et al.* (1979). As seen to the left,

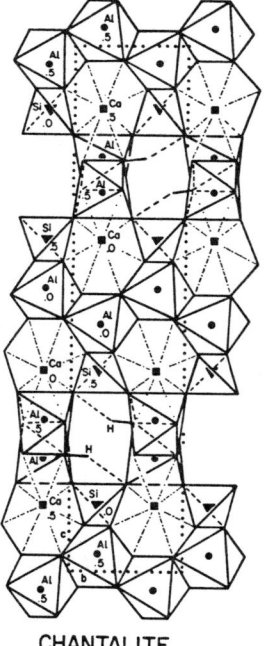

chantalite is comprised of $AlO_2(OH)_4$ octahedra sharing edges to form zig-zag chains joined by isolated SiO_4 tetrahedra and $CaO_4(OH)_4$ polyhedra. There are four of these layers in a unit cell, with each succeeding layer having its AlO_6 chains perpendicular to the next.

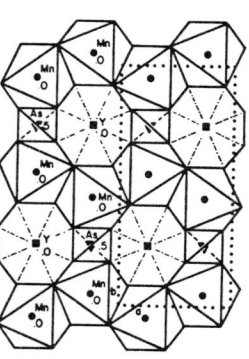

CHANTALITE   **RETZIAN**

The structure of chantalite viewed down c as compared with the structure of isoelectronic retzian built up of polyhedra of the same kind. From Liebich *et al.* (1979).

● CHANTALITE, continued

- - -

Leistner, H. and N.D. Chatterjee (1978) Wasserhaltige Minerale in System $CaSiO_3$-Al_2O_3-H_2O: Rosenhahnite $Ca_3[Si_3O_8(OH)_2]$, Vaugnatit, $CaAl[SiO_4](OH)$ and Chantalit $CaAl_2SiO_4(OH)_4$. Fortschr. Mineral. 56, 79-80.

Liebich, B.W., H. Sarp and E. Parthé (1979) The crystal structure of chantalite, $CaAl_2(OH)_4SiO_4$. Z. Kristallogr. 150, 53-63.

Moore, P.B. (1967) Crystal chemistry of the basic manganese arsenate minerals. 1. The crystal structure of flinkite, $Mn_2^{2+}M^{3+}(OH)_4(AsO_4)$ and retzian, $Mn_2^{2+}Y^{3+}(OH)_4(AsO_4)$. Am. Mineral. 52, 1603-1616.

Sarp, H., J. Deferne and B.W. Liebich (1977) La chantalite $CaAl_2SiO_4(OH)_4$, un nouveau silicate natural d'aluminium et de calcium. Bull. Suisse Minéral. Pétrogr. 57, 149-156.

● CHERALITE, $Ca_{1.08}Th_{1.15}REE_{1.62}U_{0.14}[Si_{0.34}P_{3.64}]O_4$. See Chapter 4.

● CHLORITOID, $(Fe^{+2},Mg,Mn)_2Al_4Si_2O_{10}(OH)_4$. See Chapter 6.

● CHONDRODITE, $2Mg_2SiO_4 \cdot MgF_2$. A humite mineral; see Chapter 10.

● CHRSOLITE, $(Mg,Fe)SiO_4$. An olivine; see Chapter 11.

● CLINOHUMITE, $4Mg_2SiO_4 \cdot MgF_2$. A humite mineral; see Chapter 10.

● COFFINITE, $USiO_4$. An actinide orthosilicate; see Chapter 4.

● CUPROSKLODOWSKITE. See section on *URANYL SILICATES*, this chapter.

● DATOLITE, $CaBSiO_4(OH)$.

Monoclinic, $P2_1/c$; $a = 4.832$, $b = 7.608$, $c = 9.636$ Å, $\beta = 90.4°$; $Z = 4$.

Datolite is found as a hydrothermal alteration mineral in hypabyssal and volcanic mafic igneous rocks as well as in contact-metamorphosed calcsilicate rocks. Datolites are normally close to the ideal chemistry, with only minor amounts of Fe, Al, Mg, Mn and alkalis. Semenov *et al.* (1963) reported a datolite of composition, $CaYBeBSi_2O_8(OH)_2$, indicating that the coupled substitution $Ca^{2+} + B^{3+} \rightarrow (Y,RE)^{3+} + Be^{2+}$ relates it to *GADOLINITE* (see this chapter).

The structure of datolite has been determined several times, most recently by Foit *et al.* (1973). It is a sheet structure comprised of four- and eight-membered rings of alternating SiO_4 and $BO_3(OH)$ tetrahedra in the (001) plane (see figure below). These tetrahedral sheets are joined in the [001] direction by Ca atoms coordinated by six oxygens and 2(OH) at the corners of distorted tetragonal antiprisms.

● DATOLITE, continued

BAKERITE. Based on its x-ray powder pattern and chemistry, bakerite is interpreted as a B-rich datolite with the possible formula $Ca_8B_8(BO_4)_2(SiO_4)_6(OH)_6$ ·$3H_2O$ (Kramer and Allen, 1956). Bakerite occurs as a hydrothermal or late-magmatic mineral.

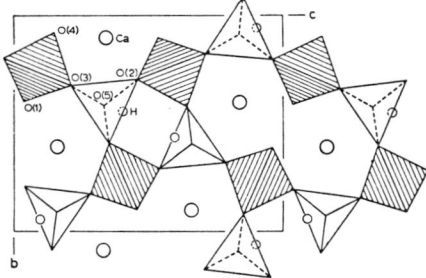

The structure of datolite projected down [100] showing the sheet of four- and eight-membered rings of alternating SiO_4 (shaded) and $BO_3(OH)$ tetrahedra (unshaded). From Foit *et al.* (1973, Fig. 1).

- - -

Deer, W.A., R.A. Howie and J. Zussman (1962) *Datolite.* In *Rock-forming minerals, Ortho- and Ring Silicates.* Longmans, London, pp. 171-175.

Foit, F.F., Jr., M.W. Phillips and G.V. Gibbs (1973) A refinement of the crystal structure of datolite, $CaBSiO_4(OH)$. Am. Mineral. 58, 909-914.

Kramer, H. and R.D. Allen (1956) A restudy of bakerite, priceite and veatchite. Am. Mineral. 41, 689-700.

● DIXENITE, $Cu^{1+}Mn^{2+}_{14}Fe^{3+}(OH)_6(As^{3+}O_3)(SiO_4)_2(As^{5+}O_4)$.

Rhombohedral, $R3$; a = 8.233, c = 37.499 Å; Z = 3.

Dixenite is a single-locality mineral from the Fe-Mn oxide deposit at Långban, Sweden, occurring in veins of hematite and dolomite with barite + pryoaurite. The crystal structure and unit formula were determined by Araki and Moore (1981) who describe this complex structure as composed of five non-equivalent layers along [001]. The Si^{4+} and As^{5+} mix over the tetrahedral sites. The As^{3+} of the $As^{3+}O_3$ trigonal pyramids are tetrahedrally coordinated about the Cu^{1+}, forming a metallic $[Cu^{1+}As^{3+}_4]$ cluster.

- - -

Araki, T. and P. B. Moore (1981) Dixenite, $Cu^{1+}Mn^{2+}_{14}Fe^{3+}(OH)_6(As^{3+}O_3)_5(Si^{4+}O_4)_2(As^{5+}O_4)$: metallic $[As^{3+}_4Cu^{1+}]$ clusters in an oxide matrix. Am. Mineral. 66, 1263-1273.

● DUMORTIERITE, $Si_3B[Al_{6.75}[]_{0.25}O_{17.25}(OH)_{0.75}]$.

Orthorhombic, *Pmcn*; a = 11.828, b = 20.243, c = 4.7001 Å; Z = 4.

Dumortierite has limited substitution of Fe^{3+}, Ti, Mg. *Holtite* is dumortierite with 12 wt % $(Ta,Nb)_2O_5$, 14 wt % Sb_2O_3 and 4.6 wt % Sb_2O_5 (Pryce, 1971). Dumortierite occurs in granitic pegmatites, hydrothermally altered rocks and in pelitic metamorphic and migmatized rocks. It was considered a hydrous oxyborosilicate of aluminum, but Claringbull and Hey (1952) proposed

392

the formula $Al_7BSi_3O_{18}$ and concluded that the small amount of water in reported analyses was sorbed. This was the accepted formula until Moore and Araki (1978) undertook a detailed structure refinement and determined that the OH was essential, attributing the low-water sums to experimental difficulties.

The crystal structure of dumortierite was solved by Golovastikov (1965) but refined in detail by Moore and Araki (1978). It is based on three types of chains built of AlO_6 octahedra extending along [100] (see the figure):

1. A column of Al(1) octahedra joined by common basal faces.
2. Two columns of edge-sharing Al(4) octahedra joined into a single chain by face-sharing.
3. A chain of Al(2,3) octahedra, based on cubic close-packing, which is 2×2 octahedra in cross-section.

The Al(2,3) chains are joined by corners to form a zigzag arrangement of the chains in a sheet parallel to (010). These sheets alternate with a layer comprising the Al(1) and Al(4) chains. The layers are joined by the SiO_4 tetrahedra and BO_3 triangles. The short cation-cation distances of 2.35 Å in the Al(1) chain, prolate thermal vibration ellipsoid for Al(1) and a site population refinement of Al(1) lead Moore and Araki (1978) to conclude that the Al(1) site has a disordered, 75% occupancy. The necessary charge balance and the larger, isotropic thermal vibration parameters for O(2) and O(7) suggested that these are, on the average, one-fourth replaced by OH.

The disordered Al(1) chain is the reason that the Moore and Araki formula differs from that of Claringbull and Hey. Ono (1981) attempted without success to synthesize dumortierites of the ideal composition, $Al_7BSi_3O_{18}$. Based on microprobe analyses, IR spectroscopy, and a crystal structure refinement of a natural dumortierite (Okamura and Ono, 1978), the material that was synthesized had a proposed formula:

$$Al_6(B_x,Al_{1-x})O_3(BO_3)[(Si_{3-y},Al_y)O_{12-y}](OH)_y$$

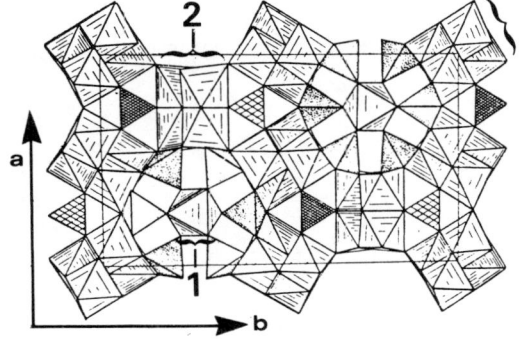

Projection of the dumortierite structure onto the *ab* plane from Golovastikov (1965). The numbers refer to the three different chains of Al-octahedra discussed in the text. The BO_3 triangles are cross-hatched; the SiO_4 tetrahedra are stippled.

393

● DUMORTIERITE, continued

- - -

Claringbull, G. F. and M. H. Hey (1958) New data for dumortierite. Mineral. Mag. 31, 901–907.

Golovastikov, N. I. (1965) The crystal structure of dumortierite. Sov. Phys. Doklady 10, 493–495.

Moore, P. B. and T. Araki (1978) Dumortierite, $Si_3B[Al_{6.75}\square_{0.25}O_{17.25} (OH)_{0.75}]$: a detailed structure analysis. N. Jb. Miner. Abh. 132, 231–241.

Okamura, F. P. and A. Ono (1978) Crystal chemical peculiarity of dumortierite from Takato. Ann. Meeting Abstr. Mineral. Soc. Japan, 145.

Ono, A. (1981) Synthesis of dumortierite in the system $Al_2O_3-SiO_2-B_2O_3-H_2O$. Japan. Assoc. Mineral. Pet. Econ. Geol. 76, 21–25.

Pryce, M. W. (1971) Holtite: a new mineral allied to dumortierite. Mineral. Mag. 38, 21–25.

● ELLESTADITE.   See section on *SILICATE APATITES*, this chapter.

● EULYTITE, $Bi_4Si_3O_{12}$.

Cubic, $I\bar{4}3d$; $a = 10.300$ Å, $Z = 4$.

Eulytite occurs with quartz and bismuth as a weathering product of bismuth ores. The location of the Bi atom in the structure was determined by Menzer (1931) in an x-ray study and the oxygen by Segal *et al.* (1966) by neutron-diffraction. The Si is on a special position. The bismuth atom is coordinated to a distorted octahedron of oxygen atoms belonging to six differ-ent, isolated SiO_4 tetrahedra.

- - -

Menzer, G. (1931) Die Kristallstruktur von Eulytin. Z. Kristallogr. 78, 136–163.

Segal, D.J., R. P. Santoro and R.E. Newnham (1966) Neutron-diffraction study $Bi_4Si_3O_{12}$. Z. Kristallogr. 123, 73–76.

● FAYALITE, Fe_2SiO_4.   An olivine; see Chapter 11.

● FERSMANITE.   See section on *NATISITE* and *FERSMANITE*, this chapter.

● FORSTERITE, Mg_2SiO_4.   An olivine; see Chapter 11.

● GADOLINITE, $Y_2Fe^{2+}Be_2O_2(SiO_4)_2$.

Monoclinic, $P2_1/a$; $a = 9.920$, $b = 7.484$, $c = 4.797$ Å; $\beta = 89°36'$; $Z = 2$.

Gadolinite is a common accessory mineral of granitoid rocks, especially pegmatites. The most common chemical variation in gadolinite is substitution of heavy rare earth elements for Y. Ito and Hafner (1974; see the figure) suggested defining boron-free gadolinites as solid-solutions of the three components: $RE_2^{3+}Fe^{2+}Be_2Si_2O_{10}$ - $CaRE_2^{3+}Fe^{3+}Be_2Si_2O_{10}$ - $H_2RE_2^{3+}Be_2Si_2O_{10}$. Appreciable amounts of B can be present in gadolinites, indicating solid solutions with either *HOMOLITE* or *DATOLITE* (see this chapter and Semenov *et al.*, 1963; Ito and Hafner, 1974).

As suggested by Ito and Mori (1953) and demonstrated by Pavlov and Belov (1959), the structure of gadolinite may be derived from that of datolite by replacing Ca by Y, B by Be, OH by O and placing the Fe at 0,0,0 and 0, 1/2, 1/2.

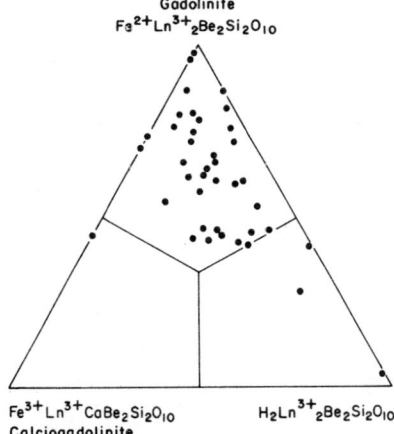

Gadolinite
$Fe^{2+}Ln^{3+}_2Be_2Si_2O_{10}$

$Fe^{3+}Ln^{3+}CaBe_2Si_2O_{10}$
Calciogadolinite

$H_2Ln^{3+}_2Be_2Si_2O_{10}$

Compositional plots of the chemical analyses of forty-four gadolinites on the triangular diagram for the system $Fe^{2+}RE_2^{3+}-$, $Fe^{3+}CaRe-$, $H_2RE^{3+}-Be_2Si_2O_{10}$, where RE = lanthanides plus Y. The plots were made in terms of molecular ratio of FeO, $FeO_{1.5}$, and H_2O in the chemical analyses. From Ito and Hafner (1974; Fig. 1).

- - -

Ito, T. and H. Mori (1953) The crystal structure of datolite. Acta Crystallogr. 6, 24-32.

Ito, J. and S.S. Hafner (1974) Synthesis and study of gadolinites. Am. Mineral. 59, 700-708.

Pavlov, P.V. and N.V. Belov (1959) The structures of herderite, datolite and gadolinite determined by direct methods. Kristallografiya 4, 324-340; Soviet Physics - Crystallogr. 4, 300-314.

Semenov, E.I., V.D. Dusmatov and N.S. Samsonova (1963) Yttrium-beryllium minerals of the datolite group. Soviet Physics - Crystallogr. 8, 539-541.

395

● GAGEITE, $(Mn,Mg)_7O(OH)_8Si_2O_6$.

Orthorhombic, *Pnnm*; $a = 13.79$, $b = 13.68$, $c = 3.279$ Å; $Z = 2$.

Gageite is a fibrous late-stage, low-temperature mineral found in fissures and cavities at Franklin, New Jersey. Moore (1969) found the structure to consist of walls of edge-sharing octahedra, three octahedra wide, connected by corner-sharing to bundles of octahedra (each 2×2 octahedra in size). See the figure. These walls and bundles of octahedra run parallel to [001], the fiber axis, forming channels. The channels have independent pairs of tetrahedral sites with a generally disordered half-occupancy by Si: if Si(1) is empty, O(9) must be empty; and if Si(2) is empty, O(8) is also. This partial occupancy of the tetrahedral site and potential partial ordering explains the faint streaks evident on long-exposure x-ray photographs requiring trebling of the *c*-axis. *Pnnm* is the average space group for the substructure.

More recently, Dunn (1979) investigated the chemistry of a number of gageite samples of differing parageneses by electron microprobe. He found the chemistry in good agreement with the original description and proposed the empirical formula $(Mn,Mg,Zn)_{40}Si_{15}O_{50}(OH)_{40}$. Mn:Mg:Zn is about 28:10:2 suggesting Mg and Zn are essential and may be ordered in the structure.

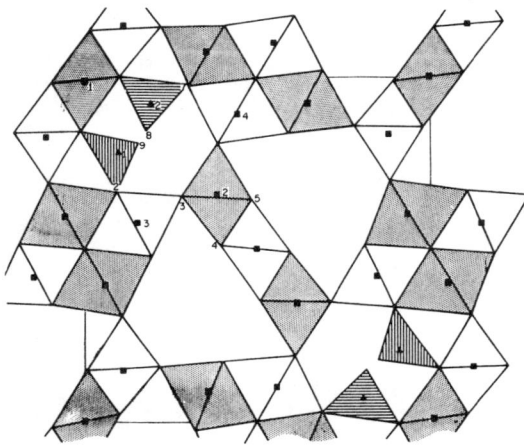

Polyhedral diagram of the gageite crystal structure down the *z* axis. Octahedra at $z = \frac{1}{2}$ are stippled. The disordered tetrahedra are drawn in two of the four equivalent open pipes. From Moore (1969, Fig. 2).

- - -

Dunn, P. J. (1979) The chemical composition of gageite: an empirical formula. Am. Mineral. 64, 1056-1058.

Moore, P. B. (1969) A novel octahedral framework structure: gageite. Am. Mineral. 54, 1005-1017.

● GLAUCOCHROITE, $CaMnSiO_4$. An olivine; see Chapter 11.

● GOLDMANITE, $Ca_3V_2Si_3O_{12}$. A garnet; see Chapter 2.

● GRANDIDIERITE, $(Mg,Fe)Al_3SiBO_9$.

Orthorhombic, *Pbnm*; $a = 10.335$, $b = 10.978$, $c = 5.760$ Å; $Z = 4$.

Grandidierite occurs in B-rich, aluminous metamorphic rocks in both contact and regional metamorphic environments (Vrána, 1979). The most important chemical variation is Fe ⇄ Mg. The structure, determined by Stephenson and Moore (1968), consists of edge-sharing Al-octahedral chains running parallel to c. The chains form a nearly square array with the orientation of the shared edge differing by 90° for every other chain (see the figure below). In the a direction, the chains are joined by BO_3 triangles and AlO_5 distorted trigonal bipyramids. In the b direction, the chains are joined by isolated SiO_4 tetrahedra and $(Mg,Fe)O_5$ distorted trigonal bipyramids. Stephenson and Moore (1968) point out a general similarity between the grandidierite and andalusite structures; however, the two are not isostructural because many of the "substituting" cations differ in coordination numbers.

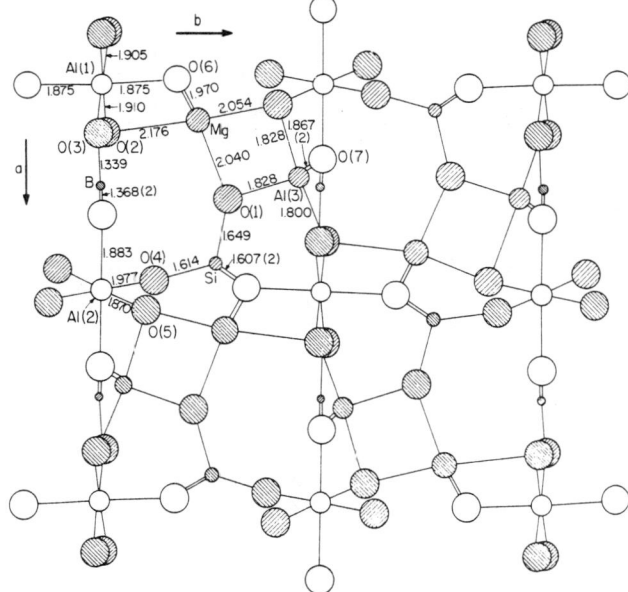

The z-axis projection of one unit cell of grandidierite. Open circles are $z \sim 0$ and 0.5, ruled circles running NE-SW are at $z = 0.25$ and running NW-SE at $z = 0.75$. Atoms in one asymmetric unit are identified and metal oxygen distances are given. From Stephenson and Moore (1968, Fig. 1).

- - -

Stephenson, D. A. and P. B. Moore (1968) The crystal structure of grandidierite, $(Mg,Fe)Al_3SiBO_9$. Acta Crystallogr. B24, 1518–1522.

Vrána, S. (1979) A polymetamorphic assemblage of grandidierite, kornerupine, Ti-rich dumortierite, tourmaline, sillimanite and garnet. N. Jahrb. Mineral., Mh., 22–33.

● GROSSULAR, $Ca_3Al_2Si_3O_{12}$. A garnet; see Chapter 2.

● HAFNON, $HfSiO_4$. Isostructural with zircon; see Chapter 3.

● HAIWEEITE.   See section on *URANYL SILICATES*, this chapter.

● HENRITERMIERITE, $Ca_3(Mn^{3+}_{1.5}Al_{0.5})[(SiO_4)_2(OH)_4]$.

Tetragonal, $I4_1/acd$; $a = 12.39$, $c = 11.91$ Å, $Z = 8$.

Henritermierite occurs in the manganese mine at Tachgagalt, Anti-Atlas, Morocco and is the manganese analog of hydrogrossular garnet (*cf.* Chapter 2, this volume). The natural material contains a small amount of ferric iron as well and has the formula $Ca_{2.97}Mn^{3+}_{1.48}Al_{0.54}Fe^{3+}_{0.06}Si_{1.93}O_{9.85} \cdot 2.05\ H_2O$ (Gaudefroy *et al.*, 1969). Aubry *et al.* (1969) determined that the structure is a distorted version of the garnet structure.

- - -

Aubry, A., Y. Dusausoy, A. Laffaille and J. Protas (1969) Détermination et étude de la structure cristalline de l'henritermiértie, hydrogrenat de Symétrie quadratique. Bull. Soc. Fr. Mineral. Cristallogr. 92, 126-133.

Gaudefroy, C., M. Orliac, F. Permingeat and A. Parfenoff (1969) L'henritermiérite, une nouvelle espèce minérale. Bull. Soc. Fr. Mineral. Cristallogr. 92, 185-190.

● HOLTITE.   See *DUMORTIERITE*, this chapter.

● HOMILITE, $Ca_2Fe^{2+}B_2O_2(SiO_4)_2$.

Monoclinic, $P2_1/a$; $a = 9.67$, $b = 7.57$, $c = 4.74$ Å; $\beta = 90°22'$; $Z = 2$.

Homilite occurs in nepheline syenites of the Langesundfiord, Norway. It can be considered the boron analogue of gadolinite, $(Y,RE)^{3+}Fe^{2+}Be_2O_2(SiO_4)_2$, by the coupled substitution $Ca^{+2} + B^{3+} = (Y,RE)^{3+} + Be^{2+}$ or an anhydrous datolite charge balanced by the addition of Fe^{2+}. Natural homilites have significant solid solutions with GADOLINITE. See *GADOLINITE* (this chapter) for references.

● HORTONOLITE, $(Fe,Mg)SiO_4$. An olivine; see Chapter 11.

● HUMITE, $3Mg_2SiO_4 \cdot Mg(F,OH)_2$. One of the humite mineral series; see Chapter 10.

● HUTTONITE, ThSiO$_4$. An actinide orthosilicate; see Chapter 4.

● HYDROANDRADITE, Ca$_3$Fe$_2$(SiO$_4$)$_{3-x}$(OH)$_{4x}$. A garnet; see Chapter 2.

● HYDROGROSSULAR, Ca$_3$Al$_2$(SiO$_4$)$_{3-x}$(OH)$_{4x}$. A garnet; see Chapter 2.

● IMOGOLITE, Al$_2$[SiO$_4$H](OH)$_3$.

Cylindrical symmetry, C_{2nh}, with n = 10, 11 or 12.

Imogolite is a gel-like hydrous aluminum silicate found in volcanic ash soils and other weathered pyroclastic deposits. Cradwick *et al.* (1972) used the electron microscope to observe that the mineral occurs in bundles of fine tubes, each ∿20 Å in diameter with repeats of 8.4 Å along the tube axis and 22-23 Å normal to the tube axis, i.e., the center-to-center separation. They deduced the structure from a gas chromatographic recognition of the [SiO$_4$] group with which they proposed replaces "the three hydroxyl groups surrounding a vacant octahedral site" in gibbsite, accounting thereby for the shortening of the 8.6 Å repeat in gibbsite to 8.4 Å in imogolite and also the curling of the gibbsite sheet to form a tube. See the figures below. X-ray intensity data were tested against cylindrical models of several diameters, and it was concluded that the 23 Å interaxial b dimension fit best.

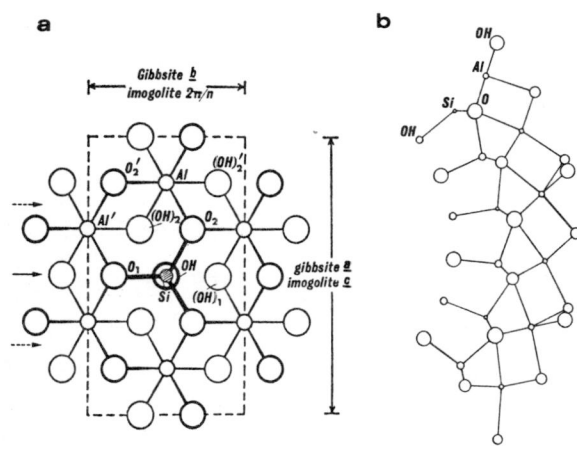

a   **b**

(a) Postulated relationship between a structural unit of imogolite and that of gibbsite. SiOH groups which would lie at the cell corners in imogolite have been omitted from the diagram. A reflection plane (solid arrow, left) and rotation-reflection planes (broken arrows) are indicated. (b) Curling of the gibbsite sheet induced by contraction of one surface to accommodate SiO$_3$OH tetrahedra: projection along the axis (imogolite c). From Chadwick *et al.* (1972, Fig. 1).

- - -
Chadwick, P. D. G., V. C. Farmer, J. D. Russell, C. R. Masson, K. Wada and
  N. Yoshinaga (1972) Imogolite, a hydrated aluminum silicate of tubular
  structure. Nature Phys. Sci. 240, 187-189.

● JERRYGIBBSITE, $Mn_9(SiO_4)_4(OH)_2$.

Orthorhombic, *Pbnm* or $P2_1nb$; $a = 4.86$, $b = 10.79$, $c = 28.30$ Å; $Z = 4$.

Jerrygibbsite was found at Franklin, New Jersey by Dunn *et al.* (1982). Its ideal composition is close to that of sonolite [see Table 4 in Chapter 10, p. 243], and its cell dimensions -- transformed to correspond to those of the humite mineral series -- are related to those of sonolite [Table 2, p. 234] with $c_{jg} = 2(c\sin\alpha)_{so}$. By analogy with ortho- and clino-pyroxene, Dunn *et al.* suggest that its structure consists of "unit-cell-twinned sonolite". If so, it is the first such polymorph in the humite or leucophoenicite families to be recognized; White and Hyde (1982b) did not report it in their HRTEM studies of Franklin sonolite (see their Fig. 3 and Chapter 10, p. 267), but their designation of this structure would be (3,2,2,2,2,2,2,3) or $(3,2^6,3)$. Winter *et al.* (1983) suggest that jerrygibbsite is more likely a member of the leucophoenicite family than the humite family.

- - -

Dunn, P.J., D.R. Peacor, W.B. Simmons, Jr. and E.J. Essene (1982) Jerrygibbsite, a new polymorph of sonolite and member of the humite-leucophoenicite groups, from Franklin, New Jersey. Am. Mineral. 67, in press.

Winter, G.A., E.J. Essene and D.R. Peacor (1983) Mn-humites from Bald Knob, North Carolina: mineralogy and phase equilibria. Am. Mineral. 68, in press.

● KANONAITE, $Mn^{3+}AlSiO_5$. The Mn-analog of andalusite; see Chapter 8.

● KASOLITE. See section on *URANYL SILICATES*, this chapter.

● KHOHARITE, $Mg_3Fe_2Si_3O_{12}$. A garnet; see Chapter 2.

● KIMZEYITE, $Ca_3(Zr,Ti)_2(Al,Fe,Si)_3O_{12}$. A garnet; see Chapter 2.

● KIRSCHSTEINITE, $CaFeSiO_4$. An olivine; see Chapter 11.

● KNEBELITE, $(Mn,Fe)SiO_4$. An olivine; see Chapter 11.

● KNORRINGITE, $Mg_3Cr_2Si_3O_{12}$. A garnet; see Chapter 2.

● LAIHUNITE, $Fe^{3+}Fe^{2+}_{0.5}SiO_4$.

Monoclinic (subcell), $P2_1/b$; $a = 4.805$, $b = 10.187$, $c = 5.801$Å; $\beta = 91.0°$; $Z = 4$.

Laihunite occurs in high-grade metamorphic rocks with magnetite, fayalite, ferrosilite, almandite and quartz. Conditions of crystallization were determined by Wang (1980) to have been 600-700°C, >15 kbar at relatively high oxygen fugacity. The rarity of these conditions accounts for the rarity of the mineral.

The chemical formula of the natural material is $Fe^{3+}_{1.00}Fe^{2+}_{0.58}Mg_{0.08}Si_{0.96}O_4$. Its crystal structure was first determined at Academica Sinica in 1976, but a more careful examination by Shen *et al.* (1982) disclosed that laihunite has a "distorted olivine-type structure," whose subcell has the dimensions listed above. Satellite reflections in x-ray and electron diffraction patterns indicate that there are two sorts of superstructures, one with $c' = 2c$ and one with $c'' = 3c$. A least-squares refinement of the latter indicated that ordering of Fe^{2+} and vacancies on the M1 octahedral site is responsible for the superstructure.

High resolution transmission electron microscopy showed laihunite to have formed by oxidation of fayalite. It commonly contains 100-200Å lamellae of precipitated magnetite and a finely dispersed amorphous phase.

- - -

Academica Sinica (1976) The crystal structure of laihunite. Ti Chi'iu Hua Hsuehi 1976, 104-106. Chem. Abstr. 85, 115090.

Shen, B., O. Tamada, M. Kitamura, and N. Morimoto (1982) Superstructure of laihunite ($Fe^{2+}_{0.5}Fe^{3+}_{1.0}SiO_4$): a nonstoichiometric olivine. I.M.A. Abstr., 381.

Wang, Sheng-Yuan (1980) Thermodynamic analysis of the stability of laihunite. Ti Chi'iu Hua Hsuehi 1980, 31-42. Chem. Abstr. 93, 50757.

Added in proof:

Academica Sinica (1982) Laihunite — a new iron silicate mineral. Geochemistry 1, 105-114.

Pingqiu, Fu, Kong Youhua and Zhang Liu (1982) Domain twinning of laihunite and refinement of its crystal structure. Geochemistry 1, 115-133.

● LARNITE, $\beta-Ca_2SiO_4$.

Monoclinic, $P2_1/n$; $a = 5.502$, $b = 6.745$, $c = 9.297$ Å; $\beta = 94.59°$; $Z = 4$.

Larnite occurs in calcsilicate rocks which have experienced high-temperature, low-pressure contact metamorphism. It is also a phase in Portland cement and slags and is important because of its hydraulic activity, i.e., its ability to harden when immersed in water.

The structure of larnite was determined by Midgley (1952) and corrected and refined by Jost *et al.* (1977); it contains isolated SiO_4 tetrahedra and 3-dimensionally connected CaO_7 and CaO_8 polyhedra (see figure). The CaO_x polyhedra are most densely packed in columns parallel to b, sharing triangular faces. The columns are cross-linked by sharing edges and corners.

a   b

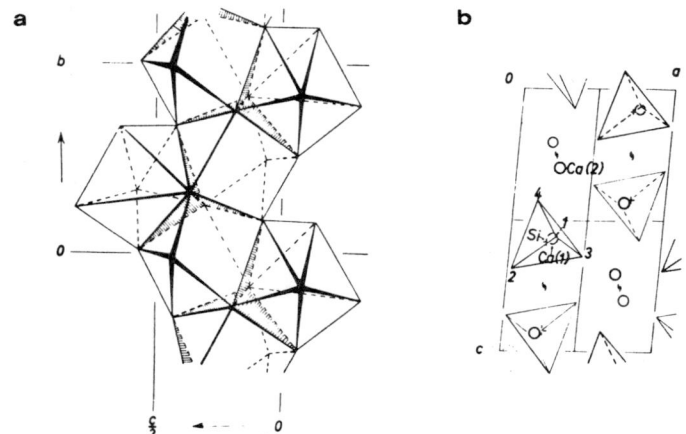

(a) Column-like structural unit in larnite made up of CaO_x polyhedra. Common faces are hatched. (b) The β-Ca_2SiO_4 structure, projected along the y-axis. Large circle: Ca; +: Si; oxygens are at the corners of tetrahedra. From Jost *et al.* (1977, Figs. 3 and 2, respectively).

- - -

Jost, K.H.. B. Ziemer and R. Seydel (1977) Redetermination of the structure of β-dicalcium silicate. Acta Crystallogr. B33, 1696-1700.

Midgley, C.M. (1952) The crystal structure of β dicalcium silicate. Acta Crystallogr. 5, 307-312.

● LEUCOPHOENICITE. $3Mn_2SiO_4 \cdot Mn(OH)_2$.

    Monoclinic, $P2_1/b$; $a = 4.826$, $b = 10.842$, $c = 11.324$ Å; $\alpha = 103.93°$; $Z = 2$. [Cell dimensions and space group transformed from Moore (1970) to conform to those of the humite minerals -- see Chapter 10.]

    Leucophoenicite occurs "as crystals in late state open hydrothermal veins and as granular masses in ore and skarn from Franklin, New Jersey, ... usually in association with green willemite, tephroite, glaucochroite and coarsely crystalline franklinite" (Moore, 1970, p. 1146; *cf*. Palache, 1935).

    Because of its composition, leucophoenicite was thought to be the Mn-isotype of humite. Palache (1935) found it to be monoclinic and not allied to the humites, and Moore (1967) reported an "O-leucophoenicite" with an orthorhombic multiple cell which he later found to be a pseudo-cell built of multiply-twinned monoclinic leucophoenicite. In fact the lattice parameters are at least superficially related to those of the Mn-humites (see Table 1 in Ch. 10, p. 234). But Moore's (1970) determination of the structure yielded the crystallochemical formula $Mn_7[SiO_4]_2[(SiO_4)(OH)_2]$. As with the humites, its structure is based on hcp oxygens stacked parallel to (100). "The octahedral populations define a new kind of kinked serrated chain ... running parallel to the z-axis, explaining the frequent twinning by reflection on {001}" (see the figures).

"The octahedrally populated chains place restrictions on the tetrahedral populations. For leucophoenicite, there is a set of fully occupied tetrahedra with point symmetry 1 [Si(2) in figure a] and a set of disordered half-occupied tetrahedra: these latter occur as edge-sharing tetrahedral pairs, with the midpoint of the common edge possessing point symmetry $\bar{1}$. This pair has an average composition $[(SiO_4)(OH)_2]$ and its presence results in unusual but explicable polyhedral distortions" (Moore, 1970, p. 1146). Bond lengths in the centrosymmetric $[Si(1)O_4(OH)_2]^{6-}$ tetrahedral group are therefore quite unique for silicates: $SiO(7)_{terminal}$ = 1.519 Å, $Si-O(7')_{basal}$ = 1.764 Å, where $O(7)...O(7')$ is the edge shared between the half-occupied Si(1) tetrahedra, and $Si-O(5)"_{basal}$ = 1.771 Å, $Si-O(4)_{basal}$ = 1.794 Å, both $[0.5Si = (OH)_{0.5}O_{0.5}]$, where *terminal* refers to the bond to an oxygen not involved in an edge shared between an octahedron and the Si(1) tetrahedron and *basal* refers to a bond to any oxygen or disordered O,OH anion that is involved in such a shared edge (see figure a). Both Si(1) and Si(2) are bonded only to oxygens because, when the Si(1) atom occupies, say the left tetrahedron, the OH ions are associated with the unoccupied right tetrahedron. This unusual stereochemistry results in a locally neutral electrostatic charge distribution in which each oxygen is coordinated by three octahedral Mn^{2+} plus one Si^{4+} and each hydroxyl is coordinated by three Mn^{2+}, just as in the Mn-humites. Thus leucophoenicite is technically an orthosilicate, albeit a most unusual one. [See also Belokoneva *et al.*, 1974.]

White and Hyde (1972a,b) describe the leucophoenicite structure (figure b) in the same terms as the humite structures (Fig. 14, Ch. 10). Like Moore (1970), they were unable to explain "additional orthorhombic reflections in X-ray powder diffraction patterns"; twinning on a unit cell scale was observed but not the periodic polysynthetic twins that would be required to rationalize those reflections. Various faulted members of the leucophoenicite family were observed to be coherently intergrown. "The possibility of ordering [of tetrahedral Si] in the [100] direction was considered but rejected because it would result in a doubled *c*-axis, for which there was no evidence in X-ray or electron diffraction patterns" (White, 1982, p. 92).

Figures a and b are on the next page.

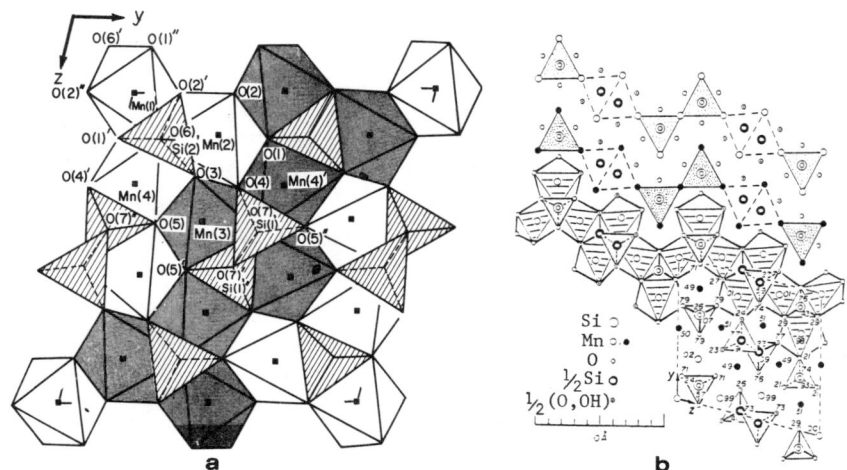

a **b**

(a) Polyhedral diagram of the structure of leucophoenicite, showing one chain of edge-sharing octahedra (shaded) and the unique pair of half-occupied edge-sharing Si(1) tetrahedra which are related by a center of symmetry half-way between the O(7) oxygens with (OH)$_{.5}$O$_{.5}$ at O(4) and O(5). The Si(2) tetrahedron is similar to those in the humites. From Moore (1970).

(b) In the upper portion of this figure the leucophoenicite structure is represented as an anion-stuffed ccp array of cations (see Chapter 10, p. 6 for full discussion). Half of the face-sharing Mn$_6$ trigonal prisms (those outlined with dashed lines) are occupied by Si in a statistical manner. The "twin formula" is $(1,2^3)$. From White (1982, Fig. 1.12, p. 17).

Belokoneva, E.L., M.A. Simonov and N.V. Belov (1974) Structures of leuco-phoenicite, Mn$_7$[SiO$_4$]$_2$[(SiO$_4$)(OH)$_2$], and synthetic Cd orthogermanate, Cd$_3$[GeO$_4$](OH)$_2$. Sov. Phys. Crystallogr. 18, 800-801.

Moore, P.B. (1967) On leucophoenicites: I. A note on form developments. Am. Mineral 52, 1226-1232.

_____ (1970) Edge-sharing silicate tetrahedra in the crystal structure on leuco-phoenicite. Am. Mineral 55, 1146-1166.

Palache, C. (1935) The minerals of Franklin and Sterling Hill, Sussex County, New Jersey. U.S. Geol. Surv. Prof. Paper 180, 104-105.

White, T.J. (1982) *An Electron Microscope Study of the Humite and Leucophoen-icite Structural Families*. Ph.D. dissertation, Australian National Univ., Canberra.

_____ and B.G. Hyde (1982a) A description of the leucophoenicite family of struc-tures and its relation to the humite family. Acta Crystallogr., in press.

_____ and _____ (1982b) An electron microscope study of leucophoenicite. Am. Mineral. (submitted).

_____

● LIEBENBERGITE, Ni$_2$SiO$_4$. An olivine; see Chapter 11.

● LUSAKITE. A cobaltan staurolite; see Chapter 7.

● MAJORITE, $Mg_3(Fe,Si,Al)_2Si_3O_{12}$.

  Cubic, *Ia3d*. a = 11.524 Å; Z = 8.

  Majorite is garnet occurring in meteorites with a composition near hypersthene. It presumably forms as a result of extraterrestrial shock metamorphism
of a pyroxene (Smith and Mason, 1970; Coleman, 1977). See also Chapter 2.

- - -

Coleman, L.C. (1977) Ringwoodite and majorite in the Catherwood meteorite.
  Canadian Mineral. 15, 97-101.

Smith, J.V. and B. Mason (1970) Pyroxene-garnet transformation in the Coorara
  meteorite. Science 168, 832-833.

● MALAYAITE, $CaSnSiO_5$. The tin analog of titanite; see Chapter 5.

● MANGANHUMITE, $3Mn_2SiO_4 \cdot Mn(OH)_2$. The Mn-analog of humite; see Chapter 10.

● MELANOCERITE, $(Ce,Ca)_5(Si,B)_3O_{12}(OH,F) \cdot nH_2O$.

  Hexagonal.

  Melanocerite occurs in nepheline-syenite pegmatites of Langesundfiord, Norway and at Wilkerforce, Ontario. It may be the same mineral as tritomite,
but with less thorium. It is mostly metamict, but single-crystal x-ray work on
the Wilkerforce material suggests that it may have the apatite structure (Erd,
1970). See the section on *SILICATE APATITES*, this chapter.

- - -

Erd, R.C. (1970) Monthly project report, listed under "Melanocerite". *In*
  M. Fleischer (1981) *The Ford-Fleischer File of Mineralogical References.*
  U.S.G.S. Open-File Report 81-1169.

● MERWINITE, $Ca_3Mg[SiO_4]_2$.

  Monoclinic $P2_1/a$; a = 13.254, b = 5.293, c = 9.328 Å; β = 91.90°; Z = 4.

  Merwinite is a major component of silicate slags used in the manufacture
of iron, steel and cement, and it is of particular value industrially because,
unlike Ca_2SiO_4, it does not go through polymorphic transformations when it is
cycled from low to high temperatures. In nature, merwinite is found only in
skarns; it was first described by Larsen and Foshag (1921) as a major constituent
at Crestmore, California. It is invariably associated with spurrite and
gehlenite, occasionally with monticellite ± spinel, and rarely with idocrase.
"Where diopside and wollastonite are abundantly associated with gehlenite, merwinite was observed to be rare or absent" (Moore and Araki, 1972, p. 1356).
Divalent Fe and Mn may substitute for Mg.

The stability of merwinite in the system $CaO-MgO-SiO_2$, so important to the ceramic and steel industries, is represented on the ternary phase equilibrium diagram compiled by Osborn and Muan (1960). Walter (1965) evaluated the equilibria of the assemblages found at Crestmore, California. Franz and Wyllie (1967) found that merwinite is stable above 755°C at 1 kbar in the system $CaO-MgO-SiO_2-CO_2-H_2$), and Kushiro and Yoder (1964) found that the assemblage merwinite + forsterite s.s. + diopside persists up to at least 38 kbar at 1500°C. Yoder (1968) concluded that merwinite may be an important mineral in Ca-rich, Si-poor regions of the upper mantle. Moore and Araki (1972) remarked that merwinite, with a volume per anion of 20.6 $Å^3$ and density 3.32 g/cc, is 10% more dense than its "olivine" counterparts, α-larnite ($3Ca_2SiO_4$) + forsterite (Mg_2SiO_4), and that its high coordination numbers for Ca (8 and 9) and density are achieved by having both O^{2-} and Ca^{2+} mixed together in the "dense-packed" layers, somewhat like the hollandite and perovskite structures.

The structure of merwinite was solved by Moore and Araki (1972 -- henceforth M & A), who first had to sort through much chemical and crystallographic misinformation. A projection down the [010] (figure a) shows some of the corner- and edge-sharing linkages amongst the SiO_4 tetrahedra, the MgO_6 octahedra and the three irregular CaO_8 and CaO_9 polyhedra. The three types of dense-packed layers, each consisting of 2 Ca + 4 oxygen atoms, are variously indicated as A, B, and P, and the stacking sequence is ...ABPABP... for a total of six layers along the a-axis. Figure b is a polyhedral diagram of a (100) slab of MgO_6 octahedra and SiO_4 tetrahedra in a "pinwheel" arrangement. The refractive indices of synthetic merwinite are α = 1.706, β = 1.711-12, γ = 1.724, $2V_γ$ = 71 ±2°. The ranges of values observed in merwinite from blast furnace slags and in nature are α = 1.705 - 1.710, β = 1.711 - 1.714, γ = 1.718 - 1.728; $2V_γ$ = 67-75° (Phemister, et al. 1942). It is the {100} cleavage that is "good" (M & A), not {010} (Tilley, 1929). This is the plane of the dense-packed slabs (figure b) and represents a direction in which only the weak Ca-O and not the stronger Mg-O or Si-O bonds need be broken. Hardness is 6. Morphologically, [011] (or [01$\bar{1}$]) prisms are predominant in synthetic merwinite, but pseudo-trigonal tabular crystals with the most prominent face being {100} are often observed. Polysynthetic twins with twin axis [001] and composition plane {110} are relatively common in natural specimens and a less common set has twin plane and composition plane {100} (Larsen and Foshag, 1921).

See discussion BREDIGITE, $Ca_{∼3.5}Mg_{∼0.5}[SiO_4]_2$, this chapter.

Figures a and b are on the next page.

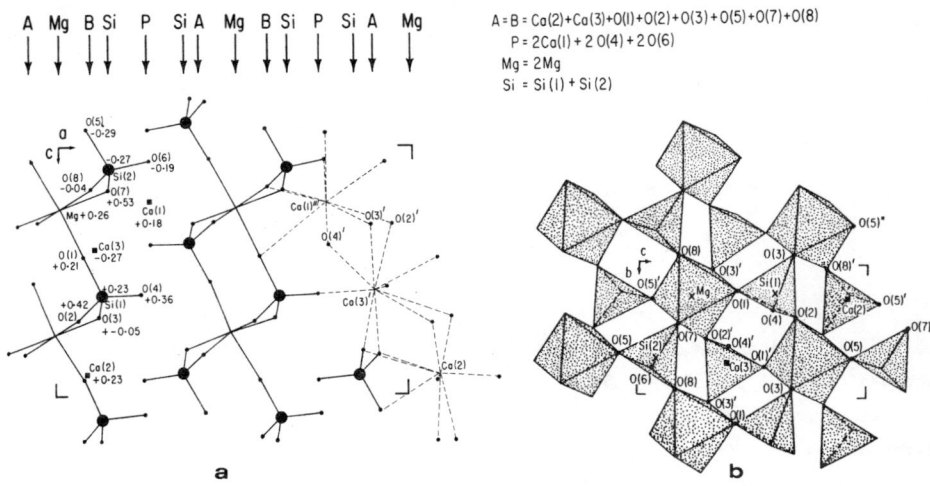

A = B = Ca(2)+Ca(3)+O(1)+O(2)+O(3)+O(5)+O(7)+O(8)
P = 2Ca(1) + 2 O(4) + 2 O(6)
Mg = 2Mg
Si = Si(1) + Si(2)

a

b

(a) Spoke diagram of the merwinite structure projected down the b-axis. The SiO$_4$ and MgO$_6$ slabs are shown at $a \backsim 0, \frac{1}{2}$ and these slabs are parallel to {100}. The Ca-O bonds are dashed. Locations of the A- and B-dense-packed layers, the P-layer, and the Mg and Si atoms along the a-axis are indicated above. Heights at atoms are in fractional coordinates along the y-direction. After Moore and Araki (1972, Fig. 3, p. 1364).
(b) Polyhedral diagram of a slab of MgO$_6$ octahedra and SiO$_4$ tetrahedra down the x^*-axis. Note the pseudo-hexagonal character of the "pinwheel" arrangement. After Moore and Araki (1972, Fig. 5, p. 1366).
- - -

Franz, G.W. and P.J. Wyllie (1967) Experimental studies in the system CaO-MgO-SiO$_2$-CO$_2$-H$_2$O. *In* P.J. Wyllie, ed., *Ultramafic and Related Rocks*, John Wiley & Sons, New York, p. 323-326.

Kushiro, I. and H.S. Yoder, Jr. (1964) Breakdown of monticellite and akermanite at high pressures. Carnegie Inst. Washington Geophys. Lab. Reo, 1963-1964, 81-83.

Larsen, E.S. and W.F. Foshag (1921) Merwinite, a new calcium magnesium ortho-silicate from Crestmore, California. Am. Mineral. 6, 143-148.

Moore, P.B. and T. Araki (1972) Atomic arrangement of merwinite, Ca$_3$Mg[SiO$_4$]$_2$, an unusual dense-packed structure of geophysical interest. Am. Mineral. 57, 1355-1374.

Phemister, J., R.W. Nurse, and F.A. Bannister (1942) Merwinite as an artificial mineral. Mineral. Mag. 26, 225-231.

Tilley, C.E. (1929) On larnite (calcium orthosilicate, a new mineral) and its associated minerals from the limestone contact zone of Scawt Hill, Co. Antrim. Mineral. Mag. 22, 77-86.

Walter, L.S. (1965) Experimental studies on Bowen's decarbonation series III: P-T univariant equilibrium of the reaction: spurrite + monticellite ⇄ merwinite + calcite and analysis of assemblages found at Crestmore, California. Am. J. Sci. 263, 64-77.

Yoder, H.S., Jr. (1968) Akermanite and related melilite-bearing assemblages. Carnegie Inst. Washington, Geophys. Lab. Report 1966-1967, 471-474.

● MONTICELLITE, $CaMgSiO_4$.   An olivine; see Chapter 11.

● NA BOLTWOODITE.   See section on *URANYL SILICATES*, this chapter.

● NATISITE, $Na_2TiOSiO_4$.

Tetragonal, $P4/nmm$; $a = 6.480$, $c = 5.107$ Å; $Z = 2$.

● FERSMANITE, $(Ca,Na)_4(Ti,Nb)_2Si_2O_{11}(F,OH)_2$.

Triclinic, $P1$ or $P\bar{1}$; $a = 7.210$, $b = 7.213$, $c = 20.451$ Å; $\alpha = 95.15°$, $\beta = 95.60°$, $\gamma = 89.04°$; $Z = 4$.

Natisite occurs in natrolite + ussingite-bearing veinlets of the Lovozero massif, Kola Peninsula, U.S.S.R (Men'shikov *et al.*, 1975). Fersmanite was de- scribed from the nepheline-bearing pegmatites of the neighboring Khibiny mas- sif (Labuntsov, 1929). An average of four microprobe analyses of natisite gave a formula of $Na_{1.99}(Ti_{0.99}Mn_{0.01}Fe_{0.01}Nb_{0.01})Si_{1.01}O_5$. An infrared spectra showed no water present. The chemistry of fersmanite is complex but was shown to be a fluorine-bearing, titanoniobosilicate of Na and Ca by Borneman- Starynkevich (1936) and Labuntsov (1933). In absence of a structure deter- mination, there have been a number of formulae proposed. The one given above is from Machin (1977) who found the main substition is CaTi = NaNb. Na ex- ceeds Nb and the charge balance is believed to be maintained by the replace- ment of some oxygen by hydroxyl.

Natisite and fersmanite are the sodium analogues of titanite, $CaTiOSiO_4$. The substitution of Na^+ for Ca^{2+} could be accomplished in several ways; among them:

$$2Na^+ = Ca^{2+}$$
$$Na^+ + (Nb,Ta)^{5+} = Ca^{2+} + Ti^{4+}$$
$$Na^+ + (OH,F)^- = Ca^{2+} + O(1)^{2-}$$

The completion of the 2Na = Ca substitution is the mineral natisite, $Na_2TiOSiO_4$, whereas the Na(Nb,Ta)(OH,F) = CaTiO(1) substitution leads to the mineral fersmanite.

The crystal structure of synthetic natisite was determined by Nikitin *et al.* (1964) and Nyman *et al.* (1978). The structure consists of layers of SiO_4 tetrahedral sharing corners with TiO_5 polyhedra alternating with layers of Na atoms (figure a). The TiO_5 polyhedron is a square pyramid (figure b) with one short Ti-O distance and four longer ones. The general structure of natisite is comparable to a group of tetragonal $AOBO_4$ compounds where A = Ta,Nb,Mo and V and B = P,S,Mo (*cf.* Longo and Arnott, 1970). In these compounds the absence

Figures a, b, and c are on the next page.

408

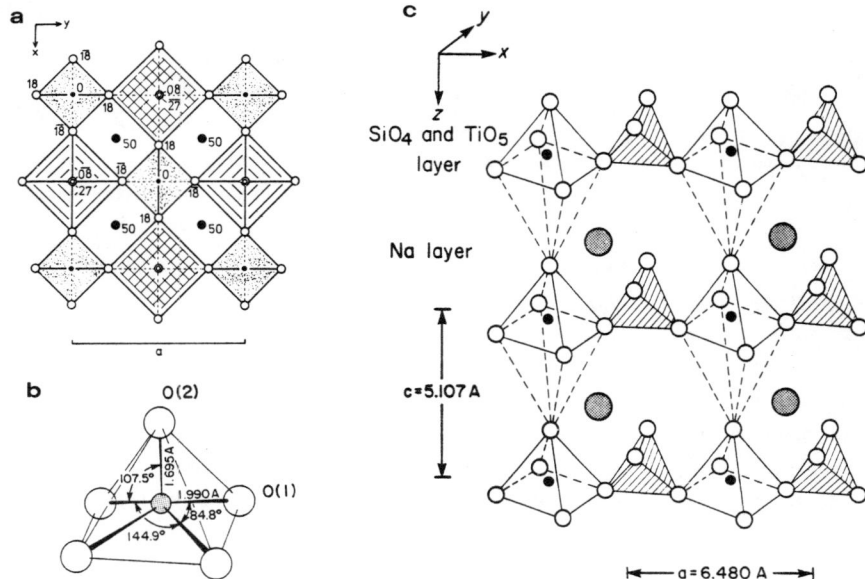

(a) The structure of synthetic natasite, Na_2TiSiO_5, projected onto 001. Large open circles are O, large filled circles Na, small open circles Ti, and small filled circles Si. Elevations are in c/100. Square pyramids with apices pointing down have basal faces cross-hatched, the others point up. From Nyman *et al.* (1978, their Fig. 1). (b) Dimensions and geometry of the TiO_5 square pyramid in synthetic natisite, Na_2TiSiO_5. Data is from the refinement of Nyman *et al.* (1978). (c) The structure of Na_2TiSiO_5 viewed on 010 showing layers of SiO_4 tetrahedra and TiO_5 square pyramids alternating with layers of Na atoms. Large open circles are oxygen, large patterned circles Na and small filled circles Ti. Dotted lines show the imagined, distorted octahedral coordination of the Ti to produce the octahedral chains found in the tetragonal $AOBO_4$ compounds.

of Na between the layers allows the layers to come closer together, converting the square pyramids to distorted octahedra. The main structural units are corner-shared octahedra forming chains parallel to *c*. The chains are coupled by tetrahedra which share the remaining corners of the octadra. Figure c shows structure of $Na_2TiOSiO_4$ drawn to show its similarity to the $AOBO_4$ compounds. This would produce alternating short and extremely long Ti-O(1) bond distances parallel to *c* : 1.695 and 3.412 Å compared to the equatorial Ti-O(2) bond distances of 1.990 Å. While clearly the Ti is not in octahedral coordination in natisite, the A-O(1) bond distances in $AOBO_4$ compounds are similar in that they alternate long and short parallel to the *c* axis.

The titanite structure is also comparable to that of natsite, containing chains of corner-shared TiO_6 octahedra which are cross-linked by SiO_4 tetrahedra (see Chapter 5, this volume). However, the octahedral chains in titanite

409

are kinked. If they are straightened and the Ca omitted, titanite would gain
a four-fold symmetry axis parallel to the chains and be isostructural with the
$AOBO_4$ compounds. Substitution of Na in the channels between the chains would
produce natisite. These structural modifications required by the $2Na^+ \rightleftharpoons Ca^{2+}$
substitution limits the solid solution of titanite and natisite.

The alternating long-short bond distances of the octahedral chains parallel to c in the $AOBO_4$ compounds results in a perfect 001 cleavage. Natisite
also has a perfect 001 cleavage, the Na atoms only weakly bonding the SiO_4 +
TiO_5 layers. The off-center displacement of the A-site cations (Ti,Nb,Ta,Mo,
V) in an octahedron is characteristic of oxygen-octahedral ferroelectrics. The
antiparallel displacements of the cations in adjacent chains in the $AOBO_4$ compounds and titanite has lead to the suggestion that these are antiferroelectric.
While the Ti octahedron in natisite is a contrivance for showing structural
similarities, the Ti is off-centered in the TiO_5 square pyramid. The antiparallel displacement in adjacent coordination polyhedra would suggest that it may
be antiferroelectric as well.

- - -

Borneman-Starynkevich, I.D. (1936): Composition of several titanosilicates from
    the Khibiny tundras. *Vernadsky Jubilee Volume*, Akad. Nauk SSSR 2, 735-755.

Laruntsov, A.N. (1929) Fersmanite, a new mineral of the Khibiny Massif. Dokl.
    Akad. Nauk, Leningrad, Ser. A, 12, 297-301.

_____ (1933) *Mineralogical Survey of the Central Part of the Khibiny Massif*
    (Deposits of Zircon, Catapleiite and Fersmanite). Khibinsky apatity, 6,
    202-208, Leningrad, Khimteoret.

Longo, J.M. and R.J. Arnott (1970) Structure and magnetic properties of $VOSO_4$.
    J. Solid State Chem. 1, 394-398.

Machin, M.P. (1977) Fersmanite $(Ca,Na)_4(Ti,Nb)_2Si_2O_{11}(F,OH)_2$: a restudy. Canad-
    ian Mineral. 15, 87-91.

Men'shikov, Yu. P., Ya. A. Pakhomovskii, E.A. Goiko, I.V. Bussen and A.N. Mer'
    kov 1975. A natural tetragonal titanosilicate of sodium, natisite. Zapiski
    Vses. Mineralog. Obshch. 104, 314-317. Am. Min. 61, 339.

Nikitin, A.V., Ilyukhin, V.V., Litvin, B.N., Kel'nikov, O.K. and Belov, N.V.
    (1964) Crystal structure of the synthetic sodium titanosilicate, $Na_2TiOSiO_4$.
    Dokl. Adad. Nauk SSSR 157, 1355-1356.

Nyman, H., O'Keefe, M and Bovin, J.O. (1978) Sodium titanium silicate, Na_2TiSiO_5.
    Acta Crystallogr. B34, 905-906.

● NELTNÉRITE, $CaMn_6SiO_{12}$.

    Tetragonal, $I4_1/acd$; a = 9.464; c = 18.854 Å; Z = 8.

    Neltnérite is part of a vein assemblage with Mn- and Ca- bearing minerals at Techgagalt, Anti-Atlas Mtns., Morocco. The mineral contains a small amount of Fe as well (Baudracco-Gritti *et al.*, 1982). Neltnérite is isostructural with braunite, the Ca occupying the distorted Mn(1) cube site (Damon *et al.*, 1966). See *BRAUNITE*, this chapter.

- - -

Baudracco-Gritti, C., R. Caye, F. Permingeat, and J. Protas (1982) La neltnérite $CaMn_6SiO_{12}$ une nouvelle espéce minérale du groupe de la braunite. Bull. Minéral. 105, 161-165.

Damon, J.L., F. Permingeat and J. Protas (1966) Étude structurale du composé $CaMn_6SiO_{12}$. C.R. Acad. Sc. Paris, 262, Sér. C. 1671-1674.

● NORBERGITE, $Mg_2SiO_4 \cdot MgF_2$. A humite mineral; see Chapter 10.

● OTTRELITE, $(Mn,Fe^{+2},Mg)_2Al_4Si_2O_{10}(OH)_4$.   The manganese analog of chloritoid; see Chapter 6.

● PARASPURRITE, $Ca_5(SiO_4)_2CO_3$.

    Monoclinic, $P2_1/a$; a = 10.473, b = 6.705, c = 27.78 Å; β = 90.58°; Z = 8.

    Paraspurrite occurs in calcsilicate rocks that have been contact-metamorphosed at high temperatures and low pressures in a roof pendant in a syenite. Its chemistry is essentially that of the ideal formula. Paraspurrite is a polymorph of spurrite differing principally by having a doubled unit cell dimension in the c^* direction. Colville and Colville (1977) have proposed a structure (see figure) based on that of spurrite by Kletsova and Belov (1961).

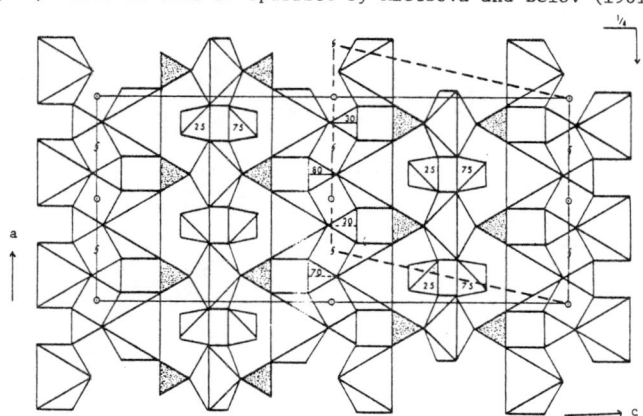

Proposed structure for paraspurrite projected onto (010) based on diagram of Kletsova and Belov (1961). The monoclinic unit cell of spurrite is outlined by dashed lines for comparison. Shaded triangles are CO_3 groups; SiO_4 tetrahedra are labelled 25 and 75 to indicate fractional coordinates along b; the other polyhedra contain Ca.

- - -

Colville, A.A. and P.A. Colville (1977) Paraspurrite, a new polymorph of
spurrite from Inyo County, California. Am. Mineral. 62, 1003-1005.

Kletsova, R.F. and N.V. Belov (1961) Crystal structure of spurrite. Soviet
Physics - Crystallogr. 5, 659-667.

● PICROTEPHROITE, $(Mg,Mn)SiO_4$. An olivine; see Chapter 11.

● PYROPE, $Mg_3Al_2Si_3O_{12}$. A garnet; see Chapter 2.

● RINGWOODITE, $(Mg,Fe)_2SiO_4$.

Cubic, $Fd3m$; a = 8.1-8.2Å; Z = 8.

A high-pressure polymorph of olivine, ringwoodite occurs in meteorites and
has practically the same composition as the coexisting olivine. Ringwoodite is
thought to be produced from olivine by shock transformation resulting from ex-
traterrestial collision. Analysed ringwoodites have compositions of approximately
$(Mg_{.75}Fe_{.25})_2SiO_4$ (Binns et $al.$, 1969; Mason et $al.$, 1968; Coleman 1977).

Ringwoodite is the spinel or γ form of $(Mg,Fe)_2SiO_4$. Variation of cell
dimension and refractive index of the synthetic material with composition is reported by Ringwood and Major (1966, see the figure). The P-T-X phase relations
of the olivine → ringwoodite transition have been studied by Akimoto and
Fujisawa (1968) and Ringwood and Major (1970). At compositions more magnesian
than Fo_{80} orthorhombic β-$(Mg,Fe)_2SiO_4$, rather than ringwoodite, is the high
pressure polymorph (Ringwood and Major, 1970). See Chapter 11, pp. 353-355 for
further discussion.

The crystal structures of synthetic Mg- and Fe-end-member "ringwoodites"
have been reported by Sasaki et $al.$ (1982) and Marumo et $al.$ (1977). They have
space group $Fd3m$, and the crystal parameters are

$$γ-Mg_2SiO_4 \quad a = 8.065Å, \; μ = 0.3685$$
$$γ-Fe_2SiO_4 \quad a = 8.234Å, \; μ = 0.3659$$

Ringwoodite has a normal spinel structure; the synthetic analogs are partially
disordered. Site occupancies are estimated to be 80-98% Si in the tetrahedral
sites and 90-99% (Mg,Fe) in the octahedral sites.

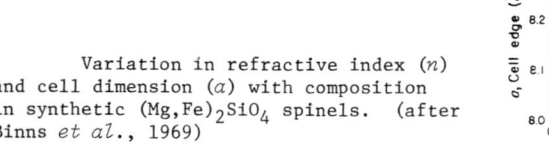

Variation in refractive index (n)
and cell dimension (a) with composition
in synthetic $(Mg,Fe)_2SiO_4$ spinels. (after
Binns et $al.$, 1969)

- - -

Akimoto, S. and H. Fujisawa (1968) Olivine - spinel solid solution equilibria in the system Mg_2SiO_4 - Fe_2SiO_4. J. Geophys. Res. 73, 1467-1479.

Binns, R.A., R.J. Davis and S.J.B. Reed (1969) Ringwoodite, natural $(Mg,Fe)_2SiO_4$ spinel in the Tenham meteorite. Nature 221, 943-944.

Coleman, L.C. (1977) Ringwoodite and majorite in the Catherwood meteorite. Canadian Mineral. 15, 97-101.

Marumo, F., M. Isobe and S. Akimoto (1977) Electron-density distributions in crystals of γ-Fe_2SiO_4 and γ-Co_2SiO_4. Acta Crystallogr. B33, 713-316.

Mason, B., J. Nelen, J.S. White, Jr. (1968) Olivine - garnet transformation in a meteorite. Science 160, 66-67.

Ringwood, A.E. and A. Major (1966) Synthesis of Mg_2SiO_4-Fe_2SiO_4 spinel solid solutions. Earth Planet Sci. Letters 1, 241-245.

Ringwood, A.E. and A. Major (1970) The system Mg_2SiO_4-Fe_2SiO_4 at high pressures and temperatures. Phys. Earth Planet. Interiors 3, 89-108.

Sasaki, S., C.T. Prewitt, Y Sato and E. Ito (1982) Single-crystal x-ray study of γ-Mg_2SiO_4. J. Geophys. Res. 87, 7829-7832.

● SCHORLOMITE, $Ca_3(Fe,Ti)_2(Si,Fe)_3O_{12}$. A garnet; see Chapter 2.

● SILICATE APATITES

ELLESTADITES and BRITHOLITES

Apatite is a calcium phosphate, $Ca_{10}(PO_4)_6(OH,F,Cl)_2$, whose structure is adaptable to numerous chemical substitutions. With reference to the general formula $A_{10}(XO_4)_6Z_2$, silicate apatites are found to be of two major types: (1) the *britholites* with A = REE and X = Si, and (2) the *ellestadites* with A primarily Ca and X = Si and S.

The structure of apatite was refined by Beevers (1946). The isolated TO_4 groups are bound together by large A cations in irregular coordination (see figure). Monovalent Z anions "just fit into a triangle of three calciums [or A cations] on each mirror plane intersecting the channel" parallel to c at the corners of the hexagonal unit cell. Silicate apatites with monoclinic symmetry are strongly pseudohexagonal and would require slight distortions of the structure to accommodate the variety of atoms at the A, X and Z sites.

Figure is on the next page.

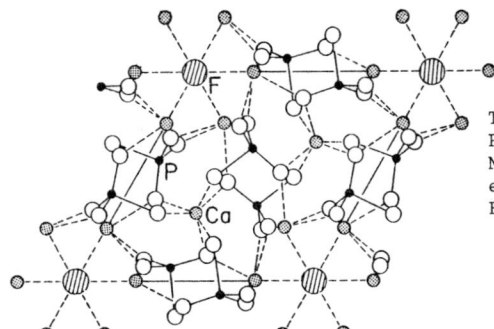

The structure of apatite, $Ca5(PO_4)_3F$. PO_4 groups are joined by solid lines. Note location of F in channels bonded to 6 Ca. From Bragg *et al.* (1965, Fig. 104, p. 144).

Ellestadites, the Silicate-Sulfate Apatites

McConnell (1937, 1938) first suggested that charge balance in phosphate apatites with the isomorphous substitution of Si but lacking rare-earth elements is maintained by the substitution

$$Si^{+4} + S^{+6} \leftrightarrows 2P^{+5}$$

This was based on the chemistry of the two silicate-bearing apatites then known, wilkeite (Eakle and Rogers, 1914) and ellestadite (McConnell, 1937). More recently, Rouse and Dunn (1982) have shown that the ratio of S:Si is essentially 1:1 for a range of compositions of silicate-sulfate apatites from Crestmore, California. They also proposed that apatites with $\sum(Si,S) > P$ be termed ellestadites. Designations of three possible end-members of $Ca_{10}(SiO_4)_3$ $(SO_4)_3 Z_2$ include those with Z = OH (hydroxylellestadite), Z = F (fluorellestadite) and Z = Cl (chlorellestadite). Deviation of the Si:S ratio from 1:1 observed in some apatite analyses is explained by Vasileva (1958) as resulting from the additional coupled substitutions

$$3P^{+5} + O^{-2} \leftrightarrows S^{+6} + 2Si^{+4} + OH^- \text{ or } Ca^{+2} + P^{+5} \leftrightarrows Na^{+1} + S^{+6}.$$

Synthetic OH, F and Cl ellestadites have been reported by Dihn and Klement (1942), Takemoto and Kato (1968) and Pliego-Cuervo and Glasser (1978).

Ellestadites are normally thought to be hexagonal. However, hydroxylellestadites from the Chichibu Mine, Saitama Prefecture, Japan have been reported with both hexagonal (Harada *et al.*, 1971) and monoclinic symmetry (Sudarsanan, 1980). The hexagonal space group is $P6_3/m$ or $P6_3$ with a = 9.49, c = 6.92 Å. The monoclinic space group is reported as either $P2_1/m$ or $P2_1$ with a = 9.476, b = 9.508, c = 6.919 Å and γ = 119.53°. This is not a standard setting for monoclinic apatites; the hexagonal to monoclinic transformation matrix is 200/001/010. [Note: The space group of monoclinic Cl-apatite is $P2_1/a$.

Sudarsanan (1980) reported on the crystal structure of natural hydroxyel-
lestadite from the Chichibu Mine refined in the space group $P2_1/a$. The structure
resembles apatite. There are three differing tetrahedral sites with average
(Si,S)-O distances of 1.52, 1.54 and 1.57 Å. These distances indicate incomplete
ordering of Si and S on the different tetrahedral sites; the mean bond length of
Si - O = 1.62 Å (Liebau, 1972) and S - O = 1.47 Å (Wuensch, 1972). Sudarsanan
also reports two groups of sites for the $Ca(2A,B,C)$ atoms; one group is
associated with F or OH,the other with Cl.

While S and Si are reported as trace constituents in apatites from a variety
of occurrences, Si- and S-rich phosphate apatites and ellestadites occur in calcareous rocks which have undergone high-temperature, low-pressure contact metamorphism (McConnell, 1937; Harada *et al.*, 1971; Vasileva, 1958).

Britholites, Rare-earth Silicate Apatites

Charge balance can be maintained with P $\leftrightarrows$ Si substitution in an apatite
structure by coupled substitution of a trivalent cation in the A site:
$$3Si^{+4} + 3(REE,Y)^{+3} + 2Ca^{+2} \leftrightarrows 3P^{+5} + 5Ca^{+2}.$$
This leads to the formula
$$(REE,Y)_3\ Ca_2\ (SiO_4)_3\ (OH,F,Cl),$$
which represents the REE silicates, called *britholites* and the yttrium-REE
silicate called *britholite-Y*. There are a number of varietal names for related,
but incompletely described substances which are discussed and referenced in
Vlasov (1966), Ito (1968) or Gay (1957). Many of the REE end-member britholites
have been synthesized as well as britholites with Ca replaced by Mg, Sr, Ba, Pb,
Mn and Cd (Cockbain and Smith, 1967; Ito, 1968; Sheyakov *et al.*, 1972). A complete solid solution between apatite and britholite-Y has been synthesized, whereas only an incomplete solid solution exists between apatite and britholite-La
(Ito, 1968).

A number of alkali- and alkaline- rare earth silicate-phosphate apatites have
been synthesized and crystallographically characterized. These include

Ca_4	REE_6	$(SiO_4)_4$	$(PO_4)_2$	O_2	Fedorov *et al.* (1975)
Ca_4	REE_6	$(SiO_4)_6$		$(OH)_2$	Cockbain (1968)
Ca_5	REE_5	$(SiO_4)_6$		OH	Cockbain (1968)
Ca_6	REE_4	$(SiO_4)_6$			Cockbain (1968)
	$REE_{4.67}$	$(SiO_4)_3$		O	Kuz'min & Belov (1966) Belokoneva *et al.* (1972)
(Li,Na,K)	REE_9	$(SiO_4)_6$		O_2	Felsche (1973); Ito (1968) Pushcharovskii *et al.* (1978)
Ca_2	REE_8	$(SiO_4)_6$		O_2	Cockbain & Smith (1967), Ito (1968)

415

● SILICATE APATITES, continued

These compounds are of interest because they suggest that deficiencies of cations or halogen site elements in analyses of britholites may represent solid solution with one of these possible end-members.

Gay (1957) found that britholites are hexagonal, space group $P6_3/m$ or $P6_3$ which is consistent with the interpretation that they are isostructural with apatite. However, britholites are often described as optically biaxial. There has been no structure refinement of a britholite, although there are several for the synthetic oxyapatites mentioned above. Cockbain and Smith (1967) did investigate the distribution of the A cations in a synthetic britholite-La: Ca and La were found to have random distributions over the 4f and 6h sites.

Britholite occurs in carbonatites and alkaline igneous rocks. Nash (1972) described enrichment in apatites of Si, Na, K, Sr and REE with increasing differentiation in the Shonkin Sag laccolith, Montana; with extreme differentiation britholites were found in the soda syenites.

- - -

Belokoneva, E.L., T.L. Petrova, M.A. Simonov and N.V. Belov (1972) Crystal structure of synthetic TR analogs of apatite $Dy_{4.67}[GeO_4]_3O$ and $Ce_{4.67}[SiO_4]_3O$. Sov. Phys. Crystallogr. 17, 429-431.

Cockbain, A.G. (1968) The crystal chemistry of the apatites. Mineral. Mag. 36, 654-660.

Cockbain, A.G. and G.V. Smith (1967) Alkaline-earth rare-earth silicate germanate apatites. Mineral. Mag. 36, 411-421.

Dihn, P. and R. Klement (1942) Isomorphe Apatitarten. Z. Electrochem. Ang. Phys. Chem. 48, 331-333.

Eakle, A.S. and A.F. Rogers (1914) Wilkeite, a new mineral of the apatite group, and okenite, its alteration product, from Southern California. Am. J. Sci., 4th Ser. 37, 262-267.

Fedorov, N.F., I.F. Andreev and N.S. Meliksetyan (1975) Growth and study of single crystals of alkali earth and rare earth oxysilicates phosphate apatites. Sov. Phys. Crystallogr. 20, 280-281.

Felsche, J. (1973) The crystal chemistry of the rare-earth silicates. Structure and Bonding 13, 99-197.

Gay, P. (1957) An x-ray investigation of some rare-earth silicates: cerite, lessingite, beckelite, britholite and stillwellite. Mineral Mag. 31, 455-468.

Harada, K., K. Nagashima, K. Nakao and A. Kato (1971) Hydroxylellestadite, a new apatite from Chichibu mine, Saitama Prefecture, Japan. Am. Mineral. 56, 1507-1518.

Ito, J. (1968) Silicate apatites and oxyapatites. Am. Mineral. 53, 890-907.

Kuz'min, E.A. and N.V. Belov (1966) Crystal Structure of the simplest silicates of La and Sm. Sov. Phys. Dokl. 10, 1009-1011.

Liebau, F. (1972) Silicon. 14-A Crystal chemistry. *In* K.H. Wedepohl, Ed., *Handbook of Geochemistry,* Vol. II/2, p. 14A1-14A32.

McConnell, D. (1937) The substitution of SiO_4- and SO_4-groups for PO_4-groups in the apatite structure; ellestadite, the end member. Am. Mineral. 22, 977-986.

_____ (1938) A structural investigation of the isomorphism of the apatite group. Am. Mineral 23, 1-19.

● SILICATE APATITES, continued

Nash, W.P. (1972) Apatite chemistry and phosphorus fugacity in a differentiated igneous intrusion. Am. Mineral. 57, 877-886.

Pliego-Cuervo, Y. and F.P. Glasser (1978) Phase relations and crystal chemistry of apatite and silicocarnotite solid solutions. Cement Concrete Res. 8, 519-524.

Pushcharovskii, D. Yu., G.I. Dorokhova, E.A. Pobedimskaya and N.V. Belov (1978) Potassium-neodyium silicate $KNd_9(SiO_4)_6O_2$ with apatite structure. Sov. Phys. Dokl. 23, 694-696.

Rouse, R.C. and P.J. Dunn (1982) A contribution to the crystal chemistry of ellestadite and the silicate sulfate apatites. Am. Mineral. 67, 90-96.

Shevylakov, A.M., I.F. Andreev, Sh. Yu. Azimov, Yu, P. Tarlakov, and N.F. Fedorov (1972) Infrared absorption spectra of synthetic britholites of the composition $Me_4^{2+}Nd_6(SiO_4)_6F_2$ where Me^{2+} is Mg, Ca, Sr or Ba. Sov. Phys. Dokl. 16, 798-799.

Sudarsanan, K. (1980) Structure of hydroxylellestadite. Acta Crystallogr. B36, 1636-1639.

Takemoto, K. and H. Kato (1968) Hydroxyl ellestadite produced by hydrothermal reaction containing calcium sulfate. Proc. 5th Intern'l. Sym. Chem. Cement, Tokyo.

Vasileva, Z.V. (1958) Sulfur-bearing apatites. Geokhimiya, 368-373 (transl. Geochemistry, 464-470, 1958).

Vlasov, K.A. (1966) Mineralogy of Rare Elements Volume II, Geochemistry and Mineralogy of Rare elements and genetic types of their deposits. Translated from Russian; published by the Israel Program for Scientific Translations, 945 pp.

Wuensch, B.J. (1972) Sulfur. 16-A Crystal chemistry. *In* K.H. Wedepohl, Ed., *Handbook of Geochemistry*, Vol. II/2, p. 16A1-16A19. Springer-Verlag, Berlin.

● SKLODOWSKITE.   See section on *URANYL SILICATES*, this chapter.

● SODDYITE.   See section on *URANYL SILICATES*, this chapter.

● SONOLITE, $4Mn_2SiO_4 \cdot Mn(OH)_2$.   The Mn-analog of clinohumite; see Chapter 10.

● SPESSARTINE, $Mn_3Al_2Si_3O_{12}$.   A garnet; see Chapter 2.

● SPURRITE, $Ca_5(SiO_4)_2CO_3$.

Monoclinic, $P2_1/a$; $a = 10.49$, $b = 6.705$, $c = 14.16$ Å, $\beta = 101.32°$; $Z = 4$.

Spurrite occurs in calcsilicate rocks which have undergone high-temperature, low-pressure contact metamorphism. It is one of the diagnostic minerals of the spurrite-merwinite facies of contact metamorphism.

The structure was determined by Smith *et al.* (1960) and Klevtsova and Belov (1961). Klevtsova and Belov (1961) describe it as layers consisting of αCa_2SiO_4 of the olivine type alternating with layers of $CaCO_3$ of aragonite (see the figure). The Ca in the "olivine layer" is 6-coordinated whereas the Ca in the "aragonite layer" is 9-coordinated.

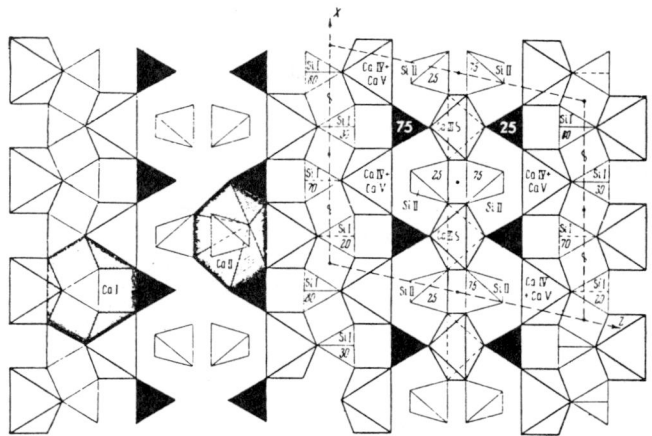

An xz projection of spurrite. The true monoclinic cell is indicated on the
pseudo-orthorhombic motif. Inside the CO3 triangles and in each of the two
kinds of SiO4 tetrahedra the heights of the central atoms are indicated by
their y coordinates. Ca III octahedra are situated along the height of the
cell in pairs one above the other with the coordinates y ± 0.05, 0.50 ± 0.05
and the Ca IV and Ca V octahedra also overlap each other with y coordinates
near to 0 and to 0.50. The Ca I and Ca II polyhedra are not shown inside
the monoclinic cell. They are distributed above and below the two types of
tetrahedra (Si I and Si II) and are illustrated on the left of the diagram.
From Kletsova *et al.* (1961).

- - -

Klevtsova, R.F. and N.V. Belov (1961) Crystal structure of spurrite. Sov.
Phys. Crystallogr. 5, 659-667.

Smith, J.V., I.L. Karle, H. Hauptman and J. Karle (1960) The crystal structure
of spurrite, $Ca_5(SiO_4)_2CO_3$. II. Description of structure. Acta Crystallogr. 13, 454-458.

● SWAMBOWITE.   See section on *URANYL SILICATES*, this chapter.

●TEPHROITE, Mn_2SiO_4. An olivine; see Chapter 11.

●THORITE, $ThSiO_4$. An actinide orthosilicate; see Chapter 4.

●TITANORHABDOPHANE. See *TUNDRITE*, this chapter.

●TOMBARTHITE, $YH[SiO_4]$. See Chapter 3.

● TUNDRITE, $Na_2Ce_2TiO_2[SiO_4](CO_3)_2$.

Triclinic, $P\bar{1}$; a = 7.560, b = 13.957, c = 5.040 Å; α = 101°07', β = 70°52'. γ = 100°01'; Z = 1.

Since its description in 1963 as the mineral titanorhabdophane, tundrite has had a number of proposed formulas. The one given is based on a crystal structure determination by Shumyatskaya *et al.* (1976). Tundrite is an alkali silico-carbonate of titanium and the rare earth elements, principally Ce and La, but Nd-rich material has been reported (tundrite-Nd). Natural tundrites exhibit the coupled substitution $Ce^{3+} + Ti^{4+} = Ca^{2+} + Nb^{5+}$. The structure of tundrite has edge-sharing TiO_6 octahedra chains parallel to c which are joined in the a direction by isolated SiO_4 tetrahedra to form layers whose outer surfaces are successively covered by Ce-polyhedra and CO_3 groups. These layers are stacked parallel to b and are joined by interlayer Na atoms. This structure explains the elongated habit of tundrite parallel to c and its perfect (010) cleavage.

– – –

Shumyatskaya, N.G., A.A. Voronkov, V.V. Ilyukhim and N.V. Belov (1976) Tundrite, $Na_2Ce_2TiO_2[SiO_4](CO_3)_2$ - refinement of the crystal structure and chemical formula. Sov. Phys. Crystallogr. 21, 399-405.

● URANYL SILICATES

The uranyl silicate group of minerals are listed in Table 1. They are interpreted as being secondary in origin, forming in oxidized, supergene environments as alteration products of other uranium minerals. Their formulas are based on crystal structure determinations or crystal chemical arguments made in the referenced papers. Their systematic mineralogy, based on earlier studies -- as well as information about other related, but incompletely characterized substances, was given by Frondel (1958); their crystal chemistry has been reviewed recently by Stohl and Smith (1981). Except for variable amounts of water, analyses of these minerals indicate nearly end-member compositions. Systematic variation of physical properties with small variations among Ca, K, Pb, Mg and Cu suggests that some substitutions occur.

The uranyl silicates are divided into three groups by Stohl and Smith (1981) on the basis of their U:Si ratios (Table 1). The basic structural unit is a chain of edge-sharing UO_7 pentagonal bipyramids (see the figures). The UO_7 coordination polyhedron has two oxygens with interatomic distances of about 1.8 Å, which are the uranyl oxygens, nearly perpendicular to the edge-sharing, equatorial, pentagonal ring of oxygens. The distorted, pentagonal dipyramid coordination polyhedra is better written $(UO_2)O_5$. The

Table 1. Members of the uranyl silicate mineral group separated on the basis of their uranium to silicon ratios. Modified from Stohl and Smith (1981).

	Space group	Unit Cell Dimensions a, Å	b, Å	c, Å	β	n	Sheet Orient'n	References
U:Si = 1:1, Uranophane Group								
URANOPHANE $Ca(H_3O)_2[(UO_2)_2(SiO_4)_2]\cdot2H_2O$	$P2_1$	15.858	6.985*	6.641	97°33'	1	(100)	Stohl & Smith (1981)
BETA-URANOPHANE $Ca[(UO_2)(UOOH)(SiO_4)(SiO_3OH)]\cdot4H_2O$	$P2_1/a$	13.898*	15.394	6.609	91°25'	1	(010)	Smith & Stohl (1972)
BOLTWOODITE $K(H_3O)[(UO_2)(SiO_4)]$	$P2_1$	7.073	7.064*	6.638	105°45'	0.5	(100)	Stohl & Smith (1981)
Na-BOLTWOODITE $(Na_{0.7}K_{0.3})(H_3O)[(UO_2)(SiO_4)]\cdot H_2O$	$P2_12_12_1$	27.40	7.02	6.65		0.5		Chernikov et al. (1975)
KASOLITE $Pb[(UO_2)(SiO_4)]\cdot H_2O$	$P2_1/c$	6.704	6.932*	13.252	104°13'	0.5	(100)	Mokeeva (1965) Rosenzweig & Ryan (1977)
SKLODOWSKITE $Mg(H_3O)_2[(UO_2)_2(SiO_4)_2]\cdot4H_2O$	$C2/m$	17.382	7.047*	6.610	105°54'	1	(100)	Mokeeva (1965)
CUPROSKLODOWSKITE $Cu[(UO_2)_2(SiO_3OH)_2]\cdot6H_2O$	$P\bar{1}$	7.052*	9.267	6.655	α=109°14' β= 89°50' γ=110°01'	1	(010)	Rosenzweig & Ryan (1975)
SWAMBOITE $U_{0.33}H_2[(UO_2)_2(SiO_4)_2]\cdot10H_2O$	$P2_1/a$	17.64	21.00	20.12	103°24'			Deliens & Piret (1981)
U:Si = 1:3, Weeksite Group								
WEEKSITE $K_2(UO_2)_2(Si_2O_5)_3\cdot4H_2O$	Pnmm Pnm2 Pm2m F2mn F222	7.106	17.90	7.087*	90°00'			Stohl & Smith (1981)
HAIWEEITE $Ca(UO_2)_2(Si_2O_5)_3\cdot5H_2O$	$P2/c$	15.4	7.05	7.10	107°52'			McBurney & Murdock (1959)
U:Si = 2:1, Soddyite								
SODDYITE $(UO_2)_2(SiO_4)\cdot2H_2O$	$Fddd$	8.297	11.219	18.661	90°00'			Belokoneva et al. (1979)

*Asterisk indicates chain axis direction.

$(UO_2)O_5$ chains are joined into larger structural units by varying numbers of SiO_4 tetrahedra, but the chains dominate the structures and account for the accicular habit of the minerals which coincide with the chain directions.

In the 1:1 uranyl silicate minerals, the chains of $(UO_2)O_5$ polyhedra are joined by isolated SiO_4 tetrahedra to form a sheet (figure a) of composition $[(UO_2)_2(SiO_4)]_n^{-4n}$, where n is the number of these units in the unit cell. Whereas the chains parallel the direction of crystal elongation, the sheets parallel the perfect (010) cleavage of the 1:1 uranyl silicate minerals. The sheet shown in figure a is that in uranophane. The apexes of the SiO_4 tetrahedra point all in the same direction where they are on the same side of the $(UO_2)O_5$ chains. In β-uranophane, the apexes of the SiO_4 tetrahedra alternate in direction. Other 1:1 uranyl silicate minerals have the same general sheet configuration as uranophane, but exhibit varying amounts of corrugation as a result of the differing interlayer cations which join the sheets. The relative amount of corrugation can be gotten from comparing the shortening of the unit cell parameter paralleling the $(UO_2)O_5$ chains from 7.064 Å in undistorted boltwoodite to 6.932 Å in kasolite. The corresponding angles between adjacent, edge-sharing pentagonal O rings in the chains are 0° in boltwoodite and 20.6° in kasolite.

The $[(UO_2)_2(SiO_4)_2]_n^{-4n}$ sheets are joined by various cations: Ca, K, Na, Pb, Mg and Cu. Except for kasolite, the interlayer cations do not have sufficent charges to balance the sheet. The charge balance is accomplished by H, either by replacing uranyl or silicate O by OH or as hydronium ions between the sheets. The location of the hydrogens are based on crystal chemical arguments, rather than being located in the structural analysis. Stohl and Smith (1981) conclude that the variable amounts of water reported in analyses is zeolitic water.

There has been a partial structural refinement of the 1:3 uranyl silicate weeksite by Stohl and Smith (1981), indicating chains of edge-sharing $(UO_2)O_5$ polyhedra which also share an edge with a SiO_4 tetrahedral (figure b). The chains are mirror images of one another and are not immediately crosslinked. Their actual linkage is by the unlocated Si and K atoms.

The only 2:1 uranyl silicate mineral is soddyite for which the structure was determined on synthetic material by Belokoneva et al. (1979). There are the basic chains of edge-sharing $(UO_2)O_5$ polyhedra, but adjacent chains are crosslinked by a single SiO_4 tetrahedra sharing two of its edges (figure c). This forms a 3-dimensional network and explains the lack of perfect cleavage as in the 1:1 uranyl silicates.

421

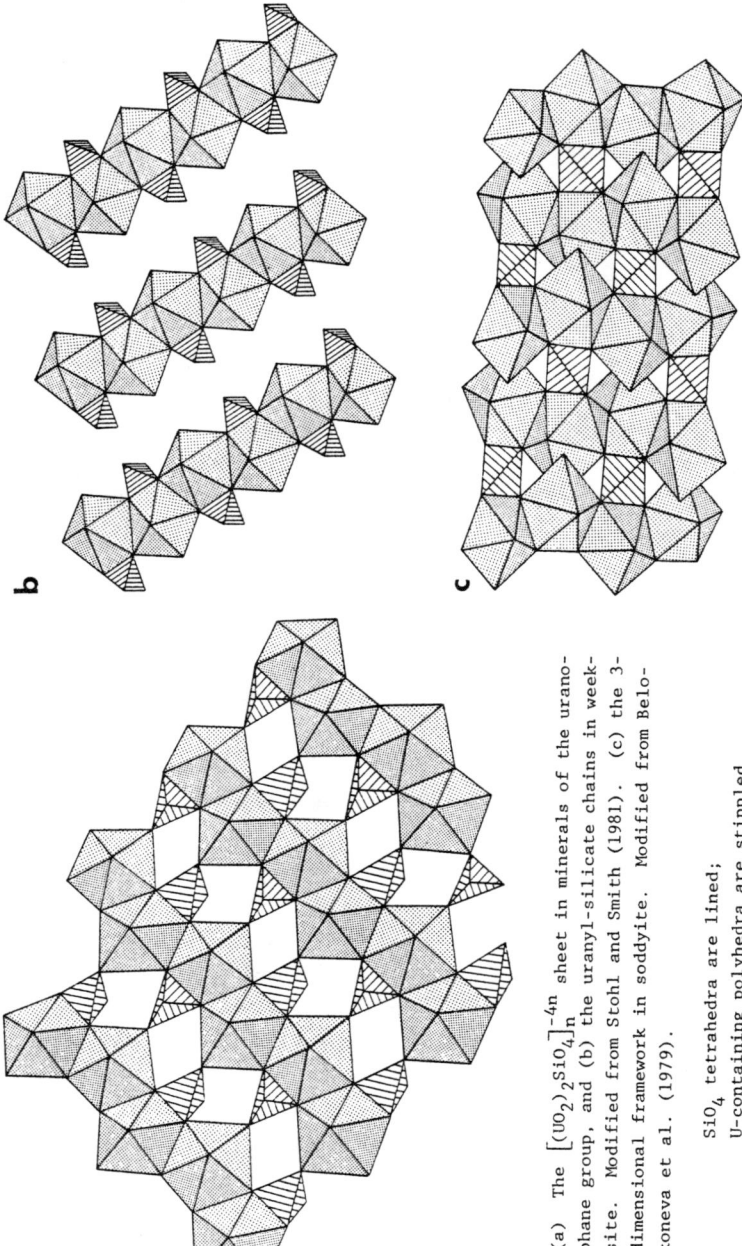

(a) The $\left[(UO_2)_2SiO_4\right]_n^{-4n}$ sheet in minerals of the uranophane group, and (b) the uranyl-silicate chains in weeksite. Modified from Stohl and Smith (1981). (c) the 3-dimensional framework in soddyite. Modified from Belokoneva et al. (1979).

SiO_4 tetrahedra are lined;
U-containing polyhedra are stippled.

- - -

Belokoneva, E.L., Mokeeva, V.I., Kuznetsov, L.M., Simonov, M.A., Makarov, E.S. and Belov, N.V. (1979) Crystal structure of synthetic soddyite $(UO_2)_2[SiO_4](H_2O)_2$. Doklady Akademii Nauk SSSR, 246, 1, 93-6.

Chernikov, A.A., Shaskin, D.P. and Gavrilova, I.N. (1975) Sodium boltwoodite. Doklady Akademii Nauk SSSR 221, 195-197. (not seen, extracted from Fleischer, M. (1976) New Mineral Names. Am. Mineral. 61, 1054-1055.

Deliens, M. and P. Piret (1981) La swamboite, nouveau silicate d'uranium hydraté du Shaba, Zaire. Canadian Mineralogist 19, 553-557.

Frondel, C. (1958) Systematic mineralogy of uranium and thorium. U.S. Geological Survey Bull. 1064, 400p.

McBurney, T.C. and Murdock, J. (1959) Haiweeite, a new uranium mineral from California. Am. Mineral. 44, 839-843.

Mokeeva, V.I. (1959) The crystal structure of sklodowskite. Doklady Akademii Nauk SSSR, 124, 578-580 (transl. Soviet Phys. Doklady, 4, 27-29, 1959).

Mokeeva, V.I. (1964) The structure of sklodowskite. Kristallografiya, 9, 2, 277-278 (transl. Soviet Physics - Crystallography, 9, 2, 217-218, 1964).

Mokeeva, V.I. (1965) The crystal structure of kasolite. Kristallografiya, 9, 5, 738-740 (transl. Soviet Phys. - Crystallography, 9, 5, 621-622, 1965).

Rosenzweig, A. and Ryan, R.R. (1975) Refinement of the crystal structure of cuprosklodowskite, $Cu[(UO_2)_2(SiO_3OH)_2] \cdot H_2O$. American Mineralogist, 60, 448-453.

Rosenzweig, A. and Ryan, R.R. (1977) Kasolite, $Pb(UO_2)(SiO_4) \cdot H_2O$. Crystal Structure Communications, 6, 617-621.

Ryan, R.R. and Rosenzweig, A. (1977) Sklodowskite, $MgO \cdot 2UO_3 \cdot 2SiO_2 \cdot 7H_2O$. Crystal Structure Communications, 6, 611-615.

Smith, D.K. and Stohl, F.V. (1972) Crystal structure of beta-uranophane. Geological Society of America Memoir, 135, 281-288.

Smith, D.K., Gruner, J.W. and Lipscomb, W.N. (1957) The crystal structure of uranophane $[Ca(H_3O)_2](UO_2)_2(SiO_4)_2 \cdot 3H_2O$. American Mineralogist, 42, 594-618.

Stohl, F.V. and D.K. Smith (1981) The crystal chemistry of the uranyl silicate minerals. Am. Mineral. 66, 610-625.

●URANOPHANE.   See section on *URANYL SILICATES*, this chapter.

●VUAGNATITE, $CaAl(OH)[SiO_4]$.

Orthorhombic, $P2_12_12_1$, a = 7.055, b = 8.543, c = 5.683 Å; Z = 4.

Discovered as a replacement product of plagioclase in rodingitized dykes in southwest Turkey, vuagnatite was named and described by Sarp *et al.* (1976). It has also been found in veins and vugs in three California localities (Pabst, *et al.*, 1976; Fitzpatrick, 1976; Pabst, 1977) and lately in Japan and Guatemala.

The structure of vuagnatite was determined independently by Fitzpatrick (1976) and McNear *et al.* (1976), after it was recognized to be isotypic with

● <u>VUAGNATITE, continued</u>

conichalcite, $CaCu(OH)[AsO_4]$. It is characterized by slightly kinked chains of edge-sharing $AlO_4(OH)_2$ octahedra parallel to c, accounting for its dominant $\{hk0\}$ forms (see inset). The octahedral chains are cross-linked and bridged by SiO_4 tetrahedra and irregular CaO_7OH polyhedra (see figure below). Due perhaps to the large number of shared polyhedral edges, vuagnatite (ρ_{calc} = 3.42 g/cc) has a high packing index and a volume per anion of 17.12 $Å^3$ which is less than that for grossular (17.34 $Å^3$; Pabst, 1977). It is biaxial nega-tive with α = 1.700, β = 1.725, γ = 1.730, $2V_\alpha$ = 48°, and dispersion $r < v$ very strong; no twinning or cleavage was observed. Fitzpatrick (1976) re-cords similar crystal data, and the lattice parameters of the crystals syn-thesized by Leistner and Chatterjee (1978) in their study of the system CaO-Al_2O_3-SiO_2-H_2O are within 2 e.s.d. of those of the natural specimens.

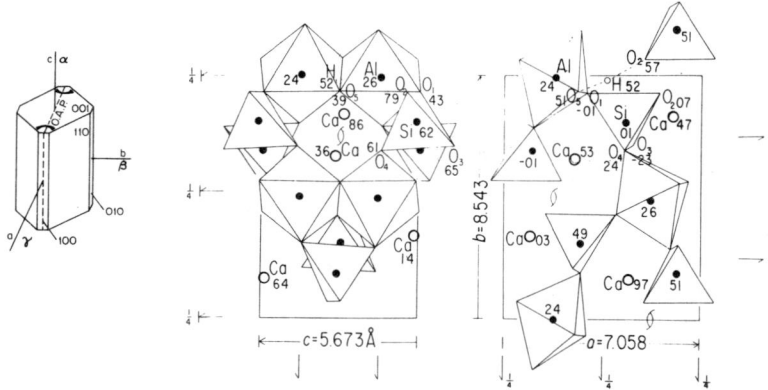

The structure of vaugnatite drawn by Pabst (1977) from the refinement by Fitz-patrick (1976).

- - -

Fitzpatrick, J.U. (1976) *Studies in the microstructure and crystal chemistry of minerals: I. Burbankite from the Green River Formation, Wyoming. II. Electron microscopy of staurolite. III. Crystal structure determination of vuagnatite, CaAlSiO₄(OH).* Ph.D. dissertation, University of California, Berkeley.

Leistner, H. and N.D. Chatterjee (1978) Wasserhaltige Minerale im System $CaSiO_3$-Al_2O_3-H_2O: Rosenhahnite $Ca_3[Si_3O_8](OH)_2$ Vaugnatit $CaAl[SiO_4](OH)$ und Chantalit $CaAl_2SiO_4(OH)_4$ Fortschr. Mineral. 56, 79-80.

McNear, E., M.G. Vincent and E. Parthé (1976) The crystal structure of vuagnatite, $CaAl(OH)SiO_4$. Am. Mineral. 61, 831-838.

Pabst, A. (1977) Über einige besonders dichte, wasserhaltige Calcium- beziehungs-weise Barium-Silicate aus der Franciscan-Formation, California (Lawsonit, Vuagnatit, Rosenhahnit, Cymrit). N. Jahrb. Mineral. Abh. 129, 1-14.

_____, R.C. Erd, L. Rosenhahn and F.E. Goff (1976) Vuagnatite from California (abstr.) 2nd Biennial MSA-FM Symposium, Tucson, Arizona.

Sarp, H., J. Bertrand and E. McNear (1976) Vaugnatite, $CaAl(OH)SiO_4$, a new natural calcium aluminum nesosilicate. Am. Mineral. 61, 825-830.

● YODERITE, $MgAl_{2.84}Fe_{0.16}O(OH)[SiO_4]_2$.

Monoclinic, $P2_1/m$; Subcell: $a = 8.022$, $b = 5.816$, $c = 7.250$ Å; $\beta = 104.9°$; $Z = 2$; supercell: $a \times 6b \times 2c$; $Z = 24$.

Yoderite has been described from Mautia Hill, Tanzania, where it exists as a major phase with quartz, kyanite, and talc (McKie, 1959; McKie and Bradshaw, 1966). Schreyer (1974) believes this 'whiteschist' assemblage formed "under water pressures in excess of some 10 kilobars and at temperatures not higher than 800-840°C." Earlier, Schreyer and Yoder (1968) found that yoderite is stable between 750°C and 875°C over a large pressure interval, even somewhat below 10 kbar.

"The average structure of $P2_1/m$ yoderite was solved by Fleet and Megaw (1962), who, in their two-dimensional Fourier refinement, deliberately neglected the weak satellite reflections reported by McKie (1959). Described in terms of coordination polyhedra, yoderite consists of chains of edgesharing $A(1)$ octahedra, $[AO_5(OH)]$, parallel to the b-axis, interconnected by isolated $[SiO_4]$ tetrahedra and two edge-sharing trigonal bipyramids, $A(2)$ and $A(3)$ of composition $[AO_4(OH)]$ and $[AO_5]$, where $A = Al, Mg, Fe^{3+}$ [figures a and b]. Thus a simplified structural formula of yoderite can be written $^{VI}(Mg, Al, Fe^{3+})_4$ $^V(Mg, Al, Fe^{3+})_4O_2(OH)_2[SiO_4]_4$. This is essentially the same as the andalusite formula with $2Mg + 2(OH) \rightarrow 2Al + 2O$. Based on recognition by McKie (1959) that the lattice parameters of yoderite are very similar to those of kyanite, Fleet and Megaw (1962; see their Fig. 3) noted that the edgesharing octahedral chains parallel to c in kyanite are linked by other $[AlO_6]$ octahedra instead of $[AO_5]$ trigonal bipyramids, as they are in yoderite. Yoderite and kyanite are frequently intergrown, and their topotaxy results from their structural similarity: $a_{yo} || b_{ky}$; $b_{yo} || c_{ky}$; $c_{yo} < a_{ky} \simeq 11°$" (Higgins et al., 1982, p. 76).

The notable feature of yoderite is that it has 'e' and 'f' satellite reflections analogous to those in plagioclase feldspars and mullite indicating an ordered, commensurate antiphase structure with an A-face-centered superlattice of dimensions $a \times 6b \times 2c$, where a, b, c are the cell parameters of the average structure (see figure c). In this model 3Al + 1Mg are presumed ordered in the edge-sharing $A(1)$ octahedral chains and 1Al + 1Mg are in the $A(2)$ trigonal bipyramid which shares an $O \cdots OH$ edge with the $A(1)$ octahedron. Al and Fe^{3+}, which occupy the $A(3)$ trigonal bipyramid in the ratio 5/6 to 1/6, may also be ordered. The $6b$ periodic modulation has conservative antiphase domain boundaries parallel to (010) with a displacement vector of $\frac{1}{2}(2c)$.

After heating for 10 hours at 800°C, the modulation is destroyed and the iron is disordered among the $A(1)$, $A(2)$ and $A(3)$ polyhedra, but the mean $A-(O, OH)$ bond lengths remain essentially unchanged.

a
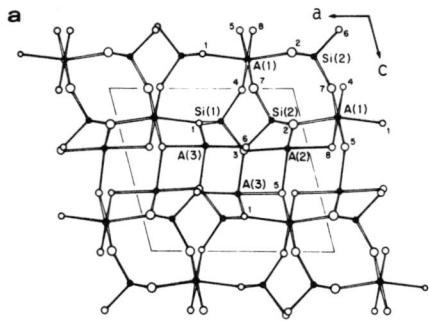

b
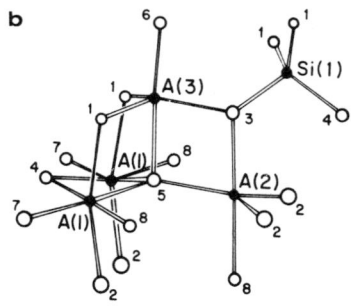

(a) A partial projection of the average structure of yoderite onto the ac plane. Oxygen atoms (open circles) are labelled with numbers only: O(8) is presumed to be the (OH) anion (Fleet and Megaw, 1962). $A(1)$ cations are octahedrally coordinated, $A(2)$ and $A(3)$ are 5-coordinated, and Si(1) and Si(2) are 4-coordinated.

(b) A partial polyhedral drawing of yoderite showing the linkage of the $A(1)$ octahedra, $A(2)$ and $A(3)$ trigonal bipyramids and one of the Si tetrahedra. The O,OH anions are identified with small numbers.

(c) A multiple-beam high-resolution TEM image of the commensurate antiphase structure of yoderite taken at 200 kV with [100] parallel to the electron beam. Below (c) is a schematic representation of the distribution in the plane normal to [100] of the two types of subcells in the antiphase structure of yoderite. Note the A-centered array of supercells with dimensions $6b \times 2c$. All figures from Higgins et al. (1982).

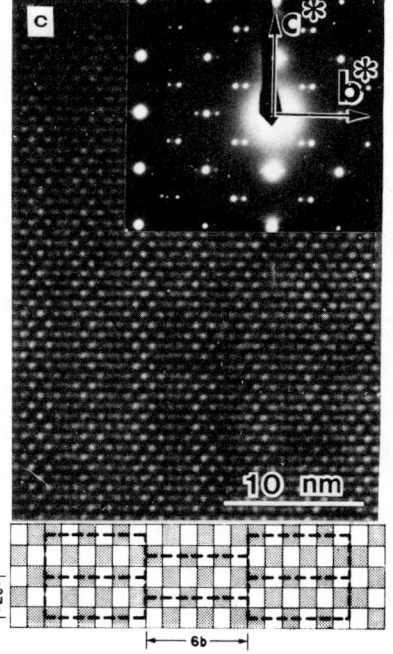

- - -

Fleet, S.G., and Megaw, H.D. (1962) The crystal structure of yoderite. Acta Crystallogr. 15, 721–728.

Higgins, J.B., Ribbe, P.H. and Nakajima, Y. (1982) An ordering model for the commensurate antiphase structure of yoderite. Am. Mineral. 67, 76–84.

McKie, D. (1959) Yoderite, a new hydrous magnesium iron aluminosilicate from Mautia Hill, Tanganyika. Mineral. Mag. 32, 282–307.

McKie, D., and Bradshaw, N. (1966) A green variety of yoderite. Nature 210, 1148.

Schreyer, W. (1974) Whiteschist, a new type of metamorphic rock formed at high pressures. Geol. Rundschau 63, 597–609.

Schreyer, W., and Yoder, H.S., Jr. (1966) Yoderite synthesis, stability and interpretation of its natural occurrence. Carnegie Inst. Washington Year Book 66, 376–380.

● UVAROVITE, $Ca_3Cr_2Si_3O_{12}$. A garnet; see Chapter 2.

● VIRIDINE, $(Al,Mn)^{3+}SiO_5$. A manganoan andalusite; see Chapter 8.

● WEEKSITE. See section on *URANYL SILICATES*, this chapter.

● WELINITE, $Mn^{+4}Mn^{+2}_3O_3(SiO_4)$.

Hexagonal, $P6_3$; $a = 8.155$, $c = 4.785$ Å; $Z = 2$.

Welinite occurs at Långban, Sweden. The above formula is idealized; natural welinite has substitution of W^{+6} and Mg for Mn^{+4} and Mn^{+2}. The presence of W (as well as OH) allows vacancies in the structure, the formula being $(Mn^{+4}_{0.08} W_{0.1})_{0.9}(Mn^{+2}_{2.0}W_{0.3})_{2.6}(O,OH)_3[SiO_4]$ (Moore, 1967). Moore (1967) found the structure to consist of two symmetry equivalent octahedral sheets related by a 2-fold screw axis (see the figure). The Si tetrahedra join the sheets, having as their bases the open part of the tetrahedral void in the octahedral sheets.

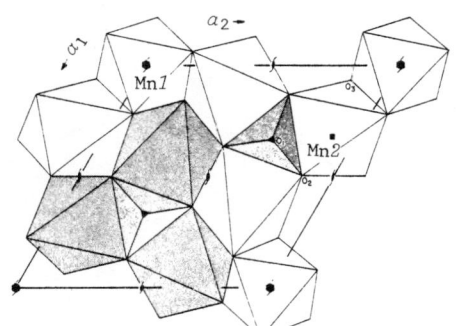

The crystal structure of welinite. The silicon atoms are between the base of the ruled tetrahedral voids and the O_1 atoms. Manganese-oxygen octahedra at $z\sim0.00$ are unshaded; octahedra at $z\sim0.50$ are stippled. The stippled octahedra at $00z$ are omitted for convenience of visualization. From Moore (1968, Fig. 1).

- - -
Moore, P.B. (1968) The crystal structure of welinite, $(Mn^{+4},W)_{<1}(Mn^{+2},W,Mg)_{<3}$ $Si(O,OH)_7$. Arkiv Mineral. 4, 459-466.

● WILKEITE. See section on *SILICATE APATITES*, this chapter.

Chapter 13

ORTHOSILICATES with SiO_4 Polymerized to Other Tetrahedral Polyanions

J.A. Speer & P.H. Ribbe

By contrast with the previous eleven chapters, Chapter 13 contains brief descriptions of the structures of all the other silicates we could locate in the major compendia which have been classified as ortho- or neso- or mono-silicates but in which "isolated" SiO_4 groups *are* polymerized into polyanions or rings or into infinite chains, sheets or frameworks of tetrahedra by corner-sharing with BeO_4, BO_4, AlO_4 or ZnO_4 groups. These include such minerals as anorthite, $CaAl_2Si_2O_8$, whose formula may be written $CaAl_2[SiO_4]_2$ and whose classification is controversial: Liebau (p. 17, this volume) considers it a "monosilicate" if completely ordered; we conventionally consider it to be a tectoaluminosilicate. Likewise larsenite, $PbZnSiO_4$, is a tectozincosilicate. Clinohedrite, $CaZnSiO_4 \cdot H_2O$, contains a sheet of corner-sharing ZnO_4 and SiO_4 tetrahedra. Sapphirine, $(Mg,Al)_8(Al,Si)_6O_{20}$, contains a single pyroxene-like aluminosilicate chain with two "wing" tetrahedra, but only one of six tetra-hedra contains only Si. In asbecasite, $Ca_3(Ti,Sn)[(As_3SiBeO_{10})_2]$, the BeO_4 and SiO_4 tetrahedra form isolated $(BeSiO_7)$ groups. Many other such examples are cited in the following text.

In an attempt to gather together the "loose ends", certain silicates which were classified at one time as orthosilicates, but whose structures are as yet unknown, have been listed alphabetically among the known structures. Other minerals that are *suspected* to be orthosilicates in the broader defi-nition of this chapter are also included. In addition a few minerals are listed whose structures, once thought to be orthosilicates, are now known to contain polymerized SiO_4 groups or chains or sheets.

● ANORTHITE, $CaAl_2Si_2O_8$.   See *DANBURITE*, this chapter.

● ASBECASITE, $Ca_3(Ti,Sn)[(As_3SiBeO_{10})_2]$.

Trigonal, $P\bar{3}c1$; a = 8.36, c = 15.30 Å; Z = 2.

Asbecasite is an alpine cleft mineral found in gneiss. The formula and crystal
structure were determined by Cannillo *et al.* (1969). As shown below, the structure is made up of layers, an A layer comprising BeO_4 tetrahedra, SiO_4 tetrahedra,
and AsO_3 pyramids, and a B layer formed by $(Ti,Sn)O_6$ octahedra and CaO_8 square
antiprisms. The A layer is made of two sublayers, related by a symmetry center,
connected by a common vertex of the BeO_4 and SiO_4 tetrahedra, forming $(BeSiO_7)$
groups.

layers

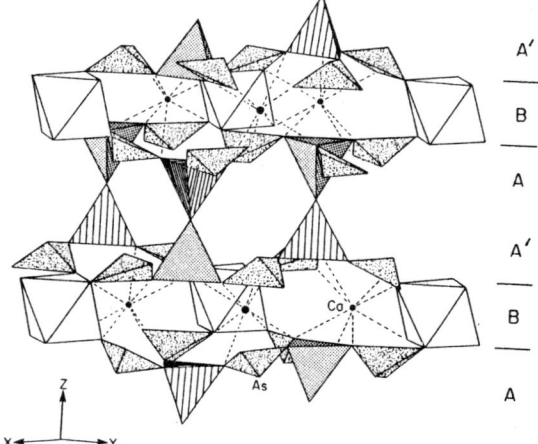

Clinographic projection of the
trigonal cell of asbecasite.
Modified from Cannillo *et al.*
(1969, Fig. 1).

- - -

Cannillo, E., G. Giuseppetti and C. Tadini (,969) The crystal structure of
    asbecasite. Accad. Naz. Lincei Ser. VIII, SLVI., 457-467.

● BAKERITE.   See *DATOLITE*, this chapter.

● BERYLLITE, $Be_3SiO_4(OH)_2 \cdot H_2O$.

Orthorhombic or monoclinic beryllite occurs in altered alkalic pegmatites of the Lovozero massif, Kola Peninsula, U.S.S.R., replacing epididymite
or sphaerobertrandite (bertrandite with high BeO and low SiO_2). Al^{3+} and Fe^{3+}
may replace Si with charge balance maintained by presence of Na and Ca.

- - -

Kuzmenko, M. V. (1954) Beryllite - a new mineral. Dokl. Akad. Nauk. S.S.R.
    99, 451-454. New Mineral Names (1955), Am. Mineral. 40, 787-788.

Vlasov, K. A. (1966) *Geochemistry and Mineralogy of Rare Elements and Genetic
    Types of Their Deposits*, Vol. 2: *Mineralogy of Rare Elements*. Israel
    Program for Scientific Translations, pp. 96-97.

● BISMUTOFERRITE.   See *CHAPMANITE*, this chapter.

● BORNEMANITE, $BaNa_4Ti_2NbSi_4O_{17}(F,OH)\cdot Na_3PO_4$.

Orthorhombic, *Ibmm* or *Ib2m*; $a = 5.48$, $b = 7.10$, $c = 48.2$ Å, $Z = 4$.

Bornemanite occurs as coatings on lomonosovite in pegmatites of the Lovozero massif, Kola Peninsula, U.S.S.R. (Men'shikov *et al.*, 1975). It may be comparable to lomonosovite (Bykova and Khomyakov, 1977).

- - -

Bykova, A. V. and A. P. Khomyakov (1977) Nature of bornemanite, a new mineral from the lomonosovite group. *In* B. I. Semenov and T. N. Chuileva (Eds.), Metod. Mineral. Issled., pp. 51-53.

Menishikov, Yu. P., I.V. Bussen, E.A. Goiko, N.I. Zabovnikova, A.N. Merkov and A.P. Khomyakov (1975) Bornemanite, a new silicophosphate of sodium, titanium, niobium and barium. Zapiski Vses. Mineral. Obshch 104, 322-326. New Mineral Data (1976), Am. Mineral. 61, 338.

● CEBOLLITE, $Ca_5Al_2Si_3O_{12}(OH)_4$.

Orthorhombic (?).

Cebollite is a low-temperature mineral formed by the hydrothermal alteration of melilite, plagioclase, etc. in carbonate-rich assemblages of carbonatites, skarns or kimberlites. Its crystal structure is unknown.

● CELSIAN, $BaAl_2Si_2O_8$.   See *DANBURITE*, this chapter.

● CHAPMANITE, $Sb^{+3}Fe_2^{+3}(SiO_4)_2(OH)$.

● BISMUTOFERRITE, $BiFe_2^{+3}(SiO_4)_2(OH)$.

Monoclinic, *Cm*. Originally considered to be orthosilicates, they have since been shown by Zhukhlistov and Zvyagin (1977) to be sheet silicates.

- - -

Zhukhlistov, A. P. and B. B. Zvyagin (1977) Determination of the crystal structures of chapmanite and bismuthoferrite by high-voltage electron diffraction. Sov. Phys. Crystallogr. 22, 419-423.

● CLINOHEDRITE, $CaZnSiO_4\cdot H_2O$.

Monoclinic, *Cc*; $a = 5.090$-5.131, $b = 15.829$-15.928, $c = 5.386$-5.422 Å; $\beta = 103.26°$-$103.39°$; $Z = 4$.

Clinohedrite occurs in the zinc deposit at Franklin, New Jersey. There have been three structure determinations of clinohedrite. This discussion is based on the two most recent by Venetopoulos and Rentzeperis (1976) and Simonov *et al.* (1977). The anionic framework of clinohedrite is a cubic close-packed array of O atoms. It has a *C*-centered arrangement of zigzag columns of edge-sharing Ca octahedra extending along c (figure a). These columns are joined by layers composed of ZnO_4 and SiO_4 tetrahedra. The Zn tetrahedra form chains along the diagonal = $a+c$ in (010) which are joined by isolated Si tetrahedra (fig. b). The Ca is in octahedral coordination, bonded to one oxygen atom, three oxygens participating in hydrogen bonding, and two oxygen atoms of H_2O molecules. The structural formula is $CaZnH[SiO_4](OH)$.

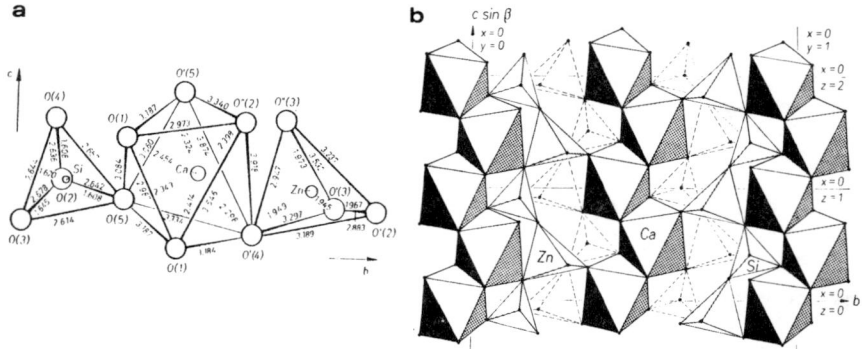

a

b

(a) Projection of the clinohedrite structure showing the edge-sharing Ca octahedral joined by SiO_4 and ZnO_4 tetrahedra.

(b) Schematic of the layer of Zn and Si tetrahedra. The chains of Zn tetrahedra run parallel to $a+c$ in (010) which is responsible for the crystal's
elongation in this direction and the name.

- - -

Simonov, M. A., E. L. Belokoneva, Yu. K. Egorov-Tismenko and N. V. Belov (1977)
The crystal structure of clinohedrite $CaZn[SiO_4]\cdot H_2O$. Sov. Phys. Dokl.
22, 614–616.

Venetopoulos, Cl. C. and P. J. Rentzeperis (1976) Redetermination of the
crystal structure of clinohedrite, $CaZnSiO_4\cdot H_2O$. Z. Kristallogr. 144,
377–392.

● DANBURITE, $CaB_2[SiO_4]_2$; ANORTHITE, $CaAl_2[SiO_4]_2$; CELSIAN and PARACELSIAN,
$BaAl_2[SiO_4]_2$; SLAWSONITE, $SrAl_2[SiO_4]_2$.

Writing the formulas in this unconventional manner gives the misleading im-
pression that these compounds are orthosilicates rather than framework (or tekto-)
borosilicate and aluminosilicates. In danburite SiO_4 tetrahedra share all four
corners with BO_4 tetrahedra, and vice versa, producing a 3-D tetrahedral network
of composition $[BSiO_4]_2^{-2}$ (Phillips *et al.*, 1974). Anorthite and celsian are, of
course, feldspars (see Volume 2 of this series); slawsonite (Griffen *et al.*, 1977)
is the Sr-analog of paracelsian (structure by Smith, 1953). Liebau would characterize these as monosilicates [≈ *orthosilicates*], but recognizes the problems associated with such a classification (see discussion on p. 17 of this volume).

Although there are topological differences among them, the aluminosilicate
structures listed here have somewhat analogous frameworks of corner-sharing SiO_4
and AlO_4 tetrahedra (see Smith and Rinaldi, 1962; Smith, 1968).

● DANBURITE, continued

Griffen, D.T., P.H. Ribbe and G.V. Gibbs (1977) The structure of slawsonite, a strontium analog of paracelsian. Am. Mineral. 62, 31-35.

Phillips, M.W., G.V. Gibbs and P.H. Ribbe (1974) The crystal structure of danburite: a comparison with anorthite, albite, and reedmergnerite. Am. Mineral. 59, 79-85.

Smith, J.V. (1953) The crystal structure of paracelsian, $BaAl_2Si_2O_8$. Acta Crystallogr. 6, 613-620.

_____ (1968) Further discussion of framework structures built from four- and eight-membered rings. Mineral. Mag. 36, 640-642.

_____ and F. Rinaldi (1962) Framework structures formed from parallel four- and eight-membered rings. Mineral. Mag. 33, 202-212.

● DAVREUXITE, $Mn_2Al_{12}[(SiO_4)_7O_3(OH)_6]$.

Monoclinic; a = 9.57, b = 5.79, c = 12.88 Å; β = 116°.

Davreuxite occurs with quartz and pyrophyllite at Ottré, Belgium (Fransolet and Bourguigon, 1976). Its structure is unknown, but Ramdohr and Strunz (1980) consider it to be an orthosilicate.

- - -

Fransolet, A.-M. and P. Bourguigon (1976) Precisions mineralogiques sur la davreuxite. Compt. Rend. Acad. Sci. (Paris) 283D, 295-297.

Ramdohr, P. and H. Strunz (1980) *Klockmanns Lehrbuch der Mineralogie*, 16 Auflage, durchgesehner Nachdruck 1980. Ferdinand Enke Verlag, Stuttgart 371 p.

● ESPERITE, $(Pb,Ca,Zn)_2SiO_4$.

Monoclinic, $P2_1/n$; a = 2×8.814, b = 8.270, c = 2×15.26 Å; β = 90°; Z = 4.
Esperite was previously known as calcium larsenite and is found at Franklin, New Jersey as replacement of massive willemite or hardystonite. It has a small range in Ca:Pb ratios of between 2.25:1 and 3:1. Ito (1968) found that esperite contained an unidentified, Pb-poor phase. Moore and Ribbe (1965) suggest that esperite is a tectozincosilicate with a possible structural formula of $Ca_3Pb[ZnSiO_4]_4$. They concluded that it is isostructural with beryllonite, where Ca,Pb correspond to Na, Zn to Be, and Si to P. A notable difference is the superstructure reflections in esperite which are presumed to result from Ca-Pb ordering. Ito (1968) synthesized an esperite (with Ca:Pb = 2.33:1) which lacked superstructure reflections.

- - -

Ito, J. (1968) Synthesis of some lead calcium zinc silicates. Am. Mineral. 53, 231-240.

Moore, P. B. and P. H. Ribbe (1965) A study of "calcium-larsenite" renamed esperite. Am. Mineral. 50, 1170-1178.

● EUCLASE, $AlBeOH(SiO_4)$.

Monoclinic $P2_1/a$, $a = 4.763$, $b = 14.29$, $c = 4.618$ Å; $\beta = 100.25°$; $Z = 4$.

Euclase shows minor substitution of Fe^{3+} and Sn for Al and F for OH. It is a low-temperature hydrothermal mineral found in alpine veins, pegmatites and greisens.

The most recent structure refinement is by Mrose and Appleman (1962; see figure below). They describe the structure as consisting of a $[Be_2(OH)_2(SiO_4)_2]_n^{-6n}$ chain extending parallel to [100] and made up of corner-sharing Be tetrahedra with Si tetrahedra on the outside edges. These chains are cross-linked by Al octahedra. Euclase crystals are elongated parallel to [100], the direction of the chain axis, and have perfect {010} cleavage, which is parallel to the chains and involves breaking a minimum number of Al-O bonds.

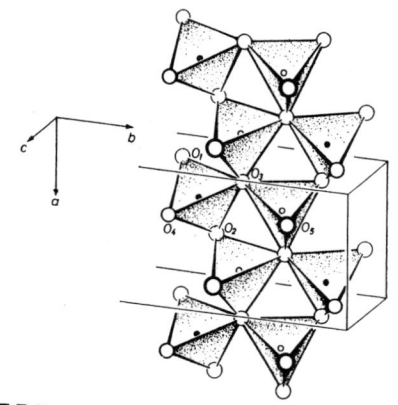

The Be,Si tetrahedral chain of euclase: large single circles, oxygen; large double circles, hydroxyl; small open circles, Be; small solid circles, Si. From Mrose and Appleman (1962, Fig. 2b).

- - -

Mrose, M. E. and D. E. Appleman (1962) The crystal structures and crystal chemistry of väyrynenite, $(Mn,Fe)Be(PO_4)(OH)$, and euclase, $AlBe(SiO_4)(OH)$. Z. Kristallogr. 117, 16-36.

● EUCRYPTITE.   See section on *PHENACITE GROUP*, this chapter.

● FRESNOITE, $Ba(TiO)[Si_2O_7]$.

Tetragonal, *P4/mbm*, *P4bm*, or *P4̄b2*; $a = 8.518$, $c = 5.211$ Å; $Z = 2$.

Before Moore and Louisnathan (1969) solved its structure, the fresnoite formula was sometimes written $BaTi[SiO_4]_2$, and it was presumed to be an ortho-silicate. But in fact it contains unusual $[TiO_5]$ tetragonal pyramids sharing basal corners with Si_2O_7 groups. The chemical bonding of the Si_2O_7 groups is similar to that in melilites.

- - -

Moore, P.B. and S.J. Louisnathan (1969) The crystal structure of fresnoite, $Ba_2(TiO)Si_2O_7$. Z. Kristallogr. 130, 438-448.

● GARRELSITE, $NaBa_3Si_2B_7O_{16}(OH)_4$.

Monoclinic, $C2/c$; $a = 14.639$, $b = 8.466$, $c = 14.438$ Å, $\beta = 114.21°$, Z =

Garrelsite occurs as authigenic crystals in dolomitic marlstone of the Green River Formation, Utah, and in borate deposits of California. The mineral formula and crystal structure were determined by Ghose *et al.* (1976). They found that garrelsite is a three-dimensional framework composed of (1) a silicoborate sheet containing a $[B_2O_{12}]^{9-}$ polyanion linked with B and Si tetrahedra and distorted $NaO_4(OH)_2$ octahedra, and (2) Ba-O polyhedra. See the figure. The $[B_2O_{12}]^{9-}$ polyanion is comprised of three B-tetrahedra and two B-triangles, as outlined in the figure below.

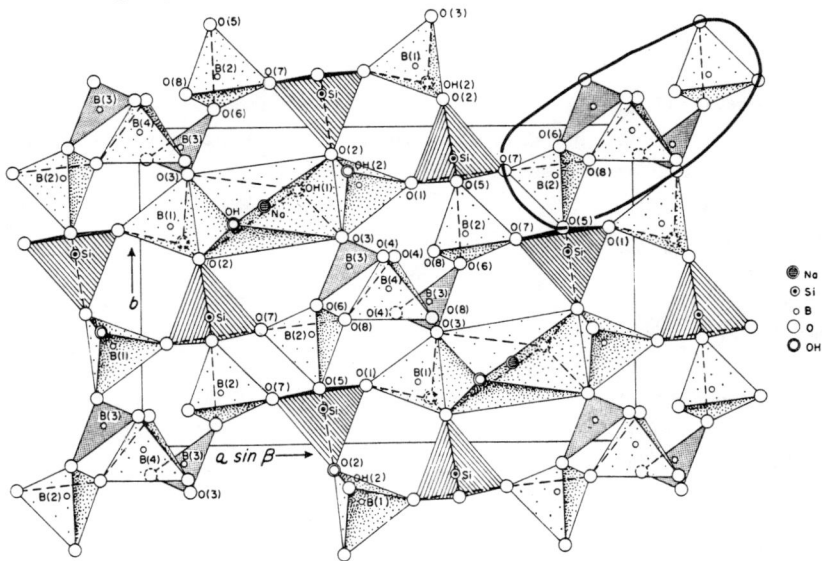

Partial projection of the garrelsite structure on (001), showing the boro-silicate sheet and the Na atom in octahedral coordination. The Ba atoms have been omitted. From Ghose *et al.* (1976, Fig. 3).

- - -

Ghose, S., C. Wan and H. Ulbrich (1976) Structural chemistry of borosilicates. I. Garrelsite, $NaBa_3Si_2B_7O_{16}(OH)_4$: a silicoborate with the pentaborate $[B_5O_{12}]^{9-}$ polyanion. Acta Crystallogr. B32, 824-832.

● GERSTMANNITE, $(Mn,Mg)Mg(OH)_2[ZnSiO_4]$.

Orthorhombic, $Bbcm$; $a = 8.185$, $b = 18.65$, $c = 6.256$ Å; $Z = 8$.

Gerstmannite occurs in hydrothermal veins at Sterling Hill, New Jersey. The mineral was described and the structure solved by Moore and Araki (1977), but as pointed out by Simonov *et al.* (1978), its structure is similar to that of clinohedrite. Both are based on cubic close-packing of oxygen atoms, but the excess (Mn,Mg) over Ca in gerstmannite occupies an additional octahedral chain forming a band of three octahedra parallel to a rather than two zigzag octahedra paralleling c as in clinohedrite. Charge balance is maintained by replacing the neutral H_2O by two (OH).

- - -

Moore, P. B. and T. Araki (1977) Gerstmannite, a new zinc silicate mineral and a novel cubic close-packed oxide structure. Am. Mineral. 62, 51–59.

Simonov, M. A., Yu. K. Egorov-Tismenko and N. V. Belov (1978) The crystal structures of clinohedrite $Ca[ZnSiO_4] \cdot H_2O$ and gerstmannite $(Mn,Mg)Mg(OH)_2[ZnSiO_4]$. Sov. Phys. Dokl. 23, 113–114.

● HARKERITE, $Ca_{24}Mg_8Al_2(SiO_4)_8(BO_3)_6(CO_3)_{10} \cdot 2H_2O$.

Cubic, $Fd3m$, pseudocell $a' = a/2 = 14.73$ Å; pseudocell Z = 2.

Harkerite occurs in calcsilicate rocks that have undergone high-temperature, low-pressure contact metamorphism and B–F–Cl metasomatism. Harkerite may contain Cl as well as variable Si, B and C contents (Davies and Machin, 1970; Barbieri. *et al.*, 1977) and has a suggested general formula:

$$Ca_{48}Mg_{16}(Al_pSi_{5-p}O_{16})_x^{m-}(BO_3)_y^{3-}(CO_3)_z^{2-}(OH,Cl)_w \cdot nH_2O$$

with m = 12 + p and mx + 3y + 2z + w = 128 with x variable between 0 and 5. The symmetry may change with chemistry as well (Barbieri *et al.*, 1977).

A structure determination by Machin and Miche (1976) shows that the tetrahedra form a pentameric anion of composition $[Al(SiO_4)_4]^{13-}$ which resembles the $[Si(SiO_4)_4]^{12-}$ pentamer found in zunyite.

- - -

Barbieri, M., D. Cozzupoli, M. Federico, M. Fornaseri, S. Merlino, P. Orlandi and L. Tolomeo (1977) Harkerite from the Alban Hills, Italy. Lithos 10, 133–141.

Davies, W.O. and M.P. Machin (1970) Isomorphous replacements in harkerite and the relation of sakhaite to harkerite. Canadian Mineral. 10, 689–695.

Machin, M.P. and G. Miehe (1976) $Al(SiO_4)_4^{13-}$ tetrahedral pentamers in harkerite. N. Jahrb. Mineral. Monatsh., 223–232.

● HELLANDITE, $[Ca_{5.5}(Y,RE)_{5.0}\Box_{1.5}](Al_{1.1}Fe^{3+}_{0.9})(OH)_4[Si_8B_8O_{40}(OH)_4]$.

Monoclinic, $P2/a$; $a = 18.99$, $b = 4.715$, $c = 10.30$ Å; $\beta = 111.4°$; Z = 1.

Hellandite is found in granitoids and granitoid pegmatites. Hogarth *et al.* (1972) made a comprehensive study of its chemistry, physical properties and crystallography. The major chemical variation is Ca ⇌ Y + REE. Hellandite is heavy rare earth enriched. There is lesser variation of Al and Fe and minor amounts of Mn, Mg, Ti and Th. The formula and crystal structure of

hellandite were determined by Mellini and Merlino (1977), who found that
it consists of five-membered SiO_4 and BO_4 tetrahedral rings forming chains
which are parallel to c and lie in (010) (see the figure). These layers of
silicoborate chains alternate with a layer comprised of Fe and Al octahedra
and Ca,Y,REE and vacant square antiprisms.

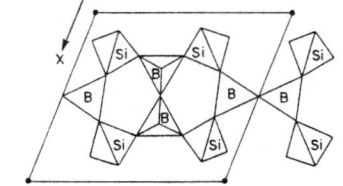

A drawing of the tetrahedral chain in
hellandite in the (010) plane. Modified
from Mellini and Merlino (1977, Fig. 1).

- - -

Hogarth, D. D., G. Y. Chao and D. C. Harris (1972) New data on hellandite.
   Canadian Mineral. 11, 760-776.

Mellini, M. and S. Merlino (1977) Hellandite: a new type of silicoborate
   chain. Am. Mineral. 62, 89-99.

● HILLEBRANDITE, $Ca_2SiO_4 \cdot H_2O$

   Monoclinic, $P2_1/a$; a = 16.60, b = 7.26, c = 11.85 Å; β = 90°; Z = 12.

   Hillebrandite occurs in high-temperature, low-pressure contact-meta-
morphic limestones and calcsilicate rocks. Its structure has been described
by Mamedov and Belov (1958) as a regular stacking of xonotlite and $Ca(OH)_2$
parallel to [001]. The silicate tetrahedra form $[Si_6O_{17}]^\infty$ chains parallel to
[010], accounting for its acicular habit. See also Heller (1953).

- - -

Heller, L. (1953) X-ray investigation of hillebrandite. Mineral. Mag. 30,
   150-154.

Mamedov, K.S. and N.V. Belov (1958) The crystal structure of hillebrandite.
   Doklady Akad. Nauk SSSR 123, 741-743. Chem. Abstracts 53, 6919.

● HODGKINSONITE, $Mn(OH)_2[Zn_2SiO_4]$.

   Monoclinic, $P2_1/a$; a = 8.171, b = 5.316, c = 11.761 Å, β = 95.25°, Z = 4.

   Hodgkinsonite is a pneumatolytic mineral from Franklin, New Jersey.
Rentzeperis (1963) determined that the structure consists of corner-sharing
ZnO_4 and SiO_4 tetrahedra forming a $[Zn_2SiO_4]$ two-dimensional network parallel
to (001), the A layer, and corner-sharing MnO_6 octahedron forming a two-
dimensional network parallel to (001), the B layer. The layers can be described as being stacked ...$ABAABA$... parallel to c.

- - -

Rentzeperis, P. J. (1963) The crystal structure of hodgkinsonite $Zn_2Mn[(OH)_2$
   $SiO_4]$. Z. Kristallogr. 119, 117-138.

● HOLDENITE, $Mn_6Zn_3(OH)_8(AsO_4)_2(SiO_4)$.

Orthorhombic, *Abma*; $a = 11.99$, $b = 31.46$, $c = 8.697$ Å; $Z = 8$.

Holdenite occurs at Franklin and Sterling Hill, New Jersey, and was originally described as a Mn-Zn arsenate. The structure and formula were determined by Moore and Araki (1977); the essential nature of silica was further supported in a chemical study by Dunn (1981). The structure of holdenite is based on a cubic close-packing of oxygens. The structural formula is complex:

$$^{VI}Mn_6(OH)_2\,^{IV}[Zn(OH)_4]\,^{IV}[Zn_2(OH)_2(AsO_4)_2(SiO_4)].$$

Silicon is in an open tetrahedral sheet of composition $[Zn_2SiO_6(OH)_2]$ in the structure.

- - -

Dunn, P. J. (1981) Holdenite from Sterling Hill and new chemical data. Mineral. Record 11, 373-375.

Moore, P. B. and T. Araki (1977) Holdenite, a novel cubic close-packed structure. Am. Mineral. 62, 513-521.

● HOWLITE, $Ca_2SiB_5O_9(OH)_5$.

Monoclinic, $P2_1/c$; $a = 12.78$, $b = 9.33$, $c = 8.60$ Å; $\beta = 104.83°$; $Z = 4$.

Howlite is found in anhydrite-gypsum evaporite deposits in Nova Scotia and the borate deposits of California. The crystal structure was determined by Finney *et al.* (1970) which confirmed the suggestion of Moenke (1960) based on IR spectra that boron occurs in both tetrahedral and triangular coordination. The structure is described as boron tetrahedra and triangles sharing corners to form six-membered boroxol rings which join to form colemanite-like chains. These chains are linked by oxygen atoms to a spiral of silicate-borate tetrahedra extending along b, forming slabs oriented approximately in (100). The slabs are held together by 8-coordinated Ca atoms.

- - -

Finney, J. J., I. Kumbasar, J. A. Konnert and J. R. Clark (1970) Crystal structure of the calcium silicoborate howlite. Am. Mineral. 55, 716-728.

Moenke, H. (1960) Die Ultrarotabsorptions-spektren wasserhaltiger. Bormineralien, des Howliths und des Danburits im Bereich von 400 bis 1800 cm^{-1}. Janaer Jahrb. 1960 I, 191-215.

● IIMORIITE, perhaps $Y_5(SiO_4)_3(OH)_3$.

Triclinic, $P\bar{1}$; $a = 11.6$, $b = 6.65$, $c = 13.1$ Å; $\alpha = 94.3°$, $\beta = 95.0°$, $\gamma = 93.6°$.

Iimoriite occurs in granitoid pegmatites in Fukushima Prefecture, Japan (Kato and Nagashima, 1970). Its structure is unknown.

- - -

Kato, A. and K. Nagashima (1970) An introduction to Japanese minerals, pp. 39 and 85-86. New Mineral Names, Am. Mineral. 58, 140.

● ILIMAUSSITE, $Ba_2Na_4CeFe^{+3}Nb_2Si_8O_{28} \cdot 5H_2O$.

Hexagonal, $P6_3/mcm$; $a = 6.19$, $c = 20.34$ Å.

Ilimaussite occurs in hydrothermal veins cutting the Ilimaussaq alkalic massif, south Greenland and in alkalic rocks of the Kola Peninsula, U.S.S.R. The structure is unknown.

- - -

Semenov, E. I., M. E. Kazakova and V. I. Bukin (1968) Ilimaussite, a new rare-earth-niobium-barium silicate from Ilimaussaq, South Greenland. Medd. Grønland 181, 3-7.

● ILMAJOKITE, $Na_2Ti[(Si,C)_3O_9] \cdot 1.5\ H_2O$.

Monoclinic $C2/c$ or Cc; $a = 39.80$, $c = 29.83$ Å; $\beta = 96.6°$.

Ilmajokite occurs in cavities in the natrolitic zone of pegmatites in the Lovozero tundra, Kola peninsula, U.S.S.R. It was considered an orthosilicate by Ramdohr and Strunz (1980), perhaps based on the original analysis with the ideal formula $Na_2Ti[Si(O,OH)_4]_3 \cdot 3\ H_2O$.

- - -

Bussen, I. V., L. F. Gannibal, E. A. Goiko, A. N. Mer'kov and A. P. Nedorezova (1972) Ilmajokite, a new mineral from the Lovozero Tundra, Zapiski Vses. Mineral. Obshch. 101, 75-79. New Mineral Names, Am. Mineral. 58, 139-140 (1973).

Goiko, E. A., I. V. Bussen, L. F. Gannibal and E. A. Lipatova (1974) Ilmajokite. Uch. Zap. Leningr. Gos. Univ. Ser. Biol. Nauk. 378, 174-181. Chem. Abstr. 84, 7517.

Ramdohr, P. and H. Strunz (1980) *Klockmanns Lehrbuch der Mineralogie*, 16 Auflage., Ferdinand Enke Verlag, Stuttgart, 371 p.

● KATOPTRITE, $(Mn_5Sb_2)(Mn_8Al_4Si_2)O_{28}$.

Monoclinic, $C2/m$; $a = 5.617$, $b = 23.02$, $c = 9.079$ Å; $\beta = 101.4°$; $Z = 2$.

● YEATMANITE, $(Mn_5Sb_2)(Mn_2Zn_8Si_4)O_{28}$.

Triclinic, $P\bar{1}$; $a = 5.604$, $b = 11.602$, $c = 9.058$ Å, $\alpha = 92.2°$, $\beta = 100.9°$, $\gamma = 88.3°$; $Z = 1$.

These minerals occur in metamorphosed Mn and Zn deposits; katoptrite in the Swedish Mn-oxide ore deposits (Långban) and yeatmanite at Franklin, New Jersey. Moore *et al.* (1976) determined the structure of katoprite and proposed a structure for yeatmanite. Both are based on dense-packing of oxygens, katoptrite with sequence ...cchh... . Mn and Sb are ordered in octahedral voids of layers which alternate with layers of ordered, corner-sharing Mn, Zn and Si tetrahedra. The stacking of these layers results in an empty octahedral sheet at $z = 1/2$ and accounts for the perfect {001} cleavage. The two minerals presumably differ in the ordering of their tetrahedral layers, (*cf.* figures a and b) and this is the reason for their different symmetries. More recently, Dunn and Leavens (1980) confirmed the formula of yeatmanite proposed by Moore et $al.$ (1976), finding evidence of limited solid solution of Mn^{2+} for Zn^{2+} in tetrahedral sites.

a
b

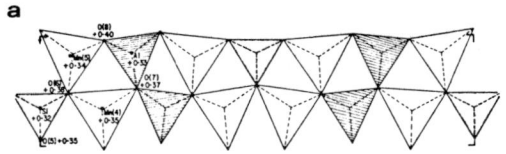 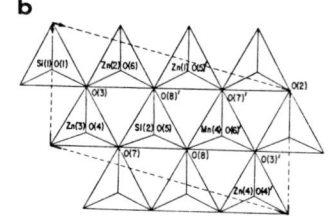

(a) Polyhedral diagram of the tetrahedral sheets above the octahedra, showing the ordering scheme in katoptrite. (b) Proposed ordering scheme of the tetrahedral layer in yeatmanite.

- - -

Dunn, P.J. and P.B. Leavens (1980) Yeatmanite: new data. Am. Mineral. 65, 196-199.

Moore, P.B., T. Araki and G.D. Brunton (1976) Katoptrite, $(Mn_5^{2+},Sb_2^{5+})^{VI}$ $(Mn_8^{2+}Al_4Si_2)^{IV}O_{28}$, a novel close-packed oxide sheet structure. N. Jahrb. Mineral. Abh. 127, 47-61.

● KORNERUPINE, $Mg_4Al_6(Si,B)_5O_{21}(OH)$.

Orthorhombic, $Cmcm$; a = 16.100; b = 13.767; c = 6.735 Å; Z = 4.

Kornerupines occur in B-rich, aluminous pelitic rocks from high metamorphic grades and show B $\rightleftarrows$ Si substitution as well as variation in Mg, Fe, Al and H contents. Crystal structure determinations of a natural (Moore and Bennett, 1968) and synthetic (Moore and Araki, 1979) kornerupine indicate that the tetrahedral atoms, Si, B and Al, comprise linear, corner-sharing T_2O_7 dimers and T_3O_{10} trimers.

- - -

Moore, P.B. and T. Araki (1979) Kornerupine: a detailed crystal-chemical study. N. Jahrb. Mineral. Abh. 134, 317-336.

—— and J.M. Bennett (1968) Kornerupine: its crystal structure. Science 159, 524-6.

● LARSENITE, $PbZnSiO_4$.

Orthorhombic $Pna2_1$; a = 8.244, b = 18.963, c = 5.06 Å; Z = 8.

Larsenite is found at Franklin and Sterling Hill, New Jersey. Prewitt et $al.$ (1967) found the structure to consist of a network of corner-sharing Zn and Si tetrahedra. The network is comprised of double chains of five-membered Zn-Si-Zn-Si-Zn tetrahedral rings which lie in the ab plane, extending along a (see the figures). These chains are interconnected by Zn-Zn-Si tetrahedral rings. The Pb occupies channels in the tetrahedral network.

● LARSENITE, continued

Larsenite was synthesized by Ito and Frondel (1967); it melts incongruently at 1000°C and 1 atm to willemite and a lead silicate liquid.

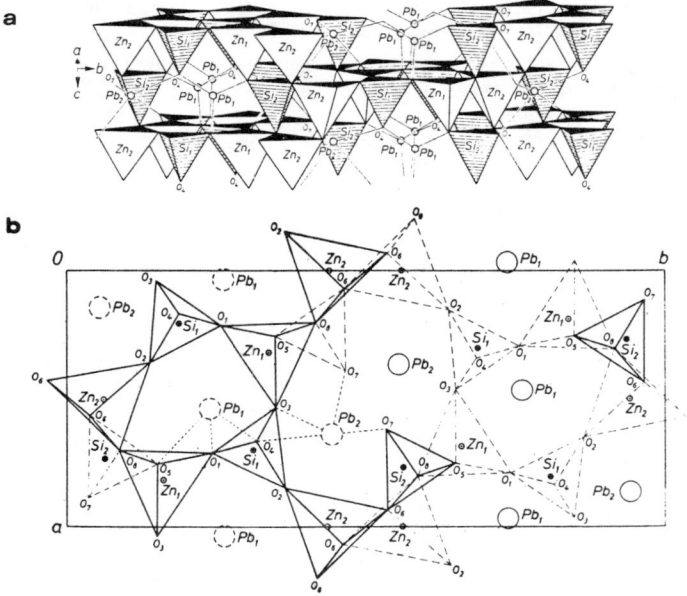

(a) Polyhedral model for larsenite viewed approximately along a and (b) along c. From Prewitt $et\ al.$ (1967, Figs. 2a and b).

- - -

Ito, J. and C. Frondel (1967) Syntheses of lead silicates: larsenite, barysilite and related phases. Am. Mineral. 52, 1077-1084.

Prewitt, C.T., E. Kirchner and A. Preisinger (1967) Crystal structure of larsenite PbZnSiO$_4$. Z. Kristallogr. 124, 115-130.

● LIBERITE, Li$_2$BeSiO$_4$.

Monoclinic, Pn; a = 4.698, b = 4.942, c = 6.104 Å; β = 90°0'; Z = 2.

Liberite is a vein mineral in tactite from the Nanling range, China. It is designated as the β phase of Li$_2$BeSiO$_4$, and the $\beta \rightarrow \gamma$ transition is at 650 ± 100°C at 1 atm. γ-Li$_2$BeSiO$_4$ is orthorhombic, $C222_1$, a = 6.853, b = 6.927, c = 6.125 Å (West, 1975).

Liberite has an ordered wurtzite structure (Chang, 1966): the oxygen atoms form a hexagonal close-packed array and the Li, Be and Si are distributed over one set of tetrahedral sites. It can also be regarded as a framework of corner-sharing tetrahedra of Li, Be and Si. γ-Li$_2$BeSiO$_4$ also has an hexagonal

● LIBERITE, continued

array of close-packed oxygens, but the cations occupy different tetrahedral
sites (Howie and West, 1974).

– – –

Chang, Han-Ching (1966) Structural analysis of liberite. Ti Chih Hsueh Pao,
     Acta Geol. Sinica 46, 76-86; Chem. Abstracts 65, 11457h.

Ch'un-Lin Chao, (1964) Liberite (Li_2BeSiO_4), a new lithium-beryllium silicate
     mineral from the Nanling Ranges, South China. Ti Chih Hsueh Pao 44, 334-
     342. Chem. Abstr. 61, 15841.

Howie, R.A. and West, A.R. (1974) The crystal structure of high (γ)-Li_2BeSiO_4
     a tetrahedral structure. Acta Crystallogr. B30, 2434-2437.

West, A.R. (1975) Crystal chemistry of liberite, Li_2BeSiO_4 and Li_2BeGeO_4.
     Bull. Soc. fr. Mineral. Cristallogr. 98, 6-10.

● MCGOVERNITE, $(Mn,Mg,Zn)_{22}(AsO_3)(AsO_4)_3(SiO_4)_3(OH)_{21}$.

     Trigonal, $R\bar{3}2/c$; a = 68.7 Å; α = 6°52'; hexagonal cell: a = 8.22, c =
203.15Å; Z = 12.

     Mcgovernite occurs in the zinc deposit at Sterling Hill, New Jersey. The
crystal structure of mcgovernite is unknown, but it has been postulated to contain structural units similar to welinite, hematolite, dixenite and kraisslite
by Wuensch (1968) and Moore and Araki (1978). The latter suggest that mcgovernite
is a generally dense-packed oxygen structure with an 84-layer repeat.

– – –

Moore, P.B. and T. Araki (1978) Hematolite: a complex dense-packed sheet struc-
     ture. Am. Mineral. 63, 150-159.

Wuensch, B.J. (1960) The crystallography of mcgovernite, a complex arseno-
     silicate. Am. Mineral. 45, 937-945.

_____ (1968) Comparison of the crystallography of dixenite, mcgovernite and
     hematolite. Z. Kristallogr. 127, 309-318.

● MULLITE, $Al(Al_{1+2x}Si_{1-2x})O_{5-x}$.   See Chapter 8.

● PARACELSIAN, $BaAl_2Si_2O_8$.   See *DANBURITE*, this chapter.

442

PHENACITE GROUP

● PHENACITE, Be_2SiO_4.

      Rhombohedral, $R\bar{3}$; $a = 7.70$ Å; $\alpha = 108°01'$; $Z = 6$;

          hexagonal cell: $a = 12.472$, $c = 8.252$ Å; $Z = 18$.

● WILLEMITE, Zn_2SiO_4.

      Rhombohedral, $R\bar{3}$; $a = 8.628$ Å; $\alpha = 107°52'$; $Z = 6$;

          hexagonal cell: $a = 13.931$, $c = 9.307$ Å; $Z = 18$.

● EUCRYPTITE, α-$LiAlSiO_4$.

      Rhombohedral, $R\bar{3}$; $a = 8.37$ Å; $\alpha = 107°52'$; $Z = 6$;

          hexagonal cell: $a = 13.53$, $c = 9.04$ Å; $Z = 18$.

PHENACITE may contain small amounts of Al,Fe^{3+},Ca,Mg,Na and K in addition to the essential elements. It occurs in granitoid pegmatites and their aureoles, greisens and hydrothermal veins. *WILLEMITE* can contain up to 12% substitution of Mn for Zn with much lesser amounts of Fe, Mg and Ca. Willemite is one of the major Zn ores in the metamorphic Franklin and Sterling Hill, New Jersey deposit. It is found sparingly in the oxidized zones of other ore deposits. *EUCRYPTITE* is usually near its ideal composition. It is found in Li-rich pegmatites.

There have been several crystal structure refinements of phenacite group minerals; among them are phenacite (Zachariasen, 1972), synthetic willemite (Hang *et al.*, 1970) and natural willemite (Simonov *et al.*, 1977). The isostructural nature of eucryptite was demonstrated by Winkler (1953, 1954). The basic unit of the crystal structure of phenacite group minerals is a column of corner-sharing 2Zn + 1Si tetrahedra about a 3_1 or 3_2 axis (figure a). These columns join 3 others by corner-sharing producing a 3-dimensional network composed of spiraled, hexagonal rings of tetrahedra (figure b). The ordering of the Si from Zn, Be and (Li + Al) reduces the symmetry.

Figures a and b are on the next page.

- - -

Hang, Chin, M.A. Simonov and N.V. Belov (1970) Crystal structures of willemite $Zn_2[SiO_4]$ and its germanium analog $Zn_2[GeO_4]$. Sov. Phys. Crystallogr. 15, 387-390.

Simonov, M.A., P.A. Sandomirskii, Yu. K. Egorov-Tismenko and N.V. Belov (1977) The crystal structure of willemite $Zn_2[SiO_4]$. Sov. Phys. Dokl. 22, 622-623.

Winkler, H.F.G. (1953) Tief-$LiAlSiO_4$ (Eukryptit). Acta Crystallogr. 6, 99.

_____ (1954) Struktur und Polymorphie des Eukryptits (Tief-$LiAlSiO_4$). Heidelberger Beitr. Mineral. Petrogr. 4, 233-242.

Zachariasen, W.H. (1972) Refined crystal structure of phenacite Be_2SiO_4. Sov. Phys. Crystallogr. 16, 1021-1025.

● PHENACITE GROUP, continued

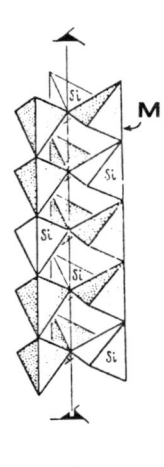 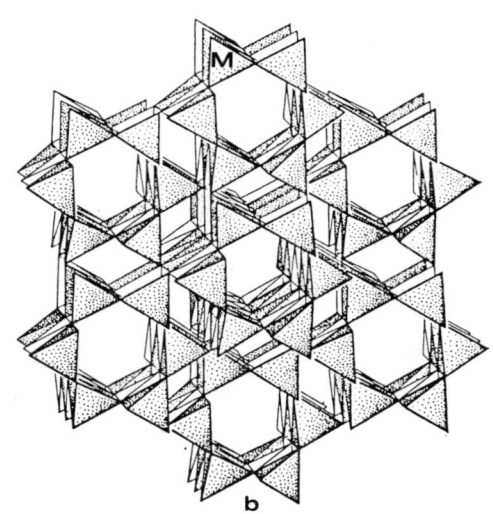

(a) The column of Si and M- cation tetrahedra in the ratio 1:2 around a 3_1 axis in a phenacite group mineral. M is Zn (willemite), Be (phenacite) or Li + Al (eucryptite). (b) The structure of willemite viewed down c. From Simonov *et al.* (1977, Figs. 2 and 4).

● ROEBLINGITE, $Pb_2Ca_7Si_6O_{14}(OH)_{10}(SO_4)_2$.

Monoclinic, Cc or $C2/c$; $a = 13.27$, $b = 8.38$, $c = 13.09$ Å, $\beta = 103.86°$, $Z = 4$.

Roeblingite occurs at Franklin, New Jersey and Långban Sweden. The structure is unknown.

Foit, F.F., Jr. (1966) New data on roeblingite. Am. Mineral. 51, 504-508.

● SAPPHIRINE, $(Mg,Al)_8(Al,Si)_6O_{20}$.

Monoclinic, $P2_1/a$; $a = 11.29$, $b = 14.44$, $c = 9.96$ Å; $\beta = 125.4°$; $Z = 4$.

Sapphirine, whose formula is given in Deer *et al.* (1962, p. 176) as $(Mg,Fe)_2Al_2O_6[SiO_4]$, is definitely misrepresented as an orthosilicate. It has a cubic closest-packed anion array with distinct octahedral and tetrahedral layers. Four tetrahedra are joined in a pyroxene-like aluminosilicate chain with two corner-sharing "wing" tetrahedra on either side (Moore, 1969). Aluminum and silicon are nonrandomly distributed amongst these six sites, but only one of them contains pure Si and one pure Al. The sapphirine formula is best written $[(Mg,Fe)_{8-x}(Al,Fe)_x]^{VI}[Al_xSi_{6-x}]^{IV}O_{20}$, where the coordination numbers of the cations are given by the Roman numerals (Higgins and Ribbe, 1979). See the figures on the following page.

444

● SAPPHIRINE, continued

(a) Schematic down [001] illustrating the layered nature of the sapphirine structure. (b) Topology of the idealized tetrahedral and (c) octahedral layers of sapphirine in [100] projection. From Higgins and Ribbe (1979, Fig. 4).

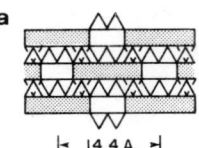

|← 14.4 Å →|

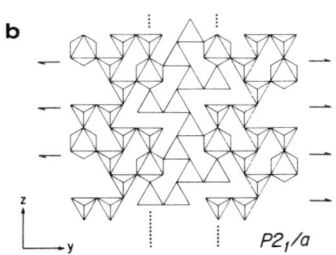

$P2_1/a$

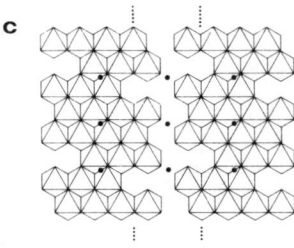

Deer, W.A., R.A. Howie and J. Zussman (1962) *Rock-Forming Minerals, Vol. 1, Ortho- and Ring Silicates.* Longmans, London, p. 176-182.

Higgins, J.B. and P.H. Ribbe (1979) Sapphirine II. A neutron and x-ray diffraction study of $(Mg,Al)^{VI}$ and $(Al,Si)^{IV}$ ordering in monoclinic sapphirine. Contrib. Mineral. Petrol. 68, 357-368.

Moore, P.B. (1969) The crystal structure of sapphirine. Am. Mineral. 59, 41-49.

● SARYARKITE, $Al_5(Ca_{.96}RE_{.76}Th_{.28})_2[(SiO_4)_{2.16}(PO_4)_{1.52}(SO_4)_{.32}]_4(OH)_{6.6} \cdot 5.6H_2O$.

Tetragonal, $P42_12$ or $P4_22_12$; $a = 8.213$, $c = 6.55$ Å; Z = 4.

Saryarkite occurs in altered granitic and rhyolitic rocks at an unspecified locality in the USSR (Krol *et al.*, 1964). The structure is unknown, but the x-ray pattern is very close to rhabdophane.

Krol, O.F., V.I. Chernov, Yu.V. Shipovalov and G.A. Khan (1964) Saryarkite, a new mineral. Zapiski Uses. Mineralog. Obsheh. 93, 147-155. New Mineral Names (1964) Am. Mineral. 49, 1775-1776.

● SILLIMANITE, Al_2SiO_5.

Sillimanite is the one aluminosilicate polymorph that is not properly described as an orthosilicate, even though its SiO_4 tetrahedra do not share corners with other SiO_4. Its fibrous habit betrays the chain-like linkage of AlO_4 and SiO_4 tetrahedra. See Chapter 8.

● SLAWSONITE, $SrAl_2Si_2O_8$.   See *DANBURITE*, this chapter.

445

● STEENSTRUPINE, $(Ce,La,Na,Mn)_6(Si,P)_6O_{18}(OH)$.

Rhombohedral; Hexagonal cell: $a = 9.47$, $c = 45$ Å, $Z = 9$.

Steenstrupine occurs in nepheline-sodalite syenites and pegmatites of the Lovozero (Kola peninsula, U.S.S.R.) and Ilimausaq (South Greenland) alkali massifs. It contains variable amounts of Th, Mn, Si, Na, P and Nb suggesting the coupled substitutions $Na^{1+}+Ce^{3+}\rightarrow 2(Ca,Mn)^{2+}$; $Na^{1+}+P^{5+}\rightarrow (Ca,Mn)^{2+}+Si^{4+}$; $Th^{4+}+Si^{4+}\rightarrow Ce^{3+}+P^{5+}$. Steenstrupine can be an important U- and Th-bearing mineral in the lujavrites of the Ilimausaq intrusion (Bohse *et al.* 1974). The structure of steenstrupine is unknown but IR spectra indicates it has isolated SiO_4 groups (Aleksandrova *et al.*, 1967).

- - -

Aleksandrova, I.T., I.I. Kupriyanov, and L.I. Rybakova (1966) Physical properties and identification of rare-earth silicates. Geol. Mestorozhd. Redk. Elementov. 26, 157-180. Chem. Abstracts 65, 1967.

Bohse, H., J. Rose-Hansen, H. Sørensen, A. Steenfelt, L. Løvborg and H. Kunzendorf (1974) On the behavior of uranium during crystallization of magmas - with special emphasis on alkaline rocks. In *Formation of Uranium Ore Deposits*, Proc. Symp. 1974, IAEA-SM-183/26, 49-60. UNIPUB: New York, N.Y.

Vlasov, K.A. (1966) Geochemistry and Mineralogy of rare elements and genetic types of their deposits, Volume II, *Mineralogy of Rare Elements*. Israel Program for Scientific Translations, 321-324.

● STILLWELLITE, $CeBO[SiO_4]$.

Hexagonal, $P3_1$; $a = 6.85$; $c = 6.70$ Å; $Z = 3$.

Stillwellite occurs in metasomatic calcareous hornfels. The predominant rare earths are Ce and La. The structure, determined by Voronkov and Pyatenko (1967), is comprised of columns of Si tetrahedra and 9-coordinated Ce-polyhedra alternating along c. The columns are laterally joined by helical chains of BO_4 tetrahedra extending in the same direction.

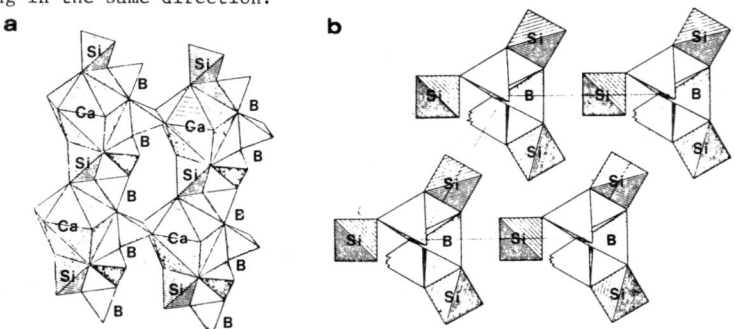

(a) Fragment of the structure of stillwellite; two translationally identical columns and chains of B tetrahedra. (b) Motif of BO_4 and SiO_4 tetrahedra in the xy projection. From Voronkov and Pyatenko (1967, Figs. 5 and 6.)

- - -

Voronkov, A.A. and Yu. A. Pyatenko (1967) X-ray diffraction study of the atomic structure of stillwellite $CeBO[SiO_9]$. Sov. Phys. Crystallogr. 12, 214-220.

● STRÄTLINGITE, $2CaO \cdot Al_2O_3 \cdot SiO_2 \cdot 8H_2O$.

Trigonal, $R3$ or $R\bar{3}$; hexagonal cell: $a = 5.747$, $c = 37.64$ Å; $Z = 3$.

Strätlingite occurs in metamorphosed limestone enclaves in basalt at Bellerberg, Mayen, Germany. Except for a small amount of iron, the mineral is near the ideal formula in chemistry (Hentschel and Kuzel, 1976). Kuzel (1976) suggests that the structure consists of a layer of composition $[Ca_2Al(OH)_6]^+$ alternating with layers of $[AlSiO_3(OH)_2 \cdot 4H_2O]^-$, which is a sheet of six-member tetrahedral rings with alternating Al and Si.

– – –

Hentschel, G. and H.J. Kuzel (1976) Strätlingite, $2\ CaO \cdot Al_2O_3 \cdot SiO_2 \cdot 8H_2O$, a new mineral. N. Jahrb. Mineral. Monatsh. 326-330.

Kuzel, H.J. (1976) Crystallographic data and thermal decomposition of synthetic gehlenite hydrate, $2\ CaO \cdot Al_2O_3 \cdot SiO_2 \cdot 8H_2O$. N. Jahrb. Mineral., Monatsh. 319-325.

● STRINGHAMITE, $CuCaSiO_4 \cdot 2H_2O$.

Monoclinic, $P2_1/c$; $a = 5.028$; $b = 16.07$; $c = 5.303$ Å; $\beta = 102.58°$.

Stringhamite occurs in a diopside-magnetite skarn at the Bawana mine, Rocky Range, Beaver Co., Utah. (Hindman, 1976). Its structure is unknown.

– – –

Hindman, J.R. (1976) Stringhamite, a new hydrous copper calcium silicate from Utah. Am. Mineral. 61, 189-192.

● SURINAMITE. $(Al_{1.36}Mg_{1.12}Fe_{0.46}Mn_{0.04})^{VI}(Si_{1.51}Al_{0.49})O_{7.36}(OH)_{0.64}$.

Monoclinic, $P2_1/a$; $a = 9.64$, $b = 11.36$, $c = 4.95$ Å; $\beta = 109.0°$; $Z = 4$.

Surinamite is found in a mylonitic mesoperthite gneiss in the Bakhuis Mtns., western Surinam, as aggregates of small crystals together with biotite, kyanite and sillimanite; intergrowths with these aluminum silicates were observed. "Surinamite resembles sapphirine optically, in X-ray powder pattern, and in structure. A proposed crystal structure involving dense-packed oxide sheets as in sapphirine satisfactorily accounts for the spinel-like substructure reflections." (P.B. Moore in de Roever *et al.*, 1976). See section on *SAPPHIRINE*, this chapter.

– – –

de Roever, E.W.F., C. Kieft, E. Murray, E. Klein, W.H. Drucker and P.B. Moore (1976) Surinamite, a new Mg-Al silicate from the Bakhuis Mountains, western Surinam. Am. Mineral. 61, 193-199.

● THAUMASITE, $[Ca_3Si(OH)_6 \cdot 12H_2O](SO_4)(CO_3)$

Hexagonal, $P6_3$; $a = 11.04$, $c = 10.39$ Å; $Z = 2$.

Thaumasite is a secondary, hydrothermal mineral occurring in cavities and fissures of mafic igneous and calcareous rocks. Edge and Taylor (1971) found that the structure of thaumasite contains unusual $Si(OH)_6^{2-}$ octahedra.

– – –

Edge, R.A. and H.F.W. Taylor (1971) Crystal structure of thaumasite, $[Ca_3Si(OH)_6 \cdot 12H_2O](SO_4)(CO_3)]$. Acta Crystallogr. B27, 594-601.

● TRANQUILLITYITE, $Fe_8^{2+}(Zr,Y)_2Ti_3Si_3O_{24}$.

Metamict.

Tranquillityite is a common, late-stage mineral found in a variety of lunar basalts (Lovering *et al.*, 1971). It contains minor amounts of Ca, Al, Mn, Cr, Nb, REE and Hf and trace amounts of U. Reheated, tranquillityite is cubic or possibly rhombohedral and may have a fluorite-related structure (Gatehouse *et al.*, 1977).

- - -

Gatehouse, B.M., I.E. Grey, J.F. Lovering and D.A. Wark (1977) Structural studies on tranquillityite and related synthetic phases. Proc. 8th Lunar Sci. Conf., 1831-1838.

● TRIMERITE, $CaMn_2(BeSiO_4)_3$.

Monoclinic, $P2_1/n$; $a = 8.098$, $b = 7.613$, $c = 14.065$ Å; $\beta = 90°$; $Z = 4$.

Trimerite occurs in manganese deposits in Harstig and Långbanshyttan, Sweden. Mg and Fe may replace Mn. The crystal structure determination of trimerite by Klaska and Jarchow (1977) shows that it consists of a 3-dimensional network of corner-sharing, ordered Be,Si tetrahedra with two different channels paralleling b which contain the Ca and Mn (see the figure). The pseudotrigonal nature of the $(BeSiO_4)$ framework explains the common trillings with composition planes (110) and ($1\bar{1}0$) observed in trimerite.

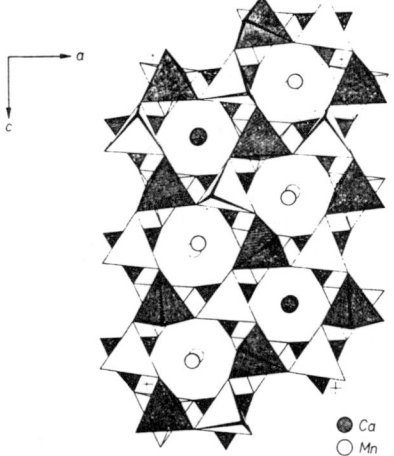

    ● Ca
    ○ Mn

The [010] projection of trimerite. From Klaska and Jarchow (1977, Fig. 1).

- - -

Klaska, K.H. and O. Jarchow (1977) Die Bestimmung der Kristallstruktur von Trimerit $CaMn_2(BeSiO_4)_3$ und das Trimeritgesetz der Verzwillingung. Z. Kristallogr. 145, 46-65.

● TRITOMITE and TRITOMITE-Y (=SPENCITE), $(Ce,La,Y,Th)_5(Si,B)_3(OH,O,F)_{13}$.

Metamict.

Tritomite is found in nepheline-syenite pegmatites of Langesundfiord near Brevik, Norway. Tritomite-Y is found in the calcite-fluorite veins of the Haliburton-Bancroft region, Ontario (Hogarth *et al*, 1973) and a plagiogranite pegmatite in Sussex Co., New Jersey (Jaffe and Molinski, 1962). Tritomites contain a variety of other elements, and when heated they give an x-ray pattern characteristic of apatite. Hogarth *et al*. (1973) suggest the parent mineral may have been *HELLANDITE* (this chapter).

- - -

Hogarth, D.D., H.R. Steacy, E.I. Semenov, E.G. Proshchenko, M.E. Kazakova and Z.T. Kataeva (1973) New occurrences and data for spencite. Canadian Mineral. 12, 66-71.

● VINOGRADOVITE, $Na_5Ti_4AlSi_6O_{24} \cdot 3H_2O$.

Monoclinic, $A2/a$; $a = 5.218$, $b = 8.692$, $c = 24.605$ Å; $\beta = 99°50'$; $Z = 2$.

Vinogradovite occurs in nepheline-syenite pegmatites of the Lovozero massif, Kola Peninsula, USSR. Analysis of the natural specimen may be on altered material which has loss of Na. It contains up to 3.5% Nb_2O_5 and 0.08% Be. Ca and K can replace Na. The mineral formula and crystallographic data for vinogradovite is based on a crystal structure determination by Simonov (1969). Unfortunately it is impossible to determine much about the structure from the 3-dimensional superposition synthesis section of vinogradovite he presents.

- - -

Semonov, E.I., E.M. Bonhtedt-Kupletskaya, V.A. Moleva and N.N. Sludskaya. (1956) Vinogradovite: a new mineral. Dokl. Akad. Nauk SSSR, 109, 617. New Mineral Names (1957) Am. Mineral. 42, 308.

Simonov, V.I. (1969) Crystal structure determination by the Fourier-transformation of the minimum function. Acta Crystallogr. B25, 1-4.

Vlasov, K.A., M.V. Kuz'menko and E.M. Es'Kova (1966) *The Lovozero Alkali Massif*. Hafner Publishing Company, pp. 379-380.

● WILLEMITE, Zn_2SiO_4. See section on *PHENACITE GROUP*, this chapter.

● YEATMANITE. See *KATOPTRITE*, this chapter.

● YFTISITE, $(Y,RE)_4(F,OH)_6TiO[SiO_4]_2$.

Orthorhombic, *Cmcm*; $a = 14.949$, $b = 10.626$, $c = 7.043$ Å.

Yftisite occurs as an accessory mineral in alkali granites, Kola Peninsula, USSR (Pletneva *et al.*, 1971). The heavy REE predominate and Sn apparently substitutes for Ti. Abstracts of crystal structure determinations of the natural (Balko and Bakakin, 1975) and synthetic (Belov and Belova, 1978) material suggest the structure is comparable to titanite, consisting of infinite chains of cornered-shared of TiO_6 octahedra cross-linked by SiO_4 tetrahedra. The (Y,RE) and (F,OH) are positioned within the openings of the resulting framework.

- - -

Balko, V.P. and V.V. Bakakin (1975) The crystal structure of the natural yttrium and rare-earth fluosilicate, $(Y,TR)_4(F,OH)_6TiO [SiO_4]_2$, (yftisite). Zh. Strukt. Khim. 16, 837–842. New Mineral Names (1977) Am. Mineral. 62, 396.

Belov, N.V. and E.N. Belova (1978) Essays on structural mineralogy. XXIX. 197. Two natural titanosilicates sphene and yftisite and their germanium (VI) and (IV) analogs. Mineral. Ab. (Lvov). 32, 5–7. Chem. Abstracts 90, 207–310.

Pletneva, N.I., A.P. Denison and N.A. Elina (1971) A new variety in the group of rare-earth fluosilicates. Materl Mineral. Kol'sk Poluostr. 8, 176–179. New Mineral Names (1977) Am. Mineral. 62, 396.